Plant Genetics: Biodiversity and Evolution

Plant Genetics: Biodiversity and Evolution

Edited by Isabelle Nickel

SYRAWOOD
PUBLISHING HOUSE

New York

Published by Syrawood Publishing House,
750 Third Avenue, 9th Floor,
New York, NY 10017, USA
www.syrawoodpublishinghouse.com

Plant Genetics: Biodiversity and Evolution
Edited by Isabelle Nickel

International Standard Book Number: 978-1-68286-755-6 (Hardback)

Cataloging-in-Publication Data

Plant genetics : biodiversity and evolution / edited by Isabelle Nickel.
    p. cm.
Includes bibliographical references and index.
ISBN 978-1-68286-755-6
1. Plant genetics. 2. Plant diversity. 3. Plants--Evolution. I. Nickel, Isabelle.
QK981 .P53 2019
581.35--dc23

TABLE OF CONTENTS

PREFACE

Plants are multicellular photosynthetic eukaryotes. There are nearly 300,000 species of green plants. They have evolved from algal mats, through bryophytes, lycopods and ferns to the highly complex angiosperms and gymnosperms. Some of the largest genomes among all organisms occur in plants. Besides the inheritance of traits, the genome of a plant is also responsible for its growth and development. Plants experience different biotic and abiotic stresses, which cause DNA damage whether directly or indirectly. Plants resist such DNA damage with a damage response, which is critical for maintaining stability of the genome. This is vital especially during seed germination, as the seed quality deteriorates with age. This book covers in detail some existing theories and innovative concepts revolving around this field. Different approaches, evaluations, methodologies and advanced studies on plant genetics with respect to biodiversity and evolution have been included herein. With state-of-the-art inputs by acclaimed experts of this field, this book targets students and professionals.

Various studies have approached the subject by analyzing it with a single perspective, but the present book provides diverse methodologies and techniques to address this field. This book contains theories and applications needed for understanding the subject from different perspectives. The aim is to keep the readers informed about the progresses in the field; therefore, the contributions were carefully examined to compile novel researches by specialists from across the globe.

Indeed, the job of the editor is the most crucial and challenging in compiling all chapters into a single book. In the end, I would extend my sincere thanks to the chapter authors for their profound work. I am also thankful for the support provided by my family and colleagues during the compilation of this book.

Editor

Genetic variation among mainland and island populations of a native perennial grass used in restoration

Kristina M. Hufford[1,2]*, Susan J. Mazer[1] and Scott A. Hodges[1]

[1] Ecology, Evolution and Marine Biology, University of California, Santa Barbara, CA, USA
[2] Present address: Ecosystem Science and Management, University of Wyoming, Laramie, WY, USA

Abstract. Genetic marker studies can assist restoration practice through selection of seed sources that conserve historical levels of gene diversity and population genetic differentiation. We examined genetic variation and structure within and among mainland and island populations of *Elymus glaucus*, a perennial bunchgrass species native to western North American grasslands that is targeted for grassland restoration. Island populations of *E. glaucus* represent sensitive sites and potentially distinctive seed sources for reintroduction, and little is known of their genetic composition. Genetic diversity and structure were estimated using amplified fragment length polymorphism markers for 21 populations and 416 individuals distributed across two coastal California mainland locations and three California Channel Islands. Eight primer combinations resulted in 166 markers, of which 165 (99.4 %) were polymorphic. The number of polymorphic bands was significantly greater among mainland populations relative to island sites, and locally common alleles were present for each sampled island and mainland location. Population structure was high (62.9 %), with most variation (55.8 %) distributed among populations, 7.1 % between mainland and island locations, and the remainder (37.1 %) within populations. Isolation by distance was only apparent among islands. Using marker data to recommend appropriate seed sources for restoration, *E. glaucus* seeds are best derived within islands with collections representing a large number of individuals from matching environments. Given the limited gene flow and prior evidence of adaptive divergence among populations of this species, regional collections are recommended in all cases to maintain diversity and to avoid long-distance introductions of highly differentiated plant material.

Keywords: AFLP markers; California Channel Islands; ecological restoration; *Elymus glaucus*; genetic drift; seed source; self-pollination; spatial genetic structure.

Introduction

Widespread anthropogenic disturbance and introductions of invasive species have resulted in the fragmentation and conversion of grassland ecosystems worldwide (D'Antonio and Vitousek 1992). Temperate grasslands originally dominated by caespitose (bunchgrass) species have been altered by introduced livestock, and are highly vulnerable to invasion by competitive annual and rhizomatous perennial exotic grasses (Mack 1989; Hayes and Holl 2003). California grasslands represent an ecosystem where plant community conversion is nearly complete (Mensing 1998). Mediterranean annual grasses and forbs introduced over the past three centuries now dominate the landscape, and native species exist as remnant populations in a matrix of exotics. Efforts are ongoing to control invasive species and to re-establish

* Corresponding author's e-mail address: khufford@uwyo.edu

native perennial bunchgrasses (e.g. Moyes *et al.* 2005; DiTomaso *et al.* 2007; Cox and Allen 2011), and guidelines are needed for the restoration of sustainable and diverse plant populations.

Fragmented populations of plant species are susceptible to environmental, demographic and genetic stochasticity (Shaffer 1981; Lande 1988). Restoration programmes commonly mitigate environmental and demographic concerns for all target species by increasing the number of both individuals and populations to minimize the probability of extirpation. In contrast, attempts to mitigate the loss of genetic variation and to minimize inbreeding have largely focused on rare and endangered species (Lande 1988), although interest in the genetic consequences of establishing or restoring populations of common species has grown over the last decade (e.g. Hufford and Mazer 2003; McKay *et al.* 2005; Bischoff *et al.* 2010). Evidence suggests that common species are subject to genetic erosion resulting from habitat fragmentation at similar or even greater rates than rare species (Honnay and Jacquemyn 2007). Genetic variation is the basis of adaptation, and the loss of genetic diversity, particularly for fitness-related traits, will impact population persistence as well as limit the ability of a population to adapt to changing environments (Frankham *et al.* 2002; Reed and Frankham 2003). Knowledge of species-level patterns of genetic diversity can inform and improve restoration protocols when population reintroduction is a restoration objective.

Primary genetic concerns for reintroduction include the maintenance of patterns of diversity within and among populations, and the introduction of genotypes adapted to environmental conditions at the restoration site (e.g. McKay *et al.* 2005). Population genetic structure is a function of a species' mating system and provides an estimate of historical levels of gene flow and connectivity among locations (Slatkin 1987). Data that provide measures of the partitioning of genetic variation, however, do not explain underlying causes of divergence, which may be a function of selection or random processes (Heywood 1991). Direct evaluations of genotypic adaptation and traits under selection can be determined in common garden and reciprocal transplant studies (Linhart and Grant 1996; Kawecki and Ebert 2004). When available, these studies provide valuable information for the identification of appropriate seed sources for reintroduction. However, data for common gardens are not available for all species, are limited in scale and may not detect all components of local adaptation (Nuismer and Gandon 2008). Thus, baseline information to describe patterns of genetic variation within and among populations remains relevant for restoration and conservation planning, and can serve as a

first step for management of genetic diversity in species reintroduction or augmentation programmes.

The genetic consequences of seed introduction during restoration may have greater impacts for populations occupying small or isolated islands relative to mainland sites. Island populations often harbour lower levels of gene diversity and higher levels of differentiation when compared with the mainland, and are at increased risk of extinction—possibly due to greater environmental and demographic stochasticity (Frankham 1997). Islands are also disproportionately vulnerable to biological invasion, and introduced species are reported to outnumber native species in grasslands of the California Channel Islands (Halvorson 1992; Schoenherr *et al.* 1999; Moody 2000). *Elymus glaucus* (blue wildrye) is a native bunchgrass species once common in California mainland and island grasslands (Holstein 2001; Bartolome *et al.* 2004). Over the past few decades, *E. glaucus* has been a target of restoration programmes due to its wide distribution, wildlife habitat value and dense root system, which prevents erosion in degraded landscapes (Knapp and Rice 1996; Erickson *et al.* 2004). Two previous reciprocal transplant studies found evidence of ecotypic variation among populations of this species as a result of adaptation to local environments over scales of 50–190 km (Hufford *et al.* 2008; Knapp and Rice 2011). These studies were conducted at mainland locations, and no information is yet available to describe genetic differentiation or adaptive variation for populations on the Channel Islands.

In this study, we used amplified fragment length polymorphisms (AFLPs) to characterize genetic diversity and structure among mainland and island populations of *E. glaucus* to address the following questions relevant for grassland restoration: (i) How much genetic variation is present in island populations, and how does this compare to mainland populations? (ii) Is genetic differentiation between island populations greater relative to nearby mainland locations? (iii) Is genetic distance correlated with geographic distance within and among islands and the mainland? Lower levels of diversity and strong genetic differentiation at island sites may indicate the reduced ability of island plants to adapt to altered environments and a greater risk of local extinction of Channel Island populations. At the same time, the geographic scale of genetic differentiation serves as an indicator of the historical rates of gene flow. These data can assist with seed provenance selection for the restoration of *E. glaucus* growing in California coastal and island grasslands by means of maintaining population genetic diversity and lowering the risk of introducing maladapted genotypes.

Methods

Study species

The genus *Elymus* includes 150 species distributed in temperate regions worldwide (Hickman 1993). *Elymus glaucus* is a perennial, non-rhizomatous bunchgrass with a broad geographic distribution from Canada to Mexico, and can be found throughout the western United States (Hickman 1993). Herbarium records for this species include much of the state of California (http://www.calflora.org), but extant populations are highly fragmented as a result of widespread land development and biological invasion (Barry *et al.* 2006). Populations of *E. glaucus* occur in diverse habitats and plant communities, and exhibit morphological variation across their range (Snyder 1951; Wilson *et al.* 2001). Polyploidy is common in the Poaceae, and *E. glaucus* is an allotetraploid ($2n = 28$) derived from *Hordeum* and *Pseudoroegneria* ancestors (Jensen *et al.* 1990). Previous studies of *E. glaucus* suggest that it is frequently self-pollinated and has a mixed mating system, allowing for some outcrossed pollination (Knapp and Rice 1996; Ie 2000; Wilson *et al.* 2001). Inflorescences are distinctive, narrow spikes, and seed dispersal is typically passive.

Study sites and collections

The Southern Channel Islands represent an archipelago of eight continental islands located at distances ranging from 13 to 61 km off the coast of mainland California (Schoenherr *et al.* 1999). Despite their proximity to the continental landmass, high numbers of endemic species (up to 47 % of native vegetation) are found in the island chain (Junak *et al.* 1995; Moody 2000). Introduced plants and animals threaten native species, and efforts to conserve and restore island ecosystems are ongoing (Halvorson 1994). In the present study, sampling sites included 21 populations distributed among two mainland locations and the three Channel Islands where *E. glaucus* is known to occur (Junak *et al.* 1997). The mainland sites were located at the University of California Sedgwick Reserve (Sedgwick) in Santa Ynez, California and Vandenberg Air Force Base (VAFB) in Lompoc, California. Offshore sites were located on Santa Rosa and Santa Cruz Islands in Channel Islands National Park, and Santa Catalina Island (Fig. 1).

Sampling locations represented diverse habitats and serpentine rock outcrops are common at the two mainland locations and Santa Catalina Island. Populations of *E. glaucus* sampled in this study occurred in oak woodland savannahs at Sedgwick and in open, coastal grasslands at VAFB. Island populations were distributed among coastal prairies and ephemeral, riparian environments (Fig. 2). The regional climate is Mediterranean, but due to the marine influence, coastal areas of VAFB and the Channel Islands experience cooler temperatures and higher humidity when compared with the interior of Santa Cruz Island and Sedgwick Ranch. Sites located at Sedgwick and in Santa Cruz Island's Central Valley are subject to greater temperature fluctuations than coastal areas, and often record seasonal temperature differences of 5 °C or more relative to the shoreline (Schoenherr *et al.* 1999).

Populations of *E. glaucus* were identified and georeferenced during several trips to each location in the spring and summer of 2002 and 2003 (Table 1). At each location, we sampled populations that represented geographically distant sites although some areas were inaccessible, limiting our sampling range. Efforts were made to sample plants separated by 0.5 m or greater within populations along 10- to 20-m walking transects. The patchy nature and small size of many *E. glaucus* populations, however, restricted the area within which we were able to collect leaf material, and in some cases we modified transect sampling to collect in a smaller radius while maintaining the separation of sampled plants. In general, population sizes ranged between 20 and 60 visible plants. At each site, leaves from 20 individuals were collected and stored in sealed bags containing silica gel for preservation. One population located on Santa Cruz Island was very small, and only allowed for 16 individual collections.

AFLP genotyping

Leaves stored in silica gel were transported to the University of California in Santa Barbara and stored at an average room temperature of 20 °C prior to AFLP genotyping. For each sampled plant, total genomic DNA was extracted from ∼20 mg of silica-dried leaf tissue using the DNeasy Plant Mini Kit (Qiagen). A total of 416 plants representing 21 *E. glaucus* populations distributed among 9 mainland and 12 island sites were included in subsequent marker analyses.

Molecular markers were amplified following the protocol of Vos *et al.* (1995) with little modification. Approximately 250 ng of DNA were digested with EcoRI and MseI restriction enzymes and ligated to corresponding adapters. The restriction–ligation mix was diluted 1 : 10 and polymerase chain reaction (PCR)-amplified with EcoRI-A and MseI-C pre-selective primers. The resulting PCR template was diluted 1 : 10 prior to amplification with selective EcoRI primers that were 5′ end labelled with IRDye 700 or 800. Eight primer combinations were selected due to clarity and repeatability of bands (Table 2). Amplification products of duplexed selective PCR reactions were denatured and separated on 7 % acrylamide gels using a LI-COR 4200 DNA sequencer (LI-COR, Lincoln, NE, USA). One or more duplicate samples were routinely included in gel runs for quality control. The

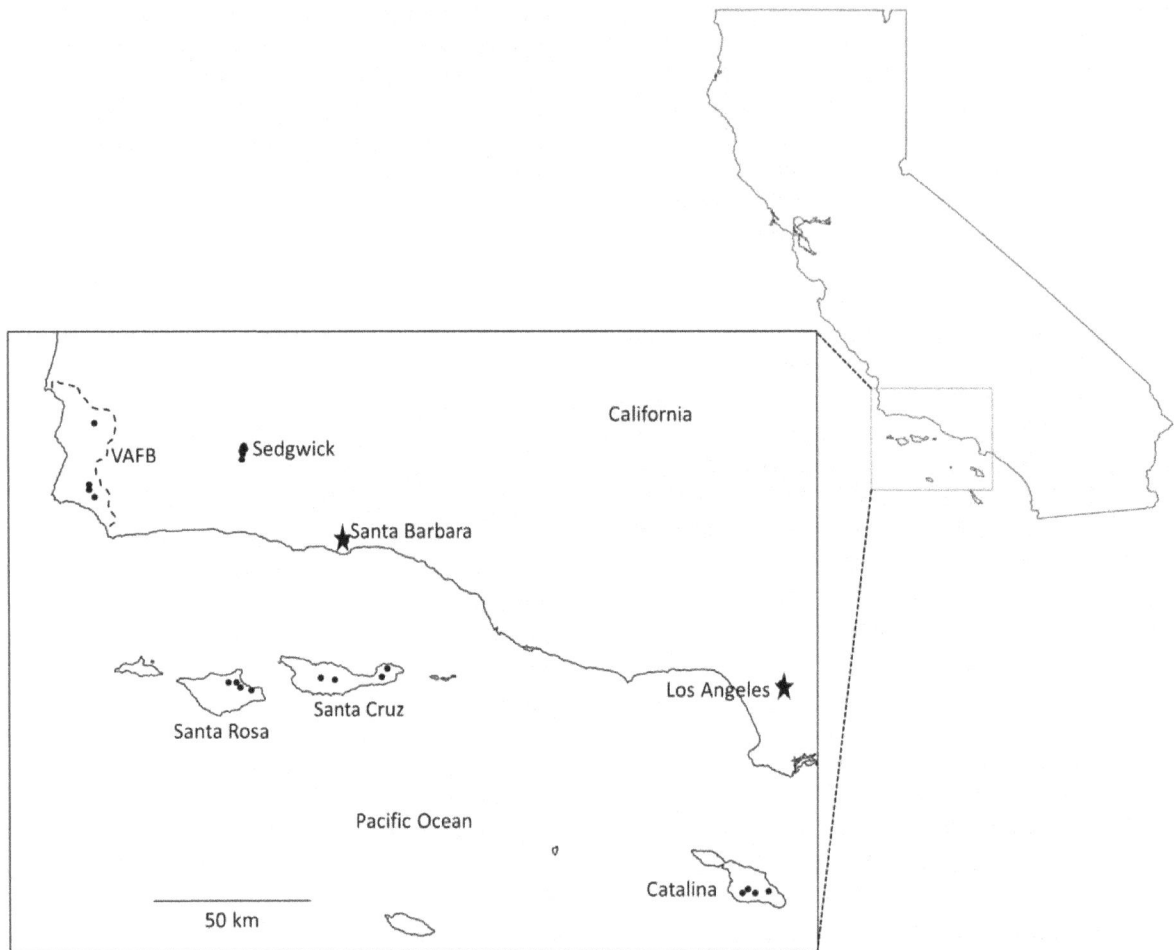

Figure 1. Map of *E. glaucus* study sites in Southern California among the three California Channel Islands and two mainland locations.

presence or absence of AFLP bands was scored manually using SAGA 2.0 MX software, and ambiguous bands were recorded as missing. Error rates (calculated as the number of mismatched genotypes divided by the number of bands) were 3–4 %.

Genetic analysis

We analysed AFLP marker data using methods previously employed for allotetraploid species, including band-based and allele-frequency metrics (e.g. Kang *et al.* 2005; Wagner *et al.* 2012). Band-based metrics compute distance-based measures of similarity within and among populations from the matrix of marker presence/absence data without inferring allele frequencies (Bonin *et al.* 2007). In contrast, allele-frequency methods estimate standard population genetic statistics; these methods were developed for diploid organisms and may not be valid for polyploid species. Allele-frequency methods were tested by Wagner *et al.* (2012) for an allotetraploid species, and estimated values correlated strongly with

band-based metrics, since allotetraploid species likely undergo meiotic segregation within each parental genome similar to diploid species (Soltis and Soltis 1993).

Genetic diversity

Using a band-based approach, we calculated the number and proportion of polymorphic bands (PPBs) within and among island and mainland sites by means of FAMD 1.2 software (Schlüter and Harris 2006). We also computed the number of fixed (or monomorphic) bands. The number of private and locally common bands (restricted to a limited area and found in ≤ 25–50 % of populations) was determined in GenAlEx 6.5 (Peakall and Smouse 2006). Differences in the number of polymorphic or fixed bands observed for island and mainland sites were compared using generalized linear models and a Poisson distribution in JMP 9 statistical software (SAS Institute Inc., Cary, NC, USA). The proportion of polymorphic loci (PLP) and expected heterozygosity (H_e) were estimated for comparison to band-based metrics

Figure 2. Riparian site representing potential *E. glaucus* habitat on Santa Rosa Island.

using a fragment-frequency approach in AFLP-SURV 1.0 (Vekemans 2002). This method assumes fixed homozygosity resulting from self-pollination and may overestimate marker frequencies for outcrossing taxa, but would meet expectations for *E. glaucus*. We subsequently compared band-based metrics with allele-frequency estimates using regression analysis. All files for genetic analyses were prepared using functions available in GenAlEx or the package AFLPdat (Ehrich 2006) for R software (R Development Core Team 2011).

Genetic differentiation among populations and regions

We conducted a hierarchical analysis of molecular variance (AMOVA) in GenAlEx to partition genetic variation within and among populations nested in two regions: island and mainland. AMOVA components of variance include Φ_{PT}, which is considered to be an analogue of F_{ST} (Wright 1951; Peakall and Smouse 2006). The significance of the different components of variance was tested with 9999 permutations. Values of Φ_{PT} were also calculated separately for populations within island or mainland regions. A Mantel test (10 000 permutations) was used to determine if there was an association between genetic

distance measured as a matrix of linearized Φ_{PT} values, and $\log_{10}$ transformed geographic distances (Rousset 1997). The Mantel test was also conducted for the two subsets of island and mainland data. The indirect rate of gene flow (N_m) was estimated following Wright (1951), and pairwise genetic distance (Φ_{PT}) values were calculated and assayed for significance with 9999 permutations.

We compared AMOVA results for population genetic structure with allele-frequency estimates calculated in HICKORY software 1.1 (Holsinger *et al.* 2002; Holsinger and Lewis 2003). HICKORY employs a Bayesian approach to estimate $\theta^{(II)}$ (comparable to F_{ST} and Φ_{PT}) using dominant markers, and does not assume Hardy–Weinberg equilibrium. We computed $\theta^{(II)}$ for three alternative models (full model, $f = 0$ model and f-free model) and selected a suitable model with the lowest deviance information criterion (DIC). Markov chain Monte Carlo (MCMC) parameters were set to default values (burn-in 50 000, sampling 250 000). To test for significant differences in population genetic structure among island and mainland locations, we ran posterior comparisons of $\theta^{(II)}$ values. If 95 % confidence intervals for the difference in paired samples included zero, we concluded that $\theta^{(II)}$ values were not significantly different.

To test the assumption that the 21 *E. glaucus* sites represented distinct populations, we used Bayesian clustering methods implemented in STRUCTURE 2.3.3 software to assign individuals to populations by employing the recessive alleles option for dominant markers (Pritchard *et al.* 2000; Falush *et al.* 2007). We ran 20 iterations of each $K = 1$–23 possible clusters using the default model that infers alpha and assumes admixture and correlated allele frequencies. Every run included a burn-in period of 150 000 MCMC cycles and 300 000 MCMC iterations (University of Oslo Bioportal; Kumar *et al.* 2009). The most likely number of clusters represented by the AFLP data was identified using the method described in Evanno *et al.* (2005) and implemented in HARVESTER software (Earl and vonHoldt 2012), which calculates ΔK as the second-order rate of change of the log probability of the data. CLUMPP 1.1.2 software (Jakobsson and Rosenberg 2007) aligned the 20 replicate runs and results were plotted with DISTRUCT 1.1 software (Rosenberg 2004).

Results

Genetic diversity

We scored clear and unambiguous AFLP bands as present or absent for each sampled individual. Of the 166 AFLP markers, 165 (99.4 %) were polymorphic for the full dataset. The proportion of missing data was calculated at 1.91 %. The average number of scorable bands generated

Table 1. Sampled locations and genetic diversity indices for 21 *E. glaucus* populations including latitude (N°) and longitude (W°), sample size (*n*), number of locally common (*f*) and fixed bands (FB), per cent polymorphic bands (PPB), per cent polymorphic loci (PLP) and expected heterozygosity (*H*$_e$) with standard errors

Location	N°	W°	Description	Population ID	*n*	*f*	FB	PPB (%)	PLP (%)	*H*$_e$ Mean	SE
Santa Catalina	33 20.81	118 26.65	Bullrush Canyon	C1	20	16	75	32.7	30.9	0.120	0.015
	33 21.43	118 25.57	Cape Canyon	C2	20	15	59	50.9	50.6	0.118	0.012
	33 20.85	118 24.32	Middle Canyon	C3	20	5	31	72.7	72.3	0.213	0.015
	33 21.19	118 21.63	Haypress	C4	20	13	40	60.6	56.0	0.157	0.013
Santa Cruz	34 00.75	119 47.77	Portezuela	SC1	20	8	82	25.5	25.3	0.095	0.014
	34 00.35	119 44.96	Valley Road	SC2	20	4	97	18.8	18.7	0.075	0.013
	34 00.94	119 35.55	End of the Line	SC3	20	2	77	28.5	28.3	0.101	0.014
	34 02.44	119 34.45	Scorpion Canyon	SC4	16	8	100	5.5	4.8	0.013	0.005
Santa Rosa	33 59.82	120 05.25	Lobo Canyon	SR1	20	4	80	23.6	22.4	0.053	0.009
	33 59.87	120 03.70	Cherry Canyon	SR2	20	3	77	30.9	30.7	0.069	0.010
	33 58.97	120 02.98	Water Canyon	SR3	20	5	58	47.3	47.0	0.117	0.013
	33 58.55	120 01.00	Box Canyon	SR4	20	7	38	61.2	59.8	0.150	0.013
Sedgwick	34 43.32	120 02.16	Figueroa 4	S1	20	4	79	23.0	22.4	0.060	0.011
	34 43.24	120 02.18	Figueroa 3	S2	20	8	55	49.1	48.8	0.089	0.009
	34 43.00	120 02.34	Figueroa 2	S3	20	6	52	51.5	51.2	0.141	0.013
	34 42.62	120 02.40	Figueroa 1	S4	20	4	97	18.2	18.1	0.035	0.008
	34 41.43	120 02.75	Ranch House	S5	20	13	36	67.3	66.9	0.211	0.015
VAFB	34 48.08	120 30.86	Campground	V1	20	8	61	47.3	47.0	0.143	0.015
	34 36.61	120 31.90	Pasture	V2	20	10	62	50.3	50.0	0.123	0.012
	34 35.64	120 31.91	San Miguelito	V3	20	15	31	65.5	65.1	0.141	0.012
	34 34.30	120 30.92	Sudden/Ave I	V4	20	13	87	24.8	24.7	0.070	0.011
				Island	236		67.8	98.8	37.2	0.107	0.015
				Mainland	180		62.2	96.4	43.8	0.113	0.018
				Mean				40.7	40.0	0.109	0.011

Table 2. Combinations of EcoRI and MseI selective primers, and the total number of bands scored and per cent polymorphism (PLP ± SE) generated by each primer combination for *E. glaucus* populations

Primer pairs	# Bands	% PLP	% SE (PLP)
E-AAC/M-CAG	17	38.38	4.93
E-ACC/M-CAG	17	37.54	4.40
E-AAG/M-CGG	22	39.39	4.65
E-AGC/M-CGG	23	43.06	5.00
E-ACA/M-CAG	25	41.33	4.82
E-AGG/M-CAG	19	37.59	5.17
E-AAC/M-CGG	23	35.40	4.32
E-ACA/M-CGG	20	46.19	5.72
Mean	20.75	39.96	4.11

by each AFLP primer combination was 20.75 (range of 17–25; Table 2). The number of polymorphic bands declined significantly for subsets of island or mainland data, reducing markers for analysis in some cases by more than half (Table 1). Mean genetic diversity for all samples was relatively high (PPB = 40.7 %), but varied considerably among populations (6–73 %). The lowest reported values for polymorphism were recorded for the population (SC4) on Santa Cruz Island with only 16 individuals, and may represent a recent founder event.

All values describing genetic diversity (PPB, PLP and *H*$_e$) were strongly correlated (pairwise comparisons, $r > 0.91$, $P < 0.0001$), indicating concordance among band-based and allele-frequency metrics. The average expected heterozygosity (*H*$_e$) within populations was low (0.1093) and varied among populations and locations (Table 1). The average expected heterozygosity within locations was greatest for the Catalina, VAFB and Sedgwick populations (0.152, 0.119 and 0.107, respectively) and declined for Santa Cruz (0.071) and Santa Rosa (0.097) Islands but was not significantly different among mainland and island locations. Comparisons of genetic diversity parameters between the two groups of mainland and island populations detected a significantly greater number of polymorphic bands among mainland populations (mean of 73 vs. 63 bands, respectively; $P = 0.007$). The mean number of fixed bands among populations did not differ significantly between mainland and island regions at the $P = 0.05$ level (67.8 vs. 62.2 bands; $P = 0.101$). Two private bands separated pooled island and mainland sites, and locally common bands were present among populations at each of the three island and two mainland locations (Table 1).

Genetic differentiation among populations and regions

Most of the variation in the AFLP profiles reported here represented variation among populations within regions. Hierarchical AMOVA for the island and mainland data partitioned 37.1 % of the variation within populations, 55.8 % among populations within regions and 7.1 % between regions (Table 3). All values were significantly different from zero ($P < 0.0001$) and Φ_{PT} summed to 0.629. The average pairwise Φ_{PT} for all sites was 0.614 with a range from 0.159 (VAFB populations, V3 and V4) to 0.91 (Sedgwick and Santa Cruz populations). All pairwise comparisons for Φ_{PT} were significant [see Supporting Information]. Population genetic structure declined within regions ($\Phi_{PT} = 0.422$ among mainland and 0.400 among island populations) relative to structure calculated for all sites. The results of the Mantel test of linearized Φ_{PT} values, and $\log_{10}$ transformed geographic distances indicated no support for isolation by distance among all island and mainland sites ($P = 0.175$). This result was repeated when

mainland sites were considered alone ($P = 0.293$). However, isolation by distance was detected among populations sampled on the three Channel Islands (matrix correlation coefficient = 0.153, $P = 0.023$). This result was limited to the groups of populations among islands and isolation by distance was not observed for populations within each island [see Supporting Information]. The value of N_m, or the indirect rate of gene flow, was calculated as 0.147, indicating that the number of migrants each generation was significantly <1 and that genetic drift plays a role in population differentiation (Slatkin 1987).

Using the Bayesian approach, two models of $\theta^{(II)}$ (full and $f = 0$) resulted in low DIC values, and we selected the full model with the expectation that the inbreeding coefficient (f) is >0 in this autogamous species (K. Holsinger, pers. comm.; Wilson et al. 2001). Under the full model, the value of $\theta^{(II)}$ for all 21 sites was 0.563, while values for the subsets of mainland or island sites were 0.557 and 0.556, respectively. Comparisons of posterior distributions detected no significant difference in population structure between island and mainland regions (difference in $\theta^{(II)}$ values: 0.001; confidence interval: -0.0379, 0.03932). Estimates of $\theta^{(II)}$ were not greater for populations located on islands relative to sampled populations on the mainland. Sedgwick and VAFB also did not differ significantly for estimates of $\theta^{(II)}$. However, Santa Catalina and Santa Rosa populations were significantly less differentiated than populations from Santa Cruz (difference in $\theta^{(II)}$ values: 0.189; confidence interval: 1349, 0.2445). Genetic structure was greater among sampled sites located on Santa Cruz Island.

Not all sampled populations were genetically distinct from one another. The highest likelihood of the number of genetic clusters represented in the AFLP dataset was consistently obtained with $K = 16$ (Fig. 3). In effect, the 21 collections represented 16 genetically distinct populations. Sites V3 and V4 corresponded to a single genetic cluster and these sites were located ~3 km apart. Individuals among the four collection sites at Santa Rosa Island were also highly similar (2–7 km apart) despite their location in separate drainages on the northeastern shore of the island. Lastly, pairs of sites within Sedgwick and within Santa Cruz Island were overlapping. Altogether, evidence of admixture was low and plants derived from no more than five to seven of the original sites appeared to have patterns of diversity representative of dispersal and gene flow within islands (Fig. 3). Essentially, these individuals shared membership in more than one population. Most populations had fixed genetic differences that did not vary widely among individuals.

Table 3. Analysis of molecular variance (AMOVA) results for E. glaucus populations located within and among the two (island and mainland) regions

Source of variation	d.f.	Sum of squares	Variance components	Variation (%)
Among regions	1	762.55	2.08	7.1
Among populations, within regions	19	336.02	16.42	55.8
Within populations	395	10.95	10.95	37.1

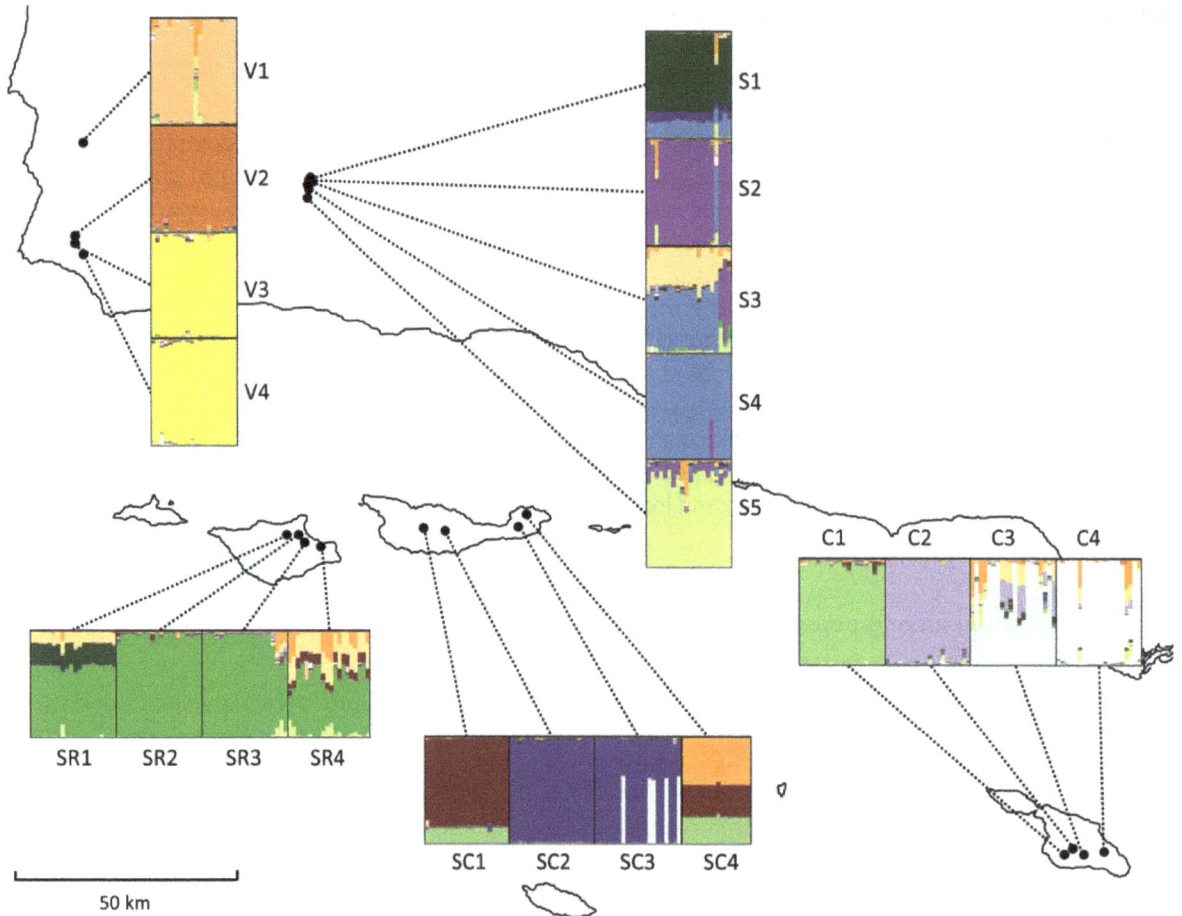

Figure 3. Estimated population structure for $K = 16$ in Bayesian analysis. Each of the study sites is represented by 21 segments and individuals within sites are designated by vertical columns. If individuals share the same pattern of genetic diversity, segments are homogeneous in colour.

Discussion

Strong levels of genetic differentiation observed in this study indicate that *E. glaucus* populations are highly self-pollinating and isolated at small distances. As a result, populations represent genetically homozygous lines and the distribution of diversity is heavily influenced by the breeding system (Schemske and Lande 1985). Similar patterns have been observed for previous studies of *E. glaucus* in North America, as well as among populations of other grass species worldwide (e.g. Jain and Allard 1960; Nevo et al. 1986; Knapp and Rice 1996). Predominantly self-pollinating species lead to different expectations for patterns of genetic diversity, and represent unique challenges for conservation that merit consideration when these species are targeted for reintroduction. In our study of *E. glaucus* among California Channel Island and mainland locations, we addressed questions regarding the levels and patterns of genetic diversity and differentiation at relatively isolated island sites, and their consequences for ecological restoration.

Genetic diversity

Despite strong evidence for self-pollination, marker data indicated that considerable genetic variation is present in blue wildrye. Measures of the mean proportion of polymorphism within studied *E. glaucus* populations were ~40 %. The distribution of genetic diversity may be the result of factors other than inbreeding, including the polyploid origin of *E. glaucus* as well as the potential, however small, for gene exchange among populations (Jain and Allard 1960). Genetic polymorphism present in alloploid grasses such as *E. glaucus* may be a consequence of segregation between homologous chromosomes within each parental genome, resulting in fixed heterozygosity in selfing lineages (Soltis and Soltis 1993). Additionally, self-pollination is rarely complete and gene exchange among lineages is possible through wind pollination in combination with seed dispersal among populations. Previous isozyme studies of *E. glaucus* growing in the northwestern United States and British Columbia also detected high levels of polymorphism at the species (77–80 %) and population level (22–31 %) (Knapp and Rice 1996;

Ie 2000; Wilson *et al.* 2001), suggesting that high levels of genetic variation are maintained across this species' range.

Relative to the entire region sampled in this study, measures of genetic variation declined significantly when data were partitioned within island or mainland locations and particularly for individual populations, many of which were genetically homogeneous for AFLP markers. Mean expected heterozygosity ($H_e = 0.109$) for Channel Island and coastal California populations was low and corresponded to values recorded for selfing species with passively dispersed seeds ($H_e = 0.097$; Hamrick and Godt 1996), but varied among sampled sites. Populations located on Santa Cruz and Santa Rosa Islands had the lowest average levels of polymorphism while populations on Catalina Island were the most diverse. Levels of genetic diversity may vary among island and mainland locations, perhaps as a result of variation in the breeding system in addition to stochastic demographic and environmental factors (Schemske and Lande 1985).

Comparisons of the number of polymorphic bands between the two regions representing islands and the mainland indicated that genetic diversity was significantly higher for mainland populations relative to the Channel Islands. These data support predictions that island populations are genetically depauperate as a result of founder events, low rates of gene flow and subsequent genetic drift (Frankham 1997). Sampled populations for this study commonly represented fewer than 60 visible plants and locating populations on the islands was difficult, probably as a result of greater geographic isolation among *E. glaucus* populations on the islands. The consequences of relatively lower levels of diversity remain unclear, however. In many taxa, genetic diversity strongly correlates with population fitness and may be useful for identification of populations at risk of decline (Reed and Frankham 2003), but this pattern requires further investigation in self-pollinating species (Leimu *et al.* 2006). High levels of inbreeding can purge deleterious alleles so that selfing species are unlikely to experience inbreeding depression, and yet heterosis is commonly reported when gene exchange occurs (Jain and Allard 1960; Schemske and Lande 1985).

Genetic differentiation among populations and regions

We discovered strong genetic differentiation in *E. glaucus* within and among island and mainland sites. Most of the variation was distributed among populations within regions and overall estimates of F_{ST} were very high ($\Phi_{PT} = 0.629$, $\theta^{(II)} = 0.563$) and similar to all prior studies of this species (Knapp and Rice 1996; Ie 2000; Wilson

et al. 2001). These data point to the dominant role of the breeding system in shaping the population structure of *E. glaucus*. At the same time, relative levels of isolation may also affect patterns of genetic differentiation if the distance among populations further limits rare episodes of gene flow.

Tests for isolation-by-distance (IBD) were not significant for populations on the mainland and IBD was only apparent between islands. Two previous mainland studies of *E. glaucus* conducted at a larger scale also failed to detect IBD (Knapp and Rice 1996; Wilson *et al.* 2001), while low levels of IBD were detected for sites sampled throughout British Columbia (Ie 2000). Overall, genetic structure was not greater among Channel Island populations than among mainland populations and IBD was not observed among populations within each island.

Bayesian cluster analysis presented some evidence for admixture in *E. glaucus* as the result of gene flow, although this evidence was limited to few individuals and populations (e.g. SR4, Santa Rosa Island). Inference of genetic structure detected 16 genetically distinct populations among 21 collections, and confirmed strong population differentiation in *E. glaucus* among islands and the mainland. The average pairwise geographic distance among sites was ~100 km. However, many sites within locations were only 2–3 km apart. We noted that the two closest collection sites (S1 and S2; 0.15 km apart) were strongly genetically differentiated. High levels of differentiation at small distances are common among self-pollinating grass species and represent challenges for their restoration in degraded ecosystems (Allard 1975).

Implications for restoration

Evidence for lower levels of polymorphism among islands suggests that island populations of blue wildrye are genetically depauperate relative to the mainland. In these cases, the introduction of diverse seed sources in restoration has been proposed to increase population viability by reduction of inbreeding depression, and creation of opportunities for hybrid vigour (Broadhurst *et al.* 2008; Pekkala *et al.* 2012). Conversely, local seed collections are often proposed to avoid disruption of local adaptation and the incidence of outbreeding depression among hybrid progeny of remnant populations and introduced plants (Hufford and Mazer 2003). How do these contrasting views apply to restoration of *E. glaucus*?

If inbreeding depression is weak in self-fertilizing species (Schemske and Lande 1985), the preservation of genetic variation to conserve population viability should no longer be a large concern in reintroduction programmes. It does not follow, however, that the diversity of lines represented in seed mixes would also be of little concern. There is strong evidence for the maintenance of heterozygotes in selfing species, and heterozygote advantage has been proposed as a mechanism to improve population fitness when outcrossing is rare (Allard and Jain 1962). Ideally, when reintroducing species such as blue wildrye, the goal would be to maintain or even increase genetic variability among plants. In this case, the introduction of seeds representing large numbers of mixed lines would be supported (Broadhurst et al. 2008). However, non-local genotypes may represent plants poorly adapted to local environmental conditions.

High levels of genetic differentiation may be the product of both genetic drift and natural selection (Slatkin 1987), and evidence for local adaptation in E. glaucus has been detected in two reciprocal transplant studies (Hufford et al. 2008; Knapp and Rice 2011). In addition, other studies of blue wildrye detected high levels of fixed variation that were correlated with morphological variation among sites (e.g. Knapp and Rice 1996; Wilson et al. 2001; Erickson et al. 2004). In light of this evidence, population genetic divergence apparent in E. glaucus may reflect not only limited gene flow but also adaptation to heterogeneous environments. We noted that genetic structure declined when the full dataset was partitioned among islands and mainland locations ($\Phi_{PT} = 0.422$ for the mainland and 0.400 for island populations relative to 0.629 among all sampled populations), and this scale corresponds to the scale of adaptive differentiation detected in field studies of this species. We consequently recommend the use of multiple seed collections specific to each Channel Island (for use at that island) to maintain diversity while avoiding long-distance introductions of non-local genotypes. Given low levels of pollen flow, these introductions are not likely to affect the viability of existing populations, and in rare cases where gene flow does occur, hybrid vigour among progeny would likely result in the formation of new selfing lines (Allard and Jain 1962; Nevo et al. 1986). Similar guidelines might apply to mainland locations, but would improve with further investigation of mainland sites.

We finish by noting that our recommendations are based on both evidence and speculation. It is unlikely that AFLP markers, which are presumed neutral, will correspond to adaptive differentiation in this species, and strategies for transfer of E. glaucus are derived from previous studies of adaptive variation as well as marker data (Kirk and Freeland 2011). Delineation of seed transfer zones, or regions within which translocation of plant materials is unlikely to result in the introduction of maladapted genotypes, is currently the best method to predict adaptive divergence and select suitable germplasm for restoration (e.g. Johnson et al. 2004). Yet, data for adaptive differentiation at the scale of contemporary grassland restoration are scarce. Additional research

(e.g. Hufford et al. 2008) to examine the scale of adaptive divergence among populations in sensitive regions would benefit restoration of temperate grasslands. In the interim, marker data can assist with the selection of germplasm to minimize the risk of introductions and yet maintain high levels of genetic diversity. Restoration sites, including large-scale seeding efforts under way across the western United States (Burton and Burton 2002), will serve as the proving ground for marker predictions of short- and long-term population viability in contemporary and changing environments.

Sources of Funding

This research was supported by the National Parks Ecological Research Fellowship Program, a programme funded by the National Park Foundation through a generous grant from the Andrew W. Mellon foundation.

Contributions by the Authors

K.M.H., S.J.M. and S.A.H. conceived and designed the study. K.M.H. and S.J.M. conducted field collections. K.M.H. conducted laboratory and data analyses and wrote the manuscript with the help of S.J.M. and S.A.H.

Acknowledgements

We thank Sarah Chaney and Dirk Rodrigues at Channel Islands National Park, Lyndal Laughrin at the Santa Cruz Island Reserve, Randy Miller at VAFB, Mike Williams at Sedgwick, and the Catalina Island Conservancy for assistance with field collections. Laura Golde assisted with DNA extractions, and Justen Whittall and Ji Yang provided assistance with genetic marker protocols.

Supporting Information

The following files are available in the online version of this article –

Table S1. Pairwise genetic distance (Φ_{PT}) values among the 21 sampled E. glaucus populations.

Figure S1. Relationship between linearized Φ_{PT} values and $\log_{10}$ values for geographic distance among populations of E. glaucus distributed on Santa Rosa, Santa Cruz and Catalina Islands off the coast of mainland California. Axes are oriented to highlight clusters of points representing each island.

Literature Cited

Allard RW. 1975. The mating system and microevolution. *Genetics* **79**:S115–S126.

Allard RW, Jain SK. 1962. Population studies in predominantly self-pollinated species. II. Analysis of quantitative genetic changes in a bulk-hybrid population of barley. *Evolution* **16**:90–101.

Barry S, Larson S, George M. 2006. California native grasslands: a historical perspective—a guide for developing realistic restoration objectives. *Grasslands* **16**:7–11.

Bartolome JW, Fehmi JS, Jackson RD, Allen-Diaz B. 2004. Response of a native perennial grass stand to disturbance in California's Coast Range grassland. *Restoration Ecology* **12**:279–289.

Bischoff A, Steinger T, Müller-Schärer H. 2010. The importance of plant provenance and genotypic diversity of seed material used for ecological restoration. *Restoration Ecology* **18**:338–348.

Bonin A, Ehrich D, Manel S. 2007. Statistical analysis of amplified fragment length polymorphism data: a toolbox for molecular ecologists and evolutionists. *Molecular Ecology* **16**:3737–3758.

Broadhurst LM, Lowe A, Coates DJ, Cunningham SA, McDonald M, Vesk PA, Yates C. 2008. Seed supply for broadscale restoration: maximizing evolutionary potential. *Evolutionary Applications* **1**: 587–597.

Burton PJ, Burton CM. 2002. Promoting genetic diversity in the production of large quantities of native plant seed. *Ecological Restoration* **20**:117–123.

Cox RD, Allen EB. 2011. The roles of exotic grasses and forbs when restoring native species to highly invaded southern California annual grassland. *Plant Ecology* **212**:1699–1707.

D'Antonio CM, Vitousek PM. 1992. Biological invasions by exotic grasses, the grass/fire cycle, and global change. *Annual Review of Ecology and Systematics* **23**:63–87.

DiTomaso JM, Enloe SF, Pitcairn MJ. 2007. Exotic plant management in California annual grasslands. In: Stromberg MR, Corbin JD, D'Antonio CM, eds. *California grasslands, ecology and management*. Berkeley, CA: University of California Press, 281–296.

Earl DA, vonHoldt BM. 2012. STRUCTURE HARVESTER: a website and program for visualizing STRUCTURE output and implementing the Evanno method. *Conservation Genetics Resources* **4**:359–361.

Ehrich D. 2006. AFLPdat: a collection of R functions for convenient handling of AFLP data. *Molecular Ecology Notes* **6**:603–604.

Erickson VJ, Mandel NL, Sorensen FC. 2004. Landscape patterns of phenotypic variation and population structuring in a selfing grass, *Elymus glaucus* (blue wildrye). *Canadian Journal of Botany* **82**:1776–1789.

Evanno G, Regnaut S, Goudet J. 2005. Detecting the number of clusters of individuals using the software STRUCTURE: a simulation study. *Molecular Ecology* **14**:2611–2620.

Falush D, Stephens M, Pritchard JK. 2007. Inference of population structure using multilocus genotype data: dominant markers and null alleles. *Molecular Ecology Notes* **7**:574–578.

Frankham R. 1997. Do island populations have less genetic variation than mainland populations? *Heredity* **78**:311–317.

Frankham R, Ballou JD, Briscoe DA. 2002. *Introduction to conservation genetics*. Cambridge: Cambridge University Press.

Halvorson WL. 1992. Alien plants at Channel Islands National Park. In: Stone CP, Smith CW, Tunison JT, eds. *Alien plant invasions in native ecosystems of Hawai'i: management and research*. Honolulu, HI: University of Hawai'i, 64–96.

Halvorson WL. 1994. Ecosystem restoration on the California Channel Islands. In: Halvorson WL, Maender JG, eds. *The Fourth California Islands Symposium: update on the status of resources*. Santa Barbara, CA: Santa Barbara Museum of Natural History, 485–490.

Hamrick JL, Godt MJW. 1996. Effects of life history traits on genetic diversity in plant species. *Philosophical Transactions of the Royal Society of London. Series B, Biological Sciences* **351**: 1291–1298.

Hayes GF, Holl KD. 2003. Cattle grazing impacts on annual forbs and vegetation composition of mesic grasslands in California. *Conservation Biology* **17**:1694–1702.

Heywood JS. 1991. Spatial analysis of genetic variation in plant populations. *Annual Review of Ecology and Systematics* **22**:335–355.

Hickman J. 1993. *The Jepson manual: higher plants of California*. Berkeley, CA: University of California Press.

Holsinger KE, Lewis PO. 2003. *Hickory: a package for analysis of population genetic data V1.0*. Storrs, CT: Department of Ecology and Evolutionary Biology, University of Connecticut.

Holsinger KE, Lewis PO, Dey DK. 2002. A Bayesian approach to inferring population structure from dominant markers. *Molecular Ecology* **11**:1157–1164.

Holstein G. 2001. Pre-agricultural grassland in Central California. *Madroño* **48**:253–264.

Honnay O, Jacquemyn H. 2007. Susceptibility of common and rare plant species to the genetic consequences of habitat fragmentation. *Conservation Biology* **21**:823–831.

Hufford KM, Mazer SJ. 2003. Plant ecotypes: genetic differentiation in the age of ecological restoration. *Trends in Ecology and Evolution* **18**:147–155.

Hufford KM, Mazer SJ, Camara MD. 2008. Local adaptation and effects of grazing among seedlings of two native California bunchgrass species: implications for restoration. *Restoration Ecology* **16**:59–69.

Ie B. 2000. *Reclaiming disturbed habitats using native grasses: the genetic story of Elymus glaucus (blue wildrye)*. MSc Thesis, University of British Columbia, Canada.

Jain SK, Allard RW. 1960. Population studies in predominantly self-pollinated species, I. evidence for heterozygote advantage in a close population of barley. *Proceedings of the National Academy of Sciences of the USA* **46**:1371–1377.

Jakobsson M, Rosenberg NA. 2007. *CLUMPP*: a cluster matching and permutation program for dealing with label switching and multi-modality in analysis of population structure. *Bioinformatics* **23**: 1801–1806.

Jensen KB, Ahang YF, Dewey DR. 1990. Mode of pollination of perennial species in the Triticeae in relation to genomically defined genera. *Canadian Journal of Plant Sciences* **70**:215–225.

Johnson GR, Sorensen FC, St. Clair JB, Cronn RC. 2004. Pacific northwest forest tree seed zones: a template for native plants? *Native Plants* **5**:131–140.

Junak S, Ayers T, Scott R, Wilken D, Young D. 1995. *A flora of Santa Cruz Island*. Santa Barbara: Santa Barbara Botanic Garden.

Junak S, Chaney S, Philbrick R, Clark R. 1997. *A checklist of vascular plants of Channel Islands National Park*. Tucson, AZ: Southwest Parks and Monuments Association.

Kang M, Ye Q, Huang H. 2005. Genetic consequence of restricted habitat and population decline in endangered *Isoetes sinensis* (Isoetaceae). *Annals of Botany* **96**:1265–1274.

Kawecki TJ, Ebert D. 2004. Conceptual issues in local adaptation. *Ecology Letters* **7**:1225–1241.

Kirk H, Freeland JR. 2011. Applications and implications of neutral versus non-neutral markers in molecular ecology. *International Journal of Molecular Sciences* **12**:3966–3988.

Knapp EE, Rice KJ. 1996. Genetic structure and gene flow in *Elymus glaucus* (blue wildrye): implications for native grassland restoration. *Restoration Ecology* **4**:1–10.

Knapp EE, Rice KJ. 2011. Effects of competition and temporal variation on the evolutionary potential of two native bunchgrass species. *Restoration Ecology* **19**:407–417.

Kumar S, Skjæveland Å, Orr RJS, Enger P, Ruden T, Mevik B, Burki F, Botnen A, Shalchian-Tabrizi K. 2009. AIR: a batch-oriented web program package for construction of supermatrices ready for phylogenomic analyses. *BMC Bioinformatics* **10**:357.

Lande R. 1988. Genetics and demography in biological conservation. *Science* **241**:1455–1460.

Leimu R, Mutikainen P, Koricheva J, Fischer M. 2006. How general are positive relationships between plant population size, fitness and genetic variation? *Journal of Ecology* **94**:942–952.

Linhart YB, Grant MC. 1996. Evolutionary significance of local genetic differentiation in plants. *Annual Review of Ecology and Systematics* **27**:237–277.

Mack RN. 1989. Temperate grasslands vulnerable to plant invasions: characteristics and consequences. In: Drake JA, Mooney HA, Di Castri F, Groves RH, Kruger FJ, eds. *Biological invasions: a global perspective*. New York: John Wiley & Sons Ltd, 155–179.

McKay JK, Christian CE, Harrison S, Rice KJ. 2005. 'How local is local?'—a review of practical and conceptual issues in the genetics of restoration. *Restoration Ecology* **13**:432–440.

Mensing SA. 1998. 560 years of vegetation change in the region of Santa Barbara, California. *Madroño* **45**:1–11.

Moody A. 2000. Analysis of plant species diversity with respect to island characteristics on the Channel Islands, California. *Journal of Biogeography* **27**:711–723.

Moyes AB, Witter MS, Gamon JA. 2005. Restoration of native perennials in a California annual grassland after prescribed spring burning and solarization. *Restoration Ecology* **13**:659–666.

Nevo E, Beiles A, Kaplan D, Storch N, Zohary D. 1986. Genetic diversity and environmental associations of wild barley, *Hordeum spontaneum* (Poaceae), in Iran. *Plant Systematics and Evolution* **153**:141–164.

Nuismer SL, Gandon S. 2008. Moving beyond common-garden and transplant designs: insight into the causes of local adaptation in species interactions. *The American Naturalist* **171**:658–668.

Peakall R, Smouse PE. 2006. GENALEX 6: genetic analysis in Excel. Population genetic software for teaching and research. *Molecular Ecology Notes* **6**:288–295.

Pekkala N, Knott KE, Kotiaho JS, Nissinen K, Puurtinen M. 2012. The benefits of interpopulation hybridization diminish with increasing divergence of small populations. *Journal of Evolutionary Biology* **25**:2181–2193.

Pritchard JK, Stephens M, Donnelly P. 2000. Inference of population structure using multilocus genotype data. *Genetics* **155**:945–959.

R Development Core Team. 2011. *R: a language and environment for statistical computing*. Vienna: R Foundation for Statistical Computing. ISBN 3-900051-07-0. http://www.R-project.org.

Reed DH, Frankham R. 2003. Correlation between fitness and genetic diversity. *Conservation Biology* **17**:230–237.

Rosenberg NA. 2004. *Distruct*: a program for the graphical display of population structure. *Molecular Ecology Notes* **4**:137–138.

Rousset F. 1997. Genetic differentiation and estimation of gene flow from *F*-statistics under isolation by distance. *Genetics* **145**:1219–1228.

Schemske DW, Lande R. 1985. The evolution of self-fertilization and inbreeding depression in plants. II. Empirical observations. *Evolution* **39**:41–52.

Schlüter PM, Harris SA. 2006. Analysis of multilocus fingerprinting data sets containing missing data. *Molecular Ecology Notes* **6**:569–572.

Schoenherr AA, Feldmeth CR, Emerson MJ. 1999. *Natural history of the islands of California*. Berkeley, CA: University of California Press.

Shaffer ML. 1981. Minimum population sizes for species conservation. *BioScience* **31**:131–134.

Slatkin M. 1987. Gene flow and the geographic structure of natural populations. *Science* **236**:787–792.

Snyder LA. 1951. Morphological variability and hybrid development in *Elymus glaucus*. *American Journal of Botany* **37**:628–636.

Soltis DE, Soltis PS. 1993. Molecular data and the dynamic nature of polyploidy. *Critical Reviews in Plant Sciences* **12**:243–273.

Vekemans X. 2002. *AFLP-SURV version 1.0*. Distributed by the author. Belgium: Laboratoire de Génétique et Ecologie Végétale, Université Libre de Bruxelles.

Vos P, Hogers R, Bleeker M, Reijans M, Van de Lee T, Hornes M, Frijters A, Pot J, Peleman J, Kuiper M, Zabeau M. 1995. AFLP: a new technique for DNA fingerprinting. *Nucleic Acids Research* **23**:4407–4414.

Wagner V, Treiber J, Danihelka J, Ruprecht E, Wesche K, Hensen I. 2012. Declining genetic diversity and increasing genetic isolation toward the range periphery of *Stipa pennata*, a Eurasian feather grass. *International Journal of Plant Sciences* **173**:802–811.

Wilson BL, Kitzmiller J, Rolle W, Hipkins VD. 2001. Isozyme variation and its environmental correlates in *Elymus glaucus* from the California Floristic Province. *Canadian Journal of Botany* **79**:139–153.

Wright S. 1951. The genetical structure of populations. *Annals of Eugenics* **15**:323–354.

Chloroplast genes as genetic markers for inferring patterns of change, maternal ancestry and phylogenetic relationships among *Eleusine* species

Renuka Agrawal[1], Nitin Agrawal[2], Rajesh Tandon[1] and Soom Nath Raina[3]*

[1] Laboratory of Cellular and Molecular Cytogenetics, Department of Botany, University of Delhi, Delhi 110007, India
[2] Cluster Innovation Centre, University of Delhi, Delhi 110007, India
[3] Present address: Amity Institute of Biotechnology, Amity University, Sector 125, Noida 201303, Uttar Pradesh, India

Abstract. Assessment of phylogenetic relationships is an important component of any successful crop improvement programme, as wild relatives of the crop species often carry agronomically beneficial traits. Since its domestication in East Africa, *Eleusine coracana* ($2n = 4x = 36$), a species belonging to the genus *Eleusine* ($x = 8, 9, 10$), has held a prominent place in the semi-arid regions of India, Nepal and Africa. The patterns of variation between the cultivated and wild species reported so far and the interpretations based upon them have been considered primarily in terms of nuclear events. We analysed, for the first time, the phylogenetic relationship between finger millet (*E. coracana*) and its wild relatives by species-specific chloroplast deoxyribonucleic acid (cpDNA) polymerase chain reaction–restriction fragment length polymorphism (PCR–RFLP) and chloroplast simple sequence repeat (cpSSR) markers/sequences. Restriction fragment length polymorphism of the seven amplified chloroplast genes/intergenic spacers (*trn*K, *psb*D, *psa*A, *trn*H–*trn*K, *trn*L–*trn*F, *16*S and *trn*S–*psb*C), nucleotide sequencing of the chloroplast *trn*K gene and chloroplast microsatellite polymorphism were analysed in all nine known species of *Eleusine*. The RFLP of all seven amplified chloroplast genes/intergenic spacers and *trn*K gene sequences in the diploid ($2n = 16, 18, 20$) and allotetraploid ($2n = 36, 38$) species resulted in well-resolved phylogenetic trees with high bootstrap values. *Eleusine coracana*, *E. africana*, *E. tristachya*, *E. indica* and *E. kigeziensis* did not show even a single change in restriction site. *Eleusine intermedia* and *E. floccifolia* were also shown to have identical cpDNA fragment patterns. The cpDNA diversity in *Eleusine multiflora* was found to be more extensive than that of the other eight species. The *trn*K gene sequence data complemented the results obtained by PCR–RFLP. The maternal lineage of all three allotetraploid species (AABB, AADD) was the same, with *E. indica* being the maternal diploid progenitor species. The markers specific to certain species were also identified.

Keywords: cpSSR; *Eleusine*; PCR–RFLP; phylogeny; Poaceae; *trn*K gene sequence.

Introduction

The wild relatives of crop species often carry beneficial alleles that are effective against various biotic and abiotic stresses. They hold the key to successful crop improvement programmes through introgression of desired genes from wild to cultivated crop species (Dida and Devos 2006). In this context, the assessment of phylogenetic relationships at the inter-specific level and the identification of gene

* Corresponding author's e-mail address: soomr@yahoo.com

pools are considered important. Based on chloroplast deoxyribonucleic acid (cpDNA) diversity, we obtained information on the molecular phylogeny of finger millet (*Eleusine coracana*) vis-à-vis its wild relatives.

The genus *Eleusine* is a member of the tribe Eragrosteae, subfamily Chloridoideae and the family Poaceae. It is a small genus of nine species, which includes six diploid ($2n = 2x = 16, 18, 20$) and three polyploid ($2n = 4x = 36, 38$) species (Bisht and Mukai 2002; Liu *et al.* 2011). These species are widely distributed in the tropical and subtropical regions of Africa, Asia and South America (Phillips 1972). East Africa is considered to be the centre of diversity for the genus (Bisht and Mukai 2002).

Eleusine coracana ($2n = 4x = 36$), commonly known as finger millet or ragi, is the only economically important species of the genus. After sorghum and pearl millet, finger millet ranks third in cereal production in the semi-arid regions of the world (Bisht and Mukai 2002). It is widely cultivated in the arid and semi-arid regions of East Africa, India, Nepal and many other Asian countries for its grain and fodder value (Verma 2009). The grain is widely used for preparing bread, cakes, soup, puddings, porridge and fermented beverages (Hilu and deWet 1976; Chandrashekar 2010; Neves 2011). Finger millet is a rich source of essential amino acids and polyphenols, and it has comparatively higher levels of calcium and iron than other known cereals (Barbeau and Hilu 1993; Chandrashekar 2010). It has a number of medicinal properties as well, particularly in controlling blood sugar levels in diabetic patients (Duke and Wain 1981; Chandrashekar 2010; Pradhan *et al.* 2010).

The assessment of genomic relationships between *E. coracana* and its allied species has been a subject of comprehensive investigations at the morphological, cytogenetic, biochemical and DNA level. Chromosome research has demonstrated the significant role of polyploidy and aneuploidy in the evolution of the genus (Chennaveeraiah and Hiremath 1974; Hiremath and Chennaveeraiah 1982; Hiremath and Salimath 1991; Bisht and Mukai 2000, 2001*a*, *b*, 2002). Biochemical, nuclear and cpDNA markers have provided valuable insight into relationships, and on the origin of the crop species (Hilu *et al.* 1978; Hilu 1988, 1995; Hilu and Johnson 1992, 1997; Werth *et al.* 1993, 1994; Salimath *et al.* 1995*a*; Neves *et al.* 2005; Dida *et al.* 2007, 2008; Liu *et al.* 2011). The $2n$ number and the genomic formula proposed on the basis of earlier studies are given in Table 1.

In spite of the enormous amount of information, there still exists considerable disagreement over the identification of the diploid ancestors of the three polyploid species and the level of speciation and evolutionary relationships among the nine species of *Eleusine* (Phillips 1972, 1995;

Hilu and Johnson 1997; Lye 1999; Bisht and Mukai 2000, 2001*a*, 2002; Devarumath *et al.* 2005; Neves *et al.* 2005; Liu *et al.* 2011). *Eleusine coracana* ($2n = 4x = 36$) (AABB) is considered to be an allotetraploid and has been domesticated from its wild progenitor, *Eleusine africana* ($2n = 4x = 36$) (AABB) (Chennaveeraiah and Hiremath 1974; Hilu and deWet 1976; Hilu *et al.* 1978; Hilu 1988, 1995; Hiremath and Salimath 1992; Werth *et al.* 1994; Hilu and Johnson 1997; Bisht and Mukai 2000, 2001*a*, *b*; Devarumath *et al.* 2005; Neves *et al.* 2005; Dida *et al.* 2007, 2008). The genetic maps generated by Dida *et al.* (2007) showed that *E. coracana* and *E. africana* are allotetraploids. There is also strong evidence to suggest that *Eleusine indica* ($2n = 2x = 18$) (AA) is the maternal genome donor of *E. coracana* and *E. africana* (Hilu 1988; Hiremath and Salimath 1992; Hilu and Johnson 1997; Bisht and Mukai 2000, 2001*a*; Neves *et al.* 2005). Bisht and Mukai (2000, 2001*a*, 2002) indicated that *Eleusine floccifolia* ($2n = 2x = 18$) could be the BB donor species of *E. coracana*. This has, however, been refuted by others (Hiremath and Salimath 1992; Neves *et al.* 2005; Devarumath *et al.* 2010; Liu *et al.* 2011). According to these authors, the BB genome donor species remains unidentified and may possibly be extinct. *Eleusine kigeziensis* ($2n = 4x = 36$ or 38) (AADD) is the third tetraploid species of the genus. *Eleusine indica* ($2n = 2x = 18$) (AA) and *E. jaegeri* ($2n = 2x = 20$) (DD) are proposed to be the wild progenitors of *E. kigeziensis* (Bisht and Mukai 2002; Devarumath *et al.* 2010). On the contrary, Neves *et al.* (2005) proposed *E. kigeziensis* to be autotetraploid, with *E. indica* being closely related to *E. kigeziensis* but not the direct genome donor to *E. kigeziensis*. Liu *et al.* (2011) have concluded that all three tetraploids (*E. coracana, E. africana* and *E. kigeziensis*) are of allotetraploid origin. They suggested independent origins of *E. kigeziensis* and *E. africana*— *E. coracana*. They are of the view that both events may have involved the diploids *E. indica* and *E. tristachya* as maternal parents, but the paternal parents remain unidentified. *Eleusine indica* and *E. tristachya* are considered to be very similar and the degree of relationship between the two remains unresolved (Hiremath and Chennaveeraiah 1982; Hiremath and Salimath 1991; Hilu and Johnson 1992; Werth *et al.* 1994; Bisht and Mukai 2001*a*, *b*; Neves *et al.* 2005; Liu *et al.* 2011).

In the present study, the cpDNA restriction site pattern variation of seven amplified chloroplast genes/intergenic spacers, the chloroplast *trn*K gene sequence and cp microsatellite polymorphism were investigated for the first time in all nine *Eleusine* species, with the objective of constructing the phylogeny of the genus *Eleusine* and identifying the maternal genome donors of the polyploid species including *E. coracana*. Both direct sequencing of

Table 1. Plant materials used in the present study. [a]Dr. Mathews M. Dida, Kenya; USDA, United States Department of Agriculture, USA; ILRI, International Livestock Research Institute, Ethopia; NBPGR, National Bureau of Plant Genetic Resources, India. [b]Bisht and Mukai (2002).

Species	Accession number	Source[a]	2n	Genome[b] formula	Growth habit
E. coracana	PI 482594	USDA	36	AABB	Annual
	PI 462778	USDA			
	PI 462779	USDA			
E. africana	EC 541535	Dida, Kenya	36	AABB	Annual
	EC 541536	Dida, Kenya			
	PI 226270	USDA			
	PI 315700	USDA			
E. tristachya	PI 477078	USDA	18	AA	Annual
	PI 331791	USDA			
E. indica	PI 442480	USDA	18	AA	Annual
E. floccifolia	PI 196853	USDA	18	BB	Perennial
E. multiflora	PI 226067	USDA	16	CC	Annual
E. jaegeri	PI 273888	USDA	20	DD	Perennial
E. kigeziensis	1112	ILRI	38	AADD	Perennial
	1079	ILRI			
E. intermedia	S. No 116	ILRI	18	AB	Perennial
D. aegyptium	IC-285214	NBPGR			

the trnK gene and polymerase chain reaction–restriction fragment length polymorphism (PCR–RFLP) of cpDNA regions and chloroplast simple sequence repeats (cpSSRs) are considered to be very good markers for detecting cpDNA variation. The chloroplast trnK gene (which also contains the matK gene within the trnK gene intron) sequence has been effectively used for the construction of grass phylogenies (Hilu and Alice 1999; Hilu et al. 1999). The matK gene sequence is one of the seven loci widely utilized for the DNA barcoding of plants (Babbar et al. 2012). Specific chloroplast genes and/or intergenic spacers can be amplified (Taberlet et al. 1991; Demesure et al. 1995; Tsumura et al. 1995, 1996; Dhingra and Folta 2005; Heinze 2005). The amplicons can be directly sequenced or restriction endonuclease digested (PCR–RFLP or cleaved amplified polymorphic sequence), and the occurrence of microsatellites (cpSSRs) in the chloroplast genome has been widely utilized for species identification, reconstruction of phylogenetic relationships, taxonomic studies and the identification of maternal parents in polyploids (Tsumura et al. 1995, 1996; Weising and Gardner 1999; Ishii and McCouch 2000; Lakshmi et al. 2000; Parani et al. 2000, 2001; Komatsu et al. 2001; Provan et al. 2001; Kishimoto et al. 2003; Nwakanma et al. 2003; Zhu et al. 2003; Asadi Abkenar et al. 2004, 2008; Van Droogenbroeck et al. 2004; Ibrahim

et al. 2007; Angioi et al. 2008; Sehgal et al. 2008; Jena et al. 2009; Liu et al. 2011; Poczai et al. 2011).

Methods

Plant materials

Seeds of the Eleusine species and one outgroup species (Dactyloctenium aegyptium) utilized in the present study were obtained from the United States Department of Agriculture (USDA) (Beltsville, MD, USA), the International Livestock Research Institute (ILRI) (Addis Ababa, Ethiopia), Dr Mathews M. Dida (Maseno University, Maseno, Kenya) and the National Bureau of Plant Genetic Resources (NBPGR) (New Delhi, India). The accession numbers and source of the seed material are given in Table 1. The seeds were grown under controlled conditions in the experimental field of the Department of Botany, University of Delhi.

DNA extraction

The total genomic DNA of the Eleusine species and D. aegyptium was isolated from fresh young leaves using the cetyl trimethyl ammonium bromide method as described by Murray and Thompson (1980) with some modifications. Instead of a CsCl–ethidium bromide ultracentrifugation step, the DNA was purified by phenol–chloroform extraction.

PCR amplification of chloroplast genes/intergenic spacers, their digestion and data analysis

Seven chloroplast genes and intergenic spacers (trnS–psbC, psaA, 16S, trnK, psbD, trnL–trnF and trnH–trnK) were amplified using previously published universal forward and reverse primers (Table 2). Amplification was carried out in 100 μL reaction mixtures containing 80 ng of template DNA, 0.1 mM dNTPs (Amersham Biosciences, UK), 2 mM MgCl₂ (Bangalore Genei, India), 1.3 μM each forward and reverse primer (Bangalore Genei), 2.5 U of Taq DNA polymerase (Bangalore Genei) and 10 μL of 10× assay buffer [100 mM Tris pH 9.0, 500 mM KCl, 0.1 % gelatin (Bangalore Genei)]. DNA amplification was performed in a My cycler (Bio-Rad, USA) programmed to 36 cycles each of 1 min (5 min for the first cycle) at 94 °C for template DNA denaturation, 1 min at the annealing temperature (63 °C for psaA, 55 °C for trnS–psbC, trnK and psbD, 60 °C for 16S, and 50 °C for trnL–trnF and trnH–trnK), and 2 min at 72 °C for primer extension, followed by a final extension cycle of 15 min at 72 °C.

Seven genes and intergenic spacers amplified from the chloroplast genome were separately restricted with 31 four-, five- and six-base cutter restriction endonucleases (AluI, AvaI, AccI, AfaI, BamHI, BalI, BglI, BglII, ClaI, DraI, EcoRV, EcoRI, HaeIII, HinfI, HindIII, HincII, MspI, KpnI, MboI, MluI, PstI, PvuII, SacI, SalI, SpeI, SphI, SmaI, SspI, TaqI, XhoI and XbaI). The 12.20 μL reaction mix contained 10 μL of the amplified gene product, 1.2 μL of enzyme buffer and 10 U of the restriction enzyme. After gentle mixing the mixture was incubated overnight at 37 °C (except for TaqI, which was incubated at 60 °C). The digestion was terminated by adding 1.5 μL of 10× loading buffer. The digested products were fractionated on 1.5 % agarose gels containing ethidium bromide (0.05 μg mL⁻¹) in 0.5× TBE buffer. A DNA ladder mix was loaded alongside the digested products to serve as size markers. After agarose gel electrophoresis, the gel was photographed in ultraviolet light. Reproducibility of the patterns was tested by repeating all the reactions at least twice.

For PCR–RFLP analysis, the presence (1) or absence (0) of a restriction fragment was recorded. Total and mean character differences between pairs of species were calculated using PAUP* 4.0 (Swofford 2002). Nei and Li's coefficient of genetic distance (Nei and Li 1979) was calculated between each pair of species after the optimality criterion was set to DISTANCE. Cluster analysis was carried out using the unweighted pair-group method using arithmetic averages (UPGMA) (Sneath and Sokal 1973) and neighbour-joining (NJ) (Saitou and Nei 1987) methods. Bootstrap values were calculated from 100 replicates using the BOOTSTRAP command in PAUP.

PCR amplification, cloning and sequencing of the trnK gene, and data analysis

The trnK gene was amplified from the nine Eleusine species and one outgroup as described above. The amplification products were separated on a 1 % agarose gel, excised from the gel and purified using a QIA quick gel extraction kit (Qiagen, Germany). The purified amplification products were cloned and sequenced with an ABI PRISM

Table 2. List of chloroplast genes/intergenic spacers amplified in the present study. ᵃSizes in source.

Genes/intergenic spacers	Primer pair	Size in bpᵃ	Source
trnS–psbC	5′-GGTTCGAATCCCTCTCTCTC-3′ 5′-GGTCGTGACCAAGAAACCAC-3′	1600	Oryza sativa (Demesure et al. 1995)
psaA	5′-AAGAATGCCCATGTTGTGGC-3′ 5′-TTCGTTCGCCGGAACCAGAA-3′	2218	Nicotiana tabacum (Shinozaki et al. 1986; Tsumura et al. 1996)
16S	5′-ACGGGTGAGTAACGCGTAAG-3′ 5′-CTTCCAGTACGGCTACCTTG-3′	1375	Nicotiana tabacum (Shinozaki et al. 1986; Tsumura et al. 1995, 1996)
trnK	5′-AACCCGGAACTAGTCGGATG-3′ 5′-TCAATGGTAGAGTACTCGGC-3′	2569	Oryza sativa (Hiratsuka et al. 1989; Tsumura et al. 1995, 1996)
psbD	5′-TATGACTATAGCCCTTGGTA-3′ 5′-TAGAACCTCCTCAGGGAATA-3′	1042	Nicotiana tabacum (Shinozaki et al. 1986; Tsumura et al. 1995, 1996)
trnL–trnF	5′-CGAAATCGGTAGACGCTACG-3′ 5′-ATTTGAACTGGTGACACGAG-3′	995	Nicotiana tabacum (Taberlet et al. 1991)
trnH–trnK	5′-ACGGGAATTGAACCCGCGCA-3′ 5′-CCGACTAGTTCCGGGTTCGA-3′	1831	Nicotiana tabacum (Demesure et al. 1995)

377 automated DNA sequencer (Applied Biosystems, Foster City, CA, USA). BLAST similarity searches were performed using the National Centre for Biotechnology Information (NCBI) BLASTN algorithm to confirm the identity of the trnK sequences. The nucleotide sequences have been submitted to GenBank with accession numbers KF357736–KF357745. The nucleotide sequence of the partial trnK gene from *Acrachne racemosa* available in GenBank (accession number JN681616.1) was also utilized for phylogenetic analysis. The consensus sequence for the trnK gene from the nine *Eleusine* taxa and one outgroup (*D. aegyptium*) was generated using the software CLC Sequence viewer 6.1.

The sequences were aligned using Clustal-X, version 1.8 (Thompson *et al.* 1997). Phylogenetic analyses were carried out with the software MEGA 5 (Tamura *et al.* 2011). Pairwise sequence divergence rates between species were calculated using the maximum composite likelihood method. Phylogeny construction was carried out using NJ, minimum evolution (ME), maximum likelihood (ML) and maximum parsimony (MP) methods. Neighbour-joining and ME trees were obtained using the maximum composite likelihood criterion while ML and MP trees were constructed using the nearest-neighbour-interchange (NNI) and tree-bisection-reconnection algorithms, respectively. In all the analyses, all positions containing gaps and missing data were eliminated from the dataset (complete deletion option). Support values of the internal branches of NJ, ME, ML and MP trees were evaluated by bootstrap (500 replicates).

PCR amplification of chloroplast microsatellites, polyacrylamide gel electrophoresis and sequencing

A total of eight primer pairs (Table 3) were used for amplification of the chloroplast microsatellites. Amplification was carried out in 25 mL reaction mixtures containing 2.5 mL of 10× reaction buffer, 1.5 mM MgCl$_2$, 200 mM dNTPs, 200 nM each primer, 0.5 U of *Taq* DNA polymerase (Bangalore Genei) and 25 ng of template DNA. DNA amplification was performed in a MyCycler™ (Bio-Rad) programmed to initial denaturation at 94 °C for 5 min followed by 35 cycles each of 1 min at 94 °C, 1 min at 55 °C (65 °C for ccmp2 and 53 °C for ccmp9) and 1 min at 72 °C, followed by a final extension cycle of 5 min at 72 °C.

An equal volume (10 mL) of formamide dye (98 % formamide, 10 mM EDTA, 0.026 g of bromophenol blue, 0.026 g of xylene cyanol) was added to each amplified product. The samples were heated for 5 min at 94 °C and immediately placed on ice. A total of 2.5 mL of each sample was loaded on a 6 % polyacrylamide gel (19 : 1 acrylamide : bisacrylamide, 7.5 M urea and 1×

TBE buffer), and electrophoresis was conducted at 55 W and 55 °C for ~2 h.

For silver staining, the gel was fixed in 10 % (v:v) acetic acid for 30 min. It was subsequently rinsed three times in de-ionized water (2 min per rinse). The gel was then kept for staining for 30 min in a 2 L solution containing 2 g of silver nitrate and 3 mL of 37 % formaldehyde (Promega, USA). The stained plate was rinsed with de-ionized water for 20 s and developed in a prechilled (10 °C) developer (2 L) solution containing 60 g of sodium carbonate, 3 mL of 37 % formaldehyde and 400 μL of sodium thiosulfate (10 mg mL^{-1}). When bands became visible, the gel was immediately transferred to 10 % acetic acid solution to stop further reaction. The gel was finally rinsed with distilled water and air dried.

The PCR products of eight primer–template combinations were characterized by direct sequencing. A PCR was performed in 100 μL volumes as described above. An aliquot of the PCR product was checked by agarose gel electrophoresis, and the remainder was purified through a QIA quick PCR clean up kit (Qiagen). Nucleotide sequencing was performed using an ABI PRISM 377 automated DNA sequencer (Applied Biosystems). The nucleotide sequences have been submitted to GenBank with accession numbers KF357730–KF357735.

Results

Chloroplast PCR–RFLP

Robust amplification products were obtained for all seven genes/intergenic spacers (trnS–psbC, psaA, 16S, trnK, psbD, trnL–trnF and trnH–trnK) from the nine *Eleusine* species and an outgroup species, *D. aegyptium*. For each gene/intergenic spacer, the amplified products appeared to be monomorphic on a 1.5 % agarose gel across the species. The approximate size of the PCR product for trnS–psbC, psaA, 16S, trnK, psbD, trnL–trnF and trnH–trnK was 1620, 2380, 1480, 2510, 1090, 1090 and 2270 bp, respectively. Aliquots of PCR products were digested separately with 31 four-, five- and six-cutter restriction endonucleases. Of the 217 amplification product–enzyme combinations, 137 did not show internal restriction sites in any of the 10 species investigated. In the remaining 80 amplification product–enzyme combinations, 1–5 restriction sites within the amplification product were obtained. Of these, 57 combinations revealed no polymorphism among the 10 species. The remaining 23 (trnK/AfaI, trnK/MspI, trnK/DraI, trnK/SpeI, trnK/SspI, trnK/KpnI, trnK/TaqI, trnK/BamHI, trnL–trnF/AluI, trnL–trnF/BglII, trnL–trnF/DraI, trnL–trnF/MboI, psaA/AluI, psaA/TaqI, psaA/MspI, psbD/HaeIII, psbD/MboI, psbD/TaqI, trnS–psbC/AfaI, trnS–psbC/AvaI, trnS–

Table 3. List of chloroplast microsatellites amplified in the present study.

Locus	Location	Repeat	Primer pair	Size in *Eleusine* (bp)	Size in source (bp)	Source
ccmp2	5′ to *trn*S	$(A)_{11}$	5′-GATCCCGGACGTAATCCTG-3′ 5′-ATCGTACCGAGGGTTCGAAT-3′	197, 200	189	*Nicotiana tabacum* (Weising and Gardner 1999)
ccmp5	3′ to *rps2*	$(C)_7(T)_{10}$ $(T)_5C(A)_{11}$	5′-TGTTCCAATATCTTCTTGTCATTT-3′ 5′-AGGTTCCATCGGAACAATTAT-3′	145, 146	103	*Nicotiana tabacum* (Weising and Gardner 1999)
ccmp6	ORF77–ORF82 intergenic	$(T)_5C(T)_{17}$	5′-CGATGCATATGTAGAAAGCC-3′ 5′-CATTACGTGCGACTATCTCC-3′	96	98	*Nicotiana tabacum* (Weising and Gardner 1999)
RCt3	Intergenic region	$(A)_{10}$	5′-TAGGCATAATTCCCAACCCA-3′ 5′-CTTATCCATTTGGAGCATAGGG-3′	113	129	*Oryza sativa* cv Nipponbare (Ishii and McCouch 2000)
RCt4	Coding region (*psb*G)	$(T)_{12}$	5′-ACGGAATTGGAACTTCTTTGG-3′ 5′-AAAAGGAGCCTTGGAATGGT-3′	131	128	*Oryza sativa* cv Nipponbare (Ishii and McCouch 2000)
RCt5	Intergenic region	$(T)_{10}$	5′-ATTTGGAATTTGGACATTTTCG-3′ 5′-ACTGATTCGTAGGCGTGGAC-3′	151	143	*Oryza sativa* cv Nipponbare (Ishii and McCouch 2000)
RCt7	Coding region (*inf*A)	$(T)_{10}$	5′-GTGTCATTCTCTAGGCGAAC-3′ 5′-AAATATGACAGAAAAGAAAAATAGG-3′	126	126	*Oryza sativa* cv Nipponbare (Ishii and McCouch 2000)
RCt8	Intron (*rpl*16)	$(T)_{17}$	5′-ATAGTCAAGAAAGAGGATCTAGAAT-3′ 5′-ACCGCGATTCAATAAGAGTA-3′	125	131	*Oryza sativa* cv Nipponbare (Ishii and McCouch 2000)

*psb*C/HaeIII, 16S/MboI, *trn*H–*trn*K/AluI) combinations were phylogenetically very informative (Table 4; Fig. 1). A total of 282 bands were scored for cluster analyses. No variation was observed across different accessions of the same taxon. The polymorphism in all the profiles was the result of site mutations.

Genetic distance (Nei and Li's) among the nine *Eleusine* species and one outgroup ranged from 0.0000 to 0.01857 (Table 5). Based on the restriction site data of the amplified gene/intergenic products, UPGMA and NJ dendrograms were generated. Both showed similar topologies. They resolved into two major clades. Clade I, supported by 100 % bootstrap support, contained *E. kigeziensis, E. indica, E. tristachya, E. coracana* and *E. africana*. Clade II was comprised of *E. jaegeri, E. floccifolia, E. intermedia* and *E. multiflora. Eleusine floccifolia* and *E. intermedia* were more closely related to each other (supported by a bootstrap value of 100 %) than either of the species was to *E. jaegeri. Dactyloctenium aegyptium* was the most diverged species among the analysed species (Fig. 2).

*trn*K sequence data

The *trn*K gene codes for tRNA-Lys^{UUU}. In rice, the chloroplast *trn*K gene is 2576 bp long, of which the exon is only 72 bp long. The exon is divided into two parts by a long intron of ~2500 bp. The 5′ exon consists of 37 bp and the 3′ exon consists of 35 bp. The *mat*K gene, which codes for maturaseK, is 1536 bp in length and is located within the chloroplast *trn*K intron. In the present study, the *trn*K gene was sequenced from the nine *Eleusine* species and *D. aegyptium*. The aligned *trn*K gene sequence of *Eleusine* corresponded to position 45 at the 5′ end to position 2531 at the 3′ end of rice. The length of the *trn*K gene in the nine *Eleusine* species ranged between 2463 (*E. jaegeri*) and 2467 bp (*E. tristachya* and *E. floccifolia*). The length of *trn*K in *A. racemosa* and *D. aegyptium* was 2466 and 2472 bp, respectively **[see Supporting Information]**. The GC content of the *trn*K gene sequence varied from 32.3 to 33 % with an average of 32.6 %. The nucleotide frequencies were 0.162 for G, 0.164 for C, 0.324 for A and 0.350 for T. The dataset including alignment gaps

Table 4. Grouping of *Eleusine* and outgroup species based on restriction fragment patterns of seven cpDNA amplified genes/intergenic spacers.

Species	trnS–psbC AfaI	trnS–psbC AvaI	trnS–psbC HaeIII	16S MboI	psbD HaeIII	psbD MboI	psbD TaqI	psaA AluI	psaA MspI	psaA TaqI	trnK AfaI	trnK DraI	trnK MspI	trnK SpeI	trnK SspI	trnK KpnI	trnK TaqI	trnK BamHI	trnH–trnK AluI	trnL–trnF AluI	trnL–trnF BglII	trnL–trnF DraI	trnL–trnF MboI
E. coracana	A1	B1	C1	D1	E1	F1	G1	H1	I1	J1	K1	L1	M1	N1	O1	P1	Q1	R1	S1	T1	U1	V1	X1
E. africana	A1	B1	C1	D1	E1	F1	G1	H1	I1	J1	K1	L1	M1	N1	O1	P1	Q1	R1	S1	T1	U1	V1	X1
E. tristachya	A1	B1	C1	D1	E1	F1	G1	H1	I1	J1	K1	L1	M1	N1	O1	P1	Q1	R1	S1	T1	U1	V1	X1
E. indica	A1	B1	C1	D1	E1	F1	G1	H1	I1	J1	K1	L1	M1	N1	O1	P1	Q1	R1	S1	T1	U1	V1	X1
E. floccifolia	A2	B1	C1	D2	E1	F2	G1	H2	I1	J1	K2	L1	M1	N1	O2	P1	Q1	R1	S1	T1	U1	V1	X1
E. intermedia	A2	B1	C1	D2	E1	F2	G1	H2	I1	J1	K2	L1	M1	N1	O2	P1	Q1	R1	S1	T1	U1	V1	X1
E. kigeziensis	A1	B1	C1	D1	E1	F1	G1	H1	I1	J1	K1	L1	M1	N1	O1	P1	Q1	R1	S1	T1	U1	V1	X1
E. jaegeri	A2	B1	C1	D2	E1	F2	G1	H2	I2	J1	K2	L1	M1	N1	O2	P1	Q1	R1	S1	T1	U1	V1	X1
E. multiflora	A2	B1	C1	D1	E2	F2	G1	H1	I1	J2	K1	L2	M2	N2	O2	P2	Q2	R2	S2	T2	U2	V2	X2
D. aegyptium	A2	B2	C2	D1	E3	F2	G2	H1	I1	J1	K2	L1	M3	N1	O1	P1	Q2	R2	S2	T2	U2	V2	X2
Combinations	I	VIII	VIII	III	IV	I	VIII	III	V	VI	VII	VI	IV	VI	II	VIII	VIII	VIII	VIII	VIII	VIII	VIII	VIII

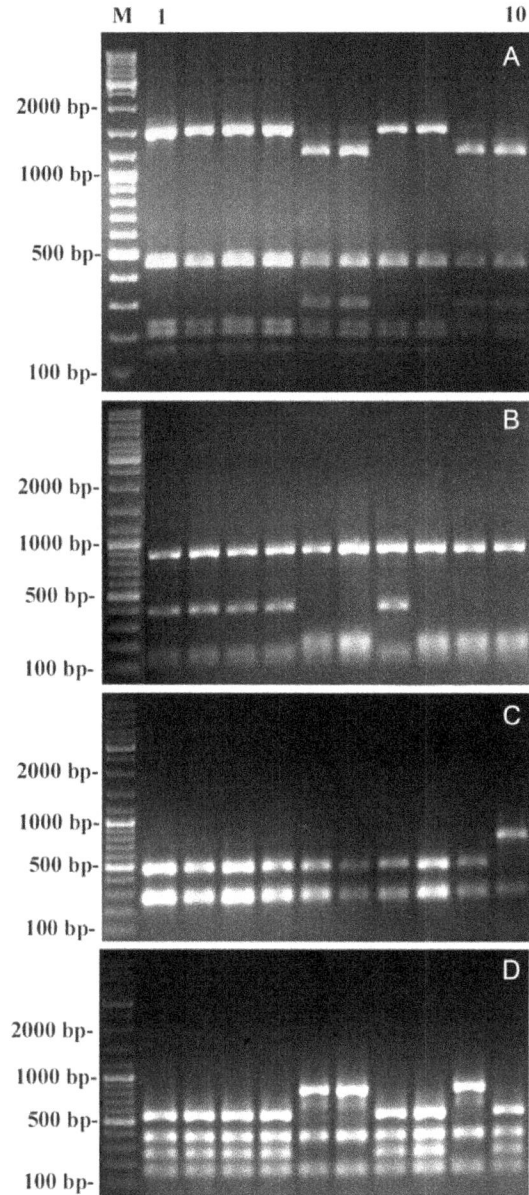

Figure 1. Restriction fragment size patterns of the amplified *trn*K gene with AfaI (A), the amplified *trn*S–*psb*C intergenic spacer with AfaI (B), the amplified *trn*L–*trn*F gene with MboI (C) and the amplified *16*S gene with MboI (D) in *E. coracana* (lane 1), *E. africana* (2), *E. tristachya* (3), *E. indica* (4), *E. floccifolia* (5), *E. intermedia* (6), *E. kigeziensis* (7), *E. multiflora* (8), *E. jaegeri* (9) and *D. aegyptium* (10). Marker DNA (M). The size of the fragments in base pairs is indicated on the left.

and missing data comprised 2486 nucleotide positions, out of which 2254 were conserved, 214 were variable and 57 were parsimony informative sites.

The *trn*K sequence data were correlated with the polymorphic PCR–RFLP profiles of the *trn*K gene. Eight amplicon–enzyme combinations for the *trn*K gene showed

Table 5. Genetic distance (Nei and Li's coefficient) matrix generated from cpDNA PCR–RFLP data of nine *Eleusine* species and an outgroup, *D. aegyptium*.

		1	2	3	4	5	6	7	8	9	10
1	*E. coracana*										
2	*E. africana*	0.00000									
3	*E. tristachya*	0.00000	0.00000								
4	*E. indica*	0.00000	0.00000	0.00000							
5	*E. floccifolia*	0.00479	0.00479	0.00479	0.00479						
6	*E. intermedia*	0.00479	0.00479	0.00479	0.00479	0.00000					
7	*E. kigeziensis*	0.00000	0.00000	0.00000	0.00000	0.00479	0.00479				
8	*E. multiflora*	0.00650	0.00650	0.00650	0.00650	0.00790	0.00790	0.00650			
9	*E. jaegeri*	0.00655	0.00655	0.00655	0.00655	0.00170	0.00170	0.00655	0.00827		
10	*D. aegyptium*	0.01443	0.01443	0.01443	0.01443	0.01815	0.01815	0.01443	0.01692	0.01857	

polymorphic profiles which revealed that polymorphism in all the profiles was the result of site mutations. The entire *trn*K gene sequence from the taxa analysed showed 214 single-nucleotide polymorphisms (SNPs). A total of 10 SNPs in the restriction sites were responsible for the eight polymorphic profiles. The polymorphic profiles produced were due to gain or loss of the restriction sites caused by SNPs. The consensus *trn*K gene sequence was generated from the taxa analysed and the locations of the polymorphic sites were marked on it (Fig. 3).

Pairwise sequence divergence ranged from 0.003 to 0.058 (Table 6). Dendrograms were generated using four different methods (NJ, ML, ME and MP). The phylogeny reconstruction through the NJ method resulted in an optimal tree with a sum of branch length of 0.095. In the NJ bootstrap consensus tree, all the *Eleusine* species were grouped into two distinct clades with 100 % bootstrap support. Clade I consisted of *E. coracana*, *E. africana*, *E. indica*, *E. kigeziensis* and *E. tristachya*. *Eleusine coracana* and *E. africana* grouped together in one subclade with a bootstrap support of 95 %, and *E. indica* and *E. kigeziensis* grouped together in another subclade with a bootstrap support of 54 % within Clade I. *Eleusine tristachya* did not group within any of the two subclades and thus was the most diverged among the five species of Clade I with a bootstrap support of 98 %. Clade II consisted of *E. intermedia*, *E. floccifolia*, *E. jaegeri* and *E. multiflora*. *Eleusine intermedia* and *E. floccifolia* were more closely related to each other than to *E. jaegeri* with a bootstrap support of 54 %. *Eleusine multiflora* represented the most diverged species within Clade II. *Acrachne racemosa* and *D. aegyptium* were the most diverged species among all the species (Fig. 4A).

The robustness of the present data lies in the fact that the dendrograms based on ME, ML and MP analysis

showed almost the same topology as that of the NJ tree (Fig. 4). Maximum parsimony analysis resulted in most parsimonious trees (Fig. 4C) with a tree length of 232. The bootstrap consensus tree had a consistency index of 0.943, a retention index of 0.859 and a rescaled consistency index of 0.811.

Chloroplast microsatellite polymorphism and sequence data

Eight consensus primer pairs were used to amplify cpDNA microsatellites from the nine species (Table 3). The primer pairs were able to amplify a product in all 10 species (Fig. 5). A major band was produced for each primer–template combination. The typical stuttering phenomenon which usually occurs upon amplification of mononucleotide- and dinucleotide-type microsatellites was observed. Size variation for the amplified products was only observed for two cpSSR loci. Two size variants were detected for ccmp2 and ccmp5. For ccmp2, all but one (*E. tristachya*) species showed a 197-bp allele. In *E. tristachya*, it was 200 bp. In the case of ccmp5, *E. jaegeri* showed a band of 146 bp, while each of the remaining nine species had a 145-bp allele (Fig. 5).

To test the presence of a repeat motif in *Eleusine*, nucleotide sequencing was carried out for amplified products from *E. coracana* with all eight cpSSR primer sets. RCt5 and RCt8 could not be successfully sequenced. ccmp2, ccmp5, ccmp6, RCt3, RCt4 and RCt7 showed sequence sizes of 197, 145, 96, 113, 131 and 126 bp, respectively (Tables 3 and 7). Repeat motifs were identified in all six amplification products at the expected positions but the degree of conservation was variable (Table 7). Among the repeat motifs, RCt4 and RCt7 were highly conserved, ccmp5 and ccmp6 were moderately conserved, and ccmp2 and RCt3 were least conserved. In the case

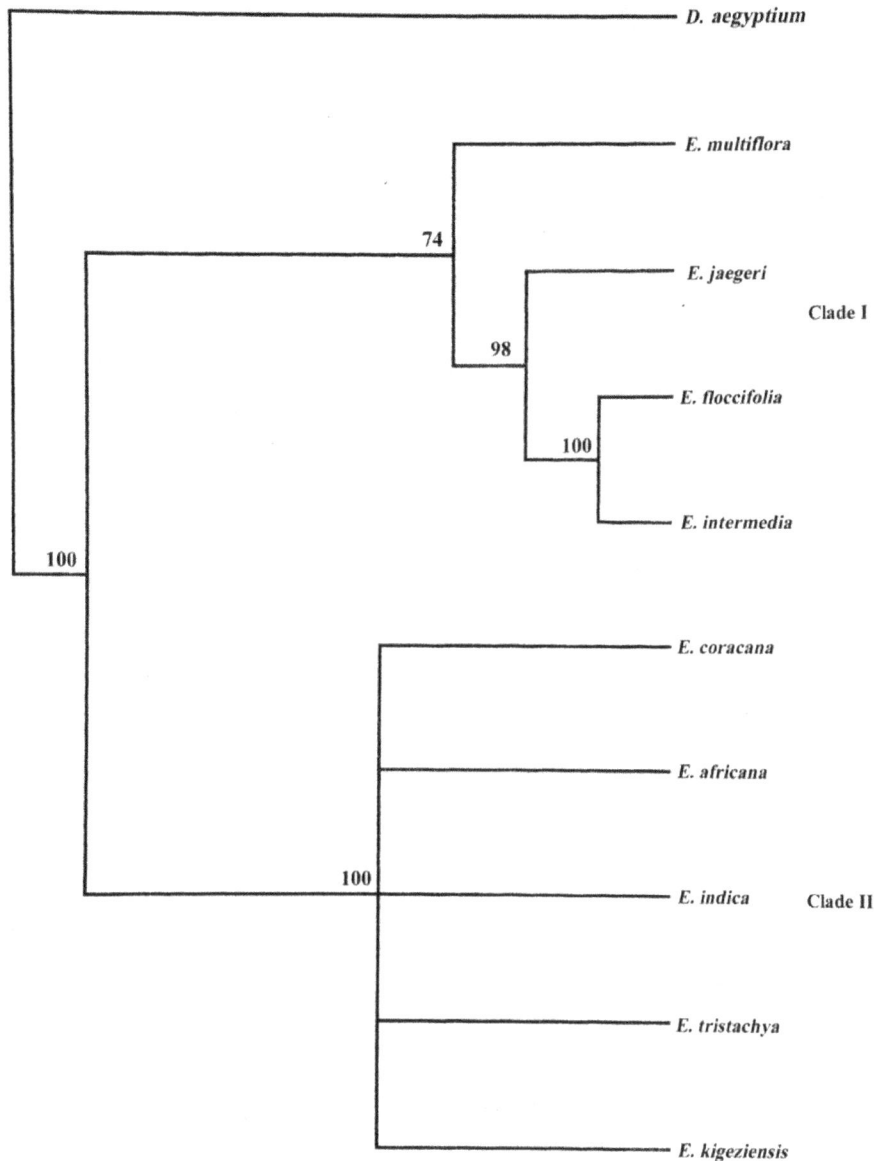

Figure 2. Unweighted pair-group method using arithmetic averages dendrogram based on the restriction fragment data of amplified chloroplast gene/intergenic spacers. Numbers at the nodes represent bootstrap probability values out of 100 replicates.

of ccmp2 and RCt3, the number of mononucleotide repeats was reduced.

Discussion

The restriction site variation identified in the amplified chloroplast gene/intergenic spacers has provided new insights into the origin and evolution of the three polyploid species, and the genetic relationships between the cultivated and wild *Eleusine* species. Species-specific markers were also identified. None of the amplicon–enzyme combinations could discriminate between *E. coracana* (AABB, $2n = 4x = 36$), *E. africana* (AABB, $2n = 4x = 36$), *E. tristachya* (AA, $2n = 2x = 18$), *E. indica* (AA, $2n = 2x = $

18) and *E. kigeziensis* (AADD, $2n = 4x = 38$) on the one hand, and between *E. floccifolia* (BB, $2n = 2x = 18$) and *E. intermedia* (AB, $2n = 2x = 18$) on the other hand. Although the maximum number of polymorphic and phylogenetically informative markers was obtained in eight *trn*K amplicon–enzyme combinations, the amount of variation that can be tapped with the PCR–RFLP method was lower than the variation tapped with sequencing of the *trn*K gene. A total of 214 SNPs were found in the *trn*K gene sequence, of which only 10 were causative to 8 polymorphic profiles. This clearly demonstrates that sequencing of the *trn*K gene is more informative than the PCR–RFLP method. Moreover, the nucleotide sequence of the chloroplast *trn*K gene fine-tuned the species

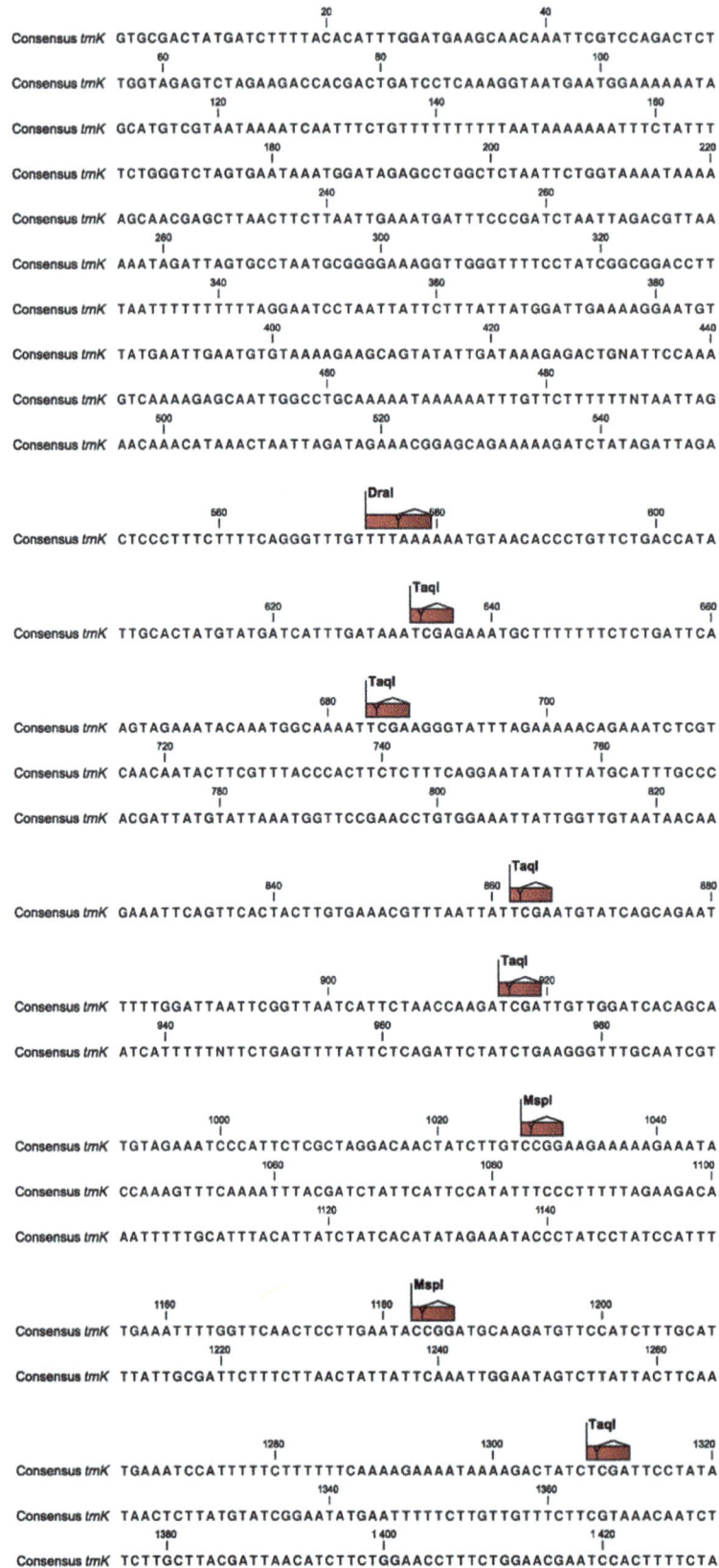

Figure 3. Consensus sequence of the *trnK* gene for nine *Eleusine* species and one outgroup (*D. aegyptium*) showing major restriction sites.

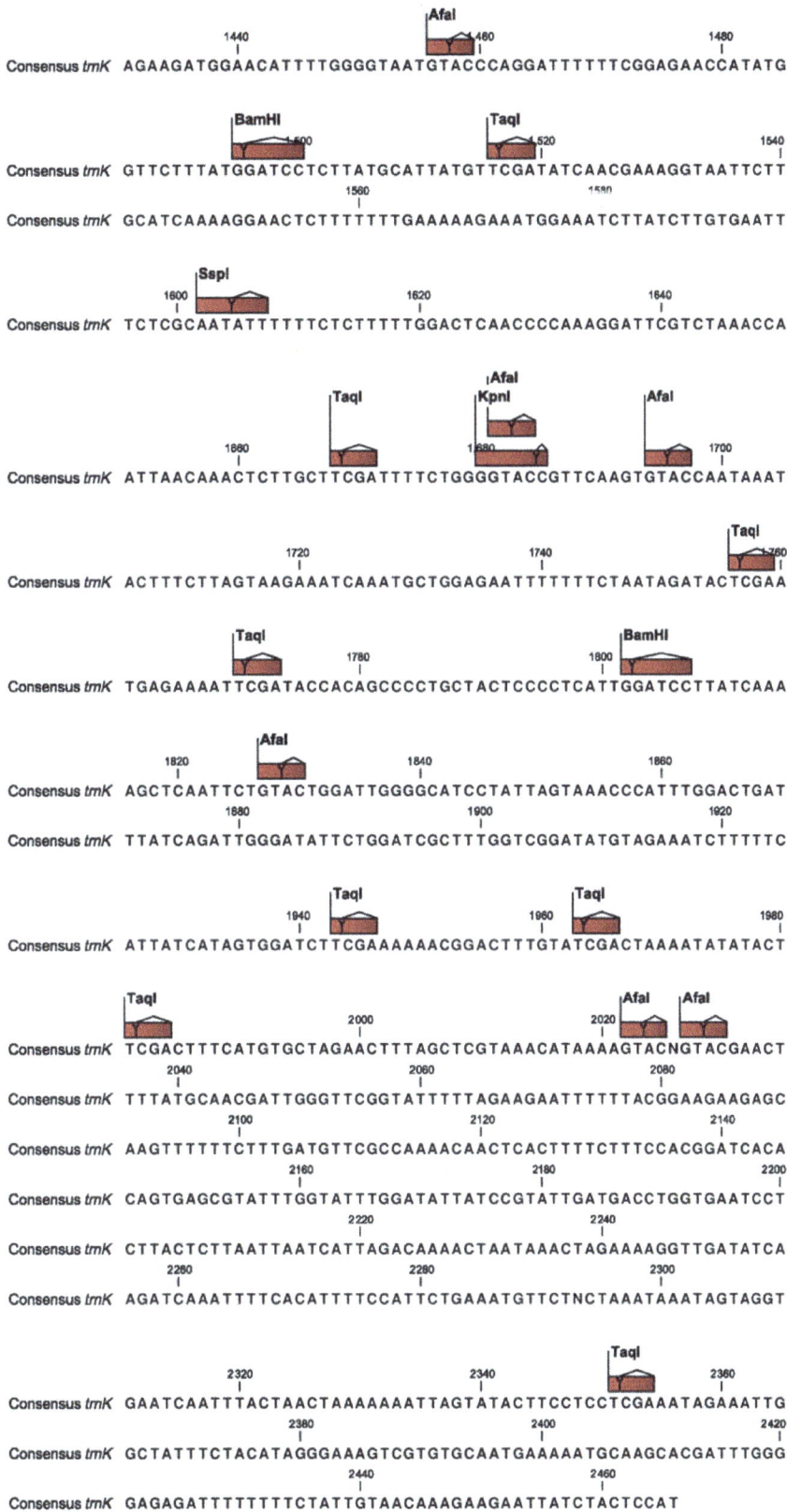

Figure 3. Continued.

Table 6. Genetic divergence (maximum composite likelihood method) from *trn*K sequence data for nine *Eleusine* species and two outgroups, *A. racemosa* and *D. aegyptium*.

1	2	3	4	5	6	7	8	9	10	11
E. floccifolia										
E. intermedia	0.008									
E. jaegeri	0.026	0.025								
E. multiflora	0.010	0.009	0.027							
E. coracana	0.012	0.012	0.026	0.012						
E. africana	0.013	0.013	0.028	0.013	0.004					
E. kigeziensis	0.014	0.014	0.030	0.014	0.007	0.008				
E. indica	0.011	0.010	0.026	0.010	0.003	0.004	0.005			
E. tristachya	0.011	0.011	0.026	0.011	0.005	0.006	0.007	0.003		
A. racemosa	0.029	0.028	0.046	0.029	0.030	0.030	0.030	0.027	0.027	
D. aegyptium	0.039	0.038	0.058	0.041	0.040	0.041	0.043	0.039	0.040	0.030

relationships. The nine species of *Eleusine* were grouped into two clades, in both PCR–RFLP and *trn*K gene sequence data analyses. The *trn*K sequence data clearly showed that the three tetraploids were closer to *E. indica* than to *E. tristachya*, as concluded from PCR–RFLP data.

The present observations based on restriction site variation and the *trn*K gene sequence of cpDNA are congruent with the conclusions reached by various authors based on the multiple marker nuclear DNA assay, chloroplast markers and chromosome research. The strong affinities between *E. coracana*, *E. africana* and *E. indica* have been highlighted in earlier studies (Hilu and Johnson 1992, 1997; Hilu 1995; Bisht and Mukai 2000, 2001a, b; Neves et al. 2005; Dida et al. 2007, 2008; Liu et al. 2011) as well. A number of studies based on chromosome research (Hiremath and Chennaveeraiah 1982; Hiremath and Salimath 1991), 2C DNA content (Hiremath and Salimath 1991), crossability data (Salimath et al. 1995b) and ribosomal DNA polymorphism (Hilu and Johnson 1992; Werth et al. 1994; Bisht and Mukai 2000), ITS sequence data (Neves et al. 2005) and cpDNA (Hilu and Johnson 1997; Neves et al. 2005; Liu et al. 2011) also support the close affinity between *E. indica* and *E. tristachya*. The allotetraploid *E. kigeziensis* exhibits the same maternal lineage as that of the two other allotetraploid species, *E. coracana* and *E. africana*. The five species share some common morphological features, such as 1–3 nerved lower glumes with a winged keel, 3–7 nerved upper glumes with a more or less winged keel, the presence of 1–3 subsidiary nerves adjacent to the central nerve of the lemma and the generally winged keel of the palea (Phillips 1972).

The present data unambiguously support the view of Chennaveeraiah and Hiremath (1974), Hilu (1988),

Hiremath and Salimath (1992), Hilu and Johnson (1997), Bisht and Mukai (2000, 2001a, b) and Neves et al. (2005) that *E. africana* is the wild progenitor of the cultivated species, *E. coracana*, and that *E. indica* is one of the diploid progenitors of the two polyploid species (Liu et al. 2011). The present results further indicate that *E. indica* with the AA genome might be the maternal parent for all three tetraploid species, viz. *E. coracana*, *E. africana* and *E. kigeziensis*. Morphologically, *E. kigeziensis* resembles *E. indica*. It is considered to be a hybrid between *E. indica* and one of the perennial species (Phillips 1972). Salimath (1990) and Bisht and Mukai (2002) have also proposed *E. indica* as one of the genome donors of *E. kigeziensis*.

The close affinity in the cpDNA PCR–RFLP profiles and *trn*K sequences between *E. floccifolia* and *E. intermedia* is also supported by the occurrence of the same chromosome number (2n = 18), 2C DNA values (Hiremath and Chennaveeraiah 1982; Hiremath and Salimath 1991) and ITS sequence data (Neves et al. 2005). Close association between *E. jaegeri*, *E. floccifolia* and *E. intermedia* suggested in earlier studies (Hiremath and Salimath 1991; Hilu and Johnson 1992; Liu et al. 2011) is further supported by our data. *Eleusine jaegeri* in all combinations grouped together with *E. floccifolia* and *E. intermedia* except in the *psa*A–MspI combination, where it shows a unique profile *vis-à-vis* the remaining species. In the case of *trn*K sequence analysis also these three species are found to be closely associated. *Eleusine multiflora* also behaves in the same fashion. The species exhibits unique profiles *vis-à-vis* the remaining species in *trn*K–DraI, *trn*K–MspI, *trn*K–SpeI, *psa*A–TaqI and *psb*D–HaeIII combinations, with the result that it represents the most diverged taxon among all the *Eleusine* species. These amplicon–enzyme

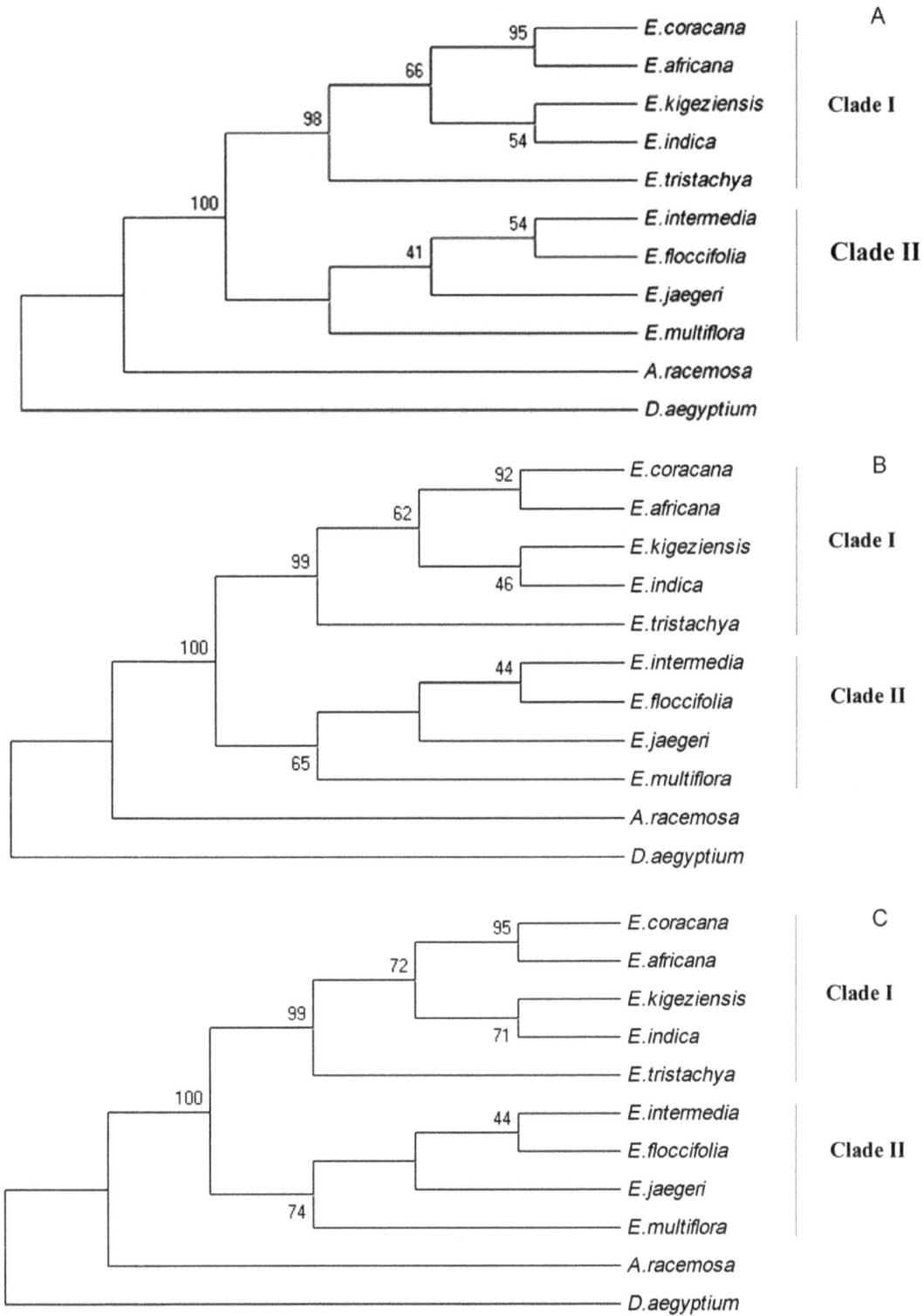

Figure 4. Dendrograms based on the sequence data of the *trnK* gene of *Eleusine* and outgroup taxa. (A) Neighbour-joining bootstrap consensus tree, (B) ML bootstrap consensus tree and (C) MP consensus tree. Numbers at the nodes represent bootstrap probability values out of 500 replicates.

combinations can therefore be used as species-specific markers for *E. jaegeri* and *E. multiflora*. The position of *E. multiflora* within the genus *Eleusine* is questionable

mainly on the basis of its unusual inflorescence morphology [considered to be intermediate between *Eleusine* and *Acrachne* (Phillips 1972; Clayton and Renvoize

1986)], distinct chemical composition (Hilu *et al.* 1978), molecular data (Hilu and Johnson 1992; Hilu 1995), chromosome number ($2n = 16$), 2C DNA value and other cytogenetic features (Mysore and Baird 1997; Bisht and Mukai 2000) including the present results. *Dactyloctenium aegyptium* in restriction site analysis, and *A. racemosa* and *D. aegyptium* in *trn*K sequence analysis indicate that they are the most diverged species *vis-à-vis Eleusine* species analysed and thus represented as outgroups.

Chloroplast microsatellites are known to have potential for phylogenetic studies (Powell *et al.* 1995*a, b*, 1996; Provan *et al.* 1997, 2001; Ishii and McCouch 2000; Angioi

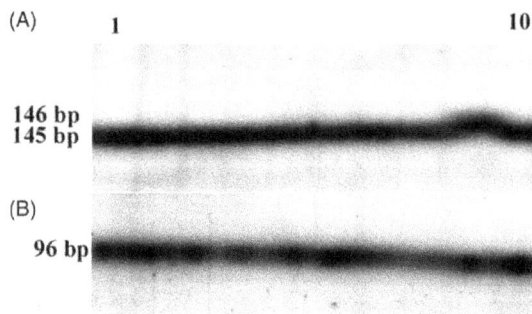

Figure 5. Silver-stained polyacrylamide gel electrophoresis patterns generated by ccmp5 (A) and ccmp6 (B) in *E. coracana* (1), *E. africana* (2), *E. tristachya* (3), *E. indica* (4), *E. floccifolia* (5), *E. intermedia* (6), *E. kigeziensis* (7), *E. multiflora* (8), *E. jaegeri* (9) and *D. aegyptium* (10). The size of the fragments in base pairs is indicated on the left.

et al. 2008). The present cpSSR data on *Eleusine* species, however, were not helpful as very little polymorphism was obtained for various amplified microsatellites. Only two cpSSR loci (ccmp2 and ccmp5) were found to be polymorphic, displaying a total of four alleles. In the case of ccmp2, the variant allele was found in *E. tristachya*, while for ccmp5 the variant allele was found in *E. jaegeri*. Overall, the sequences seem to be highly conserved. This is in strong contrast to the observations made in the other genera such as *Glycine*, *Oryza* and *Hordeum*, where different species within the genus and even within different subspecies displayed characteristic haplotypes (Powell *et al.* 1995*b*, 1996; Provan *et al.* 1997, 1999). The reason for not obtaining polymorphism in the present study could be attributed to the fact that it is very difficult to design universal primers for microsatellites that show widespread polymorphism at the inter-specific level and at the same time are able to amplify across a broad range of plant genera. Although there are primers available that amplify across wide-ranging taxa, these rarely show widespread inter-specific polymorphism. This could be due to the contrasting requirements in a short stretch of DNA of extreme stability and sequence conservation for priming sites and consistently high levels of polymorphism in the intervening region (Provan *et al.* 2001). Second, size homoplasy may not necessarily mean identical intervening microsatellite regions; it may also be due to variation in flanking regions. It is also possible that back mutations have taken place, which might result in

Table 7. Nucleotide sequences of cpDNA of *E. coracana* amplified using six cpSSR primer pairs. The location of repeats is underlined.

ccmp2 (complete sequence)

5′-GATCCCGGACGTAATCCTGGACGTATCCTGGACGTGAGGAGTAAAAATCCAAAATTTTTGGGAATTTTTTCTTACAAATTGAATTTATTTCGTACATTTATCTATGAA
AAAATCCGGGGGTTAGAATTCCTTACAATTCG<u>AAAG</u>TCCCAAACGATCCGAGGGGGCGGAAAGAGAGGGATTCGAACCCTCGGTACGAT-3′

ccmp5 (complete sequence)

5′-TGTTCCAATATCTTCTTGTCATTTTTTCCACA<u>CTTCCTTTTTTTTTTTCTTTTTTT</u>CGTCTTACCATTATGGAATTTTTTTCTTTTTG<u>AAGATTAAGAAAG</u>AGCCAAATTAT
CTTGAAATAAATAATAATTGTTCCGATGGAACCT-3′

ccmp6 (complete sequence)

5′-CGATGGATATGTAGAAAGCCCTTTTTCTAGTATTTACTAGAAAA<u>TTCATCTTTTTTTCTTCTTCT</u>CTTTCTATAGTGGAGATAGTCGCACGTAATG-3′

RCt3 (partial sequence)

5′-TTCTATCACA<u>aa</u>AATAACAT<u>AAAAA</u>CTTATAAATTGCTCCCTATGCTCCAAATGGATAAG-3′

RCt4 (complete sequence)

5′-ACGGAATTGGAACTTCTTTGGTCCAGTAACGGGAAATCCATCCAAACTTCCTGGCCGTTTTCCATGGAATCTTTTCCTTC<u>TTTTTTTTTTTTT</u>GGCGGAATATCCGGTA
AAAACCATTCCAAGGCTCCTTTT-3′

RCt7 (complete sequence)

5′-GTGTCATTCTCTAAGCGAACTCGGAACATTCCGTTGGGTAGGGCTTCCGTAACTAAACCTTCGAAAGTTACTTTTGCTTCTCTCGGG<u>TTTTTTTTTTTT</u>CTCTCCTATTTT
TTTTTTCTGTCATGTTT-3′

identical size but not identical sequence (Bryan *et al.* 1999).

Conclusions

Our results based on RFLP of the seven amplified chloroplast genes/intergenic spacers, and the *trn*K gene sequence in the nine diploid and allotetraploid *Eleusine* species and two outgroup species resulted in well-resolved phylogenetic trees. The maternal genome donor (*E. indica*, $2n = 2x = 18$) of the allotetraploid ($2n = 4x = 36$, $2n = 2x = 38$) *Eleusine* species, and the phylogenetic relationships between cultivated *E. coracana* ($2n = 4x = 36$) and wild species could be successfully resolved. The species-specific markers were also identified. The two diploid species *E. indica* and *E. tristachya* could not be resolved separately by PCR–RFLP of seven chloroplast genic/intergenic spacers, as not a single site change could be scored. However, the *trn*K gene sequence clearly demonstrated that *E. indica* is more closely related to all three allotetraploids as compared with *E. tristachya*. Therefore, *E. indica* is most likely the maternal parent to all three allotetraploids. *Eleusine multiflora* ($2n = 2x = 16$) was found to be the most diverged among all the species. The explicit identification of the maternal parent and that of the immediate wild progenitor of finger millet will be immensely useful for future genetic improvement and biotechnological programmes of the crop species.

Sources of Funding

This work was supported by the Council of Scientific and Industrial Research (CSIR), Government of India and the National Academy of Sciences, India (NASI).

Contributions by the Authors

R.A. was involved in planning and performing all the experiments, data analyses and manuscript writing. N.A. was involved in the data analysis. R.T. was involved in relevant research discussions. S.N.R. was involved in the planning and supervision of all the experimental work and in writing the manuscript. All authors have seen and agreed to the submitted manuscript.

Acknowledgements

Thanks are due to the United States Department of Agriculture (USDA), the International Livestock Research Institute (ILRI), the National Bureau of Plant Genetic Resources (NBPGR) and Dr Mathews M. Dida for supplying the seed samples. We thank Dr Vishnu Bhat and Dr Shailendra Goel for their help and support.

Accession Numbers

The nucleotide sequences of the *trn*K gene from the nine *Eleusine* species and one outgroup *D. aegyptium* have been submitted to GenBank with accession numbers KF357736–KF357745. The nucleotide sequences of six cpSSRs isolated from *E. coracana* have been submitted to GenBank with accession numbers KF357730–KF357735.

Supporting Information

The following additional information is available in the online version of this article –

File 1. Sequence data matrix of the aligned partial *trn*K gene region of cpDNA of nine *Eleusine* species and two outgroups, *A. racemosa* and *D. aegyptium*. Nucleotide sequences are displayed 5′–3′. Dots indicate the same nucleotide as in *E. coracana*; dashes indicate gaps.

Literature Cited

Angioi SA, Desiderio F, Rau D, Bitocchi E, Attene G, Papa R. 2008. Development and use of chloroplast microsatellites in *Phaseolus* spp. and other legumes. *Plant Biology* **11**:598–612.

Asadi Abkenar A, Isshiki S, Tashiro Y. 2004. Phylogenetic relationships in the 'true citrus fruit trees' revealed by PCR–RFLP analysis of cpDNA. *Scientia Horticulturae* **102**:233–242.

Asadi Abkenar A, Isshiki S, Matsumoto R, Tashiro Y. 2008. Comparative analysis organelle DNAs in acid citrus grown in Japan using PCR–RFLP method. *Genetic Resources and Crop Evolution* **55**:487–492.

Babbar SB, Raghuvanshi S, Singh HK, Parveen I, Malik S. 2012. An overview of the DNA barcoding of plants. *Phytomorphology* **62**:69–99.

Barbeau WE, Hilu KW. 1993. Protein, calcium, iron and amino acid content of selected wild and domesticated cultivars of finger millet. *Plant Foods for Human Nutrition* **43**:97–104.

Bisht MS, Mukai Y. 2000. Mapping of rDNA on the chromosomes of *Eleusine* species by fluorescence *in situ* hybridization. *Genes and Genetic Systems* **75**:343–348.

Bisht MS, Mukai Y. 2001*a*. Genomic in situ hybridization identifies genome donor of finger millet (*Eleusine coracana*). *Theoretical and Applied Genetics* **102**:825–832.

Bisht MS, Mukai Y. 2001*b*. Identification of genome donors to the wild species of finger millet, *Eleusine africana* by genomic *in situ* hybridization. *Breeding Science* **51**:263–269.

Bisht MS, Mukai Y. 2002. Genome organization and polyploid evolution in the genus *Eleusine* (Poaceae). *Plant Systematics and Evolution* **233**:243–258.

Bryan GJ, McNicoll JM, Ramsay G, Meyer RC, DeJong WQS. 1999. Polymorphic simple sequence repeat markers in chloroplast genomes of solanaceous plants. *Theoretical and Applied Genetics* **99**:859–867.

Chandrashekar A. 2010. Finger millet *Eleusine coracana*. *Advances in Food & Nutrition Research* **59**:215–262.

Chennaveeraiah MS, Hiremath SC. 1974. Genome analysis of *Eleusine coracana* (L.) Gaertn. *Euphytica* **23**:489–495.

Clayton WD, Renvoize SA. 1986. *Genera Graminum: grasses of the world.* Kew Bulletin Additional Series 13. London: Her Majesty's Stationery Office.

Demesure B, Sodzi N, Petit RJ. 1995. A set of universal primers for amplification of polymorphic non-coding regions of mitochondrial and chloroplast DNA in plants. *Molecular Ecology* **4**:129–131.

Devarumath RM, Hiremath SC, Rao SR, Kumar A, Sheelavanthmath SS. 2005. Genome interrelationship in the genus *Eleusine* (Poaceae) as revealed through heteroploid crosses. *Caryologia* **58**:300–307.

Devarumath RM, Sheelavanthmath SS, Hiremath SC. 2010. Chromosome pairing analysis in interspecific hybrids among tetraploid species of *Eleusine* (Poaceae). *Indian Journal of Genetics* **70**: 299–303.

Dhingra A, Folta MK. 2005. ASAP: amplification, sequencing & annotation of plastomes. *BMC Genomics* **6**:176.

Dida MM, Devos KM. 2006. Finger millet. In: Kole C, ed. *Genome mapping and molecular breeding in plants, Vol. 1, cereals and millets.* Heidelberg: Springer, 333–343.

Dida MM, Srinivasachary RS, Bennetzen JL, Gale MD, Devos KM. 2007. The genetic map of finger millet, *Eleusine coracana. Theoretical and Applied Genetics* **114**:321–332.

Dida MM, Wanyera N, Dunn MLH, Bennetzen JL, Devos KM. 2008. Population structure and diversity in finger millet (*Eleusine coracana*) germplasm. *Tropical Plant Biology* **1**:131–141.

Duke JA, Wain KK. 1981. *Medicinal plants of the world. Computer index with more than 85000 entries.* London: Longman Group Limited.

Heinze B. 2005. *A database for PCR primers in the chloroplast genome.* http://www.bfw.ac.at/200/1859.html (14 October 2012).

Hilu KW. 1988. Identification of the 'A' genome of finger millet using chloroplast DNA. *Genetics* **118**:163–167.

Hilu KW. 1995. Evolution of finger millet: evidence from random amplified polymorphic DNA. *Genome* **38**:232–238.

Hilu KW, Alice LA. 1999. Evolutionary implications of *MATK* indels in Poaceae. *American Journal of Botany* **86**:1735–1741.

Hilu KW, deWet JMJ. 1976. Domestication of *Eleusine coracana. Economic Botany* **30**:199–208.

Hilu KW, Johnson JL. 1992. Ribosomal DNA variation in finger millet and wild species of *Eleusine* (Poaceae). *Theoretical and Applied Genetics* **83**:895–902.

Hilu KW, Johnson JL. 1997. Systematics of *Eleusine* Gaertn. (Poaceae, Chloridoideae): chloroplast DNA and total evidence. *Annals of the Missouri Botanical Garden* **84**:841–847.

Hilu KW, deWet JMJ, Seigler D. 1978. Flavonoids and systematics of *Eleusine. Biochemical Systematics and Ecology* **6**:247–249.

Hilu KW, Alice LA, Liang H. 1999. Phylogeny of Poaceae inferred from matK sequences. *Annals of the Missouri Botanical Garden* **86**: 835–851.

Hiratsuka J, Shimada H, Whittier R, Ishibashi T, Sakamoto M, Kondo C, Honji Y, Sun CR, Meng BY, Li YQ, Kanno A, Nishizawa Y, Hirai A, Shinozaki K, Sugiura M. 1989. The complete sequence of the rice (*Oryza sativa*) chloroplast genome: intermolecular recombination between distinct tRNA genes accounts for a major plastid DNA inversion during the evolution of cereals. *Molecular and General Genetics* **217**:185–194.

Hiremath SC, Chennaveeraiah MS. 1982. Cytogenetical studies in wild and cultivated species of *Eleusine* (Gramineae). *Caryologia* **35**: 57–69.

Hiremath SC, Salimath SS. 1991. The quantitative nuclear DNA changes in *Eleusine* (Gramineae). *Plant Systematics and Evolution* **178**:225–233.

Hiremath SC, Salimath SS. 1992. The 'A' genome donor of *Eleusine coracana* (L.) Gaertn. (Gramineae). *Theoretical and Applied Genetics* **84**:747–754.

Ibrahim RIH, Azuma J-I, Sakamoto M. 2007. PCR–RFLP analysis of the whole chloroplast DNA from three cultivated species of cotton (*Gossypium* L.). *Euphytica* **156**:47–56.

Ishii T, McCouch SR. 2000. Microsatellites and microsynteny in the chloroplast genomes of *Oryza* and eight other Gramineae species. *Theoretical and Applied Genetics* **100**:1257–1266.

Jena SN, Kumar S, Nair K. 2009. Molecular phylogeny in Indian *Citrus* L. (Rutaceae) inferred through PCR–RFLP and trnL–trnF sequence data of chloroplast DNA. *Scientia Horticulturae* **119**: 403–416.

Kishimoto S, Aida R, Shibata M. 2003. Identification of chloroplast DNA variations by PCR–RFLP analysis in *Dendranthema. Journal of the Japanese Society of Horticultural Sciences* **72**:197–204.

Komatsu K, Zhu S, Fushimi H, Qui TK, Cai S, Kadota S. 2001. Phylogenetic analysis based on 18S rRNA gene and *matK* gene sequences of *Panax vietnamensis* and five related species. *Planta Medica* **67**: 461–465.

Lakshmi M, Senthilkumar P, Parani M, Jithesh MN, Parida A. 2000. PCR–RFLP analysis of chloroplast gene regions in *Cajanus* (Leguminosae) and allied genera. *Euphytica* **116**:243–250.

Liu Q, Triplett JK, Wen J, Peterson M. 2011. Allotetraploid origin and divergence in *Eleusine* (Chloridoideae, Poaceae): evidence from low-copy nuclear gene phylogenies and a plastid gene chronogram. *Annals of Botany* **108**:1287–1298.

Lye KA. 1999. Nomenclature of finger millet (Poaceae). *Lidia* **4**: 149–151.

Murray MG, Thompson KH. 1980. Rapid isolation of high molecular weight plant DNA. *Nucleic Acids Research* **8**:4321–4325.

Mysore KS, Baird V. 1997. Nuclear DNA content in species of *Eleusine* (Gramineae): a critical re-evaluation using laser flow cytometry. *Plant Systematics and Evolution* **207**:1–11.

Nei M, Li WH. 1979. Mathematical model for studying genetic variation in terms of restriction endonucleases. *Proceedings of the National Academy of Sciences of the USA* **76**:5269–5273.

Neves SS. 2011. *Eleusine.* In: Kole C, ed. *Wild crop relatives: genomic and breeding resources, millets and grasses.* Berlin: Springer, 113–133.

Neves SS, Clark GS, Hilu KW, Baird WV. 2005. Phylogeny of *Eleusine* (Poaceae: Chloridoideae) based on nuclear ITS and plastid *trnT–trnF* sequences. *Molecular Phylogenetics and Evolution* **35**: 395–419.

Nwakanma DC, Pillay M, Okoli E, Tenkouano A. 2003. Sectional relationship in the genus *Musa* L. inferred from the PCR–RFLP of organelle DNA sequences. *Theoretical and Applied Genetics* **107**:850–856.

Parani M, Lakshmi M, Ziegenhagen B, Fladung M, Senthikumar P, Parida A. 2000. Molecular phylogeny of mangroves VII. PCR–RFLP of *trnS–psbC* and *rbcL* gene regions in 24 mangrove and mangrove-associated species. *Theoretical and Applied Genetics* **100**:454–460.

Parani M, Rajesh K, Lakshmi M, Parducci L, Szmidt AE, Parida A. 2001. Species identification in seven small millet species using polymerase chain reaction–restriction fragment length polymorphism of *trnS–psbC* gene region. *Genome* **44**:495–499.

Phillips SM. 1972. A survey of the *Eleusine* Gaertn. (Gramineae) in Africa. *Kew Bulletin* **27**:251–270.

Phillips SM. 1995. Poaceae (Gramineae). In: Hedberg I, Edwards S, eds. *Flora of Ethiopia and Eritrea.* Vol. 7. Addis: Addis Ababa University and Uppsala University.

Poczai P, Cseh A, Taller J, Symon DE. 2011. Genetic diversity and relationships in *Solanum* subg. *Archaesolanum* (Solanaceae) based on RAPD and chloroplast PCR–RFLP analyses. *Plant Systematics and Evolution* **291**:35–47.

Powell W, Morgante M, McDevitt R, Vendramin GG, Rafalski JA. 1995a. Polymorphic simple sequence repeat regions in chloroplast genomes: applications to the population genetics of pines. *Proceedings of the National Academy of Sciences of the USA* **92**: 7759–7763.

Powell W, Morgante M, Andre C, McNicoll JW, Machray GC, Doyle JJ, Tingey SV, Rafalski JA. 1995b. Hypervariable microsatellites provide a general source of polymorphic DNA markers for the chloroplast genome. *Current Biology* **5**:1023–1029.

Powell W, Morgante M, Doyle JJ, McNicoll JW, Tingey SV, Rafalski JA. 1996. Genepool variation in genus *Glycine* subgenus *soja* revealed by polymorphic nuclear and chloroplast microsatellites. *Genetics* **144**:793–803.

Pradhan A, Nag SK, Patil SK. 2010. Dietary management of finger millet (*Eleusine coracana* L. Gaerth) controls diabetes. *Current Science* **98**:763–765.

Provan J, Corbett G, McNicoll JW, Powell W. 1997. Chloroplast DNA variability in wild and cultivated rice (*Oryza* spp.) revealed by polymorphic chloroplast simple sequence repeats. *Genome* **40**: 104–110.

Provan J, Russel JR, Booth A, Powell W. 1999. Polymorphic chloroplast simple sequence repeat primers for synthetic and population studies in the genus *Hordeum. Molecular Ecology* **8**:505–511.

Provan J, Powell W, Hollingsworth PM. 2001. Chloroplast microsatellites: new tools for studies in plant ecology and evolution. *Trends in Ecology and Evolution* **16**:142–147.

Saitou N, Nei M. 1987. The neighbour-joining method: a new method for reconstructing phylogenetic trees. *Molecular Biology and Evolution* **4**:406–426.

Salimath SS. 1990. *Cytology and genome relations in some species of Eleusine and its allies.* PhD Thesis, Karnataka University, Dharwad, India.

Salimath SS, de Oliveira AC, Godwin ID, Bennetzen JL. 1995a. Assessment of genome origins and genetic diversity in the genus *Eleusine* with DNA markers. *Genome* **38**:757–763.

Salimath SS, Hiremath SC, Murthy HN. 1995b. Genome differentiation patterns in diploid species of *Eleusine* (Poaceae). *Hereditas* **122**: 189–195.

Sehgal D, Rajpal VR, Raina SN. 2008. Chloroplast DNA diversity reveals the contribution of the two wild species in the origin and evolution of diploid safflower (*Carthamus tinctorius* L.). *Genome* **51**:638–643.

Shinozaki K, Ohme M, Tanaka M, Wakasugi T, Hayashida N, Matsubayashi T, Zaita N, Chunwongae J, Obokata J, Yamaguchi-Shinozaki K, Ohta C, Torazawa K, Meng BY, Sugita M, Deno H, Kamogashira T, Yamada K, Kusuda J, Takaiwa F, Kato A, Tohdoh N, Shimada H, Sugiura M. 1986. The complete nucleotide sequence of tobacco chloroplast genome: its gene organization and expression. *EMBO Journal* **5**:2043–2049.

Sneath PHA, Sokal RR. 1973. *Numerical taxonomy.* San Francisco, USA: W.H. Freeman.

Swofford DL. 2002. *PAUP*: Phylogenetic analysis using parsimony (* and other methods), Version 4.* Sunderland, MA: Sinauer Associates.

Taberlet P, Gielly L, Pautou G, Bouvet J. 1991. Universal primers for amplification of three non-coding regions of chloroplast DNA. *Plant Molecular Biology* **17**:1105–1109.

Tamura K, Peterson D, Peterson N, Stecher G, Nei M, Kumar S. 2011. MEGA5: molecular evolutionary genetics analysis using maximum likelihood, evolutionary distance, and maximum parsimony methods. *Molecular Biology and Evolution* **28**: 2731–2739.

Thompson JD, Gibson TJ, Plewniak F, Mougin FJ, Higgins DG. 1997. The Clustal X Windows interface: flexible strategies for multiple sequence alignment aided by quality analysis tools. *Nucleic Acid Research* **25**:4876–4882.

Tsumura Y, Yoshimura K, Tomaru N, Ohba K. 1995. Molecular phylogeny of conifers using RFLP analysis of PCR-amplified specific chloroplast genes. *Theoretical and Applied Genetics* **91**:1222–1236.

Tsumura Y, Kawahara T, Wickneswari R, Yoshimura K. 1996. Molecular phylogeny of Dipterocarpaceae in Southeast Asia using RFLP of PCR-amplified chloroplast genes. *Theoretical and Applied Genetics* **93**:22–29.

Van Droogenbroeck B, Kyundt T, Maertens I, Romeij-Peeters E, Scheldeman X, Romero-Motochi JP, Van Damme P, Goetghebeur P, Gheysen G. 2004. Phylogenetic analysis of highland papayas (*Vasconcellea*) and allied genera (Caricaeae) using PCR–RFLP. *Theoretical and Applied Genetics* **108**:1473–1486.

Verma V. 2009. *Textbook of economic botany.* New Delhi, India: Ane Books.

Weising K, Gardner RC. 1999. A set of conserved PCR primers for the analysis of simple sequence repeat polymorphisms in chloroplast genomes of dicotyledonous angiosperms. *Genome* **42**:9–19.

Werth CR, Hilu K, Langner CA, Baird WV. 1993. Duplicated gene expression for isocitrate dehydrogenase and 6-phosphogluconate dehydrogenase in diploid species of *Eleusine* (Gramineae). *American Journal of Botany* **80**:705–710.

Werth CR, Hilu KW, Langner CA. 1994. Isozyme of *Eleusine* (Gramineae) and the origin of finger millet. *American Journal of Botany* **81**:1186–1197.

Zhu S, Fushimi H, Cai S, Komatsu K. 2003. Phylogenetic relationship in the genus *Panax*: inferred from chloroplast *trn*K gene and nuclear 18S rRNA gene sequences. *Planta Medica* **69**:647–653.

3

Comparison of the morphogenesis of three genotypes of pea (*Pisum sativum*) grown in pure stands and wheat-based intercrops

Romain Barillot[1,2], Didier Combes[3], Sylvain Pineau[1], Pierre Huynh[1] and Abraham J. Escobar-Gutiérrez[3*]

[1] LUNAM Université, Groupe Ecole Supérieure d'Agriculture, UPSP Légumineuses, Ecophysiologie Végétale, Agroécologie, 55 rue Rabelais, BP 30748, F-49007 Angers Cedex 01, France
[2] Present address: INRA, Centre de Versailles-Grignon, U.M.R. INRA/AgroParisTech Environnement et Grandes Cultures, 78850 Thiverval-Grignon, France
[3] INRA, UR4 P3F, Equipe Ecophysiologie des plantes fourragères, Le Chêne – RD 150, BP 6, F-86600 Lusignan, France

Abstract. Cereal–legume intercrops represent a promising way of combining high productivity and agriculture sustainability. The benefits of cereal–legume mixtures are highly affected by species morphology and functioning, which determine the balance between competition and complementarity for resource acquisition. Studying species morphogenesis, which controls plant architecture, is therefore of major interest. The morphogenesis of cultivated species has been mainly described in mono-specific growing conditions, although morphogenetic plasticity can occur in multi-specific stands. The aim of the present study was therefore to characterize the variability of the morphogenesis of pea plants grown either in pure stands or mixed with wheat. This was achieved through a field experiment that included three pea cultivars with contrasting earliness (hr and HR type) and branching patterns. Results show that most of the assessed parameters of pea morphogenesis (phenology, branching, final number of vegetative organs and their kinetics of appearance) were mainly dependent on the considered genotype, which highlights the importance of the choice of cultivars in intercropping systems. There was however a low variability of pea morphogenesis between sole and mixed stands except for plant height and branching of the long-cycle cultivar. The information provided in the present study at stand and plant scale can be used to build up structural–functional models. These models can contribute to improving the understanding of the functioning of cereal–legume intercrops and also to the definition of plant ideotypes adapted to the growth in intercrops.

Keywords: Morphogenesis; *Pisum sativum*; plant architecture; plasticity; *Triticum aestivum*; wheat–pea intercropping.

Introduction

In order to ensure agriculture sustainability, efforts have been made by researchers and farmers to reduce the use of fertilizers and pesticides. This challenges the maintenance of efficient and profitable agrosystems being able to face demographic growth. Because of their ability to fix atmospheric N_2, legume species can improve the sustainability of cropping systems by helping to decrease the use of nitrogen fertilizers and favouring the diversification of crop rotations (Crews and Peoples 2004; Duc *et al.* 2010). Also, seeds or forages of legumes are among the richest sources of proteins in crops, with a high nutritional value

* Corresponding author's e-mail address: abraham.escobar@lusignan.inra.fr

for animals (Duc *et al.* 2010). However, the potential productivity of legumes has not been reached, mainly because of a strong sensitivity of these species to lodging and foliar diseases (Ney and Carrouée 2005). This is in particular the case for pea (*Pisum sativum*), which is the main source of vegetable proteins in Europe. In this context, the increasing interest in growing cereal–legume intercrops (IC), such as wheat–pea mixtures, represents an alternative for reintroducing legume species in cropping systems. Several studies reported that these mixtures can provide high and stable yields compared with pure mono-specific stands (Ofori and Stern 1987; Jensen 1996; Corre-Hellou *et al.* 2006; Hauggaard-Nielsen *et al.* 2008). Such advantages result from a balance between complementary (e.g. separate root and canopy areas) and competition processes for light, water and nitrogen that occur between intercropped species. These complex interactions depend on the pedo-climatic conditions, agricultural practices and also on the morphology and functioning of the component species (Corre-Hellou *et al.* 2006; Launay *et al.* 2009; Louarn *et al.* 2010; Naudin *et al.* 2010; Barillot 2012).

The latter point is mainly related to the choice of cultivars, which therefore appears as a determinant factor of (i) the proportion of each component species at harvest and (ii) mixture productivity. Cultivars are usually discriminated according to their earliness, sensitivity to diseases or potential yield. However, in the particular case of multi-specific stands, the above-ground architecture of a cultivar, given by its geometry, optical properties and topology of the phytoelements (Godin 2000), should also be taken into account. Indeed, plant architecture defines the plant interface with biotic (e.g. with *Mycosphaerella pinodes*; Béasse *et al.* 2000; Le May *et al.* 2009) and abiotic factors (e.g. light; Ross 1981). In the case of multi-specific stands, the complementarity between the architecture of the mixed species represents a crucial issue as it will determine their respective ability to compete for light that in turn drives the production and allocation of biomass (Varlet-Grancher *et al.* 1993; Sinoquet and Caldwell 1995).

For pea, several genes involved in the development of the above-ground architecture have been identified (for a review see Huyghe 1998). For instance, numerous *ramosus* mutants were described because of their altered branching behaviour (Arumingtyas *et al.* 1992). Plant height can also be altered through mutations made on genes involved in internode growth (Kusnadi *et al.* 1992). Genetic control of the compound leaf shape of pea has also been assessed and appears to be related to the *UNIFOLIATA* gene (Gourlay *et al.* 2000). Precocity of pea cultivars has been shown to be regulated by genes (*Hr* and *Lf*) that control the sensitivity to photoperiod for floral initiation and flowering (Murfet 1973, 1975). These studies have promoted the breeding of

several pea cultivars with contrasting architectures that therefore constitute as much as potential combinations for wheat–pea IC. Characterizing the morphogenesis (sequence of developmental and growth processes leading to plant architecture) of these various pea genotypes is therefore of major interest for improving the management of intercropped stands. Several descriptors can be used to characterize pea architecture, the most commonly used being those related to the leaf area and its spatial distribution as this strongly determines a plant's ability to compete for light interception. On a finer scale, both the amount and distribution of foliar area are related to the number and geometry of stems and leaves produced during the initiation of each phytomer by the apex. A phytomer is defined as a basic unit repeated along the stem and including an internode, a node, a leaf and one or several axillary buds (Gray 1849; White 1979). The sharing of resources within multi-specific stands also depends on the respective height reached by the component species (Sinoquet and Caldwell 1995; Schwinning and Weiner 1998; Louarn *et al.* 2010; Barillot *et al.* 2011, 2012). Although the architectural parameters involved in the leaf area and height of plants are key factors of the mixture development, they have been mainly described in mono-specific growing conditions (for a review see Munier-Jolain *et al.* 2005a). However, the morphogenesis of plants can be highly plastic when facing environmental variations; hence the question arises as to whether morphogenetic variations can occur between mono- and multi-specific stands.

The aim of the present study was therefore to characterize the variability of pea morphogenesis grown either in pure stands or mixed with wheat. In order to have a large range of plant architectures and morphogenetic responses, a field experiment was performed using three pea cultivars with contrasting growth habits. The growth and phenology of the pea cultivars were measured regularly during their growing cycle. This study provides information at both stand and plant scale in order to identify plant traits of interest that can contribute to the conception of plant ideotypes.

Methods

Plant material and growing conditions

A field experiment was carried out in 2010–11 at Brain-sur-l'Authion, western France (47°26′N, 00°26′W) in a clay soil (51 % clay, 26 % silt and 23 % sand). Daily mean air temperature, precipitation and photosynthetically active radiation (PAR) were recorded by a standard automatic agro-meteorological station located close to the field.

Winter wheat (*Triticum aestivum*) cv. Cézanne and three cultivars of winter field pea (*Pisum sativum*), cv.

Lucy (hr type), AOPH10 (hr type) and 886/01 (HR type), were sown on 28 November 2010 in sole crops (SC) and IC. Unlike hr types, flowering of the HR cultivar is sensitive to photoperiod. The sowing density of SC was 250 plants m^{-2} for wheat. Optimal densities of pea cultivars were chosen with respect to their ability for lateral development and the underlying risk of lodging. Sole crops composed of pea cultivars Lucy and AOPH10 were sown at 80 plants m^{-2} whereas cultivar 886/01 was sown at 40 plants m^{-2}. Intercrops followed a substitutive design where the two species were mixed within the row. Wheat and pea crops in IC were sown at half their respective density in pure stands. From seedbed preparation to harvest, local agronomic recommendations were followed and pest and weed were chemically controlled. Stands of sole wheat were fertilized with 14 g N m^{-2} whereas pea SC and wheat–pea IC were not supplied with external nitrogen.

Statistical analyses described below were thus performed considering two factors: (i) pea genotype with three levels (Lucy, AOPH10 and 886/01) and (ii) cropping system with two levels (SC and IC). A sole crop of wheat was added to those six treatments but was not considered for statistical analyses. These seven treatments (3 SC of pea, 3 IC and 1 SC of wheat) were arranged in experimental units of 1.2×10 m^2 within a randomized complete block design with three replicates.

Plant sampling and measurement of pea morphogenesis

On the one hand, integrated parameters defined at canopy scale (biomass, height, yield) were measured in each plot. Samplings were carried out on 0.75 m^2 in the centre of each experimental unit. The above-ground biomass and the maximal height of each SC (wheat and pea) and IC plot were measured during the growing cycle at 645, 1525 growing degree days (GDD) from emergence (base temperature 0 °C) and, lastly, at crop maturity (Table 1). The land equivalent ratio (LER) for grain yields of wheat–pea IC was also estimated according to De

Wit and Van den Bergh (1965). Land equivalent ratio is the sum of partial LER values for wheat (LER$_{wheat}$) and pea (LER$_{Pea}$):

$$LER_{wheat} = \frac{Y_{ICw}}{Y_{SCw}}, \quad LER_{pea} = \frac{Y_{ICp}}{Y_{SCp}}, \quad LER = LER_{wheat} + LER_{pea}$$

where Y_{ICw} and Y_{SCw} are yields of wheat and pea in IC, respectively, and Y_{SCw} and Y_{SCp} are yields of wheat and pea in SC, respectively. Land equivalent ratio values above 1 indicate a benefit of intercropping over sole cropping.

On the other hand, specific measurements on pea cultivars were made at plant scale. The morphogenesis of five pea plants per plot was characterized for each vegetative axis, i.e. main stems and lateral branches. Only one branch at each nodal position of the main stem was followed up. Branches were denoted according to their topological position, i.e. main stems were denoted as Axis-0, then branches that emerged from node n of the main stem were referenced as Axis-n. For each axis group, the kinetics of phytomer appearance (unfolding leaf visible to the naked eye) were measured and fitted with Schnute's non-linear model (Schnute 1981) using the least-squares method. The model is written as

$$y = \left[y_{max}^B \frac{1 - e^{-A(t)}}{1 - e^{-A(t_{max})}} \right]^{1/B} + \varepsilon_i$$

where y is the number of visible phytomers; estimated parameters are A and B which implicitly define the shape of the curve; t_{max} is the last value of the time (t) domain for which the model is fitted, corresponding to the end of the vegetative development of the stem; parameter y_{max} is the value of Y at t_{max} and ε is the residual. Parameters were optimized using the Levenberg–Marquardt iterative method (Marquardt 1963) with automatic computation of the analytical partial derivatives. The fitting procedure was performed for each axis group of each plant. The first derivatives of Schnute's adjustments were also used in order to estimate the rates of phytomer production of the pea cultivars.

Statistical analyses

Exploratory data analysis, analysis of variance and non-linear regression techniques were performed with R software (R Development Core Team 2012). Analyses of variance (ANOVAs) were performed following a two-factor linear model such that

$$y_{ijk} = \mu + B_i + G_j + C_k + (G \times C)_{jk} + \varepsilon_{ijk}$$

where Y is any dependent variable, μ the mean value of Y, B_i the effect of block i, G_j the effect of genotype j, C_k the

Table 1. Harvest time, expressed in growing degree day (DD) from emergence (base, 0 °C), of pea and wheat grown in SC and in IC.

Species	Genotype	Harvest time in SC (DD)	Harvest time in IC (DD)
Pea	Lucy	1900	2275
Pea	AOPH10	1985	2275
Pea	886/01	2130	2275
Wheat	Cézanne	2275	2275

effect of cropping system k (either SC or IC) and ε the random error of measurement ijk.

Normal distributions of the residuals of ANOVAs as well as those of Schnute's adjustments were tested using the Shapiro–Wilk test. Homoscedasticity was checked by random distribution of the residuals. Tukey's HSD tests were used for mean separation when three or more means were compared.

Results

Environmental conditions during crop growth are summarized in Fig. 1. The daily average air temperature ranged from $-4\ °C$ to $27\ °C$ on 31 January and 27 June, respectively (Fig. 1A). Irrigation supplied 30 mm of water on 21 April and 27 May 2011. The daily cumulated PAR (Fig. 1B) ranged from 2.70 to 108.70 $mol\ m^{-2}$ on 31 December 2010 and 25 June 2011, respectively.

Growth of SC and IC stands

The results described in this section are derived from measurements at the whole stand scale. First, the biomass accumulation of each crop is shown in Fig. 2. The

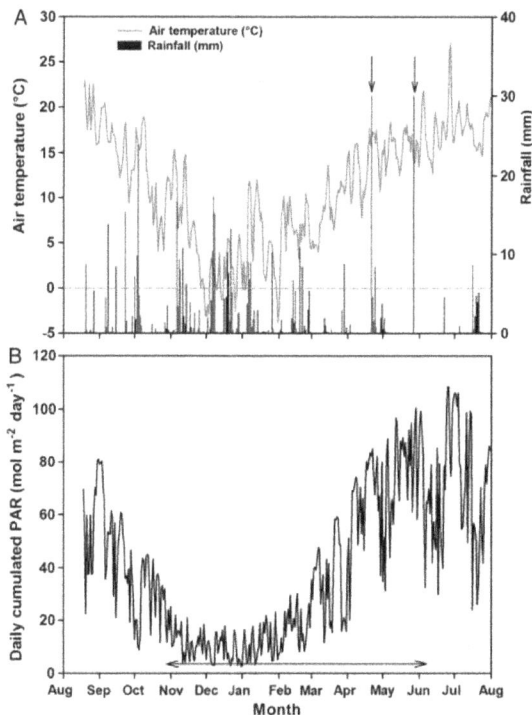

Figure 1. Meteorological conditions during the growing season 2010–11 at Brain-sur-l'Authion, France. Daily mean air temperatures and rainfall are shown in (A). Vertical arrows represent a water supply of 30 mm by irrigation. Daily cumulated PAR is shown in (B). The horizontal arrow represents the growing period of pea and wheat stands.

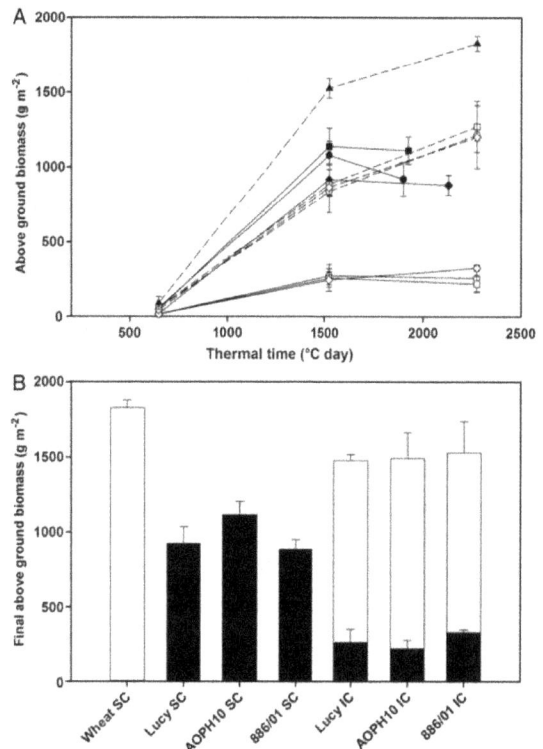

Figure 2. (A) Accumulation of above-ground biomass as a function of thermal time from emergence (base temperature = 0 °C). Sole crops are shown in closed symbols and IC in open symbols. Pea cultivars are in solid lines: Lucy is denoted by circles (filled, open), AOPH10 by squares (filled, open) and 886/01 by diamonds (filled, open). Wheat is in dotted lines with triangles (filled) in SC, and with the corresponding pea cultivar symbol in IC (open circles with 'Lucy', open squares with 'AOPH10' and open diamonds with '886/01'). (B) Contribution of wheat (white bars) and pea (black bars) in the final biomass reached by SC and IC stands.

above-ground biomass of the two species increased from 645 to 1525 GDD and then slowed down until maturity (Fig. 2A). Wheat SC showed the highest amount of biomass during the growing cycle and finally reached 1830 $g\ m^{-2}$. The final above-ground biomass accumulated by the three pea cultivars in SC ranged from 880 to 1275 $g\ m^{-2}$. On average, '886/01' exhibited the lowest biomass in SC throughout the growing cycle and finally reached $880 \pm 67\ g\ m^{-2}$. This lower biomass has to be related to its sowing density, which was 50 % of the other cultivars (see the Methods section). Intercropped wheat (IC stands) accumulated on average 1230 $g\ m^{-2}$ of biomass of all IC stands taken together. Pea grown in IC stands produced 220–325 $g\ m^{-2}$ of biomass at maturity. At this stage of development, the final biomass reached by IC stands was 1500 $g\ m^{-2}$ averaged across pea cultivars (Fig. 2B). Wheat in IC contributed to the main part of the mixture biomass (on average 3.75 times the biomass of pea). Although cultivar 886/01 exhibited the lowest biomass in SC stands, this cultivar produced more biomass when intercropped with wheat compared with 'Lucy' and 'AOPH10', despite being sown at half density.

Harvest of IC stands (Table 1) was performed when wheat reached maturity (2275 GDD). For pea grown in SC, harvest was made earlier, i.e. between 1900 and 2130 GDD depending on the cultivar. Cultivars Lucy and AOPH10 (hr types) reached their maturity earlier than '886/01'. Indeed, cultivar 886/01 (HR type) needs a longer photoperiod to reach flowering and was therefore harvested later than hr cultivars. As a result, the maturity of cultivar 886/01 and that of wheat were reached in a similar period (2130 and 2275 GDD, respectively).

Land equivalent ratios for grain yield (LERs) were estimated for IC (Fig. 3). Although wheat SC stands were fertilized with nitrogen and IC stands were not, the partial LER of wheat was systematically higher than 0.5 (0.65 on average) whatever the companion pea cultivar. Thus, wheat yields in IC were higher than those in SC when normalized by the sowing density (wheat density in IC was half that of SC but yields were reduced by <35 %). Land equivalent ratios of IC stands composed of cultivars Lucy and AOPH10 were slightly <1, meaning that the cumulative yield of wheat and pea in IC stands was lower than the sum of each SC yield. Analysis of partial LERs of 'Lucy' and 'AOPH10' (0.3 and 0.2, respectively) revealed that their yield markedly decreased in IC compared with pure stands. In contrast, the yield of IC based on cultivar 886/01 was 25 % higher than the cumulative yield of wheat and '886/01' in SC. In these IC stands, the yield of '886/01' was strongly increased compared with SC conditions ($LER_{886/01} = 0.62$ on average). Interestingly, the ratio of the final above-ground biomass of '886/01' in IC to that of '886/01' in SC (0.37, Fig. 2B) was smaller than

the grain yield ratio ($LER_{866/01}$). This means that the harvest index (ratio grain yield/above-ground biomass) was particularly high in intercropped '886/01'.

Wheat plants grown in SC and IC stands exhibited similar height and reached a maximum of 0.95 m (Fig. 4). Wheat grown in IC was taller than pea from the early stages of development and whatever the pea cultivar. The height reached by the pea cultivars was variable, depending on the cropping system (SC or IC stand). The canopy height of sole pea crops reached a maximum of 0.83–0.94 m for 'Lucy' and 'AOPH10', respectively. The height of sole pea finally decreased dramatically at the end of the growing cycle as a result of plant lodging. However, the height of pea cultivars grown in IC stands remained over 0.70 m, meaning that pea plants grown in IC were staked by wheat stems thus preventing pea lodging. The species height ratio (pea/wheat) in IC is shown in Table 2. The height ratio ranged from 0.33 to 0.48 at 700 GDD, meaning that pea cultivars were strongly dominated in the first stages of development, especially for

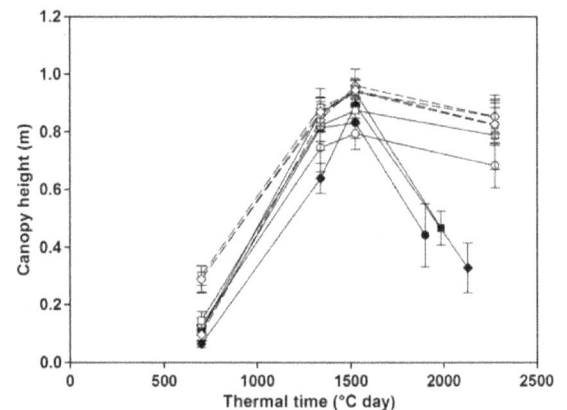

Figure 4. Observed height of canopies during the growing cycle. Sole crops are shown in closed symbols and IC in open symbols. Pea cultivars are in solid lines: Lucy is denoted by circles (filled, open), AOPH10 by squares (filled, open) and 886/01 by diamonds (filled, open). Wheat is in dotted lines with triangles (filled) in SC, and with the corresponding pea cultivar symbol in IC (open circles with 'Lucy', open squares with 'AOPH10' and open diamonds with '886/01').

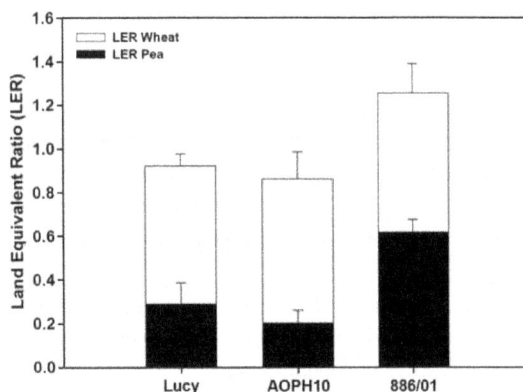

Figure 3. Land equivalent ratio of each wheat–pea mixture. Land equivalent ratio values >1 indicate a benefit of intercropping over sole cropping. N.B.: unlike pure stands of pea and IC, wheat SC were fertilized with external nitrogen.

Table 2. Height ratio of wheat–pea mixtures according to the growing degree day (GDD).

Genotype	Height ratio (pea/wheat)			
	700 GDD	1330 GDD	1525 GDD	2275 GDD
Lucy	0.45	0.87	0.85	0.80
AOPH10	0.48	0.93	0.93	0.96
886/01	0.33	0.97	1.02	1.03

'886/01'. From 1330 GDD on, the Lucy and AOPH10 cultivars were slightly shorter than wheat, whereas '886/01' reached the height of wheat.

Morphogenesis of pea cultivars

In a second step, measurements were made at plant scale in order to compare the morphogenesis of pea cultivars grown in pure stands with those intercropped with wheat.

Lateral branching of pea plants. The total number of branches (including non-flowering ones) produced by individual pea plants is shown in Fig. 5A according to their nodal position on main stems. Most lateral branches emerged from the first and second phytomer of the main stems (Axis-1 and -2) whatever the cropping system (SC and IC) and pea cultivar. Very few branches were produced on the third phytomer of the main stems (only nine branches, all cultivars and cropping systems taken together). Cultivars Lucy and AOPH10 grown in SC developed on average 3–4 branches per plant whereas cultivar 886/01 in SC was the most branching with about 6 branches per plant on average. Nevertheless, a significant effect of the genotype on the number of branches was only found for Axis-2 (ANOVA $F_{2,82} =$

13.93, $P < 0.001$). Indeed, cultivar 886/01 produced significantly more branches of type Axis-2 than Lucy and AOPH10 cultivars did (HSD $P < 0.001$). The ability of the 886/01 cultivar to develop numerous lateral branches appeared to compensate for its lower sowing density as shown by its biomass accumulation, which was similar to those of cultivars Lucy and AOPH10 (Fig. 2). Significant effects of the cropping system on the number of Axis-1 ($F_{1,82} = 14.33, P < 0.001$) and Axis-2 ($F_{1,82} = 5.48, P < 0.05$) were also found. Indeed, pea cultivars grown in IC tended to develop fewer branches than in SC. The most drastic decrease was observed for cultivar 886/01, which produced 40 % less branches in IC (HSD $P < 0.01$ and < 0.05 for Axis-1 and -2, respectively). However, for 'Lucy' (-11 %) and 'AOPH10' (-22 %) intercropped with wheat, it was not possible to detect any significant differences in the number of branches between the two cropping systems, probably because of the high variability observed.

Lateral branches emerged between 275 and 420 GDD, all cultivars and cropping systems taken together (Fig. 5B). Although they were initiated later, branches that developed on the second phytomer of the main stem (Axis-2) appeared 70 GDD earlier than those located on the first node (Axis-1). Indeed, in most cases, the first phytomer (carrying the first vestigial leaf) is located a few millimetres under the ground. This may mechanically delay the emergence of branches, depending for instance on the sowing depth or soil structure, which also affect the quantity of light perceived by the axillary bud (Jeudy and Munier-Jolain 2005). Statistical analyses showed that there was a significant effect of the genotype on the time of branch emergence ($F_{2,87} = 5.09, P < 0.01$; $F_{2,166} = 3.20, P < 0.05$ for Axis-1 and -2, respectively). Indeed, branches developed on cultivar Lucy emerged significantly earlier than those of '886/01' (HSD $P < 0.01$ and < 0.05 for Axis-1 and -2, respectively), while 'AOPH10' had an intermediary behaviour. There were however no significant differences in the time of branch emergence between SC and IC stands, meaning that this parameter of pea morphogenesis was mainly dependent on the genotype.

As shown above, pea genotypes produced several branches during the first 500 GDD. This led to complex plant architectures with a high potential number of stems per plant. For the sake of clarity, we only consider hereafter one branch at each node of the main stems. As a result, plant architectures were simplified to a main stem potentially bearing one branch on its first three nodes (see Fig. 5A). However, only a part of these stems actually grew and completed their development up to flowering. As shown in Fig. 6, very few main stems

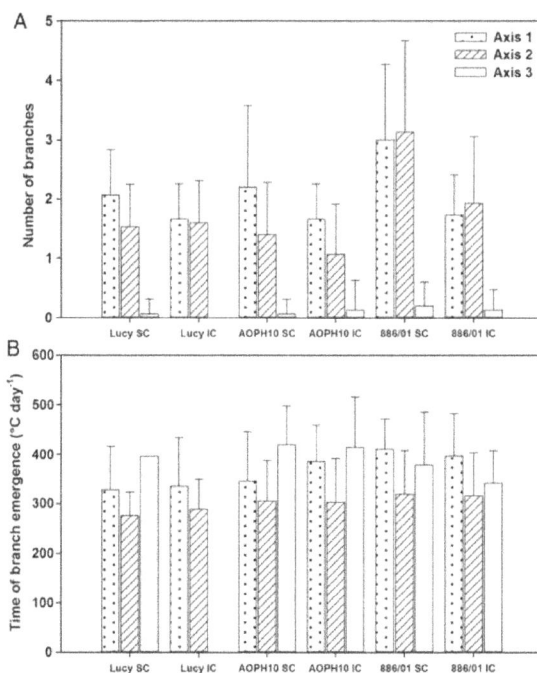

Figure 5. Lateral branching of pea cultivars ($n = 15$ plants for each condition). (A) Number of lateral branches developed by the pea cultivars Lucy, AOPH10 and 886/01 grown in SC and in IC. (B) Time of branching expressed in thermal time from crop emergence. Branches were distinguished according to their nodal position on the main stem. Axis-1: branches emerged at the first node; Axis-2: second node; and Axis-3: third node.

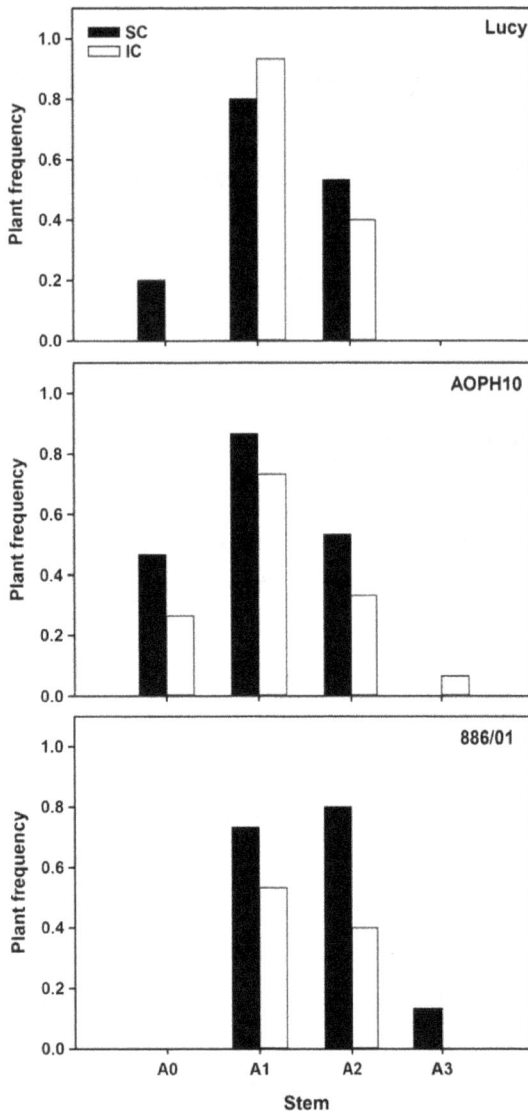

Figure 6. Frequency of pea plants grown in SC and in IC whose measured branches have reached flowering. Branches were distinguished according to their nodal position on the main stem. Axis-1: branches emerged at the first node; Axis-2: second node; and Axis-3: third node ($n = 15$ plants for each condition, only one single branch was considered at each nodal position).

of pea flowered. This was in particular the case of cultivars 886/01 (0 % of flowering main stems) and Lucy (0–20 % of flowering main stems for IC and SC, respectively). Flowering of main stems was significantly dependent on the cultivar ($F_{2,10} = 6.83$, $P < 0.05$). Indeed, the proportion of 'AOPH10' plants that had flowering main stems (25–45 % in IC and SC, respectively) was statistically higher than that of cultivar 886/01 (HSD $P < 0.05$). Nevertheless, most flowering stems of the three pea cultivars were Axis-1 and -2 branches. For cultivars Lucy and AOPH10, there were 85 % more of Axis-1 branches that carried on growing until flowering compared with Axis-2

(regardless of the cropping system). In contrast, the proportion of flowering Axis-1 of cultivar 886/01 was similar to Axis-2. Pea plants grown in SC and those intercropped with wheat exhibited similar proportions of flowering stems.

Rate of phytomer appearance and final number of phytomers. Phytomer production by stem apices followed sigmoid-type dynamics as illustrated in Fig. 7. These dynamics were fitted with Schnute's function. In order to reduce the variability on Schnute's parameters (Table 3), stems that stop growing before flowering were not taken into account in the fitting procedure. The first three parameters of Schnute's function, A, B and t_{max}, respectively, ranged from 0.36 to 4.30×10^{-3}, 0.13 to 0.56 and 1164 to 1476 GDD averaged across cultivars and cropping systems (SC and IC). The shape parameters A and B of Axis-2 branches appeared to be significantly dependent on pea cultivar ($F_{2,33} = 6.43$, $P < 0.01$; $F_{2,33} = 12.23$, $P < 0.001$ for parameters A and B, respectively). Indeed, parameters A and B of Axis-2 branches produced by cultivar 886/01 were statistically higher than those of 'Lucy' and 'AOPH10' whatever the considered cropping system (HSD $P < 0.05$ and <0.01 for A and B, respectively). We did not find any significant effect of the cropping system on parameters A and B. The duration of phytomer production (parameter t_{max}, expressed from stem emergence) appeared to be similar among the different cultivars and cropping systems. Supplementary statistical analyses were also performed in order to compare the kinetics of phytomer production of the different stems. For cultivar Lucy, the kinetics of phytomer production were statistically different between main stems and branches (HSD $P < 0.05$ and <0.01 for A and B, respectively). These analyses showed that parameter

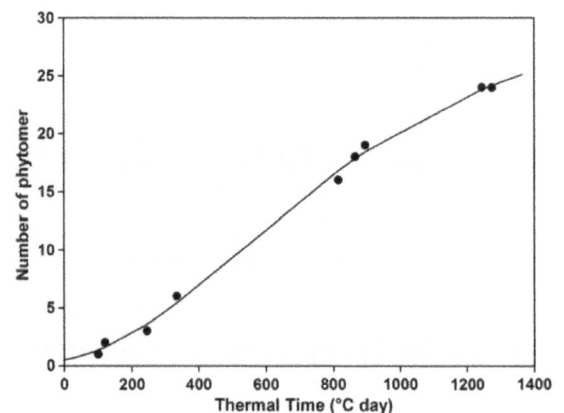

Figure 7. Typical kinetics of phytomer appearance on a vegetative stem of the pea cultivars. Observed values are in closed symbols. Non-linear adjustment (solid line) was performed using Schnute's equation.

Table 3. Parameters ($A \times 10^{-3}$, B and t_{max}) of Schnute's adjustments made on the kinetics of phytomer appearance for each pea cultivar (Lucy, AOPH10, 866/01) grown in SC or in IC. Schnute's adjustments were performed for each flowering stem (0: main stem; 1: branch developed on the first node of the main stem; 2: second node). Indicated values are the mean $\pm$ SD ($n = 15$ plants for each cultivar and cropping system).

Axis	Parameter	Stand					
		Lucy SC	Lucy IC	AOPH10 SC	AOPH10 IC	886/01 SC	886/01 IC
0	$A\ (\times 10^{-3})$	0.36 ± 0.18	–	2.50 ± 1.54	0.82 ± 1.16	–	–
	B	0.56 ± 0.04	–	0.27 ± 0.10	0.45 ± 0.17	–	–
	t_{max}	1415 ± 148	–	1378 ± 204	1257 ± 82	–	–
1	$A\ (\times 10^{-3})$	2.36 ± 1.49	2.80 ± 0.99	4.30 ± 4.74	2.48 ± 2.18	1.88 ± 0.95	2.01 ± 1.89
	B	0.19 ± 0.12	0.13 ± 0.11	0.18 ± 0.22	0.24 ± 0.21	0.16 ± 0.13	0.25 ± 0.15
	t_{max}	1294 ± 190	1288 ± 83	1164 ± 336	1288 ± 234	1279 ± 135	1336 ± 199
2	$A\ (\times 10^{-3})$	2.36 ± 0.86	2.63 ± 0.53	2.21 ± 1.38	2.49 ± 0.071	1.86 ± 1.46	2.18 ± 1.27
	B	0.18 ± 0.09	0.14 ± 0.04	0.19 ± 0.09	0.17 ± 0.08	0.34 ± 0.08	0.25 ± 0.11
	t_{max}	1362 ± 111	1352 ± 22	1292 ± 224	1378 ± 50	1476 ± 52	1365 ± 242

Table 4. Rate of phytomer production of each pea cultivar (Lucy, AOPH10, 866/01) grown in SC or in IC. Maximum rate of phytomer production (V_{max}, phytomer degree-day^{-1}) and time at which it was reached ($t_{V_{max}}$, DD) are shown (mean $\pm$ SD). Computations were carried out for each flowered stem (0: main stem; 1: branch developed on the first node of the main stem; 2: second node) ($n = 15$ plants for each cultivar and cropping system).

Axis	Parameter	Stand					
		Lucy SC	Lucy IC	AOPH10 SC	AOPH10 IC	886/01 SC	886/01 IC
0	V_{max}	0.023 ± 0.002	–	0.024 ± 0.002	0.033 ± 0.006	–	–
	$t_{V_{max}}$	1284 ± 171.3	–	641 ± 148	1004 ± 279	–	–
1	V_{max}	0.026 ± 0.005	0.030 ± 0.004	0.056 ± 0.072	0.026 ± 0.006	0.050 ± 0.020	0.030 ± 0.004
	$t_{V_{max}}$	965 ± 225	836 ± 125	1035 ± 247	950 ± 259	1141 ± 170	1121 ± 224
2	V_{max}	0.026 ± 0.004	0.030 ± 0.002	0.034 ± 0.013	0.027 ± 0.003	0.030 ± 0.001	0.030 ± 0.001
	$t_{V_{max}}$	811 ± 69	761 ± 59	929 ± 100	753 ± 47	1226 ± 207	1025 ± 93

B was significantly different between Axis-1 and -2 branches of cultivar 886/01 (HSD $P < 0.05$).

The maximum rate of phytomer production (V_{max}) and time at which it was reached ($t_{V_{max}}$) were estimated by computing the first derivative of Schnute's adjustments (Table 4). The maximum rate of phytomer production ranged from 0.023 for 'Lucy' SC to 0.056 phytomer degree-day^{-1} for 'AOPH10' SC, which means that at maximum activity one phytomer appeared each at 20–45 GDD. Parameter V_{max} of each axis group was found to be dependent neither on pea cultivar nor on cropping system. The time of maximum rate of phytomer appearance ($t_{V_{max}}$) was reached between about 640 for 'AOPH10' SC and 1285 GDD for 'Lucy' SC. Parameter $t_{V_{max}}$ of flowering main stems was significantly higher for 'Lucy' than for cultivar AOPH10 (HSD $P < 0.05$). Pea cultivars were also statistically different for parameter $t_{V_{max}}$ of Axis-1 and -2 branches ($F_{2,53} = 5.84$, $P < 0.01$; $F_{2,32} = 37.43$, $P < 0.001$,

respectively). Compared with 'Lucy', the maximal rate of phytomer appearance of cultivar 886/01 was indeed reached significantly later for Axis-1 and -2 (HSD $P < 0.01$ and <0.001, respectively). Parameter $t_{V_{max}}$ of cultivar 886/01 appeared to be significantly higher than that of 'AOPH10' but only for Axis-2 branches (HSD $P < 0.001$). For this axis group, $t_{V_{max}}$ was also found to be higher in SC (1020 GDD) than in IC (845 GDD; HSD $P < 0.05$).

The last parameter of Schnute's adjustments to be analysed is the final number of phytomers (y_{max}). The number of phytomers produced on flowering stems ranged from 9 to 32 averaged across pea cultivars and cropping (Fig. 8). A marginal proportion of flowering stems exhibited <15 phytomers (1 % for 'AOPH10', including main stems and Axis-2 branches). Stems with >15 phytomers were mainly branches developed on the first and second phytomer of main stems. This behaviour was however different in the case of cultivar AOPH10, for which $\sim$20 % of

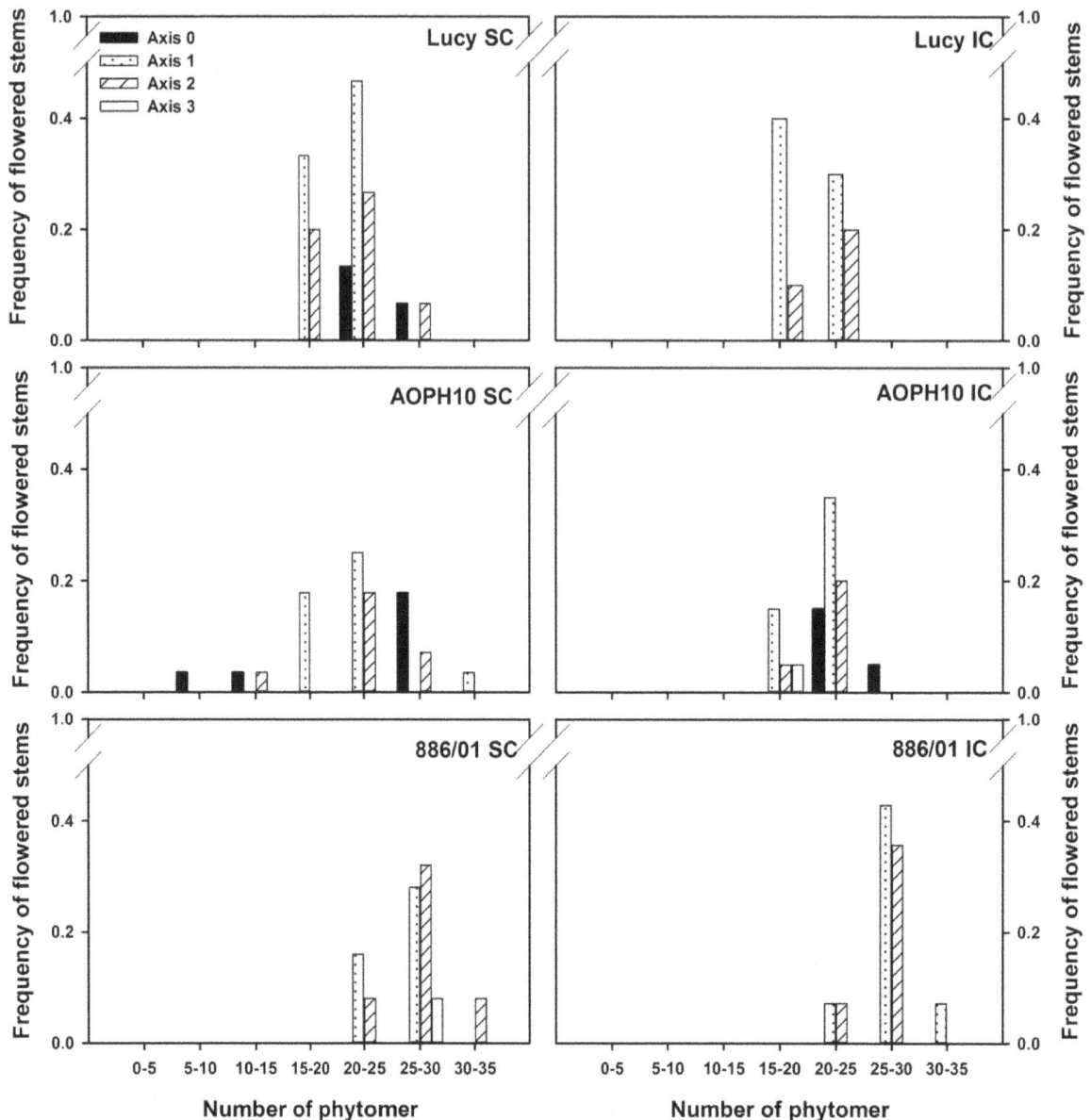

Figure 8. Frequency of flowering stems according to the final number of phytomers. Measurements were made on three pea cultivars: Lucy, AOPH10 and 886/01 grown in SC and in IC with wheat. Results are shown for stems that reached flowering. Main stems are denoted as Axis-0 and branches were distinguished according to their nodal position on the main stem. Axis-1: branches developed at the first node; Axis-2: second node; and Axis-3: third node (*n* = 15 plants for each condition).

flowering stems were main stems with >20 phytomers. Statistical analyses also showed that the final number of phytomers measured on Axis-1 and -2 branches was dependent on the cultivar ($F_{2,60} = 58.24$, $P < 0.001$; $F_{2,37} = 25.66$, $P < 0.001$ for Axis-1 and -2, respectively). Indeed, cultivar 886/01 produced significantly more phytomers than 'Lucy' and 'AOPH10' did (HSD $P < 0.001$ for both Axis-1 and -2 branches). The final number of phytomers was not statistically different for pea plants grown in pure stands or intercropped with wheat.

Flowering. The reproductive development of pea cultivars was characterized by the time of flowering of each stem, as well as the nodal position of the first flower. Flowering stage was reached between 950 and 1400 GDD for AOPH10 IC and 886/01, respectively (Fig. 9). For a given cultivar and cropping system, flowering was synchronized between main stems and branches. However, flowering of Axis-1 and -2 branches was significantly different among the pea cultivars ($F_{2,61} = 1717.29$, $P < 0.001$; $F_{2,37} = 1452.78$, $P < 0.001$ for Axis-1

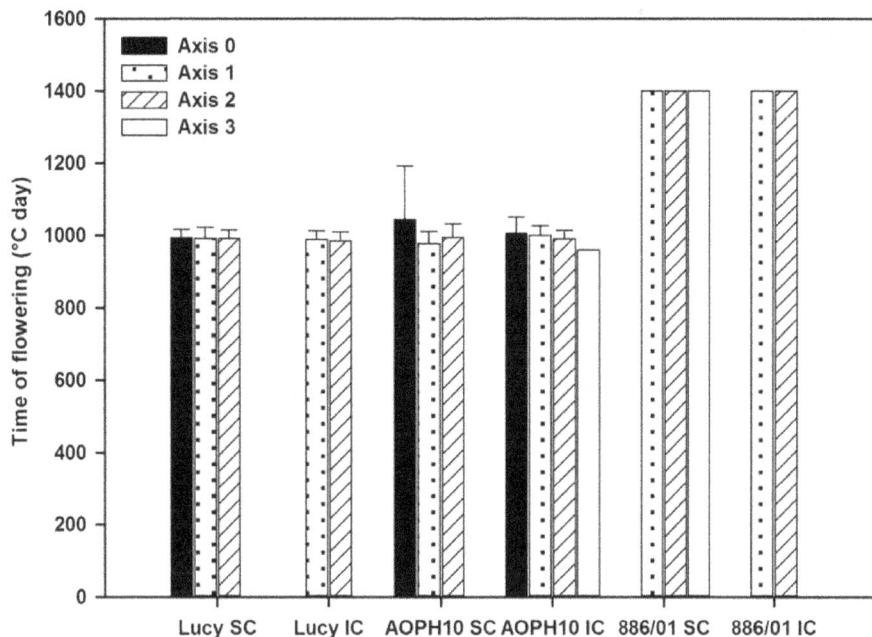

Figure 9. Time of flowering of pea cultivars Lucy, AOPH10 and 886/01 grown in SC or in IC. Main stems are denoted as Axis-0 and branches were distinguished according to their nodal position on the main stem. Axis-1: branches developed at the first node; Axis-2: second node; and Axis-3: third node ($n = 15$ plants for each condition).

and -2, respectively). Indeed, the flowering stage of cultivar 886/01 was reached significantly later (1400 GDD) than for 'Lucy' and 'AOPH10' (HSD $P < 0.001$ for both cultivars). The first flowering phytomer (Fig. 10) was located between the ninth and the 28th phytomer averaged across pea cultivars and cropping systems. First flowers of cultivar 886/01 (Axis-1 and -2 branches) were observed at higher phytomer positions (24th phytomer on average) than those measured for 'Lucy' and 'AOPH10' (HSD $P < 0.001$ for both axis group and cultivars). As observed for the time of flowering, the nodal position of the first flower was similar between pea plants grown in pure stands and those mixed with wheat whatever the cultivars.

Discussion

The aim of the present study was to address the question of the morphological responses of pea to the competition when intercropped with wheat. To this end, a field experiment was conducted on three pea cultivars that were characterized at both stand and plant scale when grown in pure stands and mixed with wheat.

Although the present study was conducted throughout 1 year only, our results on crop biomass, species height and maturity were consistent with previous studies performed on different cultivars grown under contrasting pedo-climatic conditions (e.g. Corre-Hellou et al. 2009; Naudin et al. 2010). The results presented in this study

can therefore be assumed to be representative of the conditions commonly encountered in wheat–pea mixtures. These results were obtained from unfertilized mixtures as usually performed in cereal–legume intercropping systems. Furthermore, some authors (Jensen 1996; Corre-Hellou et al. 2006) found that the contribution of the component species to the biomass of the mixture was dependent on the available nitrogen, an increase of which enhances the growth of the cereal species. The level and timing of nitrogen fertilization (Naudin 2009; Naudin et al. 2010) therefore constitute a key factor enabling one to manage the hierarchy between the mixed species. The present study also shows that the three pea cultivars exhibited a similar level of biomass, especially when grown in mixtures. This suggests that despite the genotypic differences and contrasting initial sowing densities, the three pea cultivars had a similar overall development even when they were in competition with wheat. However, this does not necessarily mean that the morphological processes of the pea cultivars had similar responses to the competition with wheat, but their integration at the stand scale leads to an equivalent growth in biomass.

Our results also show that pea was strongly affected by lodging when grown in SC. Corre-Hellou et al. (2011) reported that the high sensitivity of pea to lodging caused significant yield losses as well as an enhanced growth of weeds. Pea lodging was however strongly decreased in mixed stands, pea branches being stacked by wheat.

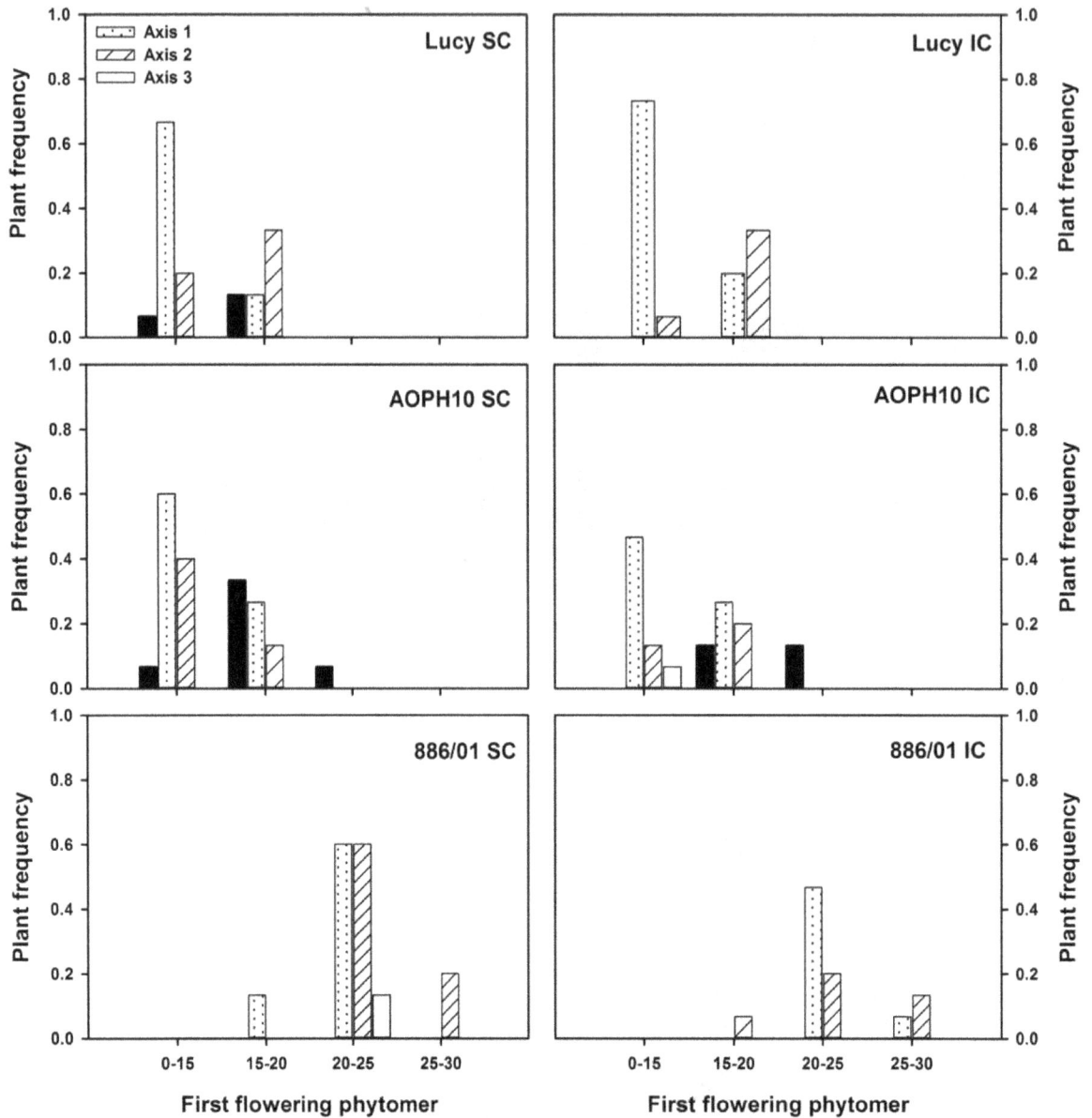

Figure 10. Plant frequency according to the first flowering phytomer of stems. Results are shown for the pea cultivars Lucy, AOPH10 and 886/01 grown in SC or in IC (*n* = 15 plants for each condition). Main stems are denoted as Axis-0 and branches were distinguished according to their nodal position on the main stem. Axis-1: branches developed at the first node; Axis-2: second node; and Axis-3: third node.

Intercropping cereals and legumes is therefore a promising way of both reintroducing legume species within agrosystems and solving the problems encountered in pure stands of legumes. Moreover, the height reached by each species in the canopy is an important feature of the stand which determines, but also emerges from, the competition processes occurring between plants. Component species height ratio has been widely shown to affect light sharing in a mixture (Sinoquet and Caldwell 1995; Louarn *et al.* 2010; Barillot *et al.* 2011, 2012) and is therefore a strong component of the inter-specific competition occurring within the mixture. The results described in the

present study illustrate that the species height ratio is not constant throughout the growing cycle; pea cultivars were much shorter than wheat until 700 GDD and then reached a similar height. Differences among pea cultivars were also observed but were not constant over time. Although '886/01' was the shorter one in the early stages of development, this cultivar finally reached the same height as that of wheat afterwards. Therefore, it seems that the competition which occurs among the inter-cropped species cannot be assessed by punctual measurements of the species height ratio (in particular during the early stages of development). This is consistent with the findings reported in a previous study (Barillot et al. 2012) where a virtual plant approach was used to demonstrate that the ability of plants to intercept light was mainly determined by the architectural parameters involved in (i) the LAI (number of branches and phytomers, leaf area) during the early stages and (ii) plant height (internode length, number of phytomers) once canopy closure was established.

The time lag between the physiological maturity of wheat and pea is a well-known issue of these mixtures. The choice of harvest timing is indeed complicated by the fact that pea generally reaches its maturity earlier than wheat. Nevertheless, physiological maturity of pea varies among the cultivars according to their earliness, which is assumed to be mainly driven by the sensitivity of flowering to the photoperiod that involves the Hr gene (Murfet 1973). The maturity of the HR cultivar (886/01) was therefore almost synchronized with that of wheat, whereas the hr cultivars had to be harvested earlier. Gaps of maturity, as encountered with hr cultivars, represent a strong practical constraint at harvest (Louarn et al. 2010); HR pea cultivars therefore appeared to be well suited to intercropping with wheat.

In order to deepen the analysis of the variability of pea morphogenesis in response to intercropping, a comparison was performed at plant scale with particular attention to pea branching, flowering, final number of phytomers and their kinetics of appearance. Branching has been shown to be dependent on several factors such as genotype, hormonal balance, environmental factors, e.g. low temperatures (Jeudy and Munier-Jolain 2005), and also plant density (Spies et al. 2010). In the present study, contrasting abilities for branching were indeed found between the genotypes [Lucy–AOPH10] and 886/01. Cultivar 886/01 was the most branching cultivar, which balanced its lower sowing density (50 % less than 'Lucy' and 'AOPH10'). Moreover, the number of branches tended to decrease in IC compared with pure stands, in particular for cultivar 886/01. Some authors like Casal et al. (1986), Ballaré and Casal (2000) or Evers et al. (2011) showed that branching of several species is affected by the quantity (PAR) and quality (red/far-red ratio) of light perceived by the axillary buds. In the present study, we can therefore hypothesize that quantity of light and/or its quality were quite similar between the respective pure stands and IC of cultivars Lucy and AOPH10. This would mean that the replacement of a 'Lucy' or an 'AOPH10' plant by a wheat one leads to similar variations of light microclimate. This could be the result of small differences in the architectural patterns of the two species in terms of leaf area, height, geometry and/or optical properties. As cultivar 886/01 has a late development (HR type), we can also hypothesize that when

branching started, wheat plants were more developed than neighbour '886/01' pea plants would have been in a pure stand. This could cause variations of the microclimate perceived by axillary buds, leading to an inhibition of branching.

The kinetics of phytomer appearance were assessed for main stems and a randomly selected branch at each node by using non-linear fittings. Our analysis showed that there were few statistical differences between the parameters belonging to the different genotypes and cropping systems. It was only found that (i) Axis-2 branches of cultivar 886/01 had kinetics different from those of 'Lucy' and 'AOPH10' and (ii) the maximum rate of phytomer appearance of '886/01' was reached later compared with the other cultivars. These results mean that the kinetics of phytomer production of different stems can be analysed/modelled by using similar Schnute's functions, at least for Lucy and AOPH10 cultivars whether they were grown in sole stands or mixed with wheat. Turc and Lecoeur (1997) also reported similar rates of leaf primordium initiation and emergence for contrasting plant growth rates, cultivars and sowing densities in spring pea. One drawback of using Schnute's function lies in the fact that some of the parameters, especially A and B, cannot be directly related to a biological meaning. It would be tempting to use linear regressions because of the reduced number of parameters and easy interpretation. However, phytomer production is not intrinsically constant and is actually characterized by a maximum rate (which can be estimated by the derivative of Schnute's functions) and a time at which development stops. These aspects cannot be handled by linear models. Supplementary statistical analyses (data not shown) showed that the residual sum of squares was significantly higher for linear regressions than that obtained with Schnute's function (HSD $P < 0.001$). These tests also indicated that the residuals of most linear regressions were not normally distributed and have means differing from zero. Nevertheless, the estimated parameters derived from Schnute's adjustments were highly variable. This variability is related to pea branching which is rather complex, particularly in the case of winter-sown cultivars. Winter conditions often cause frost damage, which induces the cessation of the development of the main stems and the initiation of numerous branches at different times (Fig. 5A and B). The result is a high variability in the characteristics of branches.

In the present study, a significant difference was observed among the pea genotypes for the final number of phytomers reached on stems. Indeed, '886/01' (HR type) was found to produce more phytomers than the other cultivars. Similar results were also reported for this particular cultivar but grown in controlled conditions and individual pots (Barillot *et al.* 2012). In contrast, the number of initiated phytomers was similar whether pea plants were grown in pure stands or mixed with wheat, whatever the genotype. Moreover, our results show that the canopy of the three pea cultivars was mainly composed of branches as main stems had stopped growing with few phytomers. As reported by Jeudy and Munier-Jolain (2005), the development of branches is increased in winter pea cultivars because of the frost damage experienced on the apex of the main stem. Such conditions were encountered during the first months of the growing cycle (December–February; Fig. 1) which corresponds to the emergence of the lateral branches (Fig. 5B).

Flowering is a crucial stage of the growing cycle that has been widely studied and used in order to model pea growth. Truong and Duthion (1993) showed that the time of flowering is a function of leaf appearance rate and position of the node bearing the first flower. The reproductive development of pea cultivars was therefore characterized by two main indicators: the nodal position of the first flower and its emergence time. As also reported by Jeuffroy and Sebillotte (1997), we found similar time of flowering between main stems and basal branches (although these were produced later) for all cultivars and cropping systems. Furthermore, the position of the first flowering node was similar among the genotypes and cropping systems. Some authors also showed that for a given genotype, the position of the first flowering node was constant over various conditions (Roche *et al.* 1998; Munier-Jolain *et al.* 2005*b*).

Finally these results highlight that in the present experiment, the morphogenesis of pea was mainly determined by the genotype and was only little affected by the competition with wheat. This suggests that the architectures of pea and wheat may be quite similar, so that the environmental conditions perceived by plants in the canopy (phylloclimate; Chelle 2005) were not strongly different between sole pea crops and wheat–pea mixtures. Functional–structural models (Vos *et al.* 2010; DeJong *et al.* 2011) are able to take into account the explicit architecture of plants and its interactions with physiological processes and environmental conditions. Such models therefore constitute suitable tools for assessing these hypotheses and can in particular be used to characterize the microclimate perceived by plants located in mono- and multi-specific stands.

Conclusion

To our knowledge, the present study is the first to compare the morphogenesis of pea grown in sole stands with that of pea grown intercropped with wheat. On the one hand, the present results show that most of the assessed parameters of pea morphogenesis (phenology, branching, final number of phytomers and their kinetics of appearance) were mainly dependent on the considered genotype. This emphasizes the importance of the selection of cultivars, in particular for intercropping systems, as this will determine the level of competition and complementarity between the component species. On the other hand, there was a low variability of pea morphogenesis between sole and mixed stands except for plant height and branching of the late cultivar 886/01. Complementary studies on wheat–pea mixtures under contrasting levels of nitrogen fertilization are now needed to provide information on how nitrogen would affect plant morphogenesis and interspecific competition. The information provided in the present study can be used for modelling pea morphogenesis in pure and mixed stands and therefore contributes to a better understanding of the functioning of cereal–legume IC. This kind of approach is also well suited for the identification of plant traits to be integrated in the definition of plant ideotypes.

Sources of Funding

This research was supported by 'La Région Pays de la Loire', France through a Ph.D. fellowship to R.B. The research of D.C. and A.E.-G. was partially funded by 'La Région Poitou-Charentes', France.

Contributions by the Authors

All authors have contributed substantially to this manuscript. R.B. completed the writing and was involved in each step of the experimentation and analysis. D.C. and A.E.G. were actively involved in the conception and design of the experiment as well as in the analyses and writing of the manuscript. S.P. was involved in the conception of the experiment and also performed the measurements. P.H. was involved in database programming and data analysis.

Acknowledgements

We gratefully acknowledge the technical staff of LEVA (Légumineuses, Ecophysiologie Végétale, Agroécologie) as well as the experimental station of Brain sur l'Authion (FNAMS, Fédération Nationale des Agriculteurs Multiplicateurs de Semences). We also thank Dr Guénaëlle Corre-Hellou (LEVA, Groupe ESA) for valuable discussions. We also thank reviewers for particularly helpful comments on this article.

Literature Cited

Arumingtyas EL, Floyd RS, Gregory MJ, Murfet IC. 1992. Branching in *Pisum*: inheritance and allelism tests with 17 *ramosus* mutants. *Pisum Genetics* **24**:17–31.

Ballaré CL, Casal JJ. 2000. Light signals perceived by crop and weed plants. *Field Crops Research* **67**:149–160.

Barillot R. 2012. *Modélisation du partage de la lumière dans l'association de cultures blé—pois (Triticum aestivum L. – Pisum sativum L.)—Une approche de type plante virtuelle*. PhD Thesis, Groupe Ecole Supérieure d'Agriculture, UPSP Légumineuses, Ecophysiologie Végétale, Agroécologie, Angers, France.

Barillot R, Louarn G, Escobar-Gutiérrez AJ, Huynh P, Combes D. 2011. How good is the turbid medium-based approach for accounting for light partitioning in contrasted grass–legume intercropping systems? *Annals of Botany* **108**:1013–1024.

Barillot R, Combes D, Chevalier V, Fournier C, Escobar-Gutiérrez AJ. 2012. How does pea architecture influence light sharing in virtual wheat–pea mixtures? A simulation study based on pea genotypes with contrasting architectures. *AoB PLANTS* **2012**:pls038; doi:10.1093/aobpla/pls038.

Béasse C, Ney B, Tivoli B. 2000. A simple model of pea (*Pisum sativum*) growth affected by *Mycosphaerella pinodes*. *Plant Pathology* **49**: 187–200.

Casal JJ, Sanchez RA, Deregibus VA. 1986. The effect of plant density on tillering: the involvement of R/FR ratio and the proportion of radiation intercepted per plant. *Environmental and Experimental Botany* **26**:365–371.

Chelle M. 2005. Phylloclimate or the climate perceived by individual plant organs: what is it? How to model it? What for? *New Phytologist* **166**:781–790.

Corre-Hellou G, Fustec J, Crozat Y. 2006. Interspecific competition for soil N and its interaction with N$_2$ fixation, leaf expansion and crop growth in pea–barley intercrops. *Plant and Soil* **282**:195–208.

Corre-Hellou G, Faure M, Launay M, Brisson N, Crozat Y. 2009. Adaptation of the STICS intercrop model to simulate crop growth and N accumulation in pea–barley intercrops. *Field Crops Research* **113**:72–81.

Corre-Hellou G, Dibet A, Hauggaard-Nielsen H, Crozat Y, Gooding M, Ambus P, Dahlmann C, von Fragstein P, Pristeri A, Monti M, Jensen ES. 2011. The competitive ability of pea–barley intercrops against weeds and the interactions with crop productivity and soil N availability. *Field Crops Research* **122**:264–272.

Crews TE, Peoples MB. 2004. Legume versus fertilizer sources of nitrogen: ecological tradeoffs and human needs. *Agriculture, Ecosystems & Environment* **102**:279–297.

DeJong TM, Da Silva D, Vos J, Escobar-Gutiérrez AJ. 2011. Using functional–structural plant models to study, understand and integrate plant development and ecophysiology. *Annals of Botany* **108**:987–989.

De Wit CT, Van den Bergh JP. 1965. Competition between herbage plants. *Journal of Agricultural Science* **13**:212–221.

Duc G, Mignolet C, Carrouée B, Huyghe C. 2010. Importance économique passée et présente des légumineuses: Rôle historique dans les assolements et les facteurs d'évolution. *Innovations Agronomiques* **11**:1–24.

Evers JB, van der Krol AR, Vos J, Struik PC. 2011. Understanding shoot branching by modelling form and function. *Trends in Plant Science* **16**:464–467.

Godin C. 2000. Representing and encoding plant architecture: a review. *Annals of Forest Science* **57**:413–438.

Gourlay CW, Hofer JMI, Ellis THN. 2000. Pea compound leaf architecture is regulated by interactions among the genes UNIFOLIATA, COCHLEATA, AFILA, and TENDRIL-LESS. *The Plant Cell Online* **12**: 1279–1294.

Gray A. 1849. On the composition of the plant by phytons, and some applications of phyllotaxis. *Proceedings of the American Association for the Advancement of Science* 438–444.

Hauggaard-Nielsen H, Jørnsgaard B, Kinane J, Jensen ES. 2008. Grain legume–cereal intercropping: the practical application of diversity, competition and facilitation in arable and organic cropping systems. *Renewable Agriculture and Food Systems* **23**:3–12.

Huyghe C. 1998. Genetics and genetic modifications of plant architecture in grain legumes: a review. *Agronomie* **18**:383–411.

Jensen ES. 1996. Grain yield, symbiotic N$_2$ fixation and interspecific competition for inorganic N in pea–barley intercrops. *Plant and Soil* **182**:25–38.

Jeudy C, Munier-Jolain N. 2005. Developpement des ramifications. In: Munier-Jolain N, Biarnes V, Chaillet I, Lecoeur J, Jeuffroy M-H, eds. *Agrophysiologie du pois protéagineux*. Paris: Inra-Quae, 51–58.

Jeuffroy MH, Sebillotte M. 1997. The end of flowering in pea: influence of plant nitrogen nutrition. *European Journal of Agronomy* **6**:15–24.

Kusnadi J, Gregory M, Murfet IC, Ross JJ, Bourne F. 1992. Internode length in *Pisum*: phenotypic characterisation and genetic identity of the short internode mutant Wt11242. *Pisum Genetics* **24**: 64–74.

Launay M, Brisson N, Satger S, Hauggaard-Nielsen H, Corre-Hellou G, Kasynova E, Ruske R, Jensen ES, Gooding MJ. 2009. Exploring options for managing strategies for pea–barley intercropping using a modeling approach. *European Journal of Agronomy* **31**: 85–98.

Le May C, Ney B, Lemarchand E, Schoeny A, Tivoli B. 2009. Effect of pea plant architecture on spatiotemporal epidemic development of ascochyta blight (*Mycosphaerella pinodes*) in the field. *Plant Pathology* **58**:332–343.

Louarn G, Corre-Hellou G, Fustec J, Lô-Pelzer E, Julier B, Litrico I, Hinsinger P, Lecomte C. 2010. Déterminants écologiques et physiologiques de la productivité et de la stabilité des associations graminées-légumineuses. *Innovations Agronomiques* **11**: 79–99.

Marquardt DW. 1963. An algorithm for least-squares estimation of nonlinear parameters. *Journal of the Society for Industrial and Applied Mathematics* **11**:431–441.

Munier-Jolain N, Biarnès V, Chaillet I, Lecoeur J, Jeuffroy M-H, Carrouée B, Crozat Y, Guilioni L, Lejeune I, Tivoli B. 2005*a*. *Agrophysiologie du pois protéagineux*, Paris: INRA-Quae.

Munier-Jolain N, Turc O, Ney B. 2005*b*. Développement reproducteur. In: Munier-Jolain N, Biarnes V, Chaillet I, Lecoeur J, Jeuffroy M-H, eds. *Agrophysiologie du pois protéagineux*. Paris: Inra-Quae, 45–50.

Murfet IC. 1973. Flowering in *Pisum*. Hr, a gene for high response to photoperiod. *Heredity* **31**:157–164.

Murfet IC. 1975. Flowering in *Pisum*: multiple alleles at the *If* locus. *Heredity* **35**:85–98.

Naudin C. 2009. *Nutrition azotée des associations Pois-Blé d'hiver (Pisum sativum L. – Triticum aestivum L.): Analyse, modélisation et propositions de stratégies de gestion*. PhD Thesis, Groupe Ecole Supérieure d'Agriculture, UPSP Légumineuses, Ecophysiologie Végétale, Agroécologie, Angers, France.

Naudin C, Corre-Hellou G, Pineau S, Crozat Y, Jeuffroy M-H. 2010. The effect of various dynamics of N availability on winter pea–wheat intercrops: crop growth, N partitioning and symbiotic N$_2$ fixation. *Field Crops Research* **119**:2–11.

44
Plant Genetics: Biodiversity and Evolution

intercrops: crop growth, N partitioning and symbiotic N_2 fixation. *Field Crops Research* **119**:2–11.

Ney B, Carrouée B. 2005. Préface. In: Munier-Jolain N, Biarnes V, Chaillet I, Lecoeur J, Jeuffroy M-H, eds. *Agrophysiologie du pois protéagineux*. Paris: Inra-Quae.

Ofori F, Stern WR. 1987. Cereal–legume intercropping systems. *Advances in Agronomy* **41**:41–90.

R Development Core Team. 2012. *R: a language and environment for statistical computing*. Vienna, Austria: R Foundation for Statistical Computing.

Roche R, Jeuffroy M-H, Ney B. 1998. A model to simulate the final number of reproductive nodes in pea (*Pisum sativum* L.). *Annals of Botany* **81**:545–555.

Ross J. 1981. Role of phytometric investigations in the studies of plant stand architecture and radiation regime. In: Ross J, ed. *The radiation regime and architecture of plant stands*. The Hague, The Netherlands: Junk, W, 9–11.

Schnute J. 1981. A versatile growth model with statistically stable parameters. *Canadian Journal of Fisheries and Aquatic Sciences* **38**:1128–1140.

Schwinning S, Weiner J. 1998. Mechanisms determining the degree of size asymmetry in competition among plants. *Oecologia* **113**: 447–455.

Sinoquet H, Caldwell MM. 1995. Estimation of light capture and partitioning in intercropping systems. In: Sinoquet H, Cruz P, eds. *Ecophysiology of tropical intercropping*. Paris: INRA Editions, 79–97.

Spies JM, Warkentin T, Shirtliffe S. 2010. Basal branching in field pea cultivars and yield–density relationships. *Canadian Journal of Plant Science* **90**:679–690.

Truong HH, Duthion C. 1993. Time of Flowering of Pea (*Pisum sativum* L.) as a function of leaf appearance rate and node of first flower. *Annals of Botany* **72**:133–142.

Turc O, Lecoeur J. 1997. Leaf primordium initiation and expanded leaf production are co-ordinated through similar response to air temperature in pea (*Pisum sativum* L.). *Annals of Botany* **80**: 265–273.

Varlet-Grancher C, Bonhomme R, Sinoquet H. 1993. *Crop structure and light microclimate*. Paris: INRA Editions.

Vos J, Evers JB, Buck-Sorlin GH, Andrieu B, Chelle M, de Visser PHB. 2010. Functional–structural plant modelling: a new versatile tool in crop science. *Journal of Experimental Botany* **61**: 2101–2115.

White J. 1979. The plant as a metapopulation. *Annual Review of Ecology and Systematics* **10**:109–145.

4

Genome size variation and evolution in allotetraploid *Arabidopsis kamchatica* and its parents, *Arabidopsis lyrata* and *Arabidopsis halleri*

Diana E. Wolf[1]*, Janette A. Steets[1,2], Gary J. Houliston[1,3] and Naoki Takebayashi[1]

[1] Department of Biology and Wildlife, Institute of Arctic Biology, University of Alaska Fairbanks, 311 Irving I, Fairbanks, AK 99775-7000, USA
[2] Present Address: Department of Botany, Oklahoma State University, 301 Physical Sciences, Stillwater, OK 74078-3013, USA
[3] Present Address: Landcare Research, Gerald St, Lincoln 7608, New Zealand

Associate Editor: W. Scott Armbruster

Abstract. Polyploidization and subsequent changes in genome size are fundamental processes in evolution and diversification. Little is currently known about the extent of genome size variation within taxa and the evolutionary forces acting on this variation. *Arabidopsis kamchatica* has been reported to contain both diploid and tetraploid individuals. The aim of this study was to determine the genome size of *A. kamchatica*, whether there is variation in ploidy and/or genome size in *A. kamchatica* and to study how genome size has evolved. We used propidium iodide flow cytometry to measure 2C DNA content of 73 plants from 25 geographically diverse populations of the putative allotetraploid *A. kamchatica* and its parents, *Arabidopsis lyrata* and *Arabidopsis halleri*. All *A. kamchatica* plants appear to be tetraploids. The mean 2C DNA content of *A. kamchatica* was 1.034 pg (1011 Mbp), which is slightly smaller than the sum of its diploid parents (*A. lyrata*: 0.502 pg; *A. halleri*: 0.571 pg). *Arabidopsis kamchatica* appears to have lost ~37.594 Mbp (3.6 %) of DNA from its 2C genome. Tetraploid *A. lyrata* from Germany and Austria appears to have lost ~70.366 Mbp (7.2 %) of DNA from the 2C genome, possibly due to hybridization with *A. arenosa*, which has a smaller genome than *A. lyrata*. We did find genome size differences among *A. kamchatica* populations, which varied up to 7 %. *Arabidopsis kamchatica* ssp. *kawasakiana* from Japan appears to have a slightly larger genome than *A. kamchatica* ssp. *kamchatica* from North America, perhaps due to multiple allopolyploid origins or hybridization with *A. halleri*. However, the among-population coefficient of variation in 2C DNA content is lower in *A. kamchatica* than in other *Arabidopsis* taxa. Due to its close relationship to *A. thaliana*, *A. kamchatica* has the potential to be very useful in the study of polyploidy and genome evolution.

Keywords: Allotetraploid; *Arabidopsis halleri* ssp. *gemmifera*; *Arabidopsis kamchatica*; *Arabidopsis lyrata*; C-value; 2C DNA content; flow cytometry; genome size; genome size variation.

Introduction

Polyploidy is one of the most important forces influencing plant diversification. Polyploidy was likely involved in 15 % of all recent angiosperm speciation events (Wood *et al.* 2009) and ancient polyploidy is apparent in all plant genomes sequenced to date (Jiao *et al.* 2011). Similarly, the majority of cultivated crops have undergone polyploidization during domestication (Otto and Whitton 2000).

* Corresponding author's e-mail address: dewolf@alaska.edu

Polyploidy influences the ecology and physiology of plants by generating novel phenotypes that may influence mating system, habitat and geographical distribution (Levin 2002). It can have major genetic and genomic effects, such as altering chromosome segregation, masking deleterious mutations, influencing levels of genetic diversity, changing gene expression, causing rearrangements, gene loss and epigenetic changes, rewiring genetic networks, and altering rates of adaptation (Levin 2002; Adams and Wendel 2005; Chen 2007; De Smet and Van de Peer 2012; Madlung 2013). Ploidy variation has the potential to promote the origin of new species, but ploidy variation within species (or species complexes) may also be an important source of genetic and phenotypic variation (Thompson and Lumaret 1992). Thus, plant biodiversity cannot be understood without understanding the processes of polyploid evolution (Lutz 1907; Stebbins 1950; Grant 1981; Madlung 2013).

Polyploids are thought to experience high levels of genomic instability and undergo massive genetic and epigenetic changes within the first few generations after formation (Chen 2007). It is likely that a great deal of genomic and phenotypic diversity is generated and the majority of early generation polyploids are unable to survive in nature. However, if one or a few stable genotypes arise that happen to reconcile genomic incompatibilities, are vigorous and are well suited to survival in the prevailing habitat, polyploids can persist (Chen 2007; Madlung et al. 2012). After this rapid 'genomic revolution', it is likely that a slow process of diploidization begins, where gene duplicates may be silenced, lost or evolve new functions (Wolfe 2001). It is thought that nearly all angiosperms have experienced at least one polyploidy event in their evolutionary history (Wolfe 2001). However, due to extensive mutation, gene loss and rearrangements, these diploidized paleopolyploids, such as *Arabidopsis thaliana*, have only recently been recognized as whole-genome sequences became available for detailed analysis (Vision et al. 2000). Both the rapid genomic revolution and gradual process of diploidization are likely to result in variation and evolution in genome size as DNA is deleted, duplicated and rearranged, and variants are subject to genetic drift and selection.

Polyploidy can arise from the duplication of genomes within a single species (autopolyploidy) or through hybridization between two species, accompanied by chromosome doubling (allopolyploidy) (Levin 2002). Either allopolyploidy or autopolyploidy may arise via a single polyploidization event, like in *Arabidopsis suecica* (Säll et al. 2003; Jakobsson et al. 2006), or may have multiple origins (Soltis and Soltis 1999), as has been suggested for *A. kamchatica* (Shimizu-Inatsugi et al. 2009). Further, variation in ploidy level is frequently found within species both within and among populations (Schmuths et al. 2004; Marhold et al. 2010), and gene flow between ploidy levels is known to occur, either via a triploid bridge or through recurrent formation of unreduced gametes by diploids (Levin 2002; Husband 2004; Henry et al. 2005, 2009; Jørgensen et al. 2011). This gene flow from diploids to polyploids is likely an important source of genetic variation in polyploids (Jørgensen et al. 2011).

Arabidopsis kamchatica is an allotetraploid plant produced through hybridization through two closely related diploid taxa, *Arabidopsis lyrata* ssp. *petraea* and *Arabidopsis halleri* ssp. *gemmifera* (Shimizu et al. 2005; Shimizu-Inatsugi et al. 2009). *Arabidopsis kamchatica* has an amphi-Beringian distribution, and the pattern of genetic diversity suggests that it migrated northward out of Japan (or near Japan) to eastern Russia, across the Bering land bridge into Alaska, and down the west coast of Canada (Shimizu-Inatsugi et al. 2009). It has been suggested that *A. kamchatica* may have multiple origins through independent hybridization and polyploidization events (Shimizu-Inatsugi et al. 2009), and/or that it may hybridize with its diploid parental taxa (Shimizu-Inatsugi et al. 2009; Wang et al. 2010). Both of these processes have the potential to give rise to genome size variation. Further, *A. kamchatica* has been suggested to contain both diploid and tetraploid individuals (Dawe and Murray 1981; Wang et al. 2010). Because *A. kamchatica* is a close relative of the model plant, *A. thaliana*, a treasure trove of molecular research is easily applied to this organism, and development of *A. kamchatica* into a model system for the evolution of polyploidy has the potential to yield a great deal of insight into the evolution of polyploid genomes.

The goal of this study was to investigate genome size variation in *A. kamchatica* using flow cytometry. We characterized the nuclear DNA content of *A. kamchatica* and its putative parental species, *A. lyrata* and *A. halleri*, in a total of 25 populations from North America, Europe and Japan. We used the results to determine whether there is variation in ploidy and/or genome size in *A. kamchatica* and its parents, and to determine how genome size has evolved in polyploids relative to their diploid parents.

Methods

Plant material

We estimated genome size from a total of 73 samples from *A. kamchatica* and its parental taxa *A. lyrata* (subspecies *A. l. lyrata* and *A. l. petraea*) and *A. halleri* ssp. *gemmifera* (Table 1, Fig. 1). All plants were germinated from seed and grown in the Institute for Arctic Biology Greenhouse at the University of Alaska Fairbanks. In populations with multiple samples, we sampled plants from different maternal families.

Table 1. Collection locations, collectors and mean ($\pm$ 1 SE) genome size of each population. [1]Assumes *Glycine max* 'Polanka' 2C DNA content of 2.5 pg (Doležel et al. 1994; Doležel and Greilhuber 2010). [2]Populations with different letters have significantly different means ($P < 0.05$) in post hoc comparisons among *A. kamchatica* populations with >2 individuals. [3]Conversion from pg to Mbp assuming Mbp = pg × 978 (Doležel et al. 2003).

Taxon	Location	Latitude	Longitude	Collector/donor	Sample size	2C DNA content (pg)[1,2]	SE	Ploidy (2C)	2C genome size (Mbp)[3]
A. h. gemmifera	Japan	34.93	133.63	Fujita Corp.	9	0.571	0.0127	2x	558.35
A. kamchatica	USA, Alaska								
	Bear Creek	65.41355	−145.62545	C. Parker	1	1.013	NA	4x	990.43
	Chena River	64.82	−147.32	N.T., D.E.W.	4	1.023[AB]	0.0035	4x	1000.06
	Fairbanks	64.83333333	−147.7	C. Parker	1	1.025	NA	4x	1002.51
	Goodnews Bay	59.11666667	−161.583333	C. Parker	5	1.016[A]	0.0056	4x	994.06
	Grant Lagoon, Kodiak Island	57.37	−154.65	C. Parker	3	1.043[B]	0.0054	4x	1020.23
	Liberty Falls	61.62	−144.55	D.E.W.	1	1.039	NA	4x	1015.72
	Portage Glacier	60.79161667	−148.9021333	N.T., D.E.W.	3	1.039[AB]	0.0104	4x	1016.02
	Parks Highway	63.25	−149.25	N.T., D.E.W.	6	1.035[AB]	0.0035	4x	1012.30
	Rainbow Ridge	63.32	−145.64	N.T., D.E.W.	3	1.032[AB]	0.0053	4x	1009.16
	Shoup Bay	61.13	−146.59	N.T., D.E.W.	4	1.033[AB]	0.0050	4x	1010.30
	Thompson Pass	61.13	−145.73	N.T., D.E.W.	1	1.033	NA	4x	1010.20
	Canada, Vancouver Island								
	Strathcona Park	49.82915	−125.8728	J.A.S., D.E.W.	15	1.027[AB]	0.0032	4x	1004.24
	Japan, Honshu Island								
	Lake Biwa, Shinbo	35.44444444	136.05	H. Marui	5	1.083[C]	0.0027	4x	1059.29
A. l. lyrata	USA, Michigan, Grand Mere	42.01	−86.54	J.A.S.	1	0.525	NA	2x	513.27
	New York	40	−74	T. Mitchell-Olds	1	0.479	NA	2x	468.72
	Pennsylvania, Presque Isle	42.14	−80.11	J.A.S.	1	0.510	NA	2x	499.19
	Pennsylvania, Raccoon Creek	40.51	−80.34	J.A.S.	2	0.502	0.0097	2x	491.38
	Wisconsin	44	−89	T. Mitchell-Olds	1	0.498	NA	2x	486.70
A. l. petraea	England, Exeter	50.72	−3.53	T. Mitchell-Olds	2	0.477	0.0082	2x	466.68
	Germany, Plech	49.65	11.47	T. Mitchell-Olds	2	0.494	0.0034	2x	482.82
	Iceland, Reykjavik, Esja Mountain	64.2	−21.7	M. Schierup	2	0.526	0.0010	2x	513.96
	Scotland, Braemer	57.01	−3.4	R. Ennos	1	0.514	NA	2x	502.68
	Austria, Mödling	48.08	16.32	S. Ansell	1	0.922	NA	4x	901.93
	Germany, Dürn	49.27	11.6	T. Mitchell-Olds	1	0.941	NA	4x	920.06

Figure 1. Map of collection localities of plants used for flow cytometry.

Ploidy determination

Chromosome counting is the traditional method for determining ploidy level of an organism; however, it is labour intensive and may be inaccurate in *Arabidopsis* species due to their very small chromosomes (ranging from 1.5 m to 2.8 μm in *A. thaliana*; Schweizer *et al.* 1987) and the high frequency of endopolyploidy (Galbraith *et al.* 1991; Melarango *et al.* 1993). Flow cytometry allows rapid analysis of thousands of nuclei per sample and high throughput of many samples (Kron *et al.* 2007). Therefore, we used flow cytometry to estimate genome size and infer DNA ploidy. Because flow cytometry reveals genome size rather than a count of chromosomes, ploidy must be verified by chromosome counts in at least a few samples. In our study, we included both diploid and tetraploid references from *Arabidopsis* locations where both flow cytometry and chromosome counts have previously been carried out (*A. kamchatica* from Japan, and *A. l. petraea* from Iceland and Austria; Table 1; Dart *et al.* 2004).

Flow cytometry

Each *Arabidopsis* sample was co-chopped and run with soybean leaf, *Glycine max* 'Polanka', as an internal reference standard. The standard was grown from the same seed stock previously quantified (Doležel *et al.* 1994). Young leaves were collected from each *Arabidopsis* plant and kept on ice until processing, which occurred within 3 h of leaf collection. For each plant, three fresh leaves were placed in a plastic Petri dish with approximately half as much fresh leaf tissue from *G. max*. Leaf tissue was chopped in the presence of 0.5 mL of cold chopping buffer using a fresh stainless-steel razor blade. The chopping buffer was modified from Otto (1990) Buffer I by adding 0.5 % v/v of Triton X-100 rather than Tween 20. When the leaves were well chopped, we added an additional 0.5 mL of cold chopping buffer. The sample was then filtered through a 30-μm Partec Cell-Trics® filter and centrifuged for 20 s at 3500 rpm. The supernatant was drawn off and 2 μL of RNase A was

added to the pellet. The pellet was resuspended in 0.2 mL of propidium iodine staining buffer. The propidium iodine staining buffer (28.65 g of dibasic sodium phosphate, 190 mL of deionized water and 10 mL of propidium iodine stock, which consists of 5 mg of propidium iodine and 10 mL of deionized water) was modified from Otto (1990). Samples were stained in the dark for 40 min prior to performing flow cytometry.

Flow cytometry was performed on a BD Biosciences FACSAria flow cytometer (BD Biosciences, San Jose, CA, USA) equipped with FACSDiva Software (BD Biosciences), using a Coherent Sapphire Solid State laser (488 nm) as the excitation source. Noise signals derived from subcellular debris were eliminated by gating. Samples were run until 5000 *Arabidopsis* nuclei were scored. Since propidium iodide was used to stain the nuclei, fluorescence was measured using the R-phycoerythrin (PE) detector, which uses the 576/26 nm bandpass filter. 2C DNA content was estimated from gated fluorescence histograms of PE area (Fig. 2). Due to endopolyploidy, the populations of plant nuclei typically gave multiple peaks of fluorescence, representing 2C, 4C and 8C nuclei (and sometimes even higher endopolyploid levels) (Galbraith *et al.* 1991; Melarango *et al.* 1993). The 2C DNA content of each sample was calculated using the smallest of the peaks, and comparing it to the *G. max* standard ((sample fluorescence/soybean fluorescence) × 2.5 pg; Doležel *et al.* 1994). All samples had a coefficient of variance (CV) for relative fluorescence among nuclei that was <10 %; however, only 48 % of samples had a CV ≤5 %, as recommended (Doležel *et al.* 2007). We believe that this is due in part to the very small *Arabidopsis* genome (Doležel *et al.* 2007), as the larger soybean standard peak had a mean CV of 3.32 %, and only 6.1 % of the samples had a CV >5 %. All soybean samples had a CV ≤5.7 %. To ensure that genome size measurements were repeatable, eight samples were repeated on different days. Differences between repeat measurements never exceeded 1.1 %, indicating that genome size measurements were highly repeatable (Doležel *et al.* 2007).

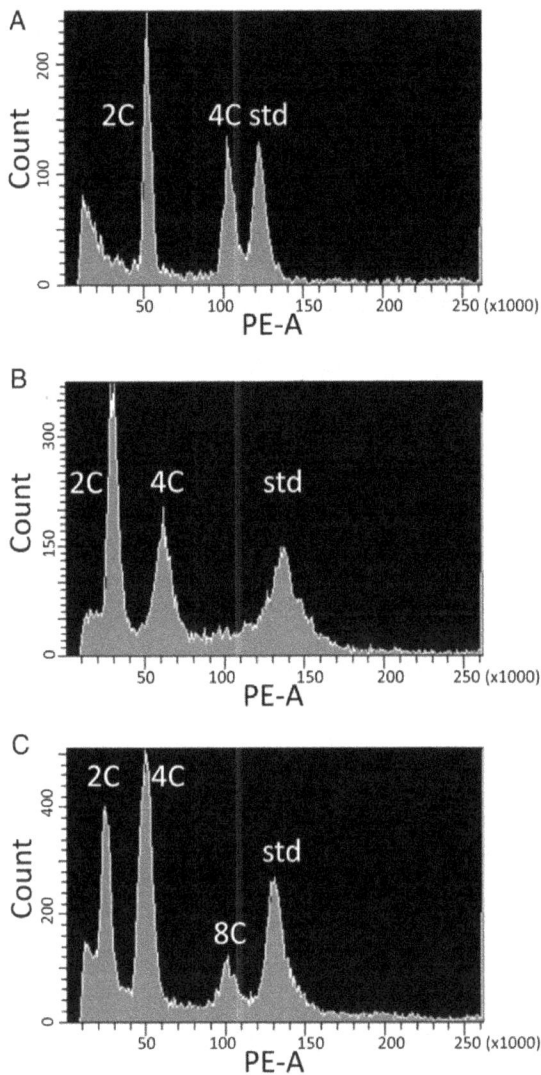

Figure 2. Fluorescence intensity histograms (PE-A) for (A) tetraploid *A. kamchatica* ($2C = 4x = 32$), (B) diploid *A. h. gemmifera* ($2C = 2x = 16$) and (C) diploid *A. lyrata* ($2C = 2x = 16$). *Arabidopsis* leaves show extensive endopolyploidy (Galbraith *et al.* 1991), and the 2C, 4C and 8C peaks are indicated, along with the soybean standard (std). The mean fluorescence of the smallest peak (2C) relative to the soybean peak was used to estimate 2C DNA content.

To determine whether the taxa differed in genome size, we used a linear mixed-effects model with species as the fixed effect, populations as the random effect and 2C DNA content (pg) as the dependent variable with the *lme4* package (Bates 2005), implemented in R. The hypothesis test of the species effect was conducted with 5000 iterations of the parametric bootstrap approach based on the likelihood ratio statistics, $D = -2 \times$ (log-likelihood ratio), of Faraway (2006). To determine which species differed from one another in 2C DNA content, we performed Tukey's multiple comparison tests with an R package, *multcomp* (Hothorn *et al.* 2008). To determine whether

populations of *A. kamchatica* differed in 2C DNA content, a second one-way ANOVA was performed with population as the fixed effect and 2C DNA content (pg) as the dependent variable. For this analysis we restricted our dataset to include only the nine *A. kamchatica* populations for which we had at least three samples. The mean number of samples per population was 5.3. Tukey's multiple comparison tests were performed to determine which *A. kamchatica* populations significantly differed from one another in genome size. In order to test the additivity of the tetraploid genome size, we examined a contrast null hypothesis, where the 1Cx genome size (i.e. the haploid genome size, *sensu* Greilhuber 2005) of *A. kamchatica* is the average of the two parental species, in the subset of data including *A. kamchatica*, *A. h. gemifera* and *A. lyrata* (two subspecies were combined). A linear mixed-effects model was fitted with 1Cx values as the dependent variable, population as a random effect and species as a fixed effect, and the linear contrast, (1Cx of *A. kamchatica*) = [(1Cx of *A. h. gemifera*) + (1Cx of *A. lyrata*)]/2, was tested with an R package, *multcomp*. For estimates of genome size diversity in each taxon, we used the CV among populations in 2C DNA content with the bias correction (Sokal and Rohlf 1995). To estimate genome size diversity in diploid *A. thaliana*, we used data from Schmuths *et al.* (2004) collected from 18 worldwide accessions using the same flow cytometry methods that we used.

Results

We found that 2C DNA content in *A. kamchatica* populations varied from 1.013 to 1.083 pg/2C, with a mean 2C DNA content of 1.034 ± 0.005 pg/2C (mean $\pm$ SE). *Arabidopsis kamchatica* and two of the *A. lyrata* ssp. *petraea* samples (Austria and Dürn, Germany) had approximately double the genome size of the other *A. lyrata* (ssp. *lyrata* and ssp. *petraea*) and *A. halleri* ssp. *gemmifera* samples (Fig. 3, Table 1). These taxa significantly differed in nuclear DNA content ($D = 136.18$, df = 1, $P < 0.0002$). These results, when taken together with chromosome counts and flow cytometry results conducted by Dart *et al.* (2004) in some of the same collections we used, suggest that the majority of *A. l. lyrata*, *A. l. petraea* and *A. h. gemmifera* are diploids, while *A. kamchatica* and two *A. l. petraea* samples are tetraploids (Fig. 3; Table 1).

There was significant variation in genome size among *A. kamchatica* populations ($F_{8,39} = 15.7$, $P < 10^{-9}$). Post hoc tests indicate that the genome size of the Japanese *A. kamchatica* population (Shinbo) was significantly larger than the North American populations. The Canadian *A. kamchatica* population did not differ in genome size from the Alaskan populations. However, two of the six Alaskan populations differed in genome size; the nuclear

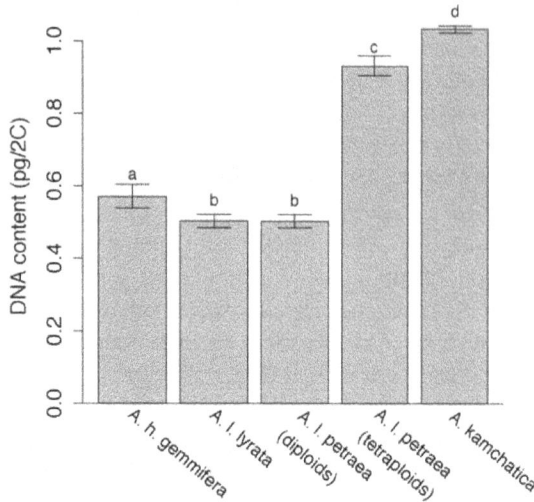

Figure 3. Estimates of 2C DNA content (pg) of each taxon and the 95 % confidence intervals of the estimates. Letters indicate significant differences ($P < 0.05$) based on Tukey's post hoc comparisons.

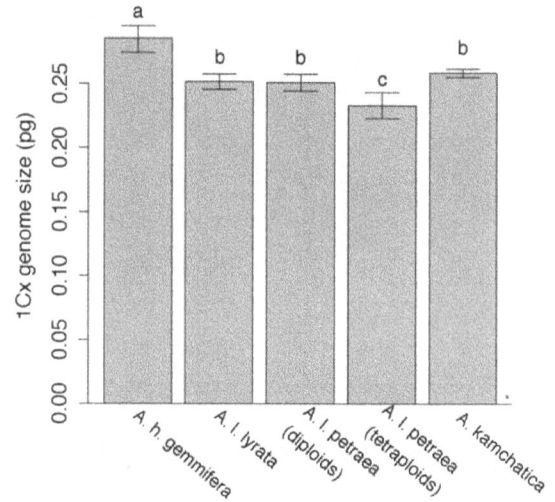

Figure 4. Estimates of 1Cx (haploid) genome size (pg) of each taxon. The error bars are 95 % confidence intervals of the estimates. Letters indicate significant differences ($P < 0.05$) based on Tukey's post hoc comparisons.

DNA content of the Goodnews Bay population was 3 % smaller than that of the Grant Lagoon population. Despite the minor amounts of variation among populations, none of the *A. kamchatica* plants sampled appear to be diploid.

The 2C DNA content of *A. l. lyrata* (0.503 pg/2C, 95 % CI [0.484, 0.522]) and diploid *A. l. petraea* (0.502 pg/2C, 95 % CI [0.484, 0.521]) did not significantly differ from one another (Fig. 3). The *A. h. gemmifera* genome (0.571 pg/2C, 95 % CI [0.539, 0.604]) was 14 % larger than *A. l. petraea* and *A. l. lyrata* (Fig. 3). We did not have enough samples/population of these taxa to analyse differences among populations.

Arabidopsis kamchatica appears to have been derived through allopolyploidy from *A. lyrata* and *A. h. gemmifera* (Shimizu-Inatsugi *et al.* 2009). Thus, if polyploidization was recent, and there were no subsequent changes in genome size, we would predict that the genome size of the allotetraploid should be equal to the sum of the two parental taxa. Further, the 1Cx genome size (i.e. the haploid genome size, *sensu* Greilhuber 2005) should be an average of its parents. However, *A. kamchatica*, on average, is slightly smaller than expected. Comparing the 1Cx genome sizes of *A. kamchatica* to its parents (Fig. 4), we can see that the *A. kamchatica* 1Cx genome size is intermediate to its parents, but less than the average of its parents (*A. kamchatica*: 0.259 pg; mean of parents: 0.268 pg, $z = -2.81$, $P = 0.0049$). Further, it is not significantly different from the smaller parent, *A. lyrata* (Fig. 4), suggesting that *A. kamchatica* may have lost DNA. *Arabidopsis kamchatica* appears to have lost ~37.594 Mbp/2C of DNA or 3.6 % of its genome. Autotetraploid *A. l. petraea* also

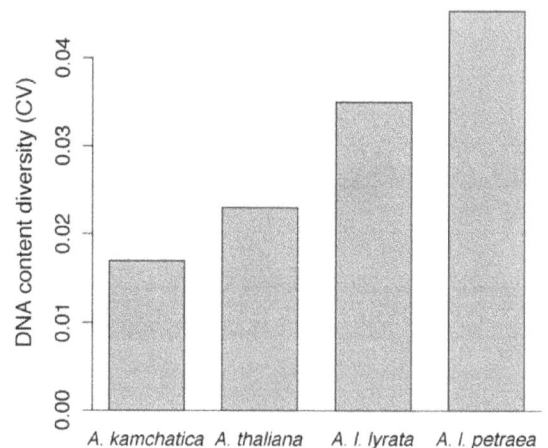

Figure 5. Genome size diversity in *Arabidopsis* taxa, measured as CV in 2C DNA content. Only diploid *A. thaliana* and *A. l. petraea* are included because there were too few tetraploids to estimate CV (two from each taxon).

appears to have lost DNA. The mean 1Cx genome size of tetraploid *A. l. petraea* (0.233 pg, 95 % CI [0.223, 0.243]) is less than the 1Cx content of diploid *A. l. petraea* (0.251 pg, 95 % CI [0.244, 0.258]), a loss of ~70.366 Mbp/2C, or 7.2 % of the genome.

We were able to estimate genome size diversity (i.e. the CV in 2C DNA content) in *A. kamchatica*, *A. l. petraea*, *A. l. lyrata* and *A. thaliana*, which were all sampled from multiple populations (*A. thaliana* data were from Schmuths *et al.* 2004). *Arabidopsis kamchatica* has the lowest diversity of all the *Arabidopsis* taxa studied (Fig. 5), including *A. thaliana* (Schmuths *et al.* 2004).

Discussion

Reliability of ploidy estimates

Our genome size estimates are very similar to those of Dart et al. (2004) for diploid and tetraploid collections in common (Table 1), suggesting that our results are reliable. Using both chromosome counting and flow cytometry, Dart et al. (2004) found that plants from Japan (Shinbo) and Austria are tetraploid ($2n = 4x = 32$) with genome sizes of 1.1 pg/2C (Japan) and 0.9 pg/2C (Austria), while plants from Iceland are diploid ($2n = 2x = 16$) with a genome size of 0.52 pg/2C. The small differences between our data and those of Dart et al. (2004) are likely due to the fact that Dart et al. (2004) used fluorescent beads as an internal size standard, whereas we used leaf tissue from G. max. While beads are sufficient for ploidy determination, leaf tissue is the preferred internal size standard for absolute genome size estimation because staining variation can be taken into account (Doležel et al. 2007).

No ploidy variation within A. kamchatica

Several previous reports have suggested that A. kamchatica contains both diploid and tetraploid individuals (Dawe and Murray 1981; Wang et al. 2010). While many species show a mix of ploidy levels, even within a population, these are likely autopolyploids (Schmuths et al. 2004; Jørgensen et al. 2011). Given that A. kamchatica is an allopolyploid, diploids spontaneously produced from tetraploids would likely have low vigour and fertility (Kerber 1964; Ladizinsky and Fainstein 1978), as allopolyploidization appears to rapidly result in gene silencing and gene loss for numerous loci (Kashkush et al. 2002; Adams and Wendel 2005). Our data from 52 A. kamchatica specimens representing most of the species' range found no evidence of diploid A. kamchatica, and we suggest that the species is likely to be entirely tetraploid. If diploids are present, they are likely to be in very low frequencies, and not maintained by selection.

Deeper investigation into previous reports also suggests that there is no good evidence for the presence of diploid A. kamchatica. Dawe and Murray (1981) report chromosome counts from three diploid ($2n = 2x = 16$) and two tetraploid A. kamchatica samples ($2n = 4x = 32$). Arabidopsis kamchatica is very difficult to morphologically distinguish from mostly diploid A. lyrata; however, molecular data suggest that the two species have distinct geographical ranges (Schmickl et al. 2010). The tetraploid counts reported by Dawe and Murray (1981) are within the species range of A. kamchatica suggested by Schmickl et al. (2010), whereas two of the three diploid counts are from plants growing north of the Brooks Range in Alaska and are probably A. l. petraea (Schmickl et al.

2010) or A. media (Mulligan 1995). One of the diploid counts (originally reported in Dawe and Murray 1979) comes from well within A. kamchatica's range in interior Alaska, near several of our collections (63°02′N, 145°29′W), and was likely taken from A. kamchatica. However, Mulligan (1995) claims that the diploid report is an error, and that the voucher in ALA indicates that $2n = 32$ (tetraploid), not $2n = 16$ (diploid). Other chromosome counts reported for A. kamchatica by Mulligan (1995) are all tetraploid, and he suggests that the species is entirely tetraploid.

Wang et al. (2010) claim to have detected both diploid and tetraploid A. kamchatica in Taiwan using flow cytometry and sequencing of nuclear DNA from 98 genes. They suggest that diploids have a 'mosaic genome' of the two parental species. Although this would be very interesting if confirmed, more complete evidence is desirable. First, their flow cytometry runs seem to lack an internal standard. The absolute value of nucleus fluorescence cannot reliably be used to estimate genome size as this value shifts due to variation in sample preparation, staining and analysis (Doležel et al. 2007). This shift can be seen by comparing Fig. S1A and S1B in Wang et al (2010), which were presented as evidence of diploid and tetraploid A. kamchatica. Further, their DNA sequence data do not provide any evidence of ploidy since only a single clone per PCR reaction was sequenced, ensuring that only a single homeologue (randomly chosen from one of the two parental genomes of tetraploids) could be obtained from each individual (Wang et al. 2010). Although we have not sampled A. kamchatica from Taiwan for our study, the 'mosaic genome' of purported diploid A. kamchatica can possibly be explained by misinterpretation of flow cytometry data and randomly sequencing only one of the two homeologues from each gene.

DNA content variation within A. kamchatica

We appear to have identified variation in the 2C DNA content among A. kamchatica populations. Greilhuber (2005) suggested that a great deal of apparent within-species, within-ploidy variation in genome DNA content estimated by flow cytometry is due to methodological artefacts. For instance, different levels of anthocyanins, tannic acid and other secondary metabolites in leaves can influence fluorescence and apparent DNA content (Loureiro et al. 2006; Bennett et al. 2008). Following best-practice recommended protocols (Doležel et al. 2007), we used an internal size standard co-chopped with each sample, we used Otto's buffer, which reduces the effects of tannic acid (Loureiro et al. 2006), and leaves were not pigmented. Further, repeated measurements of the same plant on different days produced very similar DNA content estimates ($<1.1\,\%$ variation). Thus the variation we observed should be biologically real (Schmuths *et al.* 2004). However, co-chopping two putatively different samples from different populations would further increase certainty that differences among populations are not artefactual (Greilhuber 2005).

The 2C DNA content of Japanese *A. kamchatica* appears to be slightly larger than North American *A. kamchatica*. This observed genome size difference may differentiate the two *A. kamchatica* subspecies: *A. kamchatica* ssp. *kamchatica* and *A. kamchatica* ssp. *kawasakiana*. Our Japanese *A. kamchatica* samples are from subspecies *A. k. kawasakiana*, whereas the rest of our samples represent subspecies *A. k. kamchatica* from North America. These two subspecies differ in habitat, morphology and nucleotide allele frequencies (Shimizu-Inatsugi *et al.* 2009; Higashi *et al.* 2012), and Shimizu-Inatsugi *et al.* (2009) suggested that *A. k. kawasakiana* may represent a distinct origin of *A. kamchatica*. The difference in genome size between the Japanese *A. k. kawasakiana* and North American *A. k. kamchatica* potentially supports that hypothesis. Alternatively, ongoing hybridization between *A. kamchatica* and its diploid parent, *A. h. gemmifera*, in Asia (Wang *et al.* 2010) could increase the genome size in Asia by reintroducing homeologues that may have been deleted in the allotetraploid.

Other possible explanations for the genome size differences between Japan and North America include biogeographic history and selection. It has been suggested that time-limited environments may select for a smaller genome with more rapid cell division (reviewed in Šmarda and Bureš 2010). As *A. kamchatica* expanded north out of Japan and across the cold Bering land bridge into North America (Shimizu-Inatsugi *et al.* 2009), a smaller genome may have been favoured due to the short growing season. Interestingly, despite the difference in genome size, Japanese and North American samples appear to have lost similar numbers of genes (P. L. Chang, unpubl. res.). Our sampling from Japan was very limited. A thorough investigation of genome size variation from throughout Japan, accompanied by an investigation of introgression and deletions, is needed for a thorough understanding of genome size evolution in this species.

Within-species variation in nuclear genome size may be an important source of genetic diversity, especially if it is associated with phenotypic and ecological variation (Levin 2002; Matsushita *et al.* 2012). Although we did find significant levels of genome size diversity in the allotetraploid *A. kamchatica*, levels of genome size diversity were much lower than in the diploid *Arabidopsis* taxa studied (Fig. 5). This is consistent with the low levels of nucleotide diversity in *A. kamchatica* relative to the other taxa studied (Shimizu-Inatsugi *et al.* 2009). Although nucleotide diversity is generated by point mutations, while

genome size variation is generated by indels, changes in repetitive DNA and transposon activity (Šmarda and Bureš 2010; Long *et al.* 2013), the two forms of genetic diversity are likely to be governed by many of the same population genetic processes such as mating system, biogeography and demographic history (Loveless and Hamrick 1984; Ingvarsson 2002; Glémin *et al.* 2006; Duchoslav *et al.* 2013).

Loss of DNA in tetraploids

The DNA content of tetraploid *A. kamchatica* was slightly less than expected based on the sum of the two parental taxa. It is possible that this apparent loss in DNA content could be artefactual, due to differences between species in plant secondary compounds (Greilhuber 2005). However, rapid loss of DNA after polyploidization appears to be common in polyploids, as the 1Cx genome size has been shown to decrease as the ploidy level increases (Bennett and Thomas 1991; Raina *et al.* 1994; Ozkan *et al.* 2001; Leitch and Bennett 2004; Angulo and Dematteis 2013; Duchoslav *et al.* 2013). Bennett and Thomas (1991) suggest that these changes in DNA content may have adaptive significance, perhaps because the rate of cell division is slowed considerably as genome size increases (Bennett 1972) and it may be beneficial to remove unnecessary DNA when ploidy level is high.

The majority of genome size variation within plant species at a single ploidy level is due to variation in amounts of repetitive DNA such as transposable elements, ribosomal genes and centromeric repeats (Levin 1993; Davison *et al.* 2007; Šmarda and Bureš 2010; Long *et al.* 2013). However, polyploids may also lose considerable amounts of functional DNA either because it is not necessary to have two copies or because it may allow the two parental genomes to resolve incompatibilities (Kashkush *et al.* 2002; Adams and Wendel 2005; Buggs *et al.* 2012). Whole-genome sequencing of *A. kamchatica*, and comparison to its parental taxa, suggests that each of three accessions from different geographic regions lost ~463 of more than 60 000 total genes (~2 % of assembled genes; P. L. Chang, unpubl. res.). Considering that our flow cytometry estimate of the *A. kamchatica* genome size was 3.6 % smaller than expected based on the sum of the parental genomes, the total amount of DNA lost is comparable to the percent of genes lost. This suggests that DNA was lost from both genic regions and nonfunctional regions in *A. kamchatica*.

Arabidopsis l. petraea tetraploids appear to have lost considerably more DNA than *A. kamchatica*. Although these plants are thought to be *A. l. petraea* autotetraploids, they may have experienced hybridization and introgression of DNA from *A. arenosa* (Jørgensen *et al.* 2011; Schmickl and Koch 2011), which has a genome size that is 13 % smaller than *A. l. petraea* (Jørgensen *et al.* 2011). Using DNA content numbers from Jørgensen *et al.* (2011), *A. l. petraea* tetraploid genomes are just slightly smaller than expected from the sum of diploid *A. l. petraea* and diploid *A. arenosa* genomes: observed tetraploid *A. l. petraea* relative genome size 0.44; vs expected diploid *A. l. petraea* 0.23 + diploid *A. arenosa* 0.20 = 0.43 (data are presented as a ratio of the sample peak over the internal standard peak, and cannot be converted to picograms since the 2C DNA content of the standard, *Ilex crenata*, is unknown; Jørgensen *et al.* 2011). The apparent loss of DNA in tetraploid *A. l. petraea* may thus be largely due to hybridization rather than gradual DNA loss through diploidization.

Conclusions

Contrary to some prior reports, all *A. kamchatica* plants in our samples appear to be tetraploid. We found that the allotetraploid, *A. kamchatica*, has a genome size that is just slightly less than the sum of its diploid parental taxa, *A. l. petraea* and *A. h. gemmifera*. Genome size diversity was lower in *A. kamchatica* than in other *Arabidopsis* taxa. However, there was some variation in genome size, where North American populations of *A. k. kamchatica* seem to have lost slightly more DNA than the Japanese population of subspecies *A. k. kawasakiana*. The development of *A. kamchatica* into a model system for the study of polyploidy has the potential to yield a great deal of insight, as its parental taxa have been well studied at both the ecological and genetic levels, and myriad molecular tools from *A. thaliana* are available.

Sources of Funding

This research was supported by the National Center for Research Resources (NCRR) at the National Institutes of Health (NIH) (Alaska INBRE) [RR016466 and 5P20RR016466] and the National Science Foundation (NSF) Experimental Program to Stimulate Competitive Research (Alaska EPSCoR) [0346770].

Contributions by the Authors

Plants were collected by J.A.S., D.E.W. and N.T. Methods were developed and debugged by J.A.S., G.J.H. and N.T. J.A.S. collected the data. J.A.S. and N.T. analysed the data and produced figures. D.E.W. conceived and wrote the manuscript, which was edited by J.A.S., G.J.H. and N.T.

Acknowledgements

We thank Beth Dunkel for helping with collection of the flow cytometry data. We thank J. Doležel for generously providing seeds of the genome size standard, *Glycine max* 'Polanka'. We thank Mark Wright at the UAF IAB Greenhouse for growing and maintaining plants.

Literature Cited

Adams KL, Wendel JF. 2005. Polyploidy and genome evolution in plants. *Current Opinion in Plant Biology* **8**:135–141.

Angulo MB, Dematteis M. 2013. Nuclear DNA content in some species of *Lessingianthus* (Vernonieae, Asteraceae) by flow cytometry. *Journal of Plant Research* **126**:461–468.

Bates D. 2005. Fitting linear mixed models in R. *R News* **5**:27–30.

Bennett MD. 1972. Nuclear DNA content and minimum generation time in herbaceous plants. *Proceedings of the Royal Society B Biological Sciences* **181**:109–135.

Bennett MD, Price HJ, Johnston JS. 2008. Anthocyanin inhibits propidium iodide DNA fluorescence in *Euphorbia pulcherrima*: implications for genome size variation and flow cytometry. *Annals of Botany* **101**:777–790.

Bennett ST, Thomas SM. 1991. Karyological analysis and genome size in *Milium* (Gramineae) with special reference to polyploidy and chromosomal evolution. *Genome* **34**:868–878.

Buggs RJA, Chamala S, Wu W, Tate JA, Schnable PS, Soltis DE, Soltis PS, Barbazuk WB. 2012. Rapid, repeated, and clustered loss of duplicate genes in allopolyploid plant populations of independent origin. *Current Biology* **22**:248–252.

Chen ZJ. 2007. Genetic and epigenetic mechanisms for gene expression and phenotypic variation in plant polyploids. *Annual Review of Plant Biology* **58**:377–406.

Dart S, Kron P, Mable BK. 2004. Characterizing polyploidy in *Arabidopsis lyrata* using chromosome counts and flow cytometry. *Canadian Journal of Botany* **82**:185–197.

Davison J, Tyagi A, Comai L. 2007. Large-scale polymorphism of heterochromatic repeats in the DNA of *Arabidopsis thaliana*. *BMC Plant Biology* **7**:44.

Dawe JC, Murray DF. 1979. In IOPB chromosome number reports LXIII. *Taxon* **28**:265–279.

Dawe JC, Murray DF. 1981. Atlas of chromosome numbers document for the Alaskan flora. http://www.uaf.edu/museum/collections/herb/links-and-references/chromatl.html.

De Smet R, Van De Peer Y. 2012. Redundancy and rewiring of genetic networks following genome-wide duplication events. *Current Opinion in Plant Biology* **15**:168–176.

Doležel J, Greilhuber J. 2010. Nuclear genome size: are we getting closer? *Cytometry Part A* **77A**:635–642.

Doležel J, Doleželová M, Novák FJ. 1994. Flow cytometric estimation of nuclear DNA amount in diploid bananas (*Musa acuminata* and *M. balbisiana*). *Biologia Plantarum* **36**:351–357.

Doležel J, Bartos J, Voglmayr H, Greilhuber J. 2003. Nuclear DNA content and genome size of trout and human. *Cytometry Part A* **51A**:127–128.

Doležel J, Greilhuber J, Suda J. 2007. Estimation of nuclear DNA content in plants using flow cytometry. *Nature Protocols* **2**:2233–2244.

Duchoslav M, Šafářová L, Jandová M. 2013. Role of adaptive and non-adaptive mechanisms forming complex patterns of genome size variation in six cytotypes of polyploid *Allium oleraceum* (Amaryllidaceae) on a continental scale. *Annals of Botany* **111**:419–431.

Faraway JJ. 2006. *Extending the linear model with R*. New York: Taylor and Francis.

Galbraith DW, Harkins KR, Knapp S. 1991. Systemic endopolyploidy in *Arabidopsis thaliana*. *Plant Physiology* **96**:985–989.

Glémin S, Bazin E, Charlesworth D. 2006. Impact of mating systems on patterns of sequence polymorphism in flowering plants. *Proceedings of the Royal Society B Biological Sciences* **273**:3011–3019.

Grant V. 1981. *Plant speciation*. New York: Columbia University Press.

Greilhuber J. 2005. Intraspecific variation in genome size in angiosperms: identifying its existence. *Annals of Botany* **95**:91–98.

Henry IM, Dilkes BP, Young K, Watson B, Wu H, Comai L. 2005. Aneuploidy and genetic variation in the *Arabidopsis thaliana* triploid response. *Genetics* **170**:1979–1988.

Henry IM, Dilkes BP, Tyagi AP, Lin HY, Comai L. 2009. Dosage and parent-of-origin effects shaping aneuploid swarms in *A. thaliana*. *Heredity* **103**:458–468.

Higashi H, Ikeda H, Setoguchi H. 2012. Population fragmentation causes randomly fixed genotypes in populations of *Arabidopsis kamchatica* in the Japanese Archipelago. *Journal of Plant Research* **125**:223–233.

Hothorn T, Bretz F, Westfall P. 2008. Simultaneous inference in general parametric models. *Biometrical Journal* **50**:346–363.

Husband BC. 2004. The role of triploid hybrids in the evolutionary dynamics of mixed-ploidy populations. *Biological Journal of the Linnean Society* **82**:537–546.

Ingvarsson PK. 2002. A metapopulation perspective on genetic diversity and differentiation in partially self-fertilizing plants. *Evolution* **56**:2368–2373.

Jakobsson M, Hagenblad J, Tavaré S, Säll T, Halldén C, Lind-Halldén C, Nordborg M. 2006. A unique recent origin of the allotetraploid species *Arabidopsis suecica*: evidence from nuclear DNA markers. *Molecular Biology and Evolution* **23**:1217–1231.

Jiao Y, Wickett NJ, Ayyampalayam S, Chanderbali AS, Landherr L, Ralph PE, Tomsho LP, Hu Y, Liang H, Soltis PS. 2011. Ancestral polyploidy in seed plants and angiosperms. *Nature* **473**:97–100.

Jørgensen MH, Ehrich D, Schmickl R, Koch MA, Brysting AK. 2011. Interspecific and interploidal gene flow in central european *Arabidopsis* (Brassicaceae). *BMC Evolutionary Biology* **11**:346.

Kashkush K, Feldman M, Levy AA. 2002. Gene loss, silencing and activation in a newly synthesized wheat allotetraploid. *Genetics* **160**:1651–1659.

Kerber ER. 1964. Wheat: reconstitution of the tetraploid component (AABB) of hexaploids. *Science* **143**:253–255.

Kron P, Suda J, Husband BC. 2007. Applications of flow cytometry to evolutionary and population biology. *Annual Review of Ecology, Evolution and Systematics* **38**:847–876.

Ladizinsky G, Fainstein R. 1978. A case of genome partition in polyploid oats. *Theoretical and Applied Genetics* **51**:159–160.

Leitch IJ, Bennett MD. 2004. Genome downsizing in polyploid plants. *Biological Journal of the Linnean Society* **82**:651–663.

Levin DA. 1993. S-gene polymorphism in *Phlox drummondii*. *Heredity* **71**:193–198.

Levin DA. 2002. *The role of chromosomal change in plant evolution*. Oxford: Oxford University Press.

Long Q, Rabanal FA, Meng DZ, Huber CD, Farlow A, Platzer A, Zhang QR, Vilhjálmsson BJ, Korte A, Nizhynska V, Voronin V, Korte P, Sedman L, Mandáková T, Lysak MA, Seren Ü, Hellmann I, Nordborg M. 2013. Massive genomic variation and strong selection in *Arabidopsis thaliana* lines from Sweden. *Nature Genetics* **45**:884–890.

Loureiro J, Rodriguez E, Doležel J, Santos C. 2006. Flow cytometric and microscopic analysis of the effect of tannic acid on plant nuclei and estimation of DNA content. *Annals of Botany* **98**:515–527.

Loveless MD, Hamrick JL. 1984. Ecological determinants of genetic structure in plant populations. *Annual Review of Ecology and Systematics* **15**:65–95.

Lutz AM. 1907. A preliminary note on the chromosomes of *Oenothera lamarckiana* and one of its mutants, *O. gigas*. *Science* **26**:151–152.

Madlung A. 2013. Polyploidy and its effect on evolutionary success: old questions revisited with new tools. *Heredity* **110**:99–104.

Madlung A, Henkhaus N, Jurevic L, Kahsai EA, Bernhard J. 2012. Natural variation and persistent developmental instabilities in geographically diverse accessions of the allopolyploid *Arabidopsis suecica*. *Physiologia Plantarum* **144**:123–133.

Marhold K, Kudoh H, Pak JH, Watanabe K, Španiel S, Lihová J. 2010. Cytotype diversity and genome size variation in eastern Asian polyploid *Cardamine* (Brassicaceae) species. *Annals of Botany* **105**:249–264.

Matsushita SC, Tyagi AP, Thornton GM, Pires JC, Madlung A. 2012. Allopolyploidization lays the foundation for evolution of distinct populations: evidence from analysis of synthetic *Arabidopsis* allohexaploids. *Genetics* **191**:535–547.

Melarango JE, Mehrotra B, Coleman A. 1993. Relationship between endopolyploidy and cell cize in epidermal tissue of *Arabidopsis*. *The Plant Cell* **5**:1661–1668.

Mulligan GA. 1995. Synopsis of the genus *Arabis* (Brassicaceae) in Canada, Alaska and Greenland. *Rhodora* **97**:109–163.

Otto FJ. 1990. DAPI staining of fixed cells for high-resolution flow cytometry of nuclear DNA. In: Darzynkiewickz Z, Crissman H, eds. *Methods in cell biology*. San Diego: Academic Press, 105–110.

Otto S, Whitton J. 2000. Polyploid incidence and evolution. *Annual Reviews of Genetics* **34**:401–437.

Ozkan H, Levy AA, Feldman M. 2001. Allopolyploidy-induced rapid genome evolution in the wheat (*Aegilops–Triticum*) group. *The Plant Cell* **13**:1735–1747.

Raina SN, Parida A, Koul KK, Salimath SS, Bisht MS, Raja V, Khoshoo TN. 1994. Associated chromosomal DNA changes in polyploids. *Genome* **37**:560–564.

Säll T, Jakobsson M, Lind-Halldén C, Halldén C. 2003. Chloroplast DNA indicates a single origin of the allotetraploid *Arabidopsis suecica*. *Journal of Evolutionary Biology* **16**:1019–1029.

Schmickl R, Koch MA. 2011. Arabidopsis hybrid speciation processes. *Proceedings of the National Academy of Sciences of the USA* **108**: 14192–14197.

Schmickl R, Jørgensen M, Brysting A, Koch M. 2010. The evolutionary history of the *Arabidopsis lyrata* complex: a hybrid in the amphi-Beringian area closes a large distribution gap and builds up a genetic barrier. *BMC Evolutionary Biology* **10**:98.

Schmuths H, Meister A, Horres R, Bachmann K. 2004. Genome size variation among accessions of *Arabidopsis thaliana*. *Annals of Botany* **93**:317–321.

Schweizer D, Ambros P, Gründler P, Varga F. 1987. Attempts to relate cytological and molecular chromosome data of *Arabidopsis thaliana* to its genetic linkage map. *Arabidopsis Information Service* **25**:27–34.

Shimizu KK, Fujii S, Marhold K, Watanabe K, Kudoh H. 2005. *Arabidopsis kamchatica* (fisch. Ex dc.) K. Shimizu & Kudoh and *A. kamchatica* subsp. *kawasakiana* (Makino) K. Shimizu & Kudoh, new combinations. *Acta Phytotaxonomica et Geobotanica* **56**:163–172.

Shimizu-Inatsugi R, Lihova J, Iwanaga H, Kudoh H, Marhold K, Savolainen O, Watanabe K, Yakubov VV, Shimizu KK. 2009. The allopolyploid *Arabidopsis kamchatica* originated from multiple individuals of *Arabidopsis lyrata* and *Arabidopsis halleri*. *Molecular Ecology* **18**:4024–4048.

Šmarda P, Bureš P. 2010. Understanding intraspecific variation in genome size in plants. *Preslia* **82**:41–61.

Sokal RR, Rohlf FJ. 1995. *Biometry*, 3rd edn. New York: W. H. Freeman, 58.

Soltis DE, Soltis PS. 1999. Polyploidy: recurrent formation and genome evolution. *Trends in Ecology and Evolution* **14**:348–352.

Stebbins GL. 1950. *Variation and evolution in plants*. New York: Columbia University Press.

Thompson JD, Lumaret R. 1992. The evolutionary dynamics of polyploid plants: origins, establishment and persistence. *Trends in Ecology and Evolution* **7**:302–307.

Vision TJ, Brown DG, Tanksley SD. 2000. The origins of genomic duplications in *Arabidopsis*. *Science* **290**:2114–2117.

Wang W-K, Ho C-W, Hung K-H, Wang K-H, Huang C-C, Araki H, Hwang C-C, Hsu T-W, Osada N, Chiang T-Y. 2010. Multilocus analysis of genetic divergence between outcrossing *Arabidopsis* species: evidence of genome-wide admixture. *New Phytologist* **188**: 488–500.

Wolfe KH. 2001. Yesterday's polyploids and the mystery of diploidization. *Nature Reviews Genetics* **2**:333–341.

Wood TE, Takebayashi N, Mayrose I, Barker MS, Greenspoon PB, Rieseberg LH. 2009. The frequency of polyploid speciation in vascular plants. *Proceedings of National Academy of Sciences of the USA* **106**:13875–13879.

5

Genome downsizing and karyotype constancy in diploid and polyploid congeners: a model of genome size variation

Lidia Poggio, María Florencia Realini, María Florencia Fourastié, Ana María García and Graciela Esther González*

Instituto de Ecología, Genética y Evolución (IEGEBA)-Consejo Nacional de Investigaciones Científicas y Técnicas (CONICET) and Laboratorio de Citogenética y Evolución (LaCyE), Departamento de Ecología, Genética y Evolución, Facultad de Ciencias Exactas y Naturales, Universidad de Buenos Aires, Ciudad Autónoma de Buenos Aires, Argentina

Associate Editor: Kermit Ritland

Abstract. Evolutionary chromosome change involves significant variation in DNA amount in diploids and genome downsizing in polyploids. Genome size and karyotype parameters of *Hippeastrum* species with different ploidy level were analysed. In *Hippeastrum*, polyploid species show less DNA content per basic genome than diploid species. The rate of variation is lower at higher ploidy levels. All the species have a basic number $x = 11$ and bimodal karyotypes. The basic karyotypes consist of four short metacentric chromosomes and seven large chromosomes (submetacentric and subtelocentric). The bimodal karyotype is preserved maintaining the relative proportions of members of the haploid chromosome set, even in the presence of genome downsizing. The constancy of the karyotype is maintained because changes in DNA amount are proportional to the length of the whole-chromosome complement and vary independently in the long and short sets of chromosomes. This karyotype constancy in taxa of *Hippeastrum* with different genome size and ploidy level indicates that the distribution of extra DNA within the complement is not at random and suggests the presence of mechanisms selecting for constancy, or against changes, in karyotype morphology.

Keywords: Bimodal karyotype; DNA amount variation; genome size; *Hippeastrum*; karyotype constancy; polyploids.

Introduction

The diversity of plant genomes is manifested through a wide range of chromosome number and genome size (Leitch and Leitch 2013). The partitioning of total DNA in chromosomes is a complex level of structural and functional organization of nuclear genomes. Each species has a characteristic chromosome complement, its karyotype, which represents the phenotypic appearance of somatic chromosomes. Karyotype features more commonly recorded for comparative evolutionary analysis are number and size of the chromosomes, position and type of primary and secondary constrictions, karyotype symmetry and genome size, among others. Genome size does not necessarily reflect chromosome number variation since mechanisms producing changes in total DNA amount are different for those leading to changes in chromosome number. The increases in genome size arise predominantly through polyploidy and amplification of non-coding repetitive DNA, especially retrotransposons (Bennetzen *et al.* 2005). These mechanisms are counterbalanced by processes that result in a decrease in genome size such as unequal recombination and illegitimate recombination (Leitch and Leitch 2013). Genome size changes

* Corresponding author's e-mail address: gegonzalez@ege.fcen.uba.ar

(amplification or deletions) are correlated with karyotype parameters and can affect the entire chromosome complement or they may be restricted to a subset of chromosomes.

Different patterns of distribution of DNA among chromosomes or chromosome arms, even in the absence of chromosomal rearrangements, could lead to important changes in the karyotype parameters, mainly in the asymmetry (Peruzzi et al. 2009). This parameter refers to karyotypes with a predominance of chromosomes with terminal/subterminal centromeres (intrachromosomal asymmetry) and highly heterogeneous chromosome sizes (interchromosomal asymmetry) (reviewed by Peruzzi and Eroglu 2013). It is interesting to point out that evolutionary chromosome change involving alteration in DNA amount does not always lead to changes in the morphology of the karyotype, given that in several groups of plants karyotype orthoselection has been found (White 1973), as was described in Asparagaceae, Xanthorrhoeacae (Brandham 1971; Brandham and Doherty 1998) and Vicia (Fabaceae) (Naranjo et al. 1998), among others.

The bimodal karyotype represents a special case of asymmetry and is characterized by the presence of two sharply distinct classes of chromosomes without a gradual transition. The bimodal karyotype has been reported in monocots such as Xanthorrhoeacae (Aloe, Haworthia, Gasteria), Asparagaceae (Agave, Yucca) and Amaryllidaceae (Hippeastrum, Rodophiala) (Naranjo 1969; Brandham 1971; Naranjo and Andrada 1975; Arroyo 1982; Naranjo and Poggio 1988; Brandham and Doherty 1998; Vosa 2005; Poggio et al. 2007; Weiss-Schneeweiss and Schneeweiss 2013). Taxonomic groups with bimodal karyotypes and genome size variation offer the opportunity to analyse the nature and distribution of changes between chromosome arms and among members of the haploid chromosome set.

Hippeastrum Herb. is a genus of perennial and bulbous plants of the tribe Hippeastreae of Amaryllidaceae J.St.-Hil. (Meerow et al. 2000) with ca. 60 species inhabiting tropical and subtropical America from Mexico and the Antilles to central Argentina. Their species have economic value as ornamentals and are used in the pharmaceutical industry due to their high content of alkaloids. In the genus Hippeastrum, chromosomes of about 41 species have been studied and all presented bimodal karyotypes and a basic number $x = 11$. The karyotypes consist of four short metacentric (m) chromosomes and seven large chromosomes (four submetacentric—sm and three subtelocentric—st) (Naranjo 1969; Naranjo and Andrada 1975; Arroyo 1982; Brandham and Bhandol 1997). This genus is an interesting model to analyse how and where gain or loss of DNA occurs, and how these changes affect karyotype morphology.

Poggio et al. (2007) found, in 12 Hippeastrum diploid species from South America, karyotypes similar to that previously described but significant differences in nuclear DNA content. These authors report that karyotype constancy is a product of changes in DNA content occurring in the whole-chromosome complement, and that DNA addition to the long and short sets of chromosomes varies independently. The authors state that the evolutionary changes in DNA amount are proportional to chromosome length, maintaining karyotype uniformity. They found that in diploid species with higher DNA content, the short chromosomes add equal DNA amounts to both arms, maintaining their metacentric morphology, whereas the long chromosomes add DNA only to the short arm, increasing chromosome symmetry.

Several authors reported variation in ploidy level ($3x$ to $7x$) in several species of the genus (Sato 1938; Neto 1948; Naranjo 1969; Lakshmi 1980; Arroyo 1982; Beltrao and Guerra 1990; Zou and Quin 1994). It is interesting to point out that several polyploids previously analysed were considered to be autopolyploids, because they have similar basic bimodal karyotypes to those described in diploid species (Naranjo 1969; Naranjo and Andrada 1975). The genome size of the polyploid species of Hippeastrum has not yet been reported. It has been frequently documented that the major trend in vascular plants is a decrease in the genome size per haploid genome ($1Cx$), when a polyploidization event occurs (Leitch and Bennett 2004; Leitch and Leitch 2013). This genome downsizing, which could be involved in the genetic and cytogenetic diploidization of polyploids, consists in non-random deleting of coding and non-coding sequences, changes in retroelements, chromosome reorganization, gain or loss of chromosomes or entire genomes, altered patterns of gene expression and epigenetic modifications (Feldman and Levy 2005; Ma and Gustafson 2006; Jones and Langdom 2013; Leitch and Leitch 2013).

In the present work, variation of DNA amount in species of Hippeastrum with different ploidy level is presented with the aim to evaluate if genome size per haploid genome decreases when a polyploidization event occurs. Besides, karyotype parameters are evaluated to analyse if bimodality and karyotype' constancy detected in diploids can still take place in different ploidy levels, even in the presence of genome downsizing. Finally, the variation in DNA content and correlated karyotype parameters will be discussed in the different ploidy levels studied.

Methods

Cytological studies were carried out on material cultivated at the Royal Botanic Gardens, Kew, with the exception of one specimen of Hippeastrum argentinum that

Table 1. Origin, accession numbers and ploidy level of the *Hippeastrum* species.

Species	Ploidy level	Origin	Kew accession or Herba Nt.
H. machupijchense (Vargas) Hunt	2x	Perú, Cuzco, Machupichu	376-76-03600
H. solandriflorum Herb.	2x	Argentina, Corrientes	301-79-02627
H. aulicum Herb.	2x	Brazil, Santa Catarina	434-79-04428
H. hybrid Sealy	2x	Brazil	344-79-03154
H. argentinum (Pax) Hunz.	2x	Argentina, Catamarca	ATH18258
H. psittacinum (Ker Gawl.) Herb.	2x	Brazil	088-60-08801
H. evansiae (Traub & Nels.) Moore	2x	Bolivia	302-79-02858
H. tucumanum Holmb.	2x	Argentina, Tucumán	361-75-03430
H. parodii Hunz. & Coc.	2x	Argentina, Corrientes, Três Cerros	400-76-03888
H. correiense (Bury) Worsley	2x	Brazil, Sao Paulo	419-72-03854
H. rutilum (Ker Gawl.) Herb.	2x	Brazil	501-66-50111
H. morelianum (Lamaire) Traub	2x	Brazil, Sao Paulo, Serra do Mar	419-72-03853
H. puniceum (Lamb.) Kuntze	3x	Guyana, Mt Roraina, Kako	236-80-02247
H. reginae (L.) Herb.	4x	Peru, Cuzco, Marcapata	408-53-40803
H. rutilum (Ker Gawl.) Herb.	4x	Brazil	006-69-16919
H. starkii (Nels. & Traub) Moore	4x	Bolivia	487-67-48702
H. blossfeldiae (Traub & Doran) Vam Scheepen	4x	Brazil, Sao Paulo	139-74-01555
H. scopulorum Baker	5x	Bolivia, La Paz	037-72-00389
H. rutilum (Ker Gawl.) Herb.	5x	Brazil, Pelotas	396-70-03892
H. cybister (Herb.) Benth. ex Baker	5x	Brazil	418-72-09675
H. puniceum (Lamb.) Kuntze	6x	Brazil, Sao Paulo, Araras	277-78-030023

was collected by A. T. Hunziker (ATH 18258). The sources of the materials are listed in Table 1.

Cytological analysis

For squashing, root tips were pretreated for 2.5 h in 0.002 M 8-hydroxyquinoline at 20 °C, fixed in 3 : 1 absolute ethanol : acetic acid and stained in Feulgen solution. The average of centromeric indices, for small and large chromosomes (CI$_S$ and CI$_L$), was calculated according to Poggio *et al.* (2007). The nomenclature used for chromosome morphology is that proposed by Levan *et al.* (1964). To estimate karyotype asymmetry, the coefficient of variation of chromosome length (CV$_{CL}$) and the mean centromeric asymmetry (M$_{CA}$) were calculated according to Peruzzi and Eroglu (2013). The A1 and A2 indices from Romero Zarco (1986) were also calculated for comparison with previously published data in *Hippeastrum* and related genera. Chromosomal parameters were measured using the freeware program MicroMeasure 3.3 (http://www.colostate.edu/Depts/Biology/MicroMeasure/). Mean values for the karyotype parameters were measured

from a minimum of five scattered metaphase plates in each accession.

Feulgen staining and cytophotometry

Root tips were fixed in 3 : 1 absolute ethanol : acetic acid for 1–4 days. The staining method was performed as described in Tito *et al.* (1991). The amount of Feulgen staining per nucleus, expressed in arbitrary units, was measured at a wavelength of 550 nm using the scanning method on a Vickers M85 Microspectrophotometer (Jodrell Laboratory, RBG, Kew, UK). The DNA content per basic genome expressed in picograms (pg) was calculated using *Allium cepa* var. *Ailsa Craig* as a standard (2C = 33, 55 pg; Bennett and Smith 1976). DNA content was measured in 25–50 telophase nuclei (2C) per accession.

Statistical analysis

The differences between species in 1Cx DNA content were tested through an analysis of variance (ANOVA) using generalized linear mixed models. The mean values of genome sizes were calculated and multiple contrasts were performed with the LSD Fisher method (Fisher

1932). These statistical analyses were considered significant if their P values were <0.05.

The relationship between total DNA content and ploidy level was studied by fitting a weighted least-squares linear regression. This method compensates for the variable number of DNA measurements available for each species and ploidy level (Aitken 1935).

The statistical analyses were performed using the Infostat program, FCA, National University of Córdoba (Di Rienzo et al. 2012) and the R programming language (R Development Core Team 2004).

Results

Total genome size (2C), DNA per basic genome (1Cx), karyotype formulae and karyotype parameters for diploid and polyploid species are listed in Table 2.

All the diploid and polyploid species presented $x = 11$ (Table 2 and Fig. 1). The karyotype formulae and parameters show a basic bimodal karyotype, with the presence of two distinct classes of chromosomes, long and short (Figs 1 and 2). The relative chromosome sizes and relative arm sizes per basic haploid complement ($x = 11$) are given in a diagrammatic form (Fig. 2). The volume of the short chromosomes as a percentage of the volume of all chromosomes (CV_S) is similar in all the taxa analysed (23.05–25.12) (Table 2). The centromeric indices of short chromosomes (CI_S) are very similar among diploid and polyploid taxa (42.42–46.87). On the other hand, the centromeric indices of large chromosomes (IC_L) decrease at lower genome size in diploids (19.9–26.17), while $3x$, $4x$ and $5x$ present similar values (23.18–24.37). The hexaploid differs from the rest of the species in their karyotype parameters, having a similar CI_S but a higher CI_L (Table 2). The karyotype asymmetry indices M_{CA} and CV_{CL} are given in Table 2 and are plotted against DNA content in Fig. 3. In this figure, it can be seen that Hippeastrum puniceum ($6x$), with the lowest basic DNA amount (1Cx), occupies an isolated position when compared with the rest of the Hippeastrum species. This is a consequence of its more symmetrical karyotype.

Significant differences in 1Cx DNA amount were found among the taxa ($F = 427.44$, $P < 0.0001$). They are indicated in Table 2. The total DNA content (2C) increases with ploidy level (DNA 2C: $y = 8.9x + 13.6$; $x =$ ploidy level, $R^2 = 95$ %) but the calculated regression line has a gentler slope than the line extrapolated from the diploid mean, which assumes that when the number of genomes increases DNA is added as an exact multiple of the DNA content per basic genome (Fig. 4). When DNA content per basic genome is plotted against ploidy level, a hyperbolic curve is obtained (1Cx: $y/x = 13.6/x + 8.9$) (Fig. 5).

This new formula results from rearranging the linear regression equation of Fig. 4.

Discussion

In the present work, genome size and karyotype parameters of Hippeastrum species with different ploidy level were analysed and compared with previous data.

Total DNA (2C) varies from 26.80 to 34.17 pg among diploids and increases with ploidy level, reaching a value of 64.67 pg in the hexaploid species. This genus has large genomes, since according to Leitch et al. (1998) most angiosperms actually have small 1C values (from 0.1 to 3.5 pg).

DNA per basic genome (1Cx), calculated from total DNA content, varies from 17.08 to 13.40 pg in diploids. The difference between these extreme values is significant. In polyploids there is a gradual decrease in the 1Cx value when ploidy level increases, varying from 12.90 pg in triploids to 10.78 pg in hexaploids. In Hippeastrum the polyploids studied show less DNA content per basic genome than diploids. Considering the average of basic DNA content for diploids, the triploid diminishes by 16.77 % while the decrease among $3x$–$4x$, $4x$–$5x$ and $5x$–$6x$ ploidy levels is lower, varying between 5.5 and 6.5 %. These results show that in Hippeastrum, DNA per haploid genome decreases in polyploids, the rate of variation being lower at higher ploidy levels. Many examples are found in the literature where polyploidy is associated with decreasing genome size, in terms of DNA content per haploid genome. Moreover, comparative genome studies have shown that the downsizing of the genome can take place even in a few generations and could be involved in the genetic and cytogenetic diploidization (Soltis et al. 2003; Kellogg and Bennetzen 2004; Leitch and Bennett 2004; Feldman and Levy 2005; Ma and Gustafson 2006; Leitch and Leitch 2013). While polyploidy, joined with transposable element amplification, is widely considered to play a role in generating increased genome size, mechanisms that generate small deletions such as unequal homologous recombination and illegitimate recombination could be involved in genome downsizing (Bennetzen et al. 2005; Leitch and Leitch 2013). To explain this widespread phenomenon it could be postulated that at polyploid level, the DNA elimination leads to a more adequate balance between total DNA content and certain cellular parameters. Moreover, at polyploid level, the partial elimination of DNA sequences is more easily tolerated. However, in some cases, as in genus Larrea (Zygophyllaceae) (Poggio et al. 1989) or Aloe (Xanthorrhoeacae) (Brandham and Doherty 1998), differences in 1Cx at different ploidy levels are not statistically significant.

Table 2. Chromosome numbers, genome sizes and karyotype parameters of the *Hippeastrum* species. 2C DNA, total genomic DNA; 1Cx DNA, DNA per basic genome; CI$_S$, average of centromeric index of short chromosomes; CI$_L$, average of centromeric index of long chromosomes; A1, intrachromosomal asymmetry index; A2, interchromosomal asymmetry index; M$_{CA}$, mean centromeric asymmetry; CV$_{CL}$, coefficient of variation of chromosome length; CV$_S$, volume of short chromosomes as a percentage of the volume of all chromosomes. Means with the same letter are not significantly different ($P \le 0.05$). *Data taken from Poggio *et al.* (2007), except for M$_{CA}$ and CV$_{CL}$ values.

Species	2n	2C DNA (pg) (X ± SE)	1Cx DNA (pg) (X ± SE)	CI$_S$	CI$_L$	A1	A2	M$_{CA}$	CV$_{CL}$	CV$_S$ (%)	Karyotype formula
H. machupijchense*	22	34.17 (±0.20)	17.08 (±0.10)[A]	42.42	26.17	0.50	0.30	–	30.56	23.65	[4m] + 4sm + 3st
H. solandriflorum*	22	33.77 (±0.50)	16.88 (±0.25)[AB]	42.48	24.39	0.51	0.31	36.00	31.03	23.59	[4m] + 4sm + 1sm–st + 2st
H. psittacinum*	22	31.34 (±0.23)	15.67 (±0.12)[E]	45.85	25.37	0.48	0.32	–	32.03	24.85	[4m] + 3sm + 1sm–st + 3st
H. evansiae*	22	30.92 (±0.28)	15.46 (±0.14)[EF]	46.87	23.83	0.47	0.32	36.08	32.20	23.24	[4m] + 3sm + 1sm–st + 2st + 1st–t
H. tucumanum*	22	30.64 (±0.17)	15.32 (±0.09)[FG]	43.20	24.89	0.50	0.31	39.24	31.01	24.90	[4m] + 3sm + 1sm–st + 3st
H. parodii*	22	30.21 (±0.23)	15.11 (±0.11)[G]	42.46	23.27	0.52	0.29	37.04	29.20	23.91	[4m] + 3sm + 1sm–st + 3st
H. correiense*	22	29.05 (±0.25)	14.53 (±0.13)[H]	45.58	22.78	0.51	0.29	35.46	29.04	24.44	[4m] + 2sm + 2sm–st + 1st + 2t
H. rutilum	22	27.98 (±0.28)	13.99 (±0.14)[I]	45.10	22.38	0.51	0.31	33.57	31.03	23.97	[4m] + 2sm + 1sm–st + 3st + 1t
H. morelianum*	22	26.80 (±0.19)	13.40 (±0.09)[J]	43.75	19.99	0.55	0.32	37.39	32.08	23.21	[4m] + 2sm + 1sm–st + 2st + 2t
H. puniceum	33	38.69 (±0.48)	12.90 (±0.16)[K]	44.76	23.97	0.49	0.30	31.88	30.33	24.14	[4m] + 1sm + 3sm–st + 2st + 1t
H. reginae	44	52.79 (±0.30)	13.20 (±0.08)[J]	–	–	–	–	–	–	–	–
H. rutilum	44	48.93 (±0.37)	12.23 (±0.09)[L]	42.63	23.23	0.54	0.32	39.75	32.02	23.14	[3m + 1m–sm] + 1sm + 2sm–st + 3st + 1t
H. starkii	44	47.19 (±0.30)	11.80 (±0.08)[M]	–	–	–	–	–	–	–	–
H. blossfeldiae	44	46.04 (±0.29)	11.51 (±0.07)[N]	42.85	23.18	0.53	0.32	39.30	32.01	23.05	[3m + 1m–sm] + 2sm + 1sm–st + 3st + 1t
H. scopulorum	55	58.71 (±0.26)	11.74 (±0.05)[M]	–	–	–	–	–	–	–	–
H. rutilum	55	58.20 (±0.42)	11.64 (±0.10)[MN]	45.26	24.37	0.49	0.29	35.62	29.02	24.69	[4m] + 3sm–st + 4st
H. cybister	55	56.35 (±0.38)	11.20 (±0.11)[O]	45.23	23.15	0.50	0.30	37.55	30.04	25.01	[4m] + 1sm + 3sm–st + 3st
H. puniceum	66	64.67 (±0.41)	10.78 (±0.07)[P]	44.88	34.10	0.42	0.33	28.61	33.01	25.12	4m + 3sm + 3 sm–st + 1 st

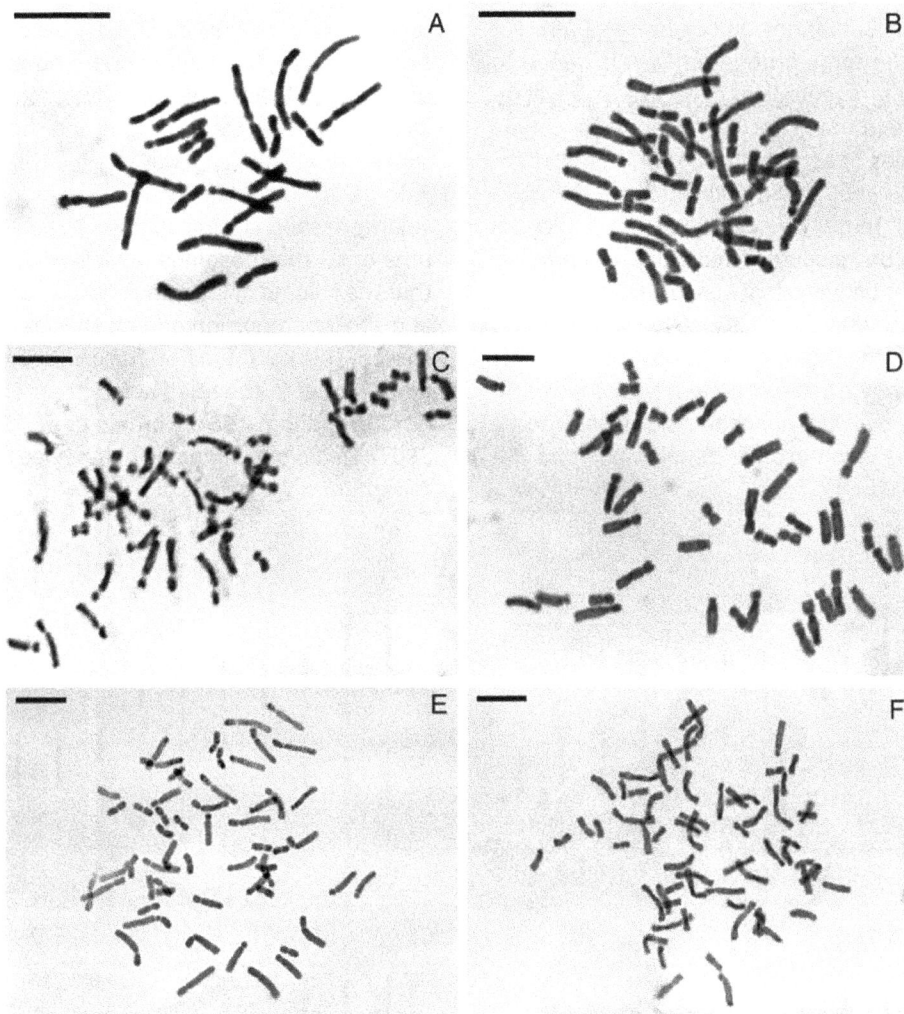

Figure 1. Mitotic metaphases of *Hippeastrum* species: (A) *H. rutilum* ($2n = 22$), (B) *H. puniceum* ($2n = 33$), (C) *H. rutilum* ($2n = 44$), (D) *H. blossfeldiae* ($2n = 44$), (E) *H. cybister* ($2n = 55$) and (F) *H. puniceum* ($2n = 66$). Scale bar: 10 μm.

The diploid and polyploid species of *Hippeastrum* here studied presented $x = 11$ and despite possessing significant differences in their genome size, all have a basic bimodal karyotype with four small m and seven large sm/t chromosomes. The constancy of the karyotype in taxa of *Hippeastrum* with different genome size and ploidy level indicates that the distribution of extra DNA within the complement is not at random and suggests the presence of mechanisms selecting for constancy, or against changes, in karyotype morphology, processes named by White (1973) as karyotype constancy or karyotype orthoselection, respectively. Several studies have shown that karyotype orthoselection in diploid species with significant differences in genome sizes involves proportional changes in all chromosomes, preserving the morphology of the complement (Brandham and Doherty 1998; Naranjo *et al.* 1998). Chromosomal parameters such as

centromeric indices and karyotype asymmetry provide some insights into how the additional DNA is distributed in the genome, between small and large chromosomes as well as between arms of individual chromosomes. In this work we use M_{CA} and CV_{CL} to estimate the intrachromosomal and interchromosomal asymmetries, respectively (Peruzzi and Eroglu 2013). Moreover, we also employ the A1 and A2 indices from Romero Zarco (1986) only for comparative purposes with previous work in the *Hippeastrum* species and related genera (Naranjo and Poggio 1988; Poggio *et al.* 2007).

Different patterns of addition of DNA amount in a chromosome complement were reviewed by Peruzzi *et al.* (2009). For 'proportional increase', the amount of DNA added to each chromosome arm is proportional to its length. This pattern does not result in a change in karyotype asymmetry when genome size changes. This pattern has been observed in several genera, including *Aloe* and *Gasteria* (Brandham and Doherty 1998). For 'equal increase', the same amount of DNA is added to each chromosome arm regardless of its size. This will result in an increase in the intrachromosomal karyotype symmetry. Examples of genera showing this pattern include *Vigna* (Parida *et al.* 1990) and *Papaver* (Srivastava and Lavania 1991). In many genera of Liliaceae, Peruzzi *et al.* (2009) found an 'unequal increase', i.e. the amount of DNA added varies between longer and shorter chromosome arms unequally.

In *Hippeastrum*, with two sets of chromosomes that differ in size and morphology, a different pattern was observed. In diploid species the evolutionary changes in DNA amount occur in the whole-chromosome complement and are proportional to chromosome length, maintaining karyotype uniformity (Poggio *et al.* 2007). These authors analysed separately the CI of short and long chromosomes and proposed a model of genome size change where the DNA increase or decrease to the long and short sets of chromosomes varies independently.

In the diploid and polyploid species analysed here, the volume of short chromosomes as a percentage of the volume of all chromosomes (CV_S) is very similar, indicating that the volume of long and short chromosomes remains in a similar proportion among species. As previously discussed, this karyotype uniformity occurs if changes are proportional to the relative length of each chromosome arm (Brandham 1983; Naranjo *et al.* 1998; Poggio *et al.* 2007). In diploid and polyploid species the CI_S are similar

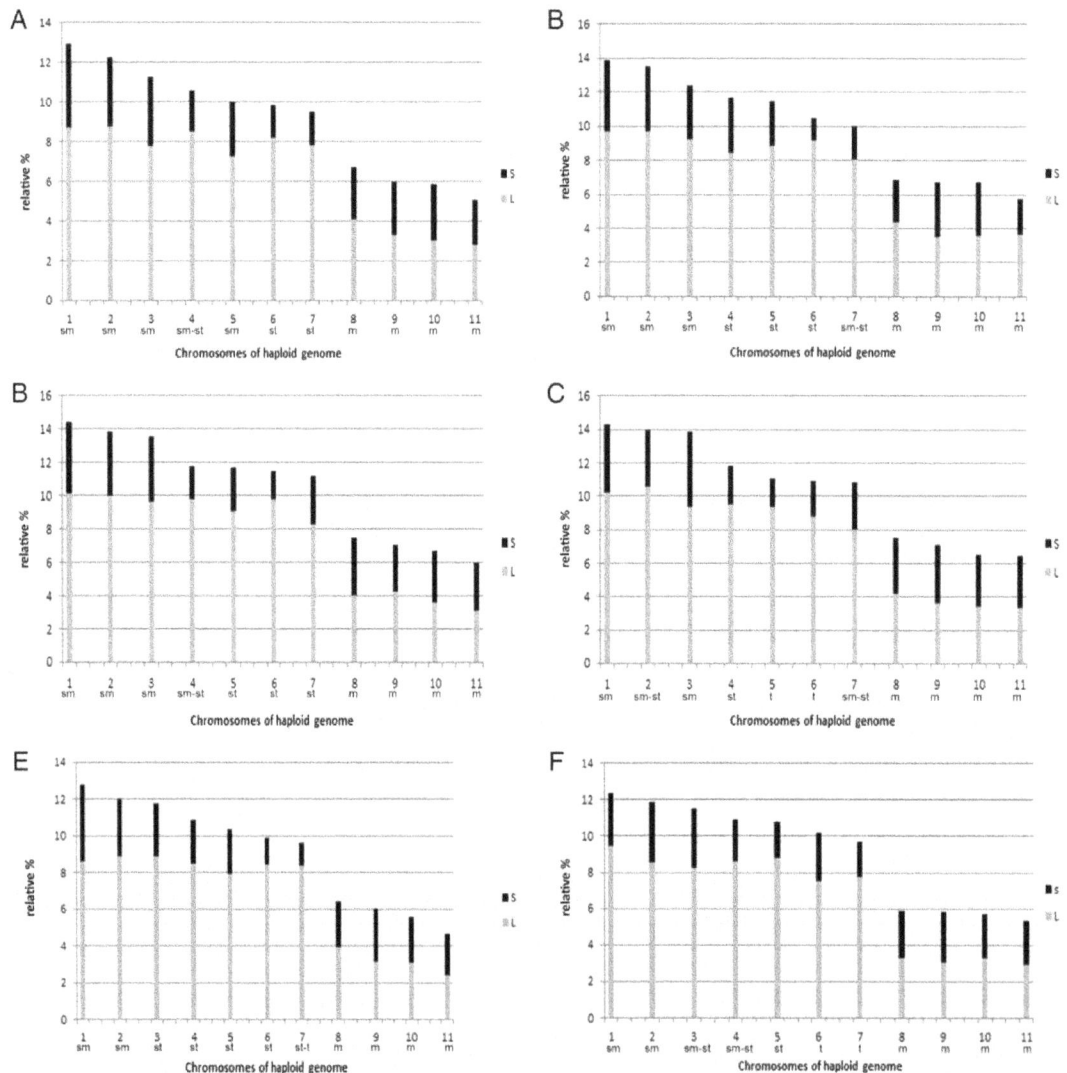

Figure 2. Relative chromosome and arm sizes per haploid complement ($x = 11$): (A) *H. solandriflorum* (2x), (B) *H. tucumanum* (2x), (C) *H. parodii* (2x), (D) *H. correiense* (2x), (E) *H. rutilum* (2x), (F) *H. morelianum* (2x), (G) *H. puniceum* (3x), (H) *H. rutilum* (4x), (I) *H. blossfeldiae* (4x), (J) *H. cybister* (5x), (K) *H. rutilum* (5x) and (L) *H. puniceum* (6x). S, short arm; L, long arm; m, metacentric; sm, submetacentric; st, subtelocentric; t, telocentric.

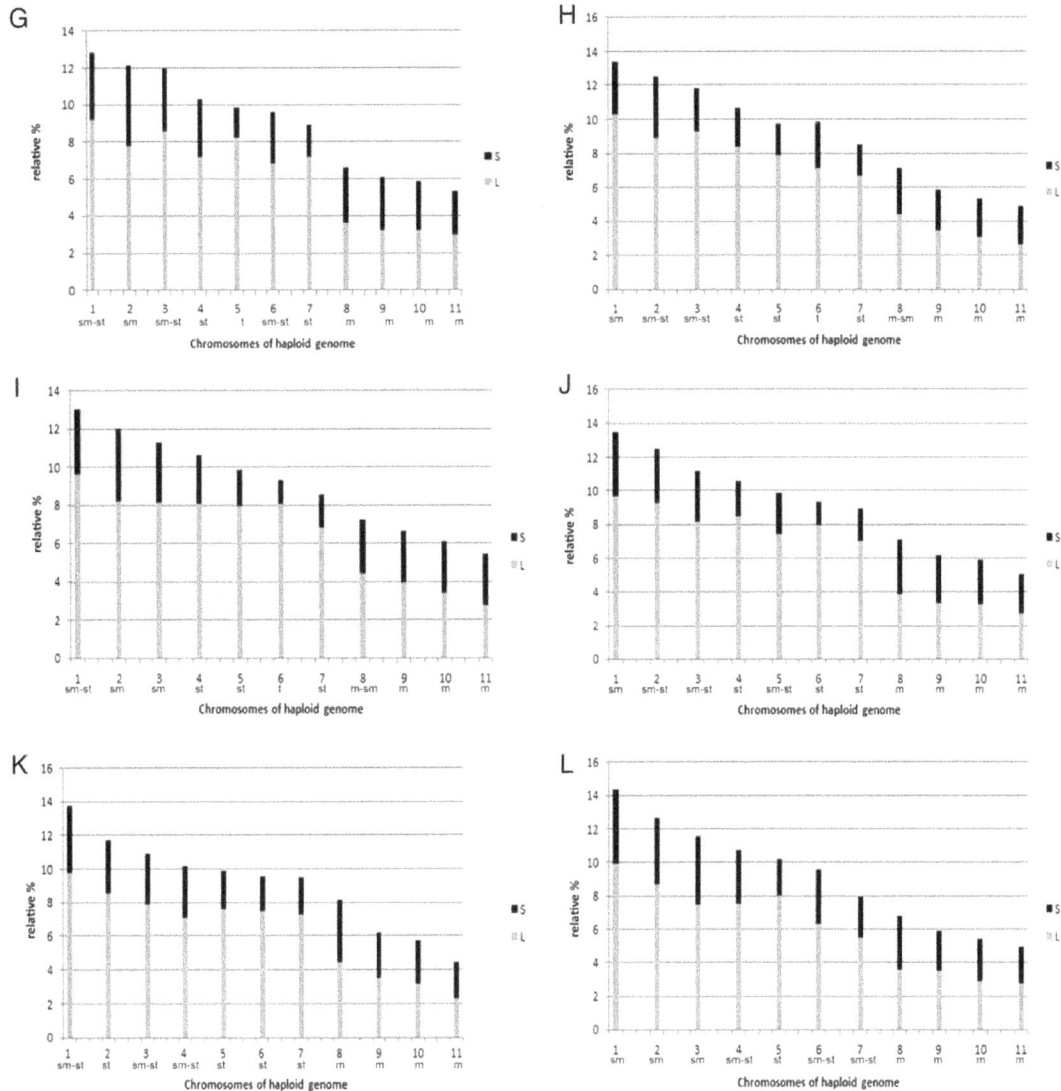

Figure 2. Continued.

and did not show any relationship with DNA amount, varying from 42.60 to 46.80. This could be explained if the short chromosomes add or lose equal DNA amounts to both arms, maintaining their metacentric morphology. Diploid species with lower DNA content have minor CI_L indices and have more asymmetric karyotypes, with a greater number of long chromosomes st or t, i.e. the changes in DNA amount in the long chromosomes affect mainly in their short arms. Among the triploids, tetraploids and pentaploids variation in CI_L was not detected, being similar to that of the diploid species with lower DNA content. This could be attributed to the lower downsizing at higher ploidy level.

In the hexaploid species analysed here, CV_S and the bimodality are maintained, and CI_S values are similar to those of the diploid and polyploid species. However, a different pattern of changes is observed in the long chromosomes of its karyotype. CI_L is greater than that of the other studied species, indicating that centromeres have a more median position. While the number of chromosomes sm–st, st and t varies from 3 to 7 from diploids to pentaploids, the hexaploids have just one st chromosome. Moreover, it is the only species with m–sm long chromosomes, i.e. the subset of long chromosomes is more symmetrical. This could be explained if there is a threshold for the distribution of changes in the larger chromosomes when the chromosome number is >55. This threshold could be related to nuclear organization at the chromosome level, arrangement of nuclear territories, interactions among genomes to sharing a nucleus and disturbances during cell division. Anyway, still very little is known about the mechanisms

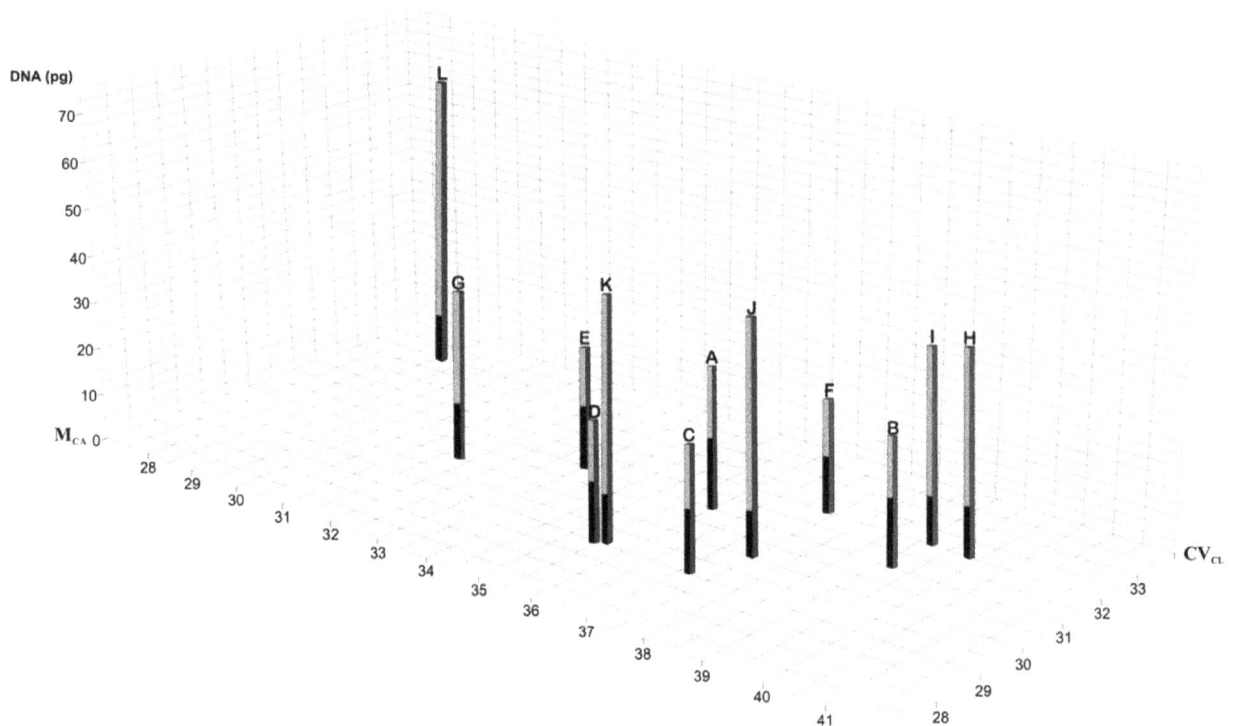

Figure 3. Asymmetry parameters (M_{CA} and CV_{CL}) plotted against DNA content. The bars represent the total DNA amount (2C) and the black zone indicates the basic DNA amount (1Cx). (A) *H. solandriflorum* (2x), (B) *H. tucumanum* (2x), (C) *H. parodii* (2x), (D) *H. correiense* (2x), (E) *H. rutilum* (2x), (F) *H. morelianum* (2x), (G) *H. puniceum* (3x), (H) *H. rutilum* (4x), (I) *H. blossfeldiae* (4x), (J) *H. cybister* (5x), (K) *H. rutilum* (5x) and (L) *H. puniceum* (6x). M_{CA}, mean centromeric asymmetry; CV_{CL}, coefficient of variation of chromosome length.

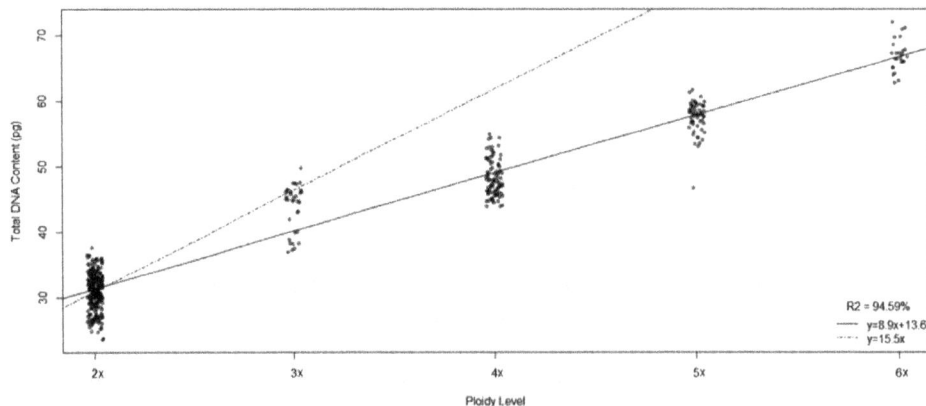

Figure 4. Total DNA content (2C) plotted against ploidy level. Solid line, linear fit; broken line, extrapolated from diploids.

and sequences involved in genome downsizing in *Hippeastrum*.

Navrátilová *et al.* (2003) reported that amplification of retroelement sequences is likely to increase the size of all chromosomes within the karyotype in an approximately equal manner. In *Hippeastrum*, the absence of notorious C and DAPI bands (unpubl. res.), joined to the presence of conserved bimodal karyotypes, even with changes in ploidy level and 1Cx value, strongly suggests that DNA

changes could occur by amplification or deletion of retroelement sequences, which are generally dispersed in the genome.

Conclusion

In the genus *Hippeastrum*, evolutionary chromosome change involves variation in DNA amount in diploids and genome downsizing in polyploids. Besides, the

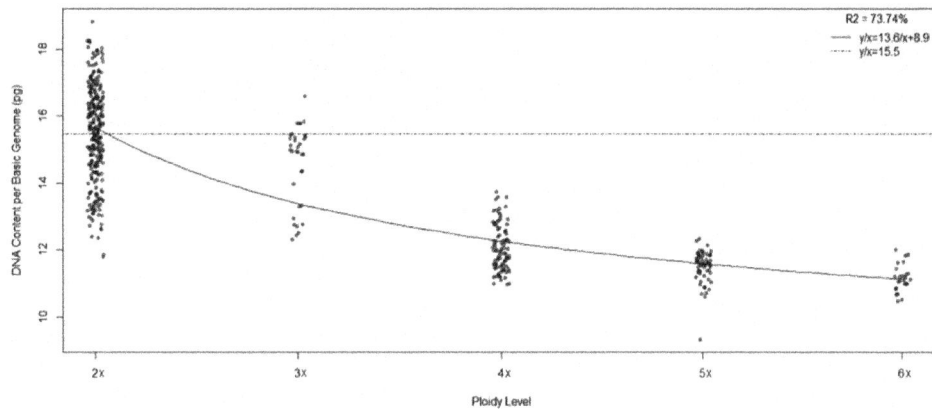

Figure 5. DNA content per basic genome (1Cx) plotted against ploidy level. Solid line, linear fit/ploidy level (*x*); broken line, extrapolated from diploids/ploidy level (*x*).

bimodal karyotype is preserved maintaining the relative proportions of members of the haploid chromosome set by karyotype orthoselection. The presence of conserved karyotypes, even with changes in ploidy level and DNA content per basic genome, is strongly susceptible to an adaptive interpretation, suggesting the existence of mechanisms that select for constancy in karyotype morphology.

Sources of Funding

Funding was provided by grants from the Consejo Nacional de Investigaciones Científicas y Técnicas (CONICET-PIP 00342), Universidad de Buenos Aires (UBACYT 20020100100859) and Agencia Nacional de Producción Científica y Tecnológica—SECyT (PICT 2010-1665).

Contributions by the Authors

All authors contributed to the experimental design, data analysis and manuscript preparation.

Acknowledgements

We thank Agr. Bch. M.F. Schrauf for statistical contributions and Mr D. Fink for assistance with the image analysis. We are also grateful to the anonymous referees and the associate editor of the journal *AoB PLANTS* for their helpful comments.

Literature Cited

Aitken AC. 1935. On least squares and linear combinations of observations. *Proceedings of the Royal Society of Edinburgh* **55**:42–48.

Arroyo SC. 1982. The chromosomes of *Hippeastrum*, *Amaryllis* and *Phycella* (Amaryllidaceae). *Kew Bulletin* **37**:211–216.

Beltrao GT, Guerra M. 1990. Cytogenetica de angiospermas coletadas em Pernambuco. *Ciência e Cultura* **42**:839–845.

Bennett MD, Smith JB. 1976. Nuclear DNA amounts in angiosperms. *Proceedings of the Royal Society of London* **274**:227–274.

Bennetzen JL, Jianxin MA, Devos KM. 2005. Mechanisms of recent genome size variation in flowering plants. *Annals of Botany* **95**: 127–132.

Brandham PE. 1971. The chromosomes of the Liliaceae II. Polyploidy and karyotype variation in the Aloineae. *Kew Bulletin* **25**: 381–399.

Brandham PE. 1983. Evolution in a stable chromosome system. In: Brandham PE, Bennet MD, eds. *Kew Chromosome Conference II*. London: George Allen & Un Win, 251–260.

Brandham PE, Bhandol PS. 1997. Chromosomal relationships between the genera *Amaryllis* and *Hippeastrum* (Amaryllidaceae). *Kew Bulletin* **52**:973–980.

Brandham PE, Doherty MJ. 1998. Genome size variation in the Aloaceae, an angiosperm family displaying karyotypic orthoselection. *Annals of Botany* **82**:67–73.

Di Rienzo JA, Casanoves F, Balzarini MG, Gonzalez L, Tablada M, Robledo CW. InfoStat versión 2012. Grupo InfoStat, FCA, Universidad Nacional de Córdoba, Argentina. http://www.infostat.com.ar.

Feldman M, Levy AA. 2005. Allopolyploidy a shaping force in the evolution of wheat genomes. *Cytogenetic and Genome Research* **109**:250–258.

Fisher RA. 1932. *Statistical methods for research workers*, 4th edn. Edinburgh: Oliver and Boyd.

Jones RN, Langdom T. 2013. The plant nucleus at war and peace: genome organization in the interphase nucleus. In: Leitch IJ, Greilhuber J, Dolezel J., Wendel JF, eds. *Plant genome diversity*, Vol. 2. New York: Springer, 13–31.

Kellogg EA, Bennetzen JL. 2004. The evolution of nuclear genome structure in seed plants. *American Journal of Botany* **91**: 170–1725.

Lakshmi N. 1980. Cytotaxonomical studies in eight genera of Amaryllidaceae. *Cytologia* **45**:663–773.

Leitch IJ, Bennett MD. 2004. Genome downsizing in polyploid plants. *Biological Journal of the Linnean Society* **82**:651–663.

Leitch IJ, Leitch AR. 2013. Genome size diversity and evolution in land plants. In: Leitch IJ, Greilhuber J, Dolezel J, Wendel JF, eds. *Plant genome diversity*, Vol. 2. Vienna: Springer, 307–322.

Leitch IJ, Chase MW, Bennett MD. 1998. Phylogenetic analysis of DNA C-value provides evidence for a small ancestral genome size in flowering plants. *Annals of Botany* **82**:85–94.

Levan A, Fredga K, Sandberg AA. 1964. Nomenclature for centromeric position on chromosomes. *Hereditas* **52**:201–220.

Ma XF, Gustafson JP. 2006. Timing and rate of genome variation in *Triticale* following allopolyploidization. *Genome* **49**:950–958.

Meerow AW, Charles LG, Qin-Bao L, Si-Lin Y. 2000. Phylogeny of the American Amaryllidaceae based on nrDNA ITS sequences. *Systematic Botany* **25**:708–726.

Naranjo CA. 1969. Cariotipo de nueve especies argentinas de *Rhodophiala, Hippeastrum, Zephyrantes, Habranthus* (Amaryllidaceae). *Kurtziana* **5**:67–87.

Naranjo CA, Andrada AB. 1975. El cariotipo fundamental del género *Hippeastrum* Herb. (Amaryllidaceae). *Darwiniana* **19**:556–582.

Naranjo CA, Poggio L. 1988. A comparison of karyotype, Ag-NOR bands and DNA contents in *Amaryllis* and *Hippeastrum* (Amaryllidaceae). *Kew Bulletin* **42**:317–325.

Naranjo CA, Ferrari MR, Palermo AM, Poggio L. 1998. Karyotype, DNA content and meiotic behaviour in five South American species of *Vicia* (Fabaceae). *Annals of Botany* **82**:757–764.

Navrátilová A, Neumann PA, Macas J. 2003. Karyotype analysis of four *Vicia* species using *in situ* hybridization with repetitive sequences. *Annals of Botany* **91**:921–926.

Neto EM. 1948. Numeros de cromossomos dos genero *Hippeastrum* Herb. *Boletim da Sociedade Brasileira de Agronomia* **8**:383–388.

Parida A, Raina SN, Narayan RKJ. 1990. Quantitative DNA variation between and within chromosome complements of *Vigna* species (Fabaceae). *Genetica* **82**:125–133.

Peruzzi L, Eroglu HE. 2013. Karyotype asymmetry: again, how to measure and what to measure? *Comparative Cytogenetics* **7**:1–9.

Peruzzi L, Leitch IJ, Caparelli KF. 2009. Chromosome diversity and evolution in Liliaceae. *Annals of Botany* **103**:459–475.

Poggio L, Burghardt AD, Hunziker JH. 1989. Nuclear DNA variation in diploid and polyploid taxa of *Larrea* (Zygophyllaceae). *Heredity* **63**:321–328.

Poggio L, González GE, Naranjo CA. 2007. Chromosome studies in *Hippeastrum* (Amaryllidaceae): variation in genome size. *Botanical Journal of the Linnean Society* **155**:171–178.

R Development Core Team. 2004. *R: a language and environment for statistical computing.* Vienna, Austria: R Foundation for Statistical Computing. http://www.R-project.org.

Romero Zarco C. 1986. A new method for estimating karyotype asymmetry. *Taxon* **35**:526–530.

Sato D. 1938. Karyotype alteration and phylogeny. IV. Karyotypes in Amaryllidaceae with special reference to the SAT-chromosome. *Cytologia* **9**:203–242.

Soltis DE, Soltis PS, Tate JA. 2003. Advances in the study of polyploidy since plant speciation. *New Phytologist* **161**:173–191.

Srivastava S, Lavania UC. 1991. Evolutionary DNA variation in *Papaver. Genome* **34**:763–768.

Tito CM, Poggio L, Naranjo CA. 1991. Cytogenetic studiesin genus *Zea* 3. DNA content and heterochromatin in species and hybrids. *Theoretical and Applied Genetics* **83**:58–64.

Vosa CG. 2005. On chromosome uniformity, bimodality and evolution in the tribe Aloineae (Asphodelaceae). *Caryologia* **581**: 83–85.

Weiss-Schneeweiss H, Schneeweiss GM. 2013. Karyotype diversity and evolutionary trends in angiosperms. In: Leitch IJ, Greilhuber J, Dolezel J, Wendel JF, eds. *Plant genome diversity*, Vol. 2. Vienna: Springer, 209–230.

White MJD. 1973. *Animal cytology and evolution.* Cambridge: Cambridge University Press, 961.

Zou QL, Quin ZX. 1994. The karyotype analysis of *Hippeastrum rutilum. Guihaia* **14**:37–38.

6

White-tailed deer are a biotic filter during community assembly, reducing species and phylogenetic diversity

Danielle R. Begley-Miller[1]*, Andrew L. Hipp[2], Bethany H. Brown[2], Marlene Hahn[2] and Thomas P. Rooney[3]

[1] Department of Ecosystem Science and Management, The Pennsylvania State University, University Park, PA 16802, USA
[2] The Morton Arboretum, Lisle, IL 60532, USA
[3] Department of Biological Sciences, Wright State University, Dayton, OH 45435, USA

Associate Editor: J. Hall Cushman

Abstract. Community assembly entails a filtering process, where species found in a local community are those that can pass through environmental (abiotic) and biotic filters and successfully compete. Previous research has demonstrated the ability of white-tailed deer (*Odocoileus virginianus*) to reduce species diversity and favour browse-tolerant plant communities. In this study, we expand on our previous work by investigating deer as a possible biotic filter altering local plant community assembly. We used replicated 23-year-old deer exclosures to experimentally assess the effects of deer on species diversity (H'), richness (SR), phylogenetic community structure and phylogenetic diversity in paired browsed (control) and unbrowsed (exclosed) plots. Additionally, we developed a deer-browsing susceptibility index (DBSI) to assess the vulnerability of local species to deer. Deer browsing caused a 12 % reduction in H' and 17 % reduction in SR, consistent with previous studies. Furthermore, browsing reduced phylogenetic diversity by 63 %, causing significant phylogenetic clustering. Overall, graminoids were the least vulnerable to deer browsing based on DBSI calculations. These findings demonstrate that deer are a significant driver of plant community assembly due to their role as a selective browser, or more generally, as a biotic filter. This study highlights the importance of knowledge about the plant tree of life in assessing the effects of biotic filters on plant communities. Application of such knowledge has considerable potential to advance our understanding of plant community assembly.

Keywords: Browsing; herbivory; phylogenetic clustering; phylogenetic community ecology; plant–animal interactions; species diversity.

Introduction

During the community assembly process—the formation of local communities from a regional species pool—most available species are 'filtered out' of local communities on the basis of genotypes, dispersal limitations or sets of traits that are least suited to a particular habitat (Keddy 1992; HilleRisLambers *et al.* 2012). The species found in a local community are those that can pass through environmental (abiotic) and biotic filters and successfully compete. Herbivores may act as a biotic filter (Augustine and McNaughton 1998; Suzuki *et al.* 2013) by preventing a species that is otherwise well adapted to the abiotic conditions in a local community from persisting over time. Herbivory is thus expected to produce local communities consisting of species with traits that confer resistance to or tolerance of herbivory. If herbivory resistance or

* Corresponding author's e-mail address: begley.danielle@gmail.com

tolerance evolves on the plant tree of life (i.e. if it is phylogenetically heritable), we would expect herbivory to alter phylogenetic diversity within communities. Such changes are important from both a theoretical and applied perspective: understanding shifts in the phylogenetic structure of plant communities in response to experimental removal from herbivory can help elucidate how and to what extent herbivory is a biotic filter shaping plant communities. Phylogenetic diversity needs to be better investigated as a tool for more targeted conservation efforts and for understanding the maintenance of biodiversity in conservation areas (Faith 1992).

White-tailed deer (*Odocoileus virginianus*) overabundance is a conservation issue throughout parts of eastern North America, because browsing alters community structure, composition and diversity of forests (Horsley *et al.* 2003; Côté *et al.* 2004). Changes in community composition reflect the selective browsing strategy of white-tailed deer. Deer consume palatable, nutrient-rich species when available and lesser quality browse when high-quality sources are depleted (Beals *et al.* 1960; Balgooyen and Waller 1995; Waller and Alverson 1997; Côté *et al.* 2004). Unpalatable, browse-tolerant and non-preferred plant species are commonly observed in heavily browsed areas (Tremblay *et al.* 2006; Rooney 2009; Martin *et al.* 2010; Royo *et al.* 2010; Goetsch *et al.* 2011). It is thus not surprising that deer browsing has been linked to both plant population extirpations and reductions in forest understorey species richness (SR) (Rooney and Dress 1997; Horsley *et al.* 2003; Rooney *et al.* 2004; Martin *et al.* 2010).

Here, we examine a community in which deer have been experimentally removed for two decades and investigate how this has affected the phylogenetic community structure. In this study, we surveyed vascular plant taxa in successive years in paired control and deer exclosure plots. We had three main goals of our analysis. We first determined the effects of deer browsing on community structure by comparing both species and phylogenetic diversity in control and exclosed areas. We were particularly interested in whether phylogenetic data provided additional information not contained in species diversity measures (Vellend *et al.* 2011), and whether deer browsing altered the degree of phylogenetic relatedness within each local community. We next tested for phylogenetic patterns in two categories of traits associated with vulnerability to deer browsing: browse type (Rooney 2009) and pollination mode (Rooney *et al.* 2004). We then developed a deer-browsing susceptibility index (DBSI) to quantitatively separate vulnerable from non-vulnerable species at our study site. By applying the tools of phylogenetic diversity to an applied study of the conservation impacts of deer herbivory, we refine our understanding of how deer serve as a biotic filter in plant community assembly.

Methods

Study site

This study was conducted at the 2500 ha Dairymen's Club in Wisconsin, USA (46.15°N, 89.68°W). The site is privately owned and managed for conservation, recreation and scientific research. The climate is continental, with average yearly precipitation between 550 and 780 mm and a mean temperature range of $-20\,°C$ in winter to $32\,°C$ in summer (Rooney *et al.* 2004). The landscape is heterogeneous, including lakes, sedge meadows and mixed conifer–hardwood forest. Dairymen's Club purchased the property in 1925, and all hunting has been prohibited since then (Rooney 2006). In the absence of hunting, the deer population grew quickly. Growth was further fuelled by a supplemental deer-feeding programme from 1950 to 2000. In feeding areas, local concentrations exceeded 100 deer km^{-2}. Forests are the predominant land cover type in the area, and dominant canopy trees include *Acer saccharum*, *Tsuga canadensis* and *Betula alleghaniensis*.

Deer exclosure experiment

In 1990, four deer exclosures were constructed within 500 m of feeding areas on the property to protect vulnerable plant species from continuous browsing. These long-term exclosures are 1.8 m tall, range in size from 196 to 720 m^2 (Rooney 2009) and are constructed of 2.5 × 7.5 cm wire mesh. The deer densities at this site throughout the 20th century were much higher than were found throughout the northern Wisconsin region (Rooney 2006). To understand the long-term effects of these prolonged deer densities on forest understorey plant communities, three permanent ground-vegetation transects were established inside and outside each exclosure in 2006 (Rooney 2009). Each transect totals 10 m in length and extends 5 m into an adjacent unfenced area (control) and 5 m into its paired exclosure (separated by the exclosure fence). The unique history of deer population dynamics at this study site, combined with the construction of these exclosures decades ago, allows us to better assess the long-term effects of deer as a driver of plant community assembly.

Vegetation data collection

Per cent cover data were collected from the permanent transects during the first or second week of June each year from 2006 to 2012, except 2007. The line-intercept method was used to obtain cover data. All plants ≤ 1 m tall were identified to species. Each exclosure and control area was sampled equally, regardless of size of exclosure. Along each transect, a measuring tape was laid on the ground beneath the vegetation. If any part of the organ of a plant (i.e. leaf, stem, flower) intercepted the transect the plant was identified to species, and length (to the nearest cm) of the tape covered was recorded. Per cent cover for the ith species in a plot was calculated as $(\Sigma n_i)/1500$, where n is the length of the tape covered by each occurrence of species i (to the nearest cm) along that transect. The denominator is the length in centimetres of three 5 m transects. Because multiple species can intercept the same transect segment at different heights, the total per cent cover can exceed 100 %. Per cent cover of the ith species across all four plots within a treatment is $(\Sigma n_i)/6000$ (Rooney 2009).

Diversity metrics and analysis

Phylogeny. A rooted phylogenetic tree was created using DNA sequences of three gene regions. The tree is site specific, in that we only obtained gene sequences from species found at the study site. We did not sequence species from the regional species pool not found in our study plots. Of these DNA regions, one (the 5′ end of the chloroplast *rbc*L gene) is highly conserved across angiosperms. It is a widely used DNA barcoding gene (Kress and Erickson 2007) that has been the workhorse of broad-scale phylogenetics across higher plants (e.g. Chase *et al.* 1993). This gene aligns unambiguously across green plants and provides solid information on genetic relationships across the samples we studied. The other two DNA regions are more rapidly evolving and used widely in fine-scale phylogenetics in flowering plants: the chloroplast intergenic spacer between the 3′ end of the *trn*L exon and the 5′ end of the *trn*F exon (hereafter in the paper referred to as the *trn*L–*trn*F region) (Taberlet *et al.* 1991), and nuclear ribosomal internal transcribed spacer regions (ITS1 and ITS2), including the embedded 5.8S gene (hereafter in the paper referred to collectively as the ITS region) (Baldwin *et al.* 1995). These genes, however, were not fine-scale enough to determine intraspecific differences between individuals. All sequences were used to determine genetic differences at the species level, and to confirm species identification in cases of uncertainty. Gene sequences for 18 of the 36 species at our study site were obtained from NCBI GenBank (Benson *et al.* 2010) **[see Supporting Information, bolded]**, and the other 18 were sequenced from material collected during the 2012 field season **[see Supporting Information, italicized]**. Sequences for this study used the following PCR primers: ITS-I (Urbatsch *et al.* 2000) and ITS-4 (White *et al.* 1990); *rbc*La-F (Levin *et al.* 2003) and *rbc*La-R

(Kress *et al.* 2009) and for the *trn*L–*trn*F intergenic spacer, Taberlet *et al.* (1991) primers e and f. PCR reactions were conducted as in Hipp *et al.* (2006), with the following cycling regimens: ITS: 94.0° for 5:00; 35 cycles of: 94.0° for 0:30, 48.0° for 1:00, 72.0° for 1:30; 72.0° for 7:00. *rbc*L: 94.0° for 5:00; 35 cycles of: 94.0° for 0:30, 52.0° for 1:00, 72.0° for 1:30; 72.0° for 7:00. *trn*L–*trn*F: 95.0° for 3:00; 50 cycles of: 95.0° for 0:20, 45.0° for 0:30, 52.0° for 4:00; 72.0° for 7:00. PCR products were cycle sequenced using BigDye reaction kits and the PCR primers, and unincorporated dye terminators were removed using CleanSEQ magnetic beads (Agencourt, Beckman Coulter). PCR was conducted at The Morton Arboretum, and sequencing was conducted on an ABI 3730 capillary sequencer in The Pritzker Lab of the Field Museum. Double-stranded DNA sequence contigs were cleaned manually in Sequencher 3.0 (Gene Codes Corporation, Ann Arbor, MI, USA) and exported as text for analysis.

DNA sequences were aligned using Muscle v. 3.8.31 (Edgar 2004a, b) and manually adjusted. Data for the *rbc*L region were globally aligned without ambiguities, including all taxa. Global alignment of all taxa simultaneously for the ITS and *trn*L–*trn*F regions, however, produced alignments that were riddled with ambiguities. To address this, data matrices were first aligned by APGIII order. Then, profile-to-profile alignments were utilized, in which the alignment within each order is held fixed but nucleotide positions are allowed to shift among orders. Profile-to-profile alignments were conducted among most closely related orders, moving progressively up the tips to the root of the green plants tree of life, using the Angiosperm Phylogeny Group tree (APG III 2009) as updated in APG Web (Stevens 2001 onwards).

Multiple alignments were then concatenated and analysed under likelihood in RAxML v.7.2.6 (Stamatakis 2006), using the GTRCAT nucleotide substitution model, using the multithreading option on a 4-core Intel processor (Stamatakis and Ott 2008). Analysis was conducted using 200 bootstrap replicates. Branch lengths were optimized on the resulting tree using penalized likelihood (Sanderson 2002) as implemented in the ape package (Paradis *et al.* 2004) of R v.2.13.1 (R Development Core Team 2011). Smoothing parameters from 10 to 0.001 were tried and found to have no appreciable effect on the branch lengths on the tree. The reported tree (Fig. 1) utilizes a smoothing parameter of 1.0. All DNA sequences generated for this study are deposited in NCBI GenBank **[see Supporting Information]**.

Phylogenetic diversity. Using the site-specific phylogenetic tree, we analysed phylogenetic signal, diversity and patterns of community structure. To test for phylogenetic

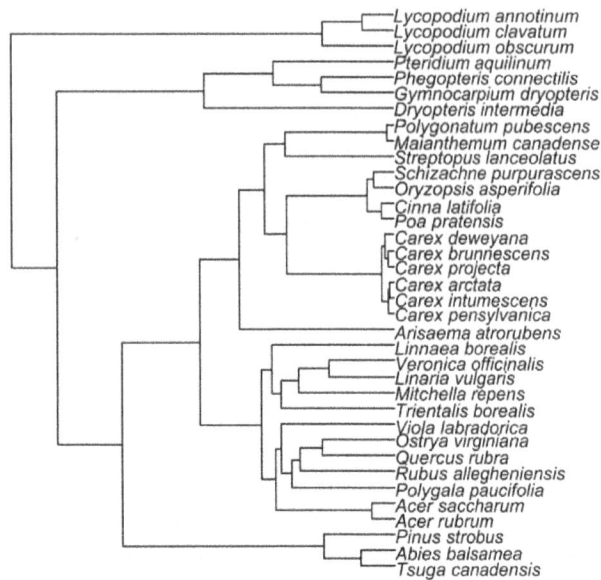

Figure 1. Rooted phylogenetic tree (using ITS, *rbcL* and *trnL* – *trnF* regions) of all species present at the study site. Phylogeny was estimated using maximum likelihood in RAxML, as described in the Methods.

signal we used two different approaches with two different character traits: browse type (woody, broadleaf forb, fern, grass, sedge or lycopod) and pollination mode (biotic or abiotic). Based on previous research, we predicted species persisting in browsed areas would be abiotically pollinated grasses and sedges (graminoids) (Rooney *et al.* 2004; Rooney 2009). We estimated the phylogenetic signal of pollination mode (Rooney *et al.* 2004), a binary trait, using Fritz and Purvis' (2010) D statistic in the caper package in R (Orme 2013). A value of $D = 0$ indicates a trait consistent with a Brownian threshold model, while a value of $D = 1$ indicates a trait following a random distribution. Values can fall outside of this range, with those significantly less than 0 indicating high phylogenetic conservatism, and values significantly higher than 1 indicating phylogenetic overdispersion. Significance is assessed by comparing observed trait distributions with expected distributions simulated under a Brownian motion model or by random permutation of the original tip states. For the multistate trait browse type (Rooney 2009), we used Mesquite (Maddison and Maddison 2011) to calculate the minimum number of character transitions needed to observe trait distribution under maximum parsimony on our site-specific maximum likelihood tree (Fig. 1), and compared the observed value with a null distribution calculated over 1000 permutations of the tip states. The Type I error rate (P) value was estimated as the number of permutations

for which the parsimony score (minimum number of character steps) was less than or equal to the parsimony score for the browse type data.

As a general metric of phylogenetic diversity we used mean pairwise phylogenetic distance (MPD) as implemented in the R package picante (Kembel *et al.* 2013). We chose MPD because it is less correlated with SR than Faith's (1992) phylogenetic diversity (Yessoufou *et al.* 2013) and is more sensitive to changes between distantly related taxa than mean nearest taxon distance. To test for phylogenetic patterns of community structure we used the net relatedness index (NRI, also in picante), which compares the phylogenetic structure of our measured communities with randomly permuted trees. Specifically, we tested if control or exclosure communities were phylogenetically autocorrelated (clustered), showing less phylogenetic diversity than expected at random. Large positive NRI values indicate clustering, while a large absolute value derived from a negative NRI indicates phylogenetic overdispersion.

Species diversity. Species richness was defined as the total number of species encountered along transects at each plot. Species diversity indices were calculated using the Shannon – Wiener (Shannon and Weaver 1949) standard diversity metric (H'). We used H' because it is weighted for abundance and is less correlated with SR than Simpson's diversity index (D). Per cent cover data for each species from each control and exclosure plot was used for abundance.

Statistical analysis

We compared MPD, SR and H' of controls and exclosures using a linear mixed effects model with site as a random effect, and treatment, year and treatment/year interaction as fixed effects. The random effect of site was retained within the model if $P < 0.10$. Significant fixed effects were reported at the $P < 0.05$ level. For NRI, we pooled all sites together by treatment and analysed by year, comparing phylogenetic distance of communities across the most parsimonious tree with simulated ones with shuffled tip labels over 999 permutations. This shuffling simulates a null expectation of no effect of phylogeny. Resulting P values were used to identify which years and which exclosures were most influenced by phylogenetic structure.

Deer-browsing susceptibility index

To evaluate plant species susceptibility to deer browsing and identify those reliant on exclosures for persistence, we developed a DBSI. It compares relative cover of plant species inside and outside exclosures, and scales from 0 to 1 for each species. The DBSI value for each species represents the fraction of that species' per cent cover inside versus outside the exclosure. A score of 0 indicates a species is only found outside exclosures, while a score of 1 indicates a species is only found inside exclosures. To exclude rare species that would skew DBSI calculations, we only included species present in more than two exclosures or controls in two or more years. Deer-browsing susceptibility index was calculated separately for each species in each year using the following equation:

$$\text{DBSI} = \left(\sum C_e\right) / \left(\sum C_e + \sum C_c\right)$$

where C_e is the per cent cover inside the exclosure, C_c is the per cent cover outside the exclosure and $\Sigma(s = 1,2,...,n)$ is the sum of C_e (or C_c) across all exclosures (or controls). In the results, we report a single mean DBSI value across all years for each species.

Results

Species and phylogenetic diversity

Species richness and H' responded similarly to browsing from deer. Browsing caused a 17 % reduction in SR ($F_{(1,43)} = 5.81$, $P = 0.02$; Fig. 2A) and 12 % reduction in H' ($F_{(1,42.99)} = 4.43$, $P = 0.04$; Fig. 2B). Neither metric showed a year (SR: $F_{(5,38)} = 0.189$, $P = 0.96$; H':

$F_{(5,37.99)} = 0.658$, $P = 0.66$), or treatment/year interaction effect (SR: $F_{(5,33)} = 0.913$, $P = 0.48$; H': $F_{(5,32.99)} = 0.468$, $P = 0.7971$). Site was significant in both models (SR: $\chi^2_{(1,48)} = 11.48$, $P = 0.0007$; H': $\chi^2_{(1,48)} = 7.52$, $P = 0.0061$).

Browsing significantly reduced phylogenetic diversity (MPD) by 63 % ($F_{(1,42.98)} = 42.36$, $P < 0.00001$; Fig. 2C). The effect of year ($F_{(5,37.98)} = 0.24$, $P = 0.94$) and the interaction of year and treatment ($F_{(5,32.98)} = 0.46$, $P = 0.80$), however, were not significant in browsed or unbrowsed plots. Site was a significant random effect in the model at the $P < 0.10$ level ($\chi^2_{(1,48)} = 2.72$, $P = 0.099$). Analysis of NRI showed significantly higher relatedness in browsed areas than expected by chance in each year sampled (range: $+1.53$ to $+2.51$, $P < 0.05$; Fig. 3). Exclosure areas did not show structured phylogenetic response (range: -0.60 to $+0.01$, $P > 0.05$).

Phylogenetic signal

Pollination mode exhibits clustering ($D = -0.672$; Table 1) relative to a random null model (two-tailed $P < 0.001$; 1000 permutations of the tip states). Clustering is also stronger than expected under Brownian motion null distribution, but not significantly so (two-tailed $P = 0.14$; 1000 Brownian motion simulations of the tip states under a threshold model; Fritz and Purvis 2010). Browse type (woody, broadleaf herb, fern, grass, sedge or lycopod) also showed significant phylogenetic clustering relative to a phylogenetically neutral null model (Table 1). The number of evolutionary steps in the maximum parsimony reconstruction of browse type was 8,

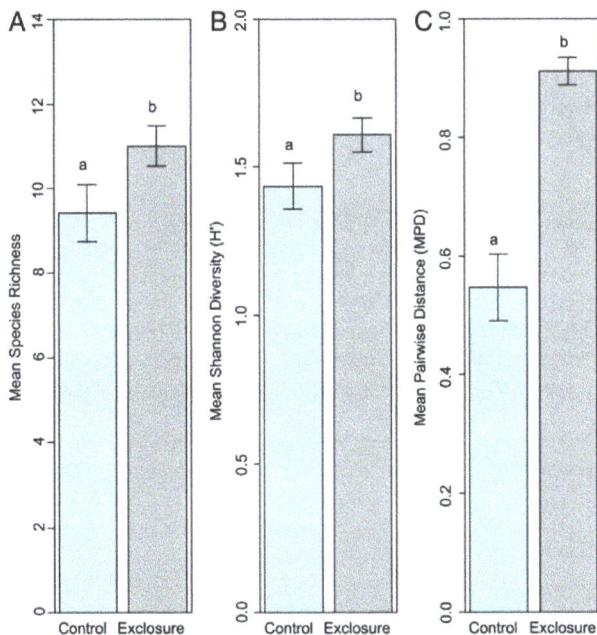

Figure 2. Mean species richness (A), Shannon–Weiner diversity (B) and mean pairwise phylogenetic distance (C) in exclosure (grey bars) and control (blue bars) areas across all years (2006–2012, excluding 2007). Different letters indicate statistical significance between groups at the $P < 0.05$ level, as tested using ANOVA in a linear mixed effects model. Error bars are ± 1 SE.

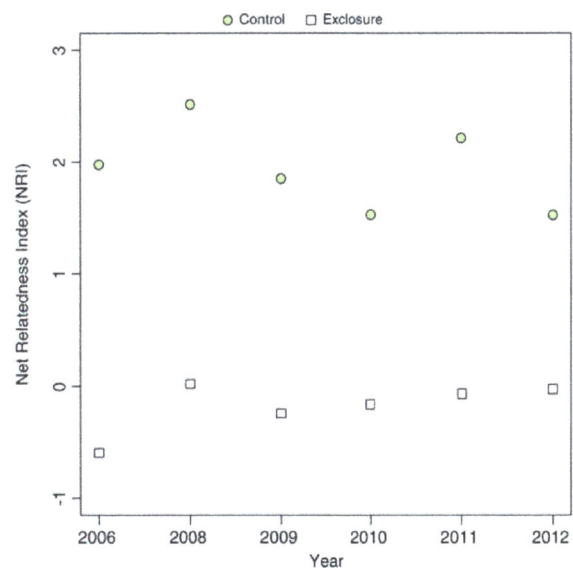

Figure 3. Net relatedness index of exclosure areas (squares) and control areas (circles) from 2006 to 2012 (excluding 2007). Points highlighted in green indicate NRI values significantly greater than expected by chance ($P < 0.05$, based on 999 random permutations of the tip states on the phylogeny).

while the mean number of evolutionary steps in 1000 simulated trees with permutated tips was 22.1 (95 % CI [19, 25], $P < 0.001$).

Deer-browsing susceptibility index

The two most deer-browse-susceptible species (*Polygala paucifolia* and *T. canadensis*) were found exclusively inside exclosures (DBSI = 1.0; Fig. 4). Both species represent very different growth patterns, as *P. paucifolia* is a perennial broadleaf herb and *T. canadensis* is an evergreen woody-browse species. All species in the low susceptibility category were graminoids or club mosses, with *Schizachne purpurascens* being the least susceptible of all species analysed (DBSI = 0.02; Fig. 4).

Discussion

Species and phylogenetic diversity

Excluding deer for two decades significantly increased species diversity, richness and phylogenetic diversity of plant communities. These results are consistent with previous studies documenting the loss of species diversity of plant communities due to selective browsing by deer (Gill and Beardall 2001; Horsley *et al.* 2003; Rooney *et al.* 2004; Suzuki *et al.* 2013). These results also reveal a substantial loss of phylogenetic diversity in browsed plots. If we set the basal node of our ultrametric tree to a depth of 432 million years, based on a recent estimate of the age of the Tracheophyta (Smith *et al.* 2010), the observed loss of only a few branches represents an average difference in phylogenetic diversity between browsed and unbrowsed plots of 372.5 million years.

We found significant changes to the phylogenetic community structure of browsed plant communities. These findings demonstrate the ability of deer to shape northern plant communities by filtering out species that share a suite of phylogenetically heritable traits contributing to their browse susceptibility. This conclusion rests on two findings. First, pollination mode and browse type

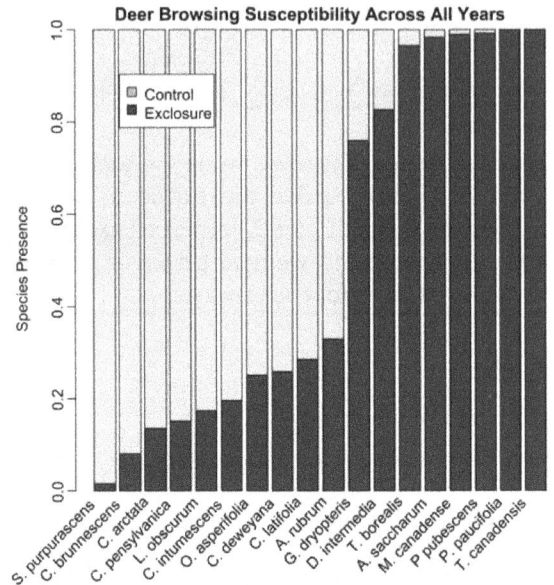

Figure 4. A stacked bar graph representing the mean calculated DBSI of each species present in more than one exclosure in two or more years. Light grey bars indicate the proportion of the species present in control areas, while dark grey bars indicate the proportion of the species present in exclosures. Species are ordered left to right from the least susceptible (*S. purpurascens*) to most susceptible (*T. canadensis*).

Table 1. Tests for phylogenetic signal for pollination mode and browse type using Fritz and Purvis' *D* statistic and phylogenetic autocorrelation, respectively. To test for phylogenetic signal for the binary trait pollination mode, we used Fritz and Purvis' *D* statistic. For this test, significance is assessed by comparing observed trait distributions with expected distributions simulated under a Brownian motion model or by random permutation of the original tip states. For the multistate trait browse type we calculated the maximum parsimony of browse type on our site-specific maximum likelihood tree, and compared the observed value with a null distribution generated by 1000 permutations of the tip states. The Type I error rate (*P*) value for this test was estimated as the minimum number of simulated trees with parsimony reconstruction of less than or equal to the number of steps in the observed reconstruction. Significant *P* values (*P* < 0.05) for both tests indicate that traits are phylogenetically clustered (i.e. not phylogenetically independent).

Parameter	Phylogenetic autocorrelation	Parameter	Fritz and Purvis' *D* statistic
Character type	Categorical (browse type)	Character type	Binary (pollination mode)
Number of permutations	1000	Number of permutations	1000
Difference in no. of evolutionary steps (MP−mean shuffled)	−14.1	*D* statistic	−0.674
95 % confidence interval	LCI: 19 UCI: 25	Probability of *D* given Brownian phylogenetic structure	0.144
P value	<0.001	Probability of *D* given random phylogenetic structure	<0.001

(two traits linked to plant-browsing susceptibility) exhibit significant phylogenetic signal. Deer are thought to visually cue in on biotically pollinated plants due to their conspicuous flowers (Wiegmann and Waller 2006), and since ungulate browse type classes are taxonomically clustered, we expected this trait to exhibit a strong phylogenetic signal. As we observe in this study, many other studies report increases in the relative abundances of the grass and sedge browse types following increases in deer density (Halls and Crawford 1960; Kie *et al.* 1980; Horsley *et al.* 2003; Rooney *et al.* 2004). We did not, however, find an increase in the relative abundance of the fern browse type in response to deer browsing, contrasting findings elsewhere in North America and New Zealand (Royo and Carson 2006).

Second, plant community phylogenetic patterns showed significant phylogenetic clustering in browsed plots across all years, indicating species in these communities were much more related than expected by chance. Unbrowsed plots showed no phylogenetic pattern relative to the suite of species observed. We interpret phylogenetic clustering in browsed communities as arising from a biotic filtering process during community assembly. Because each pair of browsed and unbrowsed plots is spatially adjacent in homogeneous environments, it is unlikely that abiotic environmental filtering accounts for phylogenetic clustering in browsed plots (Mayfield and Levine 2010). It is more likely that any species from our species pool can establish in browsed plots, but that many are competitively excluded due to fitness inequalities arising from their low browse tolerance or resistance (Chesson 2000; Mayfield and Levine 2010). Deer browsing favours species from the flowering plant clade Poales, represented in our study by the Poaceae and Cyperaceae, thus filtering out much phylogenetic diversity. This effect is reflected both in the phylogenetic patterns of trait conservatism and in our NRI values. The consequence is the loss of representative evolutionary history in the forest understorey layer.

Deer-browsing susceptibility index

In this study, species in the low deer-browsing susceptibility category exhibited characteristics of tolerant or resistant species. Eight species were of either the grass or sedge (graminoid) browse type with abiotic pollination mode. Collectively they are considered browse tolerant, because they have basal meristems and are able to regrow following browsing (Coughenour 1985). As expected, species categorized as broadleaf herbs and woody browse types were classified as susceptible, and most of these species exhibit biotic pollination. The more-susceptible broadleaf herbaceous plant species are

broadly distributed through the phylogeny, consistent with our NRI values.

Generally, our findings are consistent with other studies showing the ability of deer to promote browse tolerant and unpalatable species (Horsley *et al.* 2003; Côté *et al.* 2004; Royo and Carson 2006). In this study, however, DBSI is based on species presence inside and outside exclosures. It does not directly measure a species' susceptibility to deer browsing based on chemical composition or deer preference. As a result, DBSI could reflect a species' response to certain environmental conditions (e.g. shading and competition). To the extent that deer exclusion creates a more favourable microhabitat for species, DBSI can reflect differences not directly due to deer herbivory.

Conclusions

In our study area, deer herbivory acts as a biotic filter. Deer reduce species diversity, SR and phylogenetic diversity by filtering out species that have browse-susceptible traits. Deer have suppressed browse-intolerant species and promoted the coexistence of closely related browse-tolerant species. Phylogenetic diversity indices have been previously used as a way to assess effects of disturbance (Cavender-Bares and Reich 2012) or environmental gradients (Pellissier *et al.* 2012) on community structure. Our study is the first to identify white-tailed deer as a significant driver of plant community assembly using phylogenetic methods.

We gain two fundamental insights from applying the tools of phylogenetic community ecology to this classic study system. First, evolutionary history shapes plant responses to herbivory. The traits we measured—and, presumably a host of other unmeasured traits that affect browse-susceptibility—have high phylogenetic heritability. Second, the phylogenetic heritability of these traits shapes the effect of browsing on phylogenetic diversity and community structure. Thus, as has been shown in other studies of herbivory (e.g. Pearse and Hipp 2009, 2012), phylogenetic heritage integrates over a large number of traits and may thus be a better predictor of herbivore susceptibility than even suites of measured traits. There are hundreds of published studies that have used exclosures to examine the influence of deer herbivory on plant community composition. A re-analysis of data from these studies using the framework of phylogenetic community ecology may provide us even stronger evidence about the utility of phylogeny for predicting plant community responses to management and disturbance. It is our expectation that the resulting increased ability to identify species at risk will enable more effective conservation management and further advances in our understanding of plant community assembly.

Sources of Funding

Our work was funded by Dairymen's Inc. (USA) and a Research Incentive Fund from the Wright State University.

Contributions by the Authors

D.R.B.-M. collected data, analysed data, drafted first version of the manuscript. A.L.H. conducted analysis and collaborated in interpretation of results and manuscript preparation. B.H.B. and M.H. conducted DNA sequencing and phylogenetic analysis and helped with manuscript preparation. T.P.R. secured funding, collected data, helped with data analysis and contributed to writing the manuscript.

Acknowledgements

We thank the Dairymen's Club for building and maintaining deer exclosures. Additionally, we thank Dan Larkin for providing helpful feedback on an earlier draft of this manuscript, and two anonymous reviewers for their suggestions to improve the quality of this publication.

Accession Numbers

All novel sequences for nucleic acids have been submitted and accepted into GenBank as of 18 December 2013. A table of accession numbers for these sequences is included in the supporting information section of this manuscript.

Supporting Information

The following **Supporting Information** is available in the online version of this article –

File 1. Accession numbers as approved by GenBank on 18 December 2013. Herbarium voucher numbers pending. NA indicates the gene was not sequenced, bold numbers indicate the gene was obtained from GenBank and numbers in italics indicate the gene was sequenced by The Morton Arboretum.

Literature Cited

APG III. 2009. An update of the Angiosperm Phylogeny Group classification for the orders and families of flowering plants: APG III. *Botanical Journal of the Linnean Society* **161**:105–121.

Augustine D, McNaughton S. 1998. Ungulate effects on the functional species composition of plant communities: herbivore selectivity and plant tolerance. *The Journal of Wildlife Management* **62**: 1165–1183.

Baldwin B, Sanderson M, Porter J, Wojciechowski M, Campbell C, Donoghue M. 1995. The ITS region of nuclear ribosomal DNA: a valuable source of evidence on angiosperm phylogeny. *Annals of the Missouri Botanical Garden* **82**:247–277.

Balgooyen C, Waller D. 1995. The use of *Clintonia borealis* and other indicators to gauge impacts of white-tailed deer on plant communities in Northern Wisconsin, USA. *Natural Areas Journal* **15**:308–318.

Beals E, Cottam G, Vogl R. 1960. Influence of deer on vegetation of the Apostle Islands, Wisconsin. *The Journal of Wildlife Management* **24**:68–80.

Benson D, Karsch-Mizrachi I, Lipman D, Ostell J, Sayers E. 2010. GenBank. *Nucleic Acids Research* **38**:D46–D51.

Cavender-Bares J, Reich P. 2012. Shocks to the system: community assembly of the oak savanna in a 40-year fire frequency experiment. *Ecology* **93**:S52–S69.

Chase M, Soltis D, Olmstead R, Morgan D, Les D, Mishler B, Duvall M, Price R, Hills H, Qui Y, Kron K, Rettig J, Conti E, Palmer J, Manhart J, Sytsma K, Michael H, Kress W, Karol K, Clark W, Hendren M, Gaut B, Jansen R, Kim K, Wimpee C, Smith J, Furnier G, Smith J, Strauss S, Xiang Q, Plunkett G, Soltis P, Swensen S, Williams S, Gadek P, Quinn C, Eguiarte L, Golenberg E, Learn G, Graham S, Barrett S, Dayanandan S, Albert V. 1993. Phylogenetics of seed plants: an analysis of nucleotide sequences from the plastid gene rbcL. *Annals of the Missouri Botanical Garden* **80**:528–580.

Chesson P. 2000. Mechanisms of maintenance of species diversity. *Annual Review of Ecology and Systematics* **31**:343–366.

Côté S, Rooney T, Tremblay J, Dussault C, Waller D. 2004. Ecological impacts of deer overabundance. *Annual Review of Ecology, Evolution, and Systematics* **35**:113–147.

Coughenour M. 1985. Graminoid responses to grazing by large herbivores: adaptations, exaptations, and interacting processes. *Annals of the Missouri Botanical Garden* **72**:852–863.

Edgar R. 2004a. MUSCLE: a multiple sequence alignment method with reduced time and space complexity. *BMC Bioinformatics* **5**:113.

Edgar R. 2004b. MUSCLE: multiple sequence alignment with high accuracy and high throughput. *Nucleic Acids Research* **32**:1792–1797.

Faith D. 1992. Conservation evaluation and phylogenetic diversity. *Biological Conservation* **61**:1–10.

Fritz S, Purvis A. 2010. Selective in mammalian extinction risk and threat types: a new measure of phylogenetic signal strength in binary traits. *Conservation Biology* **24**:1042–1105.

Gill R, Beardall V. 2001. The impact of deer on woodlands: the effects of browsing and seed dispersal on vegetation structure and composition. *Forestry* **74**:209–218.

Goetsch C, Wigg J, Royo A, Ristau T, Carson W. 2011. Chronic over browsing and biodiversity collapse in a forest understory in Pennsylvania: results from a 60 year-old deer exclusion plot. *Journal of the Torrey Botanical Society* **138**:220–224.

Halls L, Crawford H. 1960. Deer-forest habitat relationships in North Arkansas. *Journal of Wildlife Management* **24**:387–395.

HilleRisLambers J, Adler P, Harpole W, Levine J, Mayfield M. 2012. Rethinking community assembly through the lens of coexistence theory. *Annual Review of Ecology, Evolution and Systematics* **43**: 227–248.

Hipp A, Reznicek A, Rothrock P, Weber J. 2006. Phylogeny and classification of *Carex* section *Ovales* (Cyperaceae). *International Journal of Plant Sciences* **167**:1029–1048.

Horsley S, Stout S, DeCalesta D. 2003. White-tailed deer impact on the vegetation dynamics of a Northern Hardwood forest. *Ecological Applications* **13**:98–118.

Keddy P. 1992. Assembly and response rules: two goals for predictive community ecology. *Journal of Vegetation Science* **3**:157–164.

Kembel S, Ackerly D, Blomberg S, Cornwell W, Cowan P, Helmus M,

Morlon H, Webb C. 2013. Picante: R tools for integrating phylogenies and ecology. cran.r-project.org/web/packages/picante/picante.pdf.

Kie J, Drawe D, Scott G. 1980. Changes in diet and nutrition with increased herd size in Texas white-tailed deer. *Journal of Range Management* 33:28–34.

Kress W, Erickson D. 2007. A two-locus global DNA barcode for land plants: the coding rbcL gene complements the non-coding trnH-psbA spacer region. *PLoS ONE* 2:e508.

Kress W, Erickson D, Jones F, Swenson N, Perez R, Sanjur O, Bermingham E. 2009. Plant DNA barcodes and a community phylogeny of a tropical forest dynamics plot in Panama. *Proceedings of the National Academy of Sciences of the USA* 106:18621–18626.

Levin R, Wagner W, Hoch P, Nepokroeff M, Pires J, Zimmer E, Sytsma K. 2003. Family-level relationships of Onagraceae based on chloroplast rbcL and ndhF data. *American Journal of Botany* 90:107–115.

Maddison W, Maddison D. 2011. *Mesquite: a modular system for evolutionary analysis*. Version 2.75 http://mesquiteproject.org.

Martin J, Stockton S, Allombert S, Gaston A. 2010. Top-down and bottom-up consequences of unchecked ungulate browsing on plant and animal diversity in temperate forests: lessons from a deer introduction. *Biological Invasions* 12:253–271.

Mayfield M, Levine J. 2010. Opposing effects of competitive exclusion on the phylogenetic structure of communities. *Ecology Letters* 13:1085–1093.

Orme D. 2013. The caper package: comparative analysis of phylogenetics and evolution in R. http://cran.r-project.org/web/packages/caper/caper.pdf.

Paradis E, Claude J, Strimmer K. 2004. APE: analyses of phylogenetics and evolution in R language. *Bioinformatics* 20:289–290.

Pearse IS, Hipp AL. 2009. Phylogenetic and trait similarity to a native species predict herbivory on non-native oaks. *Proceedings of the National Academy of Sciences of the USA* 106:18097–18102.

Pearse IS, Hipp AL. 2012. Global patterns of leaf defenses in oak species. *Evolution* 66:2272–2286.

Pellissier L, Alvarez N, Espíndola A, Pottier J, Dubuis A, Pradervand J, Guisan A. 2012. Phylogenetic alpha and beta diversities of butterfly communities correlate with climate in the western Swiss Alps. *Ecography* 35:001–010.

R Development Core Team. 2011. *R: A language and environment for statistical computing*. Vienna, Austria: R Foundation for Statistical Computing. ISBN 3-900051-07-0. http://www.R-project.org/.

Rooney T. 2006. Deer density reduction without a 12-Gauge shotgun (Wisconsin). *Ecological Restoration* 24:205–206.

Rooney T. 2009. High white-tailed deer densities benefit graminoids and contribute to biotic homogenization of forest ground-layer vegetation. *Plant Ecology* 202:103–111.

Rooney T, Dress W. 1997. Species loss over sixty-six years in the ground layer vegetation of heart's content, an old-growth forest in Pennsylvania, USA. *Natural Areas Journal* 17:297–305.

Rooney T, Wiegmann S, Rogers D, Waller D. 2004. Biotic impoverishment and homogenization in unfragmented forest understory communities. *Conservation Biology* 18:787–798.

Royo A, Carson W. 2006. On the formation of dense understory layers in forests worldwide: consequences and implications for forest dynamics, biodiversity, and succession. *Canadian Journal of Forest Research* 36:1345–1362.

Royo A, Collins R, Adams M, Kirschbaum C, Carson W. 2010. Pervasive interactions between ungulate browsers and disturbance regimes promote temperate forest herbaceous diversity. *Ecology* 91:93–105.

Sanderson M. 2002. Estimating absolute rates of molecular evolution and divergence times: a penalized likelihood approach. *Molecular Biology and Evolution* 19:101–109.

Shannon C, Weaver W. 1949. *The mathematical theory of communication*. Urbana: The University of Illinois Press.

Smith S, Beaulieu J, Donoghue M. 2010. An uncorrelated relaxed-clock analysis suggests an earlier origin for flowering plants. *Proceedings of the National Academy of Sciences of the USA* 107:5897–5902.

Stamatakis A. 2006. RAxML-VI-HPC: maximum likelihood-based phylogenetic analyses with thousands of taxa and mixed models. *Bioinformatics* 22:2688–2690.

Stamatakis A, Ott M. 2008. Efficient computation of the phylogenetic likelihood function on multi-gene alignments and multi-core architectures. *Philosophical Transactions of the Royal Society B: Biological Sciences* 363:3977–3984.

Stevens P. 2001. onwards. Angiosperm Phylogeny Website, Version 12, July 2012 [and more or less continuously updated since]. http://www.mobot.org/MOBOT/research/APweb/ (August 2012).

Suzuki M, Miyashita T, Kabaya K, Ochiai K, Asada M, Kikvidze Z. 2013. Deer herbivory as an important driver of divergence of ground vegetation communities in temperate forests. *Oikos* 122: 104–110.

Taberlet P, Gielly L, Pautou G, Bouvet J. 1991. Universal primers for amplification of three non-coding regions of chloroplast DNA. *Plant Molecular Biology* 17:1105–1109.

Tremblay J, Hout J, Potvin F. 2006. Divergent nonlinear responses of the boreal forest layer along an experimental gradient of deer densities. *Oecologia* 150:78–88.

Urbatsch L, Baldwin B, Donoghue M. 2000. Phylogeny of the coneflowers and relatives (Heliantheae: Asteraceae) based on nuclear rDNA internal transcribed spacer (ITS) sequences and chloroplast DNA restriction site data. *Systematic Botany* 25:539–565.

Vellend M, Cornwell W, Magnuson-Ford K, Mooers A. 2011. Measuring phylogenetic biodiversity. In: Magurran A, McGill B, eds. *Biological diversity: frontiers in measurement and assessment*. Oxford: Oxford University Press, 193–206.

Waller D, Alverson W. 1997. The white-tailed deer: a keystone herbivore. *Wildlife Society Bulletin* 25:217–226.

White T, Bruns T, Lee S, Taylor J. 1990. Amplification and direct sequencing of fungal ribosomal RNA genes for phylogenetics. In: Innis M, Gelfand D, Sninsky J, White T, eds. *PCR protocols: a guide to methods and applications*. New York: Academic Press, Inc. 315–322.

Wiegmann S, Waller D. 2006. Fifty years of change in northern upland forest understories: identity and traits of 'winner' and 'loser' plant species. *Biological Conservation* 129:109–123.

Yessoufou K, Davies J, Maurin O, Kuzmina M, Schaefer H, van der Bank M, Savolainen V. 2013. Large herbivores favour species diversity but have mixed impacts on phylogenetic community structure in an African savannah ecosystem. *Journal of Ecology* 101:614–625.

7

Genetic diversity of high-elevation populations of an endangered medicinal plant

Akshay Nag[1,2], Paramvir Singh Ahuja[1] and Ram Kumar Sharma[1*]

[1] Biotechnology Division, CSIR—Institute of Himalayan Bioresource Technology, Post Box 6, Palampur, 176061 Himachal Pradesh, India
[2] Academy for Scientific and Innovative Research (AcSIR), CSIR—Institute of Himalayan Bioresource Technology, Post Box 6, Palampur, 176061 Himachal Pradesh, India

Associate Editor: Kermit Ritland

Abstract. Intraspecific genetic variation in natural populations governs their potential to overcome challenging ecological and environmental conditions. In addition, knowledge of this variation is critical for the conservation and management of endangered plant taxa. Found in the Himalayas, *Podophyllum hexandrum* is an endangered high-elevation plant species that has great medicinal importance. Here we report on the genetic diversity analysis of 24 *P. hexandrum* populations (209 individuals), representing the whole of the Indian Himalayas. In the present study, seven amplified fragment length polymorphism (AFLP) primer pairs generated 1677 fragments, of which 866 were found to be polymorphic. Neighbour joining clustering, principal coordinate analysis and STRUCTURE analysis clustered 209 individuals from 24 populations of the Indian Himalayan mountains into two major groups with a significant amount of gene flow ($N_m = 2.13$) and moderate genetic differentiation $F_{st}(0.196)$, $G'_{st}(0.20)$. This suggests that, regardless of geographical location, all of the populations from the Indian Himalayas are intermixed and are composed broadly of two types of genetic populations. High variance partitioned within populations (80 %) suggests that most of the diversity is restricted to the within-population level. These results suggest two possibilities about the ancient population structure of *P. hexandrum*: either all of the populations in the geographical region of the Indian Himalayas are remnants of a once-widespread ancient population, or they originated from two types of genetic populations, which coexisted a long time ago, but subsequently separated as a result of long-distance dispersal and natural selection. High variance partitioned within the populations indicates that these populations have evolved in response to their respective environments over time, but low levels of heterozygosity suggest the presence of historical population bottlenecks.

Keywords: AMOVA; amplified fragment length polymorphism (AFLP); Baker's rule; genetic structure; Indian Himalayas; *Podophyllum hexandrum*; self-pollination.

Introduction

Throughout history, one of the many ways in which humans have benefited from plant diversity is as a source of traditional medicines. According to the World Health Organization (WHO), as many as 80 % of the world's populations depend on traditional medicine for their primary health-care needs (WHO 1993). Most traditional therapy involves the use of plant extracts or their active principles. In the present era, unprecedented growth in global population has led to subsequent increase in human demands and overexploitation of the earth's plant resources (Gadgil and Meher-Homji 1986; Işik

* Corresponding author's e-mail address: mrk_sharma@yahoo.com, ramsharma@ihbt.res.in

2011). Most plausible scenarios today suggest that we are likely to lose a large part of our traditional wealth of medicinal plants in the near future if critical steps are not taken to conserve them (Ehrlich and Daily 1993; Badola and Aitken 2003). Currently, large numbers of medicinally important plant resources face serious threat of extinction and severe genetic loss, but detailed information is lacking. For most of these endangered medicinal plant species, effective conservation plans are minimal and very little material is available in genebanks. Further, a major emphasis on discovering new drug molecules from plant resources has contributed to the loss of natural genetic resources. Nearly 25 % of the estimated 250 000 species of vascular plants in the world may become extinct within the next 40 years, if proper conservation measures are not undertaken (Kala 2000).

Knowledge of genetic variation within species, coupled with information about their reproductive biology, is very important when establishing any conservation and management programme (Newton et al. 1999; Juan et al. 2000; Frankham 2003; Silva et al. 2011) aimed at preserving genetic variation within and among populations (Eriksson 2001; Silva et al. 2011). Knowledge of genetic diversity patterns is also important in understanding the evolutionary history of a species and in the assessment of future risks to diversity (Neel and Ellstrand 2003). With regard to endangered species, measuring genetic variation among different populations is important for prioritization of sites and management choices for future conservation programmes. For example, greatly diverse or differentiated populations could be targeted for conservation, while genetically penurious populations might be targeted for management plans to restore diversity (Godt et al. 1996; Petit et al. 1998).

In the present study, we quantified the patterns of genetic diversity within Podophyllum hexandrum, an endangered plant species of great medicinal importance. Using amplified fragment length polymorphism (AFLP) markers, we have examined 209 individuals of 24 natural populations of P. hexandrum, representing the wider geographical area of the entire Indian Himalayas ranging from the states of Jammu and Kashmir, and including the Zanskar region, Himachal Pradesh, Uttarakhand to Sikkim. The genetic diversity of P. hexandrum has not been studied for the entire of the Indian Himalayas, and such a large geographical area with a greater number of samples has been analysed for the first time. The present study will comprehensively reveal the overall genetic diversity prevailing in these populations and will also aid in understanding the genetic dynamics of the species. Further, this will also throw light on how these populations are persisting despite their having a small chromosome number and self-pollinating reproductive behaviour.

Methods

Study species

Podophyllum hexandrum (Himalayan mayapple; syn: Sinopodophyllum hexandrum, Podophyllum emodi) is a species of great medicinal importance. It is confined to the alpine regions of Afghanistan, Pakistan, Nepal, Bhutan, South West China and India (Airi et al. 1997; Choudhary et al. 1998). Despite its wider distribution in the entire Indian Himalayan range, from Ladakh to Sikkim at an elevation of 3000–4200 m, the current status of P. hexandrum is now endangered. The rhizomes and roots of P. hexandrum contain anti-tumour lignans such as podophyllotoxin, 4′-dimethyl podophyllotoxin and podophyllotoxin 4-O-glucoside (Tyler et al. 1988; Broomhead and Dewick 1990). Among these lignans, podophyllotoxin or podophylloresin is most important for its use in the semi-synthesis of anti-cancer drugs etoposide and teniposide (Issell et al. 1984; Canel et al. 2000). Podophyllotoxin acts as an inhibitor of microtubule assembly. These drugs are used in the treatment of lung cancer, testicular cancer, neuroblastoma, hepatoma and other tumours. It also shows antiviral activities by interfering with some critical viral processes (Giri and Narasu 2000). The podophyllotoxin content of Himalayan mayapple is quite high (4.3 %) compared with other species of Podophyllum, notably Pelargonium peltatum (0.25 %), the most common species in the American subcontinent (Jackson and Dewick 1984). However, the percentage of resin varies greatly at different growth phases, with age of the plant, seasonal variation and different geographical sites (Purohit et al. 1999).

The life cycle of P. hexandrum is 5–6 years. Flowers blossom before the leaves grow out. According to the latest report, occasional cross-pollination has been observed in P. hexandrum (Xiong et al. 2013); however, its morphological and biological characteristics are adapted to self-pollination and effective sexual reproduction. The self-pollination mechanism of the plant is very unique. When the flower is under blossom or just in blossom, the position of the gynoecium is upright; however, when it reaches full blossom stage, the gynoecium takes a full turn and because of this, the entire gynoecium gets closer to an anther to become pollinated. Fruit bearing is almost 100 %. Thus it appears the plant shows considerable fitness (Shaobin et al. 1997; Xu et al. 1997). The important point here is that most of the members of Berberidaceae are cross-pollinated including P. peltatum, the North American counterpart of Himalayan Mayapple. The disjunction between the two species is estimated to have happened ~6.52 ± 1.89 million years ago (Liu et al. 2002). Lack of pollinators and the nectarless character (Crants 2008) of the flower might have been responsible for the evolution of self-pollination in this plant.

In natural conditions, the dispersal of seeds is facilitated primarily by herbivores, mainly Himalayan grazers that travel great distances; hence the seed dispersal distance of *P. hexandrum* is reasonably good (Rajkumar and Ahuja 2010). *Podophyllum hexandrum* has a wide region of distribution; however, within that region, it appears primarily in valleys with secondary vegetation. In any given population, the plant shows a clumping distribution pattern (Ma and Hu 1996).

Traditionally, *P. hexandrum* has been used in folk medicine in small quantities by local healers as a cure for ulcers, cuts, wounds and skin diseases (Negi *et al.* 2011), but commercialization of this plant in recent years has increased the demand and consequent exploitation of the species. Owing to habitat fragmentation (Young *et al.* 1996), overexploitation, long dormancy, low rate of natural regeneration and overgrazing, it has been classified as an endangered species (Kala 2005). There is an urgent need to conserve the genetic diversity of this prized medicinal plant, which may become extinct if its reckless exploitation continues. Earlier studies of genetic diversity in Himalayan populations have been restricted to a relatively limited geographic area (Xiao *et al.* 2006*a, b*; Alam *et al.* 2008; Naik *et al.* 2010; Li *et al.* 2011). As has been suggested by its pollination mechanism (namely, self-pollination), populations are expected to be genetically structured. Further, owing to a small chromosome number ($2n = 12$) (Nag and Rajkumar 2011) with a very large genome (*C* value $= 16.075$ Gb) (Nag *et al.* 2011), *P. hexandrum* might be experiencing severe evolutionary pressure against adaptation as suggested by the large genome size constraint hypothesis (Knight *et al.* 2005).

Plant materials

A total of 24 different geographical locations, ranging from Kashmir to the Sikkim Himalayas and representing most of the Indian Himalayas, were visited for sample collection in the present study during 2008 and 2012 (Table 1, Fig. 1). The Himalayan mountain ranges included were the Dhauladhar range, the Pir Panjal range, the Shivalik/Garhwal range, the Greater Himalayan range, the Zanskar range and the Kangchenjunga Himal section. The Zanskar range, which is a Trans-Himalayan range, and the Kangchenjunga Himal section are subranges of the Greater Himalayas. Young leaves of the Himalayan mayapple were collected in silica gel by changing the gel periodically, until the sample was completely dried. The minimum distance between sample plants within a population was kept at ~5 m. The extent of exploitation of the plant is such that at some locations, the number of plants per quadrat (1 m × 1 m) was 0.6 and hence we kept 5 as the minimum sample size in the study; however, we sampled up to 25 plants per population. The total

number of samples collected was 224, of which 209 were chosen for analysis based on the presence of good-quality DNA profiles.

DNA isolation and molecular analysis

Total genomic DNA was isolated following the CTAB method (Doyle and Doyle 1987; Doyle 1990) with minor modifications. Deoxyribonucleic acid concentrations were determined using a Nanodrop spectrophotometer (Thermo Scientific) followed by a quality check on ethidium bromide-stained agarose gels, using known amounts of uncut λ DNA as a standard.

The AFLP protocol was carried out following the procedure described by Vos *et al.* (1995) with minor modifications. Genomic DNA (250 ng) was restricted with *Eco*RI/ *Mse*I enzyme mix and ligated to standard adapters using the T4 DNA ligase. The adapter-ligated DNA served as a template for preamplification, with PCR parameters of 20 cycles at 94 °C for 30 s, 56 °C for 1 min and 72 °C for 1 min. After screening 36 primer pairs for four individuals from four populations, seven primer pairs (Table 2) were chosen for the full survey because they resulted in clear and reproducible bands **[see Supporting Information]**. Selective amplification was carried out with 2.5 μL of these diluted products using *Eco*RI primers (fluorescently labelled with NED, FAM and JOE) and *Mse*I primers, Taq polymerase, PCR buffer, $MgCl_2$, each dNTPs and deionized water in a final volume of 10 μL. The first selective amplification cycle consisted of 94 °C for 30 s, 65 °C for 30 s and 72 °C for 1 min. The annealing temperature was lowered by 0.7 °C per cycle during the next 12 cycles, followed by 23 cycles at 94 °C for 30 s, 56 °C for 30 s and 72 °C for 1 min. All PCRs were performed on the i-cycler PCR system (Bio-Rad, Australia). 0.5 μL of each selective PCR product was mixed with 0.3 μL of Gene Scan-500 ROX size standard (Applied Biosystems) and 9.2 μL of highly deionized formamide. This mixture was denatured at 94 °C for 5 min, followed by immediate chilling on ice and these denatured products were loaded on an ABI 3730xl automated DNA Analyser (Applied Biosystems, Hitachi) to visualize the amplified fragments.

The software program GeneMapper 3.7 (Applied Biosystems) was used to analyse electropherograms generated by automated genotyping using the ABI 3730xl automated DNA analyser. The large amount of data generated by the automated DNA analyser was checked manually a number of times to exclude unreliable detection and to improve the quality of data. The size range of amplified fragments, peak height threshold in terms of relative fluorescence units (rfu) and bandwidth were considered to be the most important scoring parameters; different sets of parameters were tested, and the parameter set that was optimized for the best fit was used for

Table 1. Details of the locations from where samples of *P. hexandrum* were collected for this study. *N*, number of individuals in a population.

Location	Himalayan range	State	Altitude (in metres MSL)	Latitude	Longitude	N
Bairagarh	Pir Panjal	Himachal Pradesh	2292	32.9064	76.1616	9
Chholmi	Garhwal Himalaya	Uttarakhand	2899	31.0269	78.8704	5
Dharali	Garhwal Himalaya	Uttarakhand	3005	31.0283	78.7983	12
Diankund	Dhauladhar	Himachal Pradesh	2154	32.5417	76.0275	13
Gulaba	Pir Panjal	Himachal Pradesh	2994	32.3188	77.2035	12
Gulmarg	Pir Panjal	Jammu and Kashmir	2156	34.0614	74.3876	5
Jalsu	Dhauladhar	Himachal Pradesh	3359	32.3063	77.1429	10
Kasol	Pir Panjal	Himachal Pradesh	2770	31.9919	77.3392	13
Koksar	Greater Himalayas	Himachal Pradesh	3136	32.4138	77.2349	17
Pehelgam	Pir Panjal	Jammu and Kashmir	2218	34.0149	75.3106	5
Prashar	Dhauladhar	Himachal Pradesh	2315	31.7644	77.0897	16
Purthi	Dhauladhar	Himachal Pradesh	2978	32.9225	76.474	5
Sangla Kanda	Greater Himalayas	Himachal Pradesh	2915	31.4201	78.2574	5
Sansha	Greater Himalayas	Himachal Pradesh	3273	32.6125	76.9493	5
Shopian	Pir Panjal	Jammu and Kashmir	2182	33.785	74.7942	5
Sikkim	Greater Himalayas (Kangchenjunga Himal section)	Sikkim	2741	27.3807	88.2551	5
Sissu	Greater Himalayas	Himachal Pradesh	3127	32.4833	77.1299	7
Sonamarg	Pir Panjal	Jammu and Kashmir	3031	34.2954	75.2919	5
Sural Pangi	Greater Himalayas	Himachal Pradesh	2715	33.1209	76.3788	5
Tral	Pir Panjal	Jammu and Kashmir	2433	33.8787	75.1356	5
Trilokinath	Greater Himalayas	Himachal Pradesh	2910	32.6837	76.6962	9
Tungnath	Garhwal Himalaya	Uttarakhand	3448	30.4883	79.2162	5
Zanskar	Greater Himalaya	Jammu and Kashmir	3810	33.587	76.696	25
Dharali	Garhwal Himalaya	Uttarakhand	3005	31.0283	78.7983	12

our analyses. Amplified fragments of 50–500 base pairs having present (1) and absent (0) peaks were extracted using GeneMapper 3.7. The resulting binary matrix was exported in the form of comma-separated text for data analysis.

Data analysis

Calculations for genetic distance, pairwise population matrix of Nei's genetic identity, allele frequency by population, Mantel test for correlation of genetic and geographic distance and principal coordinate analysis (PCoA) were conducted using GenAlEx 6.501 (Peakall and Smouse 2006, 2012). STRUCTURE version 2.3.4 (Pritchard *et al.* 2000) was used to infer the genetic structure so as to obtain an estimate of the likely number of population genetic clusters (*K*). The numbers of clusters of the populations (*K*) were identified by performing six iterations and setting the value of *K* from 1 to 25 with a burn-in period of 100 000 and 100 000 number of the Markov Chain Monte Carlo (MCMC) repeats after burn-in. The maximal value of LnP(D), the posterior probability of data as per Evanno *et al.* (2005), was obtained using STRUCTURE HARVESTER (Earl and vonHoldt 2012). To further confirm the number of genetic clusters, the value of *K* was estimated through analysis of molecular variance (AMOVA)-based clustering using kMeans software (Meirmans 2012). To infer the partitioning of the diversity, G_{st} & G'_{st} software package GenoType/GenoDive (Meirmans and Van Tienderen 2004) was used. In the first step GenoType detects the genotyping errors and prepares an input file for GenoDive. In the second step GenoDive calculates the parameters of diversity and diversity partitioning. AFLPSURV

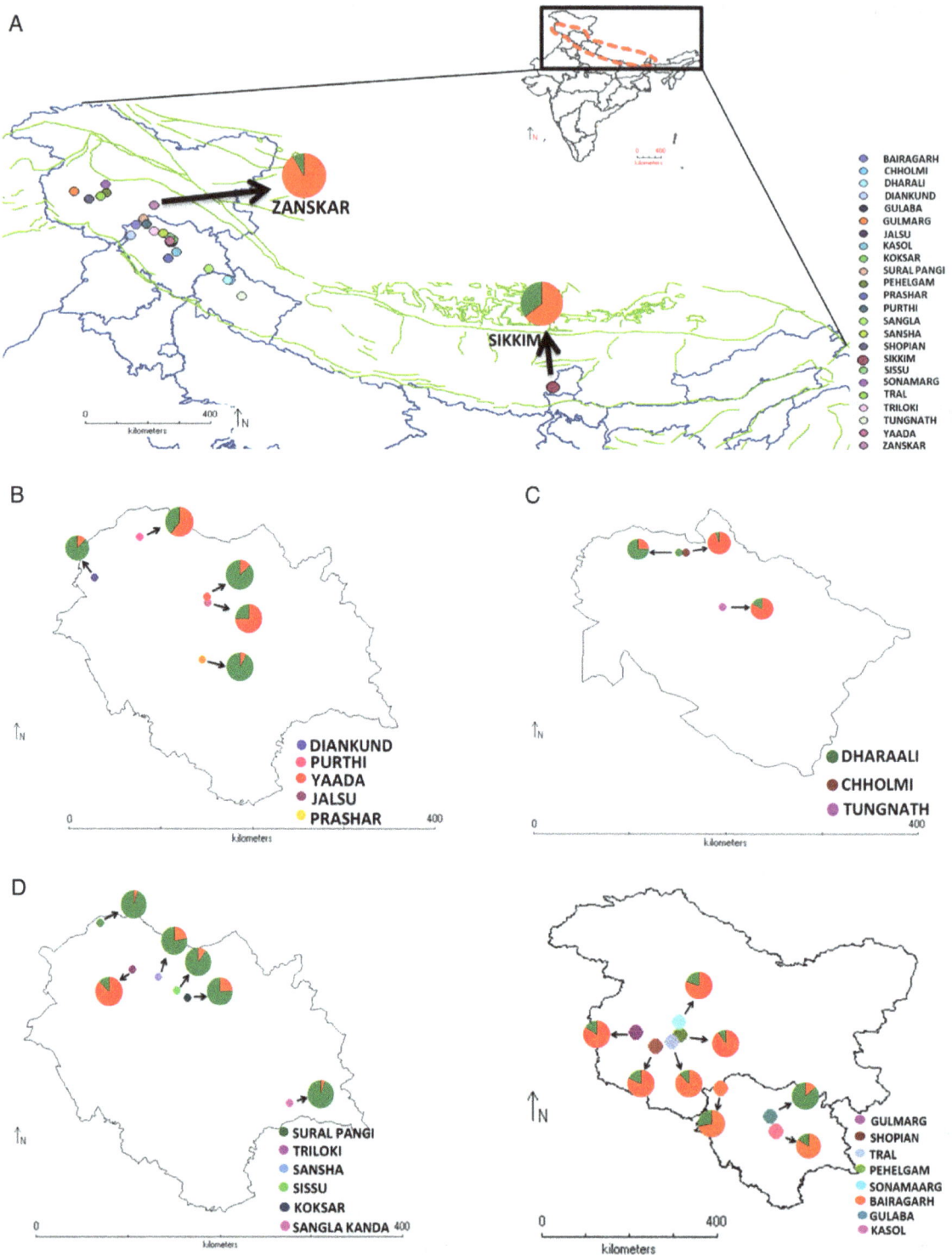

Figure 1. Geographic distribution of sampled populations of *P. hexandrum* from the Indian Himalayas with pie charts representing the percentage of the two genetic pools from each of the populations. (A) Map representing all the sampled locations, (B) Dhauladhar range, (C) Shivalik/Garhwal Himalayas, (D) Greater Himalayas, (E) Pir Panjal range.

(Vekemans *et al.* 2002), which follows a Bayesian method with non-uniform prior distribution (Zhivotovosky 1999), was used to infer the genetic relationships among populations by calculating Nei's unbiased genetic distance (Lynch

and Milligan 1994) among all possible pairs of populations from allele frequencies. Hierarchical AMOVA and F_{st} was conducted using ARLEQUIN 3.5.1.2 (Excoffier and Lischer 2010). The input file for ARLEQUIN was prepared using

Table 2. List of AFLP primer pairs used in the study.

S. no.	Primer combination	No. of bands	No. of polymorphic bands	Percentage polymorphism (%)
1	E-ACA + M-CTGC	325	135	41.53
2	E-AAC + M-CAG	256	92	35.93
3	E-AAG + M-CTAG	166	128	77.10
4	E-ACC + M-CAT	257	103	40.07
5	E-ACT + M-CAG	229	126	55.02
6	E-ACC + M-CAG	261	154	59.03
7	E-AGG + M-CTT	183	128	69.94

the program CONVERT (Glaubitz 2004). The dendrogram was computed by using the neighbour joining (NJ) clustering with DARwin5 version 5.0.158 (Perrier and Jacquemoud-Collet 2006).

Results

AFLP analysis and polymorphism

Scoring the sampled material of *P. hexandrum* (209 individuals, 24 populations) for seven AFLP primer combinations resulted in 1677 unambiguous fragments in the size range of 50–500 bp, of which 866 (51.65 %) were polymorphic (Table 2). The mean number of fragments per individual was found to be 105.5. The maximum number of polymorphic bands was found in the Zanskar population (40.42 %), followed by the Koksar population (37.99 %), whereas the Chholmi population had the minimum number of polymorphic bands (7.27 %). The number of private alleles in each population ranged from 0 to 40, and comprised 25.08 % of the total bands. The population of Zanskar contained a maximum 40 private alleles, followed by the Triloki population (28 private alleles), whereas the Tral population had only one private allele. We did not find any private alleles in the Pehelgam, Chholmi, Sural Pangi and Sansha populations. Overall heterozygosity (Nei's unbiased diversity, *uh*) estimates recorded in the AFLP analysis were found to be very low, with the maximum observed heterozygosity (0.155 ± 0.008) in the Eastern Himalayan population collected from Sikkim and the lowest heterozygosity in the Sural Pangi population (0.043 ± 0.005). Shannon's information index (*I*) values also complemented these findings (Table 3).

Genetic differentiation and partitioning of populations

Overall F_{st} revealed in Arlequin analysis was 0.196 (Table 4). G'_{st} was found to be 0.20, whereas $G_{st} = 0.19$. Pairwise F_{st} analyses (Table 5) showed that the populations from Sural Pangi (Greater Himalayan range) and

Sonamarg (Pir Panjal range) and Sural Pangi (Greater Himalayan range) and Shopian (Pir Panjal range) were found to be most divergent ($F_{st} = 0.58$) of the populations, whereas the minimum F_{st} (0.04) was observed between the Gulaba (Pir Panjal range) and Diankund (Dhauladhar range) populations, and the Dharali (Garhwal Himalaya) and Diankund (Dhauladhar range) populations.

Analysis of molecular variance analysis revealed that the majority of the variance was restricted to within-population variation (80 %), whereas variance partitioned among population was 20 %. Significant gene flow (N_m) was recorded between the populations. On the basis of $F_{st}[N_m = (1/F_{st} - 1)/4]$, N_m was found to be 1.02, whereas the value of gene flow on the basis of Gst [$N_m = G_{st}(1 - G_{st})/G_{st}$] was found to be 2.13, which indicated a considerable intermixing and low genetic differentiation among populations.

Population genetic structure and cluster analysis

AFLP-based genetic diversity analysis in the 24 populations of *P. hexandrum* was carried out using three different but complementary approaches, factorial analysis or PCoA, neighbour Joining (NJ)-based hierarchical clustering and Bayesian model-based clustering. Principal coordinate analysis (Fig. 2) complemented NJ (Fig. 3) cluster analysis in providing an overall view of genetic diversity in the natural populations of *P. hexandrum*. The first three axes effectively captured the entire diversity (91 %) in the *P. hexandrum* populations and revealed two major groups. A Mantel test between the genetic and geographic distances showed no correlation **[see Supporting Information]**. This was further confirmed by the AMOVA analysis between the mountain ranges, which showed that majority of variance was found within mountain ranges (91 %) **[see Supporting Information]**. The dendrogram (Fig. 3) obtained by NJ analysis showed that populations from the Zanskar and Kashmir Valleys (Gulmarg, Sonamarg, Pehelgam, Tral and Shopian) were clustered in one group along with populations from Bairagarh, Jalsu, Kasol, Triloki, Purthi, Tungnath and Sikkim

Table 3. Population genetic parameters of the 24 populations comprising 209 individuals of *Podophyllum hexandrum* from the Indian Himalayas. N, number of individuals in a population; uh, Nei's unbiased diversity; I, Shannon's information index.

Location	N	No. of polymorphic alleles	No. of private alleles	% Polymorphism	uh	I
Bairagarh	9	288	9	33.3	0.122 ± 0.006	0.166 ± 0.008
Chholmi	5	75	0	8.7	0.052 ± 0.006	0.058 ± 0.006
Dharali	12	247	10	28.5	0.087 ± 0.006	0.124 ± 0.008
Diankund	13	284	12	32.8	0.093 ± 0.005	0.137 ± 0.007
Gulaba	12	227	7	26.2	0.079 ± 0.005	0.113 ± 0.007
Gulmarg	5	187	3	21.6	0.118 ± 0.008	0.135 ± 0.009
Jalsu	10	281	13	32.4	0.111 ± 0.006	0.154 ± 0.008
Kasol	13	319	12	36.8	0.107 ± 0.006	0.156 ± 0.008
Koksar	17	331	24	38.2	0.102 ± 0.005	0.154 ± 0.008
Pehelgam	5	145	0	16.7	0.086 ± 0.007	0.098 ± 0.008
Prashar	16	290	7	33.5	0.084 ± 0.005	0.127 ± 0.007
Purthi	5	272	6	31.4	0.148 ± 0.008	0.176 ± 0.009
Sangla Kanda	5	196	8	22.6	0.082 ± 0.006	0.097 ± 0.007
Sansha	5	170	0	19.6	0.087 ± 0.007	0.103 ± 0.008
Shopian	5	205	1	23.7	0.091 ± 0.007	0.104 ± 0.008
Sikkim	5	298	17	34.4	0.155 ± 0.008	0.185 ± 0.009
Sissu	7	206	3	23.8	0.080 ± 0.006	0.103 ± 0.007
Sonamarg	5	194	2	22.4	0.085 ± 0.007	0.099 ± 0.008
Sural Pangi	5	127	0	14.7	0.043 ± 0.005	0.048 ± 0.006
Tral	5	179	2	20.7	0.108 ± 0.008	0.122 ± 0.009
Trilokinath	9	290	28	33.5	0.123 ± 0.007	0.167 ± 0.009
Tungnath	5	254	8	29.3	0.138 ± 0.008	0.164 ± 0.009
Yada	6	229	7	26.4	0.098 ± 0.006	0.123 ± 0.008
Zanskar	25	350	40	40.4	0.090 ± 0.005	0.142 ± 0.007

Table 4. Table of analysis of molecular variance along with F_{st} value, as calculated in ARLEQUIN.

Source	Degrees of freedom	Sum of squares	Variance components	Percentage variation	F_{st}
Among populations	23	2991.200	10.247	20	0.196
Within populations	185	7797.613	42.149	80	

(group I). The majority of the populations of the Pir Panjal range (six out of seven populations) remained together in Group I, along with two populations from the Dhauladhars and a single population each from the Zanskar, Shivalik/Garhwal, Greater Himalayas and Kangchenjunga Himal section (Table 6). The second group is composed of majority of the populations from the Greater Himalayan trail (Sansha, Koksar, Sural Pangi, Sangla Kanda and Sissu) and the Dhauladhar range (Prashar, Yada and Diankund) along with two populations from Garhwal Himalayas (Chholmi and Dharali) and a single population from Pir Panjal (Gulaba). Clustering is quite distinct with only 10 out of 209 individuals recorded as intermixing between the groups. Surprisingly, individuals from a single population did not cluster together into a single subgroup. Bayesian model-based STRUCTURE analyses showed that the maximum likelihood of clustering of the AFLP data [LnP(D)] was obtained when samples

Table 5. Pairwise F_{st} and Nei's genetic distances between 24 populations. Values above the diagonal represent Nei's genetic distances and values below the diagonal represent F_{st}.

		1	2	3	4	5	6	7	8	9	10	11	12	13	14	15	16	17	18	19	20	21	22	23	24
1	Bairagarh	0	0.034	0.028	0.029	0.024	0.032	0.028	0.028	0.027	0.020	0.020	0.020	0.023	0.035	0.051	0.058	0.051	0.046	0.046	0.048	0.073	0.045	0.042	0.069
2	Sonamrg	0.19	0	0.037	0.057	0.046	0.040	0.047	0.045	0.031	0.043	0.055	0.045	0.052	0.050	0.058	0.072	0.059	0.060	0.060	0.075	0.092	0.058	0.057	0.076
3	Shopian	0.14	0.25	0	0.051	0.042	0.044	0.037	0.039	0.039	0.039	0.042	0.035	0.041	0.053	0.065	0.072	0.066	0.066	0.064	0.077	0.096	0.061	0.059	0.078
4	Tral	0.14	0.33	0.29	0	0.029	0.042	0.043	0.046	0.043	0.025	0.025	0.031	0.032	0.049	0.063	0.061	0.060	0.055	0.055	0.052	0.075	0.055	0.053	0.078
5	Gulmarg	0.1	0.27	0.24	0.14	0	0.027	0.034	0.039	0.030	0.018	0.022	0.023	0.023	0.035	0.051	0.051	0.046	0.043	0.042	0.038	0.066	0.043	0.043	0.065
6	Pehelgam	0.17	0.28	0.29	0.25	0.14	0	0.045	0.052	0.033	0.023	0.037	0.036	0.028	0.054	0.069	0.071	0.059	0.063	0.059	0.047	0.090	0.062	0.063	0.087
7	Sikkim	0.11	0.22	0.17	0.18	0.13	0.22	0	0.031	0.035	0.034	0.032	0.025	0.035	0.040	0.049	0.054	0.048	0.046	0.045	0.061	0.074	0.044	0.041	0.058
8	Tungnath	0.12	0.23	0.2	0.22	0.17	0.27	0.09	0	0.037	0.042	0.038	0.032	0.043	0.044	0.053	0.059	0.054	0.049	0.051	0.069	0.078	0.047	0.047	0.066
9	Purthi	0.1	0.15	0.19	0.19	0.11	0.16	0.11	0.13	0	0.029	0.034	0.027	0.037	0.036	0.043	0.054	0.045	0.042	0.044	0.052	0.073	0.040	0.044	0.058
10	Triloki	0.09	0.23	0.21	0.11	0.06	0.11	0.14	0.19	0.12	0	0.016	0.018	0.013	0.041	0.059	0.059	0.053	0.052	0.048	0.036	0.071	0.049	0.049	0.075
11	Kasol	0.1	0.31	0.24	0.13	0.11	0.22	0.15	0.19	0.17	0.08	0	0.015	0.013	0.042	0.060	0.056	0.055	0.050	0.047	0.037	0.068	0.049	0.046	0.072
12	Jalsu	0.1	0.26	0.2	0.16	0.11	0.21	0.1	0.15	0.12	0.09	0.08	0	0.016	0.034	0.050	0.049	0.046	0.043	0.040	0.037	0.063	0.038	0.039	0.062
13	Zanskar	0.15	0.32	0.27	0.21	0.15	0.19	0.21	0.26	0.22	0.08	0.09	0.11	0	0.041	0.063	0.057	0.053	0.051	0.047	0.033	0.074	0.049	0.046	0.076
14	Koksar	0.2	0.29	0.3	0.28	0.2	0.31	0.2	0.23	0.19	0.23	0.25	0.2	0.27	0	0.014	0.019	0.017	0.011	0.012	0.038	0.035	0.013	0.017	0.033
15	Prashar	0.31	0.37	0.39	0.37	0.31	0.41	0.28	0.31	0.25	0.34	0.35	0.31	0.38	0.1	0	0.012	0.014	0.011	0.010	0.056	0.033	0.010	0.019	0.024
16	Sissu	0.32	0.44	0.43	0.37	0.31	0.43	0.28	0.32	0.29	0.32	0.33	0.29	0.36	0.12	0.08	0	0.012	0.014	0.013	0.051	0.029	0.014	0.023	0.030
17	Yada	0.26	0.35	0.38	0.33	0.25	0.35	0.22	0.27	0.22	0.27	0.3	0.26	0.32	0.09	0.09	0.09	0	0.016	0.014	0.047	0.036	0.013	0.022	0.028
18	Diankund	0.27	0.36	0.38	0.32	0.25	0.37	0.25	0.27	0.23	0.29	0.29	0.26	0.33	0.07	0.07	0.09	0.06	0	0.007	0.044	0.029	0.008	0.018	0.027
19	Gulaba	0.28	0.39	0.4	0.35	0.27	0.39	0.26	0.31	0.26	0.29	0.3	0.26	0.32	0.08	0.08	0.09	0.09	0.04	0	0.037	0.027	0.008	0.017	0.028
20	Chholmi	0.29	0.51	0.5	0.37	0.27	0.38	0.33	0.39	0.3	0.22	0.24	0.25	0.24	0.25	0.39	0.4	0.35	0.31	0.31	0	0.058	0.042	0.050	0.072
21	Sural_Pangi	0.4	0.58	0.58	0.48	0.43	0.57	0.4	0.44	0.4	0.39	0.39	0.38	0.44	0.24	0.27	0.28	0.3	0.23	0.24	0.54	0	0.028	0.044	0.046
22	Dharaii	0.27	0.36	0.37	0.33	0.26	0.38	0.24	0.27	0.23	0.28	0.3	0.24	0.32	0.08	0.07	0.09	0.08	0.04	0.05	0.32	0.23	0	0.015	0.024
23	Sansha	0.22	0.36	0.36	0.31	0.24	0.39	0.19	0.24	0.22	0.26	0.27	0.22	0.29	0.09	0.13	0.16	0.13	0.11	0.12	0.39	0.38	0.09	0	0.035
24	Sangla Kanda	0.35	0.45	0.45	0.42	0.35	0.48	0.28	0.33	0.29	0.36	0.38	0.34	0.42	0.21	0.18	0.22	0.18	0.18	0.22	0.5	0.4	0.17	0.24	0

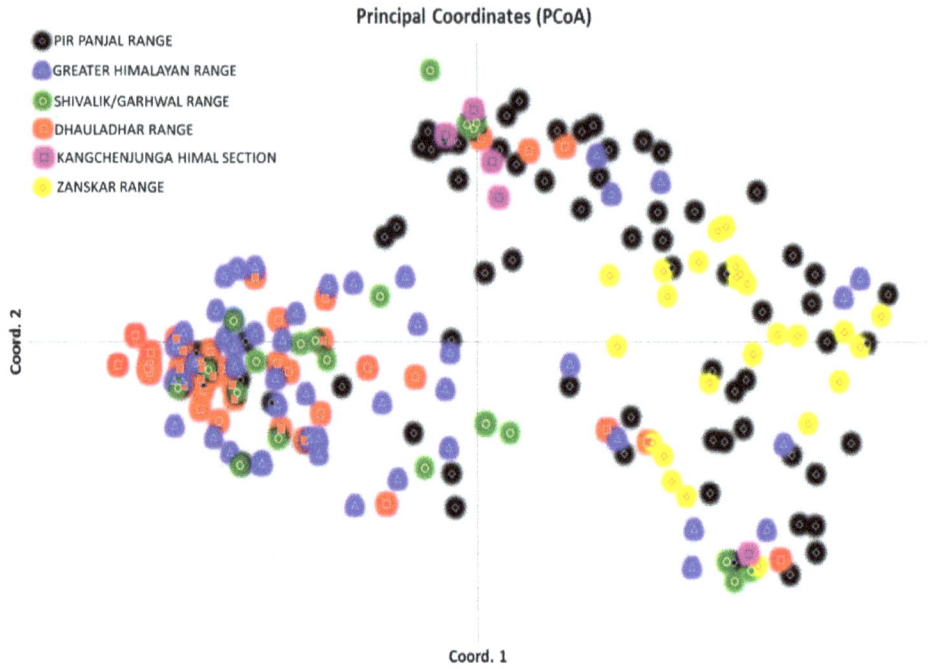

Figure 2. Principal coordinate analysis showing the differentiation of 209 individuals of *P. hexandrum* from 24 populations from the Indian Himalayas according to the mountain ranges.

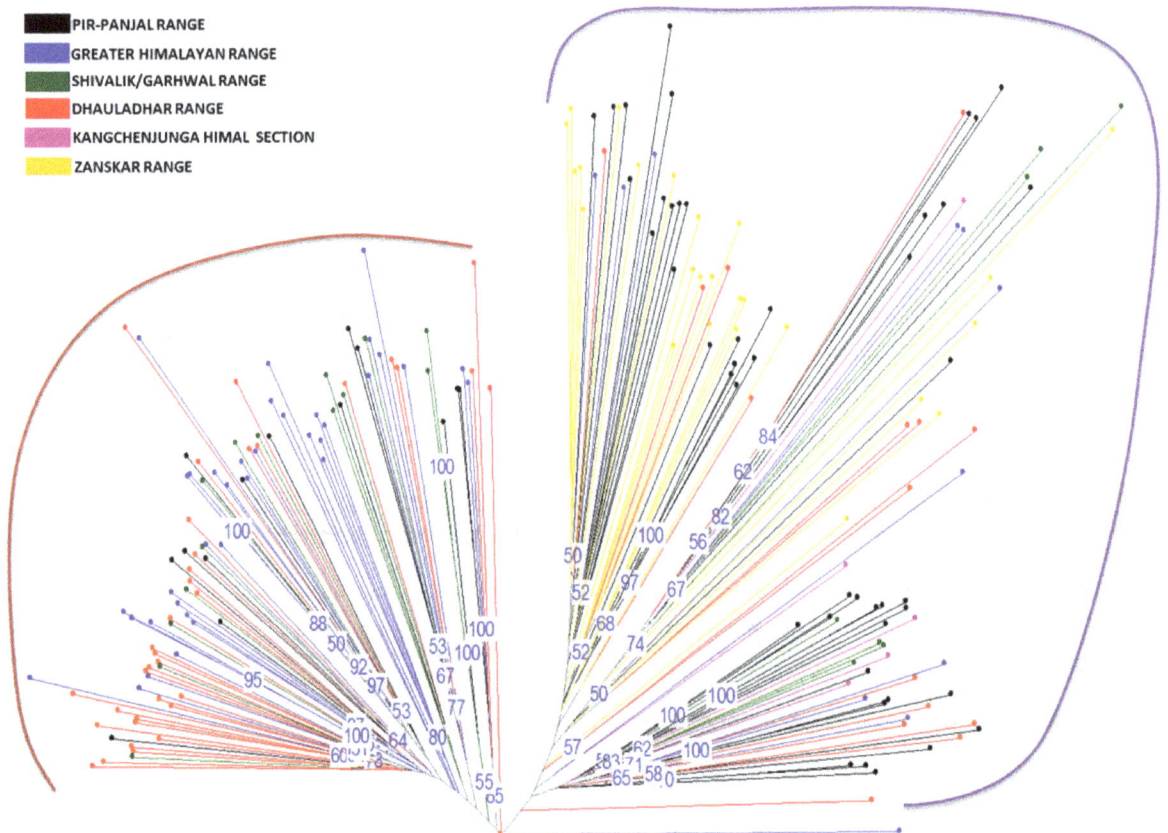

Figure 3. Neighbour-joining tree based on genetic distances of 209 individuals of *Podophyllum hexandrum* from the Indian Himalayas. Numbers above branches indicate bootstrap values >50 % (1000 replicates).

Table 6. Cluster analysis results for populations, with mountain ranges identified.

Group I	Group II
Balragarh (Pir Panjal)	Chholmi (Shivalik or Garhwal Himalayas)
Pehelgam (Pir Panjal)	Dharali (Shivalik or Garhwal Himalayas)
Sonamarg (Pir Panjal)	
Gulmarg (Pir Panjal)	Sansha (Greater Himalayas)
Shopian (Pir Panjal)	Koksar (Greater Himalayas)
Tral (Pir Panjal)	Sangla Kanda (Greater Himalayas)
Kasol (Pir Panjal)	Sissu (Greater Himalayas)
Leh (Zanskar)	Sural Pangi (Greater Himalayas)
Jalsu (Dhauladhar)	Diankund (Dhauladhar)
Purthi (Dhauladhar)	Yada (Dhauladhar)
Tungnath (Shivalik or Garhwal Himalayas)	Prashar (Dhauladhar)
Triloki (Greater Himalayas)	Gulaba (Pir Panjal)
Sikkim (Kangchenjunga Himal section)	

were clustered into two groups ($K = 2$). This confirms that among the populations included in the study, two types of gene pools/genetic populations are found in the Indian Himalayas (Fig. 4). The percentage of individuals with pure grouping was found to be 73.4 %, whereas 26.6 % individuals showed mixed grouping at various levels.

Discussion

Amplified fragment length polymorphism markers have been widely used to study the genetic diversity and population structure of various endangered plant species like *Leucopogon obtectus* (Gardens *et al.* 2001), *Eryngium alpinum* (Gaudeul *et al.* 2000), *Limonium dufourii* (Palacios *et al.* 1999), *Silene tatarica* (Tero *et al.* 2003) and others. The efficiency of a DNA marker system for analysing diversity relies upon the extent of polymorphism detected by uncovering a large number of markers spanning the whole genome (Luikart *et al.* 2003). Amplified fragment length polymorphism markers are ideal for detecting polymorphisms, as the variable regions detected are based on restriction enzyme sites and thus essentially reveal through whole-genome scans even minor genetic variations within any given organism (Mueller and Wolfenbarger 1999). We were able to obtain sufficient numbers of polymorphic markers (866; 51.65 %) for the estimation of population genetic parameters, in

accordance with the criteria for the critical number of dominant markers suggested by Staub *et al.* (2000); (80 bands) and Mariette *et al.* (2002); (100–200 bands) for reliable estimation of population genetic parameters. Additionally, three primer combinations having detected a large number of bands suggest the potential of these markers for future population biology studies in this species.

Genetic differentiation

Generally, taxa with self-pollinating behaviour have the majority of variance partitioned among populations (Loveless and Hamrick 1984). This species, therefore, might be expected to have a diverse distribution of AFLP variation among populations, but the present study does not support this *a-priori* expectation in overall levels of inter-population differentiation across the 24 populations surveyed (within population 80 % and among population 20 %, with $F_{st} = 0.196$ and $G'_{st} = 0.20$). However, significant population differentiation reported previously in allozymes (Bhadula *et al.* 1996) and various DNA fingerprinting marker studies (Xiao *et al.* 2006a, b; Alam *et al.* 2008; Naik *et al.* 2010) might have resulted due to the fact that the targeted populations were locally restricted, which is also congruent with differentiation inferences in the subpopulation in the current study. Current inferences are based on sampled populations from a wide geographical range covering the whole of the Indian Himalayas. Moreover, low population size also leads to lower levels of genetic variation, which is a general trend in endangered plants (Loveless and Hamrick 1984). The number of plants of *P. hexandrum* is very small as compared with other non-endangered plants found in any of the locations we sampled. The overall genetic differentiation between the populations in this study was found to be moderate ($F_{st} = 0.196$, $G'_{st} = 0.20$), which is evident from the cluster analysis, as all the individuals from a population remain clustered in either of the two groups, confirming that genetic structure, although weak, is present in these populations. Low values of unbiased heterozygosity ($uh = 0.043$–0.155) and Shannon's information index ($I = 0.048$–0.185) suggest that population bottlenecks resulted due to small population size. This also accounts for reduced genetic variation among the sampled populations in the present study.

Formation of Himalayas and subsequent evolution of *P. hexandrum* populations

Although self-pollinated, *P. hexandrum* is also capable of occasional cross-pollination, and this phenomenon accounts for another source of low genetic variation among populations. The seed set in the cross-pollinated plants is found to be almost the same as that in

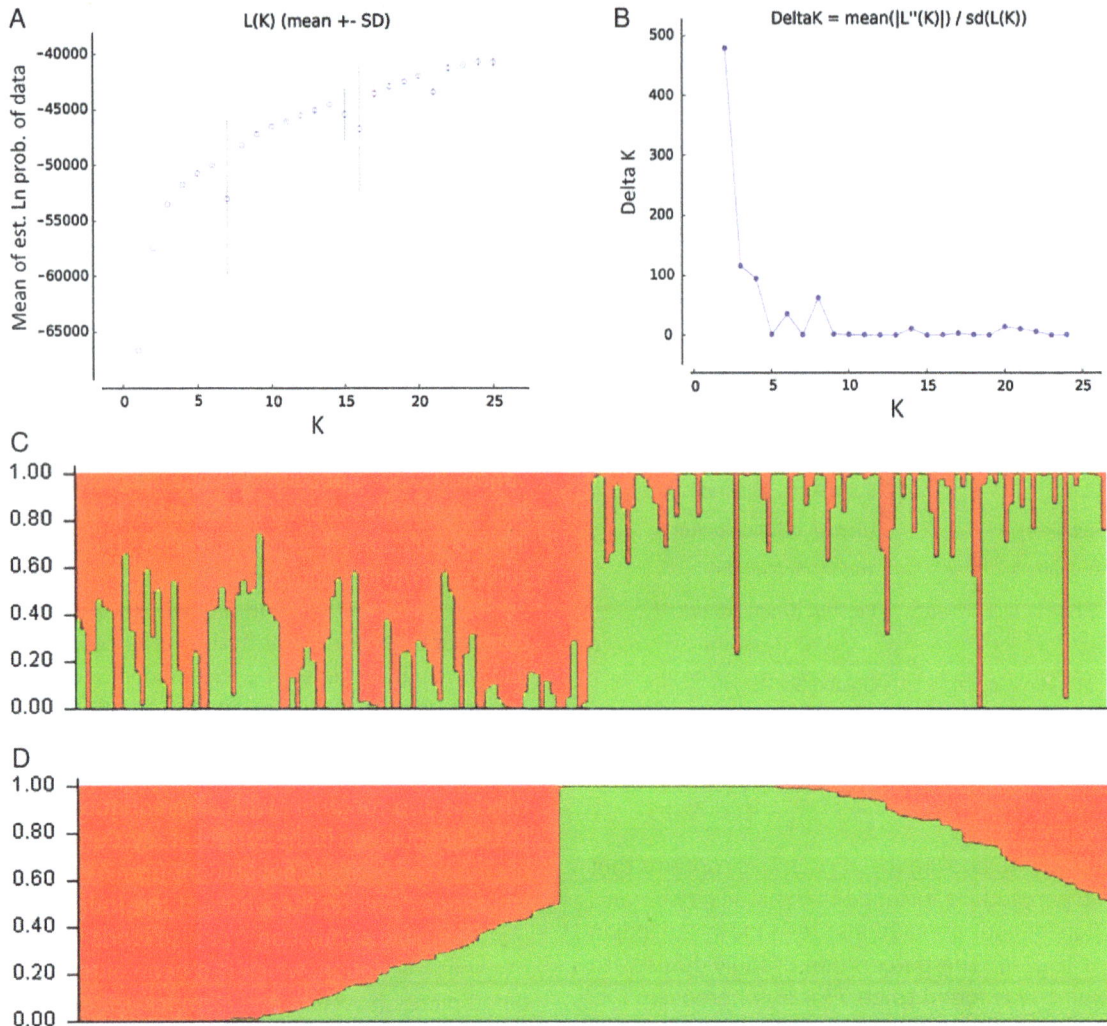

Figure 4. STRUCTURE inferences of *P. hexandrum* populations based on AFLP genotyping. (A) log likelihood, LnP(D), (B) changes in the log like-lihood, $\Delta(K)$, for different number of groups. (C) Bar plots represent STRUCTURE inferences of individual assignments ($K = 2$) as inferred in the Structure Harvester web v. 0.6.93. Each vertical bar represents one individual. (D) The bar plot represents individuals arranged according to its most likely ancestry. Each colour represents the most likely ancestry of the cluster from which the genotype or partial genotype was derived.

self-fertilized plants (Xiong *et al.* 2013) and it is believed that self-pollinated species are almost always derived from cross-fertilizing ancestors (Stebbins 1957; Wyatt 1988). *Podophyllum peltatum*, the species found in the North American subcontinent, is cross-pollinated (i.e. self-incompatible; SI) whereas *P. hexandrum* is self-compatible (SC). The sister relationship between the two species is well documented, and it is estimated that these species became separated ~6.94 ± 3.94 million years ago (Liu *et al.* 2002). The disjunction between the two species coincides with the time of upsurge of the Himalayas, and it appears that the self-pollinating mode of reproduction has evolved from the cross-fertilizing *P. peltatum* and has also been proven phylogenetically (Wang *et al.* 2007). The last rapid upsurge of the Hima-layas began ~4–3 million years ago in the late Miocene.

This suggests that a shift from SI to SC might have evolved as a result of this geographical development. The pollin-ator fauna are known to decline in terms of both species and number with rise in elevation. Thus, the uplifted habitat of the plant must have resulted in the scarcity of pollinators in early spring. In the process of evolution, the flower of *P. hexandrum* has adapted to delayed self-ing, i.e. it tends to allow earlier cross-pollination to pre-dominate when pollinators are available, which seems to be a reproductive strategy in response to the scarcity of the pollinators. Further, attractive, open, cup-shaped showy flowers with large anthers of *P. hexandrum* are characters of a cross-pollinating species. This suggests that the self-pollination mode of reproduction has evolved in this plant only to counter pollinator scarcity, although cross-pollination has not been eradicated.

Self-pollination might have been responsible for the dispersal of populations across the Himalayas following Baker's rule, which suggests that following long-distance dispersal, a solitary propagule is much more likely to reproduce and generate a sexually reproducing population if it is capable of self-fertilization (Baker 1955). If the new colony thus established is well adapted to its new environment, it can spread throughout the area, where favourable conditions are found, even though its capacity for genetic variation is greatly reduced (Stebbins 1957) and some of the traits become fixed due to genetic drift in the populations that are capable of self-pollination. These facts lead us to one of the two possible inferences derived from our study that populations from the Indian Himalayas are relics of a once-widespread ancestral stock, which subsequently became fragmented during the course of evolution. Another inference might be that all the populations prevailing in the Indian Himalayas have originated from two types of genetic populations fixed due to natural selection a long time ago. This is also evident from the fact that in the cluster analysis, most of the individuals from the similar genetic population remained in the same group regardless of their geographical location. It also suggests that during the course of evolution, genotypes favoured by natural selection have been dispersed in the Himalayan region and have maintained themselves as constant, genetically similar lines for many generations. Although there is no distinct geographical barrier shown in the cluster analysis, six out of seven populations from the Pir Panjal range cluster into one group and three out of four populations from the Dhauladhar ranges cluster into another group. Out of four populations from the Garhwal Himalayas, two populations cluster into either group, as is the case of populations from the Greater Himalayan range which also cluster into both the groups.

Gene flow between populations

Dispersal of pollens and migration of seeds determine the patterns of gene dispersion within and among populations after reproduction (Loveless and Hamrick 1984). The low levels of variance partitioned among populations suggest that a good level of gene flow is present among the populations of the Indian Himalayas, which is aptly confirmed by gene flow calculations ($N_m = 2.13$). The phenology of P. hexandrum suggests that pollen dispersal cannot be a factor accounting for the gene flow, which means that high gene flow is a result of seed dispersal. The fruits of P. hexandrum contain numerous seeds (80–120) **[see Supporting Information]**. A significant amount of gene flow was recorded across all the populations of the Indian Himalayas. The fruit is a berry which is not edible initially, but becomes edible as it ripens, and

Himalayan birds and grazing animals feed on these fruits, thus facilitating seed dispersal (Rajkumar and Ahuja 2010). Further, the seed dispersal distance depends on various factors including the flight range of the birds and migration status of the grazers. All of the sites from which these populations were collected were situated in Himalayan regions that are accessed annually by numbers of tourists through the trekking trails of the Himalayas. Collectively, these short trails form a network known as "The Greater Himalayan Trail" which ranges from Nanga Parbat in Pakistan to Namcha Barwa in Tibet and includes the Himalayan mountains falling in the vicinity of Pakistan, India, Nepal, Bhutan and part of Tibet (Harris 1992; Choegyal 2011). In the cluster analysis, the population from the Sikkim region has been clustered along with the populations from the Pir Panjal range. The Sikkim population has been situated in the Kangchenjunga Himal section of the Greater Himalayas, which is located in the northeastern part of the Himalayas. This clustering suggests that these regions experience a significant amount of anthropogenic interference, and this activity also plays a major role in the dispersal of the germplasm along these trails, helping to increase the gene flow. High gene flow results in dampening of the local adaptation due to its homogenizing effect, which prevents population differentiation.

Furthermore, various reports suggest the unsustainable extraction of various medicinal plants from the Western Himalayas (Kala 2000, 2005; Uniyal et al. 2002; Kala and Farooquee 2004; Kala et al. 2006; Larsen and Olsen 2007; Larsen 2014). One such report has been published regarding the exploitation of Picrorhiza kurroa, another endangered medicinal plant (Uniyal et al. 2011) found in habitats like those of P. hexandrum. A similar kind of exploitation occurs for P. hexandrum. One can easily find seeds and roots of P. hexandrum by visiting the local healers and traditional medicinal practitioners at high elevations. Drug dealers follow the same Greater Himalayan Trail for trading of the raw material and drugs, also facilitating seed dispersal.

The number of seeds produced per plant is large, but the germination percentage is quite low (7–45 %). The seeds remain dormant for up to 3 years (Sreenivasulu et al. 2009). The production of a large number of seeds might be an adaptive strategy, so that a few, if not all, might germinate and establish new individuals, and as mentioned earlier, a solitary propagule is potent enough to produce a population if it is capable of self-pollination. This might have resulted in a population with the same genetic pool as that of the seed which was established after long-distance dispersal. The phenomenon of establishment of genetically similar populations following long-distance seed dispersal seems to have resulted in the two genetic populations present today.

Conclusions

Based on the comprehensive molecular analysis of natural populations of *P. hexandrum,* it can be assumed that all the populations found in the Indian Himalayas are descendants of either one parent population from which two types of genotypes diverged or two different parent populations from different regions, whose dispersal to other regions was facilitated by humans, animals and birds. High variance partitioned within populations indicates that these populations are well sustained, but high levels of anthropogenic interference and habitat fragmentation are major threats for the sustainability of this plant in nature. Low levels of genetic diversity also pose a concern about the survival of the populations against ecological bottlenecks in the future. Moreover, stern conservation measures and laws need to be implemented urgently to limit the unauthorized uprooting and illegal trade of the rhizomes. The extent of overexploitation is such that during our frequent visits to the field, we found a significant amount of loss in the number of individuals within a season or two (e.g. between 2009 and 2011), the average number of plants per quadrat (1 m^2) decreasing from 2 to 0.6 at Prashar (data not shown). If effective conservation measures are not undertaken soon, we are likely to lose the invaluable genetic resources of this important medicinal plant. Inferences derived from the current study will help to guide management and conservation policies. Moreover, high-throughput sequencing efforts are required to study environmental effects on the adaptation mechanism. This type of study will be extremely helpful in understanding the genes involved in divergent selection and local adaptation.

Source of Funding

This research work was funded by the Council of Scientific and Industrial Research (CSIR), Government of India.

Contributions by the Authors

R.K.S. and A.N. designed the study. A.N. collected the populations and conducted the AFLP and data analysis. P.S.A. helped in the coordination of the study. A.N. and R.K.S. wrote and approved the final version of the manuscript. All authors have read and approved the final manuscript.

Acknowledgement

A.N. is thankful to AcSIR at CSIR-IHBT for registration in the PhD programme. The authors extend their special thanks to Dr Qazi Parvaiz Hassan for providing samples from the Kashmir region. They are also grateful to Dr S. Rajkumar for collecting samples from Kinnaur and Uttarakhand. Critical suggestions from the anonymous reviewers are greatly acknowledged. This is CSIR-IHBT communication number 3650.

Supporting Information

The following Supporting Information is available in the online version of this article –

Table S1. Analysis of molecular variance (AMOVA) between the different mountain ranges.

Figure S1. Representative AFLP profile of *P. hexandrum* samples revealed by E-ACA + M-CAG primer combination using from the automated DNA analyser (3730xl). (A) Lane window screen shot 1–96: Different *P. hexandrum* samples, (B) green fragments represent the detected fragments while red peaks are the marker fragments indicating size.

Figure S2. Pictorial representation of *P. hexandrum* plant in nature and its fruit showing numerous seeds in it.

Figure S3. Mantel test showing no correlation between genetic and geographic distance on the basis of AFLP data in 24 populations of *P. hexandrum.*

Literature Cited

Airi S, Rawal RS, Dhar U, Purohit AN. 1997. Population studies on *Podophyllum hexandrum* Royle: a dwindling, medicinal plant of the Himalaya. *Genetic Resources Newsletter* **110**:29–34.

Alam A, Naik PK, Gulati P, Gulati AK, Mishra GP. 2008. Characterization of genetic structure of *Podophyllum hexandrum* populations, an endangered medicinal herb of Northwestern Himalaya, using ISSR-PCR markers and its relatedness with podophyllotoxin content. *African Journal of Biotechnology* **7**:1028–1040.

Badola HK, Aitken S. 2003. The Himalayas of India: a treasury of medicinal plants under siege. *Biodiversity* **4**:3–13.

Baker HG. 1955. Self-compatibility and establishment after 'long-distance' dispersal. *Evolution* **9**:347–349.

Bhadula SK, Singh A, Lata H, Kuniyal CP, Purohit AN. 1996. Genetic resources of *Podophyllum hexandrum* Royle, an endangered medicinal species from Garhwal Himalaya, India. *Plant Genetic Resources Newsletter* **106**:26–29.

Broomhead AJ, Dewick PM. 1990. Tumour-inhibitory aryltetralin lignans in *Podophyllum versipelle, Diphylleia cymosa* and *Diphylleia grayi. Phytochemistry* **29**:3831–3837.

Canel C, Moraes RM, Dayan FE, Ferreira D. 2000. Podophyllotoxin. *Phytochemistry* **54**:115–120.

Choegyal L. 2011. The Great Himalaya trail: a new Nepal tourism product with both trek marketing and development rationale. *Nepal Tourism and Development Review* **1**:71–76.

Choudhary DK, Kaul BL, Khan S. 1998. Cultivation and conservation of *Podophyllum hexandrum*—an overview. *Journal of Medicinal and Aromatic Plant Sciences* **20**:1071–1073.

Crants JE. 2008. *Pollination and pollen limitation in mayapple* (Podophyllum peltatum *L.), a nectarless spring perrenial.* PhD Thesis, University of Michigan, USA.

Doyle JL. 1990. Isolation of plant DNA from fresh tissue. *Focus* **12**:13–15.

Doyle JJ, Doyle JL. 1987. A rapid DNA isolation procedure for small quantities of fresh leaf tissue. *Phytochemical Bulletin* **19**:11–15.

Earl D, vonHoldt B. 2012. STRUCTURE HARVESTER: a website and program for visualizing STRUCTURE output and implementing the Evanno method. *Conservation Genetics Resources* **4**:359–361.

Ehrlich PR, Daily GC. 1993. Population extinction and saving biodiversity. *Ambio* **22**:64–68.

Eriksson G. 2001. Conservation of noble hardwoods in Europe. *Canadian Journal of Forest Research* **31**:577–587.

Evanno G, Regnaut S, Goudet J. 2005. Detecting the number of clusters of individuals using the software STRUCTURE: a simulation study. *Molecular Ecology* **14**:2611–2620.

Excoffier L, Lischer HEL. 2010. Arlequin suite ver 3.5: a new series of programs to perform population genetics analyses under Linux and Windows. *Molecular Ecology Resources* **10**:564–567.

Frankham R. 2003. Genetics and conservation biology. *Comptes Rendus Biologies* **326**:22–29.

Gadgil M, Meher-Homji VM. 1986. Localities of great significance to conservation of India's biological diversity. *Proceedings of the Indian Academy of Sciences* **November**:165–180.

Gardens B, Authority P, Park K, Garden B, Perth W, Nutrition P. 2001. Conservation genetics of the rare and endangered *Leucopogon obtectus* (Ericaceae). *Molecular Ecology* **10**:2389–2396.

Gaudeul M, Taberlet P, Till-Bottraud I. 2000. Genetic diversity in an endangered alpine plant, *Eryngium alpinum* L. (Apiaceae), inferred from amplified fragment length polymorphism markers. *Molecular Ecology* **9**:1625–1637.

Giri A, Narasu ML. 2000. Production of podophyllotoxin from *Podophyllum hexandrum*: a potential natural product for clinically useful anticancer drugs. *Cytotechnology* **34**:17–26.

Glaubitz JC. 2004. CONVERT: a user-friendly program to reformat diploid genotypic data for commonly used population genetic software packages. *Molecular Ecology Notes* **4**:309–310.

Godt MJW, Johnson BR, Hamrick JL. 1996. Genetic diversity and population size in four rare southern Appalachian plant species. *Conservation Biology* **10**:796–805.

Harris N. 1992. The geological exploration of Tibet and the Himalaya. *The Alpine Journal* **96**:66–74.

Işik K. 2011. Rare and endemic species: why are they prone to extinction? *Turk Journal of Botany* **35**:411–417.

Issell BF, Muggia FM, Carter SK. 1984. Etoposide (VP-16): current status and new developments. London: Academic press, 233–243.

Jackson DE, Dewick PM. 1984. Aryltetralin lignans from *Podophyllum hexandrum* and *Podophyllum peltatum*. *Phytochemistry* **23**:1147–1152.

Juan C, Emerson BC, Oromí P, Hewitt GM. 2000. Colonization and diversification: towards a phylogeographic synthesis for the Canary Islands. *Trends in Ecology and Evolution* **15**:104–109.

Kala CP. 2000. Status and conservation of rare and endangered medicinal plants in the Indian trans-Himalaya. *Biological Conservation* **93**:371–379.

Kala CP. 2005. Indigenous uses, population density, and conservation of threatened medicinal plants in protected areas of the Indian Himalayas. *Conservation Biology* **19**:368–378.

Kala CP, Farooquee NA. 2004. Prioritization of medicinal plants on the basis of available knowledge, existing practices and use value status in Uttaranchal, India. *Biodiversity and Conservation* **13**:453–469.

Kala CP, Dhyani PP, Sajwan BS. 2006. Developing the medicinal plants sector in northern India: challenges and opportunities. *Journal of Ethnobiology and Ethnomedicine* **15**:32.

Knight CA, Molinari NA, Petrov DA. 2005. The large genome constraint hypothesis: evolution, ecology and phenotype. *Annals of Botany* **95**:177–190.

Larsen HO. 2014. Commercial medicinal plant extraction in the hills of Nepal: local management system and ecological sustainability. *Environmental Management* **29**:88–101.

Larsen HO, Olsen C. 2007. Unsustainable collection and unfair trade? Uncovering and assessing assumptions regarding Central Himalayan medicinal plant conservation. In: Hawksworth D, Bull A, eds. *Plant conservation and biodiversity SE—8*. The Netherlands: Springer, 105–123.

Li Y, Zhai S-N, Qiu Y-X, Guo Y-P, Ge X-J, Comes HP. 2011. Glacial survival east and west of the 'Mekong-Salween Divide' in the Himalaya-Hengduan Mountains region as revealed by AFLPs and cpDNA sequence variation in *Sinopodophyllum hexandrum* (Berberidaceae). *Molecular Phylogenetics and Evolution* **59**:412–424.

Liu J, Chen Z, Lu A. 2002. Molecular evidence for the sister relationship of the eastern Asia-North American intercontinental species pair in the *Podophyllum* group (Berberidaceae). *Botanical Bulletin of Academia Sinica* **43**:147–154.

Loveless JL, Hamrick MD. 1984. Ecological determinants of genetic structure in plant populations. *Annual Review of Ecology and Systematics* **15**:65–95.

Luikart G, England PR, Tallmon D, Jordan S, Taberlet P. 2003. The power and promise of population genomics: from genotyping to genome typing. *Nature Reviews Genetics* **4**:981–994.

Lynch M, Milligan BG. 1994. Analysis of population genetic structure with RAPD markers. *Molecular Ecology* **3**:91–99.

Ma S, Hu Z. 1996. Preliminary studies on the distribution pattern and ecological adaptation of *Sinopodophyllum hexandrum* (Royle) Ying (Berberidaceae). *Journal of Wuhan Botanical Research* **14**:47–54.

Mariette S, Cottrell J, Csaikl UM, Goikoechea P, Konig A, Lowe AJ, Dam BC, Van, Barreneche T, Bodenes C, Streiff R, Burg K, Groppe K, Munro RC, Tabbener H, Kremer A. 2002. Comparison of levels of genetic diversity detected with AFLP and microsatellite markers within and among mixed *Q. petraea* (Matt.) Liebl. and *Q. robur* L. Stands. *Silvae Genetica* **51**:2–3.

Meirmans PG. 2012. AMOVA-based clustering of population genetic data. *The Journal of Heredity* **103**:744–750.

Meirmans PG, Van Tienderen PH. 2004. Genotype and genodive: two programs for the analysis of genetic diversity of asexual organisms. *Molecular Ecology Notes* **4**:792–794.

Mueller UG, Wolfenbarger LL. 1999. AFLP genotyping and fingerprinting. *Trends in Ecology and Evolution* **14**:389–394.

Nag A, Rajkumar S. 2011. Chromosome identification and karyotype analysis of *Podophyllum hexandrum* Roxb. ex Kunth using FISH. *Physiology and Molecular Biology of Plants* **17**:313–316.

Nag A, Chanda S, Rajkumar S. 2011. Estimation of nuclear genome size of important medicinal plant species from Western Himalaya using flow cytometry. *Journal of Cell and Plant Sciences* **2**:19–23.

Naik PK, Alam A, Singh H, Goyal V, Parida S. 2010. Assessment of genetic diversity through RAPD, ISSR and AFLP markers in *Podophyllum hexandrum*: a medicinal herb from the Northwestern Himalayan region. *Physiology and Molecular Biology of Plants* **16**:135–148.

Neel M, Ellstrand N. 2003. Conservation of genetic diversity in the endangered plant *Eriogonum ovalifolium* var. *vineum* (Polygonaceae). *Conservation Genetics* **4**:337–352.

Negi VS, Maikhuri RK, Vashishtha DP. 2011. Traditional healthcare practices among the villages of Rawain valley, Uttarkashi, Uttarakhand, India. *Indian Journal of Traditional Knowledge* **10**: 533–537.

Newton AC, Allnutt TR, Gillies ACM, Lowe AJ, Ennos RA. 1999. Molecular phylogeography, intraspecific variation and the conservation of tree species. *Trends in Ecology and Evolution* **14**:140–145.

Palacios C, Kresovich S, Gonzalez-Candelas F. 1999. A population genetic study of the endangered plant species *Limonium dufourii* (Plumbaginaceae) based on amplified fragment length polymorphism (AFLP). *Molecular Ecology* **8**:645–657.

Peakall R, Smouse PE. 2006. genalex 6: genetic analysis in Excel. Population genetic software for teaching and research. *Molecular Ecology Notes* **6**:288–295.

Peakall R, Smouse PE. 2012. GenAlEx 6.5: genetic analysis in Excel. Population genetic software for teaching and research—an update. *Bioinformatics* **28**:2537–2539.

Perrier X, Jacquemoud-Collet JP. 2006. Dissimilarity analysis and representation for windows. http://darwin.cirad.fr/.

Petit RJ, El Mousadik A, Pons O. 1998. Identifying populations for conservation on the basis of genetic markers. *Conservation Biology* **12**:844–855.

Pritchard JK, Stephens M, Donnelly P. 2000. Inference of population structure using multilocus genotype data. *Genetics* **155**: 945–959.

Purohit MC, Bahuguna R, Maithani UC, Purohit AN, Rawat MSM. 1999. Variation in podophylloresin and podophyllotoxin contents in different populations of *Podophyllum hexandrum*. *Current Science* **77**:1078–1079.

Rajkumar S, Ahuja PS. 2010. Developmental adaptation of leaves in *Podophyllum hexandrum* for effective pollination and dispersal. *Current Science* **99**:1518–1519.

Shaobin M, Zhengrao X, Zhihao H. 1997. A contribution to the reproductive biology of *Sinopodophyllum hexandrum* (Royle) Ying (Berberidaceae). *Acta Botanica Boreali-Occidentalia Sinica* **1**:49–55.

Silva L, Elias R, Moura M, Meimberg H, Dias E. 2011. Genetic variability and differentiation among populations of the Azorean endemic gymnosperm *Juniperus brevifolia*: baseline information for a conservation and restoration perspective. *Biochemical Genetics* **49**: 715–734.

Sreenivasulu Y, Chanda SK, Ahuja PS. 2009. Endosperm delays seed germination in *Podophyllum hexandrum* Royle—an important medicinal herb. *Seed Science and Technology* **37**:10–16.

Staub JE, Danin-poleg Y, Fazio G, Horejsi T, Reis N, Katzir N. 2000. Comparative analysis of cultivated melon groups (*Cucumis melo* L.) using random amplified polymorphic DNA and simple sequence repeat. *Euphytica* **115**:225–241.

Stebbins GL. 1957. Self-fertilization and population variability in the higher plants. *The American Naturalist* **XCI**:337–354.

Tero N, Aspi J, Siikamaki P, Jakalaniemi A, Tuomi J. 2003. Genetic structure and gene flow in a metapopulation of an endangered plant species, *Silene tatarica*. *Molecular Ecology* **12**: 2073–2085.

Tyler VE, Brady LR, Robbers JE. 1988. Pharmacognosy, 9th edn. Philadelphia: Lea and Fabiger.

Uniyal A, Uniyal SK, Rawat GS. 2011. Commercial extraction of *Picrorhiza kurrooa* Royle ex Benth. in the Western Himalaya. *Mountain Research and Development* **31**:201–208.

Uniyal SK, Awasthi A, Rawat GS. 2002. Current status and distribution of commercially exploited medicinal and aromatic plants in upper Gori valley, Kumaon Himalaya, Uttaranchal. *Current Science* **82**:1246–1252.

Vekemans X, Beauwens T, Lemaire M, Roldán-Ruiz I. 2002. Data from amplified fragment length polymorphism (AFLP) markers show indication of size homoplasy and of a relationship between degree of homoplasy and fragment size. *Molecular Ecology* **11**: 139–151.

Vos P, Hogers R, Bleeker M, Reijans M, van de Lee T, Hornes M, Friters A, Pot J, Paleman J, Kuiper M, Zabeau M. 1995. AFLP: a new technique for DNA fingerprinting. *Nucleic Acids Research* **23**:4407–4414.

Wang W, Chen Z, Liu Y, Li R, Li J. 2007. Phylogenetic and biogeographic diversification of Berberidaceae in the northern hemisphere. *Systematic Botany* **32**:731–742.

WHO. 1993. *Guidelines on the conservation of medicinal plants*. Switzerland: IUCN, WHO & WWF.

Wyatt R. 1988. Phylogenetic aspects of the evolution of self-pollination. In: Gottlieb L, Jain S, eds. *Plant evolutionary biology*. The Netherlands: Springer, 109–131.

Xiao M, Li Q, Guo L, Luo T, Duan W, He W, Wang L, Chen F. 2006a. AFLP analysis of genetic diversity of the endangered species *Sinopodophyllum hexandrum* in the Tibetan region of Sichuan Province, China. *Biochemical Genetics* **44**:47–60.

Xiao M, Li Q, Wang L, Guo L, Li J, Tang L, Chen F. 2006b. ISSR analysis of the genetic diversity of the endangered species *Sinopodophyllum hexandrum* (Royle) Ying from Western Sichuan Province, China. *Journal of Integrative Plant Biology* **48**: 1140–1146.

Xiong Y-Z, Fang Q, Huang S-Q. 2013. Pollinator scarcity drives the shift to delayed selfing in Himalayan mayapple *Podophyllum hexandrum* (Berberidaceae). *AoB PLANTS* **5**: plt037; doi:10.1093/aobpla/plt037.

Xu ZY, Ma SB, Hu CP, Yang CY, Hu ZH. 1997. The floral biology and its evolutionary significance of *Sinopodophyllum hexandrum* (Royle) Ying (Berberidaceae). *Journal of Wuhan Botanical Research* **15**: 223–227.

Young A, Boyle T, Brown T. 1996. The population genetic consequences of habitat fragmentation for plants. *Trends in Ecology and Evolution* **5347**:413–418.

Zhivotovsky LA. 1999. Estimating population structure in diploids with multilocus dominant DNA markers. *Molecular Ecology* **8**: 907–913.

8

Genetic diversity and population structure of an extremely endangered species: the world's largest Rhododendron

Fu Qin Wu[1,†], Shi Kang Shen[1*†], Xin Jun Zhang[1], Yue Hua Wang[1] and Wei Bang Sun[2]

[1] Present address: School of Life Sciences, Yunnan University, Kunming No. 2, Green Lake North Road, Kunming, Yunnan 650091, The People's Republic of China
[2] Kunming Botanical Garden, Kunming Institute of Botany, Chinese Academy of Sciences, Kunming 650201, The People's Republic of China

Associate Editor: Kermit Ritland

Abstract. Comprehensive studies on the genetic diversity and structure of endangered species are urgently needed to promote effective conservation and management activities. The big tree rhododendron, *Rhododendron protistum* var. *giganteum*, is a highly endangered species with only two known endemic populations in a small area in the southern part of Yunnan Province in China. Unfortunately, limited information is available regarding the population genetics of this species. Therefore, we conducted amplified fragment length polymorphism (AFLP) analysis to characterize the genetic diversity and variation of this species within and between remaining populations. Twelve primer combinations of AFLP produced 447 unambiguous and repetitious bands. Among these bands, 298 (66.67 %) were polymorphic. We found high genetic diversity at the species level (percentage of polymorphic loci = 66.67 %, $h = 0.240$, $I = 0.358$) and low genetic differentiation ($G_{st} = 0.110$) between the two populations. Gene flow between populations (N_m) was relatively high at 4.065. Analysis of molecular variance results revealed that 22 % of the genetic variation was partitioned between populations and 78 % of the genetic variation was within populations. The presence of moderate to high genetic diversity and low genetic differentiation in the two populations can be explained by life history traits, pollen dispersal and high gene flow ($N_m = 4.065$). Bayesian structure and principal coordinate analysis revealed that 56 sampled trees were clustered into two groups. Our results suggest that some rare and endangered species are able to maintain high levels of genetic diversity even at small population sizes. These results will assist with the design of conservation and management programmes, such as *in situ* and *ex situ* conservation, seed collection for germplasm conservation and reintroduction.

Keywords: AFLP markers; big tree rhododendron; conservation strategies; genetic diversity; *Rhododendron protistum* var. *giganteum*; small population.

Introduction

Genetic diversity is one aspect of biological diversity that is extremely important for conservation strategies (Kaljund and Jaaska 2010; Gordon *et al.* 2012). It is well known

that preserving the genetic diversity of endangered species can significantly affect their long-term survival and evolution in changing environments (Frankham *et al.* 2002). Therefore, knowledge of the genetic diversity and

* Corresponding author's e-mail address: shikang168@yahoo.com
† These authors contributed equally to this work.

population structure of endangered plant species is crucial for their conservation and management (Frankham 2003; Gordon *et al.* 2012; Lopes *et al.* 2014). Population size is considered an important factor for maintaining genetic variation. Small populations are more vulnerable than large ones to extinction because of environmental stochasticity, genetic drift and inbreeding. Genetic drift decreases heterozygosity and eventual fixation of alleles, and inbreeding increases homozygosity within populations (Frankham 2005). In general, a drop in population size may cause the decline of genetic diversity by genetic drift and inbreeding. In the longer term, diminished genetic diversity may cause a loss of fitness and evolutionary capacity to adapt to environmental changes (Lande 1993; Kaljund and Jaaska 2010). Therefore, quantifying patterns of genetic variability and diversity within and among different populations is very important for small population species conservation and management planning.

Big tree rhododendron, *Rhododendron protistum* var. *giganteum*, exhibits a very limited distribution, with only two populations found in the Gaoligong Mountains of northwestern Yunnan Province in China (Fig. 1). In addition, only ~1500 individual plants have been found (Ma *et al.* 2012). Because of this situation, big tree rhododendron has been included in the Red List of Critically Endangered Species in China (Fu 1992) and protected under the Conservation Programme for Wild Plants with Extremely Small

Population in China (2012–15 operational plan) (Ma *et al.* 2013a). However, this species is still at risk of extinction because of continued habitat disturbance. Thus, genetic data on big tree rhododendron are urgently needed to inform current and future conservation activities.

In this study, we investigated the genetic diversity and structural patterns of big tree rhododendron both within and between the only known two natural populations. We used amplified fragment length polymorphism (AFLP) markers because this technique has been successfully employed in other studies that evaluated the genetic diversity of other *Rhododendron* species (Escaravage *et al.* 1998; Chappell *et al.* 2008; Erfmeier and Bruelheide 2011; Zhao *et al.* 2012a; Li *et al.* 2012a). Our study aimed to (i) characterize the level of genetic diversity in big tree rhododendron; (ii) reveal the distribution of genetic variation within and between the two remaining populations; and (iii) discuss possible implications of these population genetic data for management and conservation. We hypothesized that genetic diversity level would be low because of the small size of the remaining populations of big tree rhododendron.

Methods

Study species

Rhododendron is the largest woody plant genus in the Ericaceae family with at least 1025 species and one of

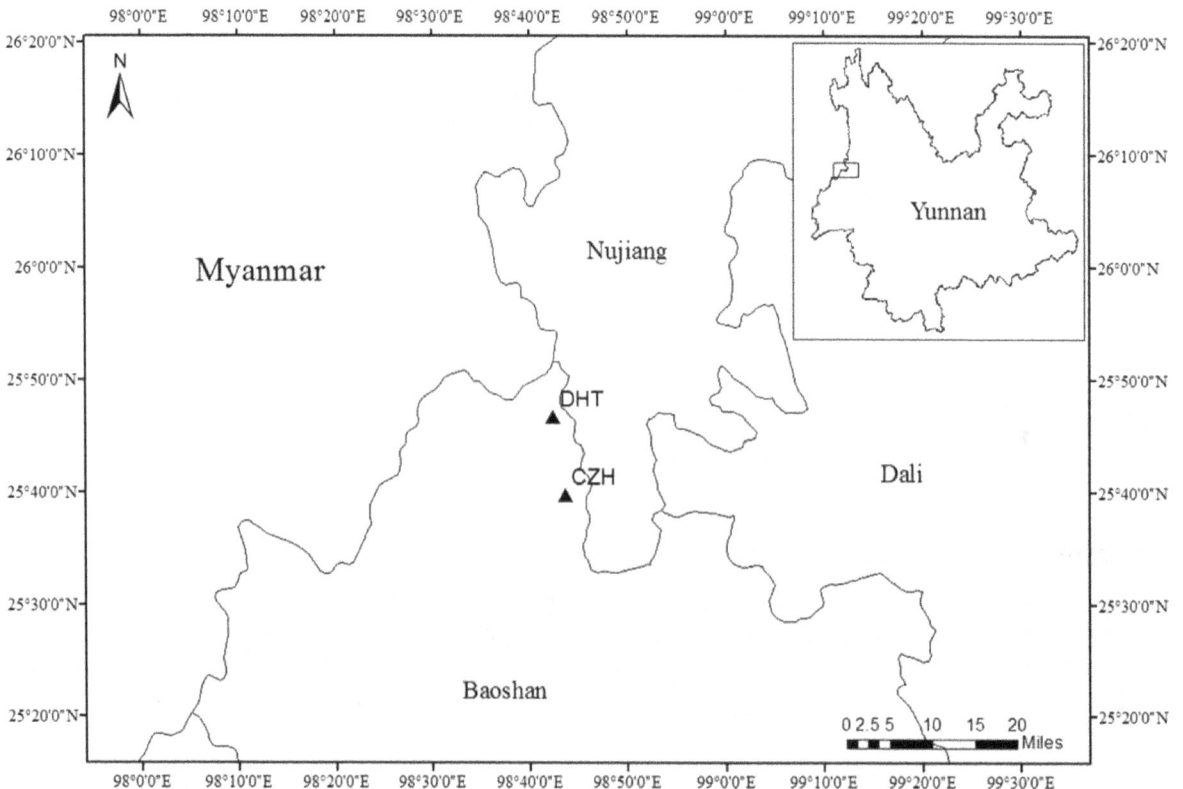

Figure 1. Location of the two populations of big tree rhododendron included in this study. CZH and DHT are population codes.

the most common woody plants that are distributed across the northern temperate zone, tropical southeast Asia and northeastern Australia (Chamberlain *et al.* 1996). Approximately 70 % of >500 *Rhododendron* species are endemic in China, and most of these species thrive in the northwestern part of Yunnan province. Thus, northwestern Yunnan has been recognized as one of the diversification and differentiation centres of modern *Rhododendron* (Ma *et al.* 2013*b*). *Rhododendron* species are not only the major composition of alpine and sub-alpine vegetation, but also the world-wide famous woody ornamental plants.

Big tree rhododendron, *R. protistum* var. *giganteum*, was first identified and named by George Forrest in 1919. It belongs to subgenus *Hymenanthes* and subsection *Grandia* (Fang *et al.* 2005). This species is one of the tallest and most ancient rhododendron trees, reaching 30 m in height and 1 m in basal diameter. Big tree rhododendron is an evergreen tree. It is characterized by large, deep and purple-red flowers and oblong-lanceolate to oblong-oblanceolate leaves with continuous and loose indumentums in the abaxial surface. Flowering stage occurs during January to March when flowers abundantly bloom. Fruiting period happens between October and December. Mature seeds are small with thin membranous wings (Fang *et al.* 2005). This plant is a very important germplasm source with high ornamental value.

Plant materials

According to our previous field investigation, two remaining big tree rhododendron populations are distributed in the Gaoligong Mountain National Nature Reserve, Tengchong County. The Gaoligong Mountains are part of the Hengduan Mountain chain that belongs to a global hotspot for biodiversity. This area is found in the border area between southwestern China and northern Myanmar between 24°40′ and 28°30′N latitude. These mountains cover a total area of 111 000 km². Approximately 3990 species of seed plants with 117 species of *Rhododendron* thrive in this region (Li *et al.* 2000; Ma *et al.* 2013*a*). In November 2012, 56 big tree rhododendron individuals were collected from the two natural populations in which 30 of these samples were collected from the Cizhuhe (CZH) population, whereas 26 samples were collected from the Dahetou (DHT) population. The distance between the collected individual samples was at least 15 m. Fresh young leaves were removed from shoots, dried in silica gel and stored at −20 °C until DNA was extracted. The detailed information regarding locations and population codes of the samples is shown in Fig. 1.

DNA extraction

Genomic DNA was extracted from dried leaves by using a modified cetyltrimethylammonium bromide protocol (Doyle and Doyle 1987). Total purified DNA was detected by 1.0 % agarose gel electrophoresis and stored at −20 °C until use.

AFLP fingerprinting

Amplified fragment length polymorphism fingerprinting was performed in accordance with the method described by Vos *et al.* (1995) with minor modifications.

Genomic DNA (150–450 ng) was double-digested using restriction endonucleases EcoRI (5 U) and MseI (5 U) (New England Biolabs, USA) in a total reaction volume of 20 μL for 4 h at 37 °C. This reaction was deactivated at 70 °C for 20 min. The digested DNA (4 μL) was added to 16 μL of the ligation mixture with 50 pmol MseI adaptor, 5 pmol EcoRI adaptor and 2 U T4 DNA ligase (New England Biolabs). This mixture was incubated at 16 °C for 16 h, and then deactivated at 65 °C for 10 min.

The ligated DNA (2 μL) was pre-selected using 33 ng of the primers for MseI and EcoRI adaptors (Table 1) with 5 nmol dNTPs, 10× Taq buffer (2.5 μL with Mg²⁺-free medium) and 1.5 U DNA Taq polymerase (Transgen Biotech, Beijing, China) to yield a total volume of 25 μL. Pre-selective polymerase chain reaction (PCR) amplification profiles are listed as follows: 94 °C for 3 min; 30 cycles of 30 s denaturing at 94 °C, 30 s annealing at 56 °C and

Table 1. Polymorphism and primer informativeness of 12 AFLP primer combinations. PPL, percentage of polymorphic loci.

Selective nucl.	Amplification bands	Polymorphism bands	PPL (%)
M-CAA/E-AGC	29	14	48.28
M-CAC/E-AAG	37	27	72.97
M-CAC/E-ACA	44	22	50.00
M-CAC/E-ACC	43	32	74.42
M-CAC/E-AGG	50	28	56.00
M-CAG/E-AAG	31	24	77.42
M-CAG/E-ACC	40	21	52.50
M-CAG/E-ACG	37	29	78.38
M-CAG/E-AGC	36	28	77.78
M-CAT/E-ACG	35	28	80.00
M-CTA/E-ACG	34	27	79.41
M-CTA/E-AGC	31	18	58.06
Total	447	298	–
Mean	37.25	24.83	67.10

60 s elongation at 72 °C; and holding at 72 °C for 5 min. The pre-selective PCR products were then diluted 10-fold, and the template was used for selective amplifications.

An initial screening was performed using two individuals from each area by using 64 primer combinations for selective amplifications. A total of 12 primer combinations (Table 1), which generated clear and abundant bands, were chosen for selective PCR. Selective amplification was performed in a 25-μL reaction volume by using selective primer pairs with 5 nmol dNTPs, $10 \times$ Taq buffer (2.5 μL with Mg^{2+}-free medium) and 1.5 U DNA Taq polymerase (Transgen Biotech). The selective PCR amplification profile was obtained by denaturation at 94 °C for 30 s, annealing at 65 °C for 30 s, temperature decrease by 0.9 °C/cycle and extension at 72 °C for 1 min for 12 cycles; followed by 94 °C for 30 s, 56 °C for 30 s, 72 °C for 1 min for 23 cycles and 72 °C for 5 min.

Amplified DNA products were mixed with 98 % formamide loading buffer (10 μL), heated at 95 °C for 7 min and immediately cooled in an ice bath for 30 min. The products were resolved by 6 % polyacrylamide gel electrophoresis in $0.5 \times$ Tris–borate–EDTA buffer by using a 100-bp DNA ladder marker (Transgen Biotech) for 4 h at 1500 V and then stained with 0.1 % silver nitrate.

Data analysis

All individuals were manually scored as either '1' or '0' corresponding to the presence or absence of AFLP bands (100–700 bp), respectively, to construct a binary matrix. The binary matrix was edited using GenALEx version 6.4.1 (Peakall and Smouse 2006). The percentage of polymorphic loci (PPL), effective number of alleles (N_e), Nei's genetic diversity (h), Shannon's information index (I), level of gene flow (N_m), total gene diversity (H_t), variability within populations (H_s) and coefficient of genetic differentiation (G_{st}) were calculated using POPGENE version 1.32 (Yeh et al. 1999) and GenALEx version 6.4.1 (Peakall and Smouse 2006) with manual corrections.

Analysis of molecular variance (AMOVA) was conducted to calculate the partitioning of genetic variation between and within the two populations by using GenALEx version 6.4.1 (Peakall and Smouse 2006).

We conducted a Bayesian analysis of the population structures by using STRUCTURE version 2.2 (Pritchard et al. 2000). A total of 10 independent runs were performed for each set with K ranging from 2 to 20, a burn-in of 1×10^5 iterations and 1×10^5 subsequent Markov Chain Monte Carlo steps. The combination of admixture and correlated allele frequency models was also analysed. The second-order rate of change in the log probability of the data with respect to the number of clusters (ΔK) was also used to estimate the probable likely number of genetic clusters (Evanno et al. 2005). The best-fit number of groupings was evaluated using ΔK by STRUCTURE HARVESTER version 0.6.8 (Earl and von Holdt 2012). Furthermore, principal coordinate analysis (PCoA) was also employed to examine the genetic relationships between the detected populations by using GenALEx version 6.4.1 (Peakall and Smouse 2006).

Results

Genetic diversity

Among 64 previously published primer pairs, 12 could amplify well-distributed fragments with good distinction, were highly polymorphic and ranged from 100 to 700 bp. A total of 12 AFLP primers were used in the entire dataset in this study. This dataset generated 447 clear and quantifiable fragments that ranged from 100 to 700 bp in 56 individuals from the two wild populations. Among 447 loci, 298 (66.67 %) were polymorphic. The total number of fragments of each primer combination ranged from 29 (M-CAA/E-AGC) to 50 (M-CAC/E-AGG) with an average of 37.25. Percentage polymorphism varied from 48.28 to 80.00 % with an average of 67.10 % per primer combination (Table 1).

Nei's gene diversity (h) and Shannon's information index (I), in the combined data matrix of all 12 primers, were 0.240 and 0.358, respectively. The genetic diversities within species (H_t) and within populations (H_s) were 0.238 and 0.212, respectively (Table 2). The genetic differentiation between the populations (G_{st}) was 0.110. Based on the G_{st} value, the level of gene flow (N_m) was estimated as

Table 2. Genetic diversity, differentiation parameters of two wild populations of big tree rhododendron. PPL, percentage of polymorphic loci; h, Nei's (1973) gene diversity; I, Shannon's information index; H_t, total variability; H_s, variability within populations; G_{st}, coefficient of genetic differentiation; N_m, estimate of gene flow.

Name	PPL (%)	h	I	H_t	H_s	G_{st}	N_m
CZH	63.31	0.234	0.346	–	–	–	–
DHT	51.90	0.190	0.281	–	–	–	–
Species level	66.67	0.240	0.358	0.238	0.212	0.110	4.065

Table 3. Analysis of molecular variance based on AFLP markers for the two populations of big tree rhododendron.

Source of variation	d.f.	Sum of squares	Variation components	Percentage of variation (%)
Among populations	1	355.866	11.318	22.00
Within populations	54	2191.313	40.580	78.00
Total	55	2547.179	51.898	

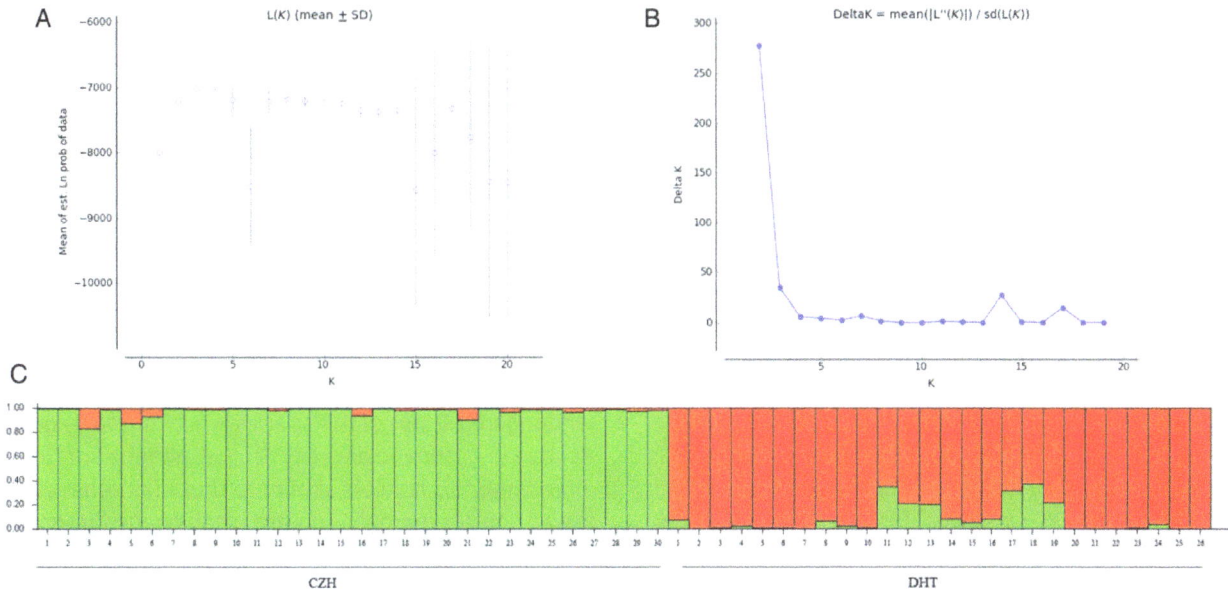

Figure 2. Results of the Bayesian model-based clustering STRUCTURE analysis of 56 individuals of big tree rhododendron. (A) The probability of the data ln $P(D)$ ($\pm$ SD) against the number of K clusters, and increase of ln $P(D)$ given K, calculated as $(LnP(D)k - LnP(D)k - 1)$. (B) ΔK values from the mean log-likelihood probabilities from STRUCTURE runs where inferred clusters (K) ranged from 1 to 20. (C) Estimated genetic clustering ($K = 2$) obtained with the STRUCTURE program for 56 individuals. Individuals are separated according to the population, and the black vertical line in the bar chart is population identifier.

4.065 ($N_m > 1$). These results indicated high gene flow and low differentiation between extant populations.

At the population level, the CZH population (PPL = 63.31 %, $h = 0.234$, $I = 0.346$) showed a higher genetic diversity level than the DHT population (PPL = 51.90 %, $h = 0.190$, $I = 0.281$; Table 2).

Genetic structure

Analysis of molecular variance results revealed that 22.00 % of the genetic variation was partitioned between populations and 78.00 % was observed within populations based on AFLP markers (Table 3). These results indicated low genetic variation levels between the two populations.

The genetic structure of big tree rhododendron was analysed on the basis of AFLP markers by using STRUCTURE and PCoA. The STUCTURE analysis based on the ΔK method revealed that ΔK was 277.7 for $K = 2$ and ΔK was <36 for all of the values of K (ranging from 3 to 20) (Fig. 2A and B). Therefore, the optimal ΔK for $K = 2$ showed that the best-fit model for the sampled 56

individuals of big tree rhododendron revealed two clusters (Fig. 2C). These 56 individuals formed a clear separation between CZH and DHT populations except a few admixed individuals, indicating weak differentiation.

The existence of two groups was also supported by the PCoA (Fig. 3). Two-dimensional PCoA separated the 56 samples into two distinct clusters along the two axes. The F1 axis separated the DHT population, whereas the F2 axis further resolved the CZH population. The first and second principal coordinates accounted for 37.74 and 17.89 % of the total genetic variation, respectively.

Discussion

Genetic diversity

The genetic diversity of species in small populations is lower than that in large populations because of genetic drift and inbreeding (Willi *et al.* 2006; Li *et al.* 2012b). Therefore, rare and endangered species with narrow geographical distributions likely maintain lower genetic

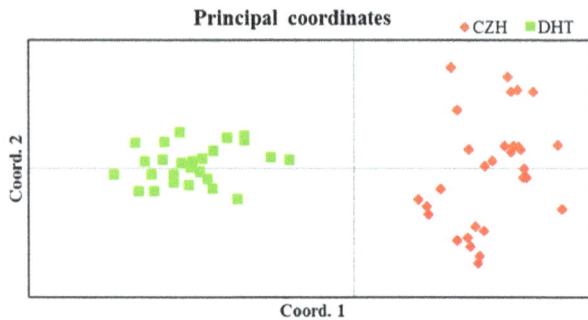

Figure 3. A two-dimensional plot of the PCoA of 56 individuals of big tree rhododendron. The first and second principal coordinates account for 37.74 and 17.89 % of total variation, respectively.

diversity than similar species with widespread geographical distributions (Hamrick and Godt 1989). In the present study, genetic diversity within big tree rhododendron was detected using AFLP markers. Unexpectedly, our study showed that big tree rhododendron showed a higher level of genetic diversity (PPL = 66.67 %, $h = 0.240$, $I = 0.358$) at the species level than other critically endangered tree species such as *Metrosideros bartlettii* (PPL = 44 %) (Drummond et al. 2000), *Abies yuanbaoshanensis* (PPL = 50.96 %, $h = 0.151$, $I = 0.1735$) (Wang et al. 2003), *Metrosideros boninensis* (PPL = 12.9 %, $h = 0.024$, $I = 0.039$) (Kaneko et al. 2008) and *Ostrya rehderiana* (PPL = 29.90 %) (Li et al. 2012b). In contrast, we found that this rhododendron exhibited a lower level of genetic diversity than reported for *Litsea szemaois* (PPL = 80.8 %) (Ci et al. 2008). However, a moderate level of genetic diversity was detected in big tree rhododendron compared with genetic parameters estimated using AFLP markers from other *Rhododendron* species (Chappell et al. 2008; Erfmeier and Bruelheide 2011; Zhao et al. 2012a). The results did not support our hypothesis that big tree rhododendron would exhibit low levels of genetic diversity because of its small populations. Our results generally supported the view that some rare and endangered species can maintain high levels of genetic diversity even at small population sizes (Rossetto et al. 1995; Ci et al. 2008; Gordon et al. 2012; Zhao et al. 2012b).

The maintenance of high genetic diversity in big tree rhododendron, including its mating system, life form and natural selection, can be explained by several possible reasons. In general, biological traits, reproductive mode and breeding system have often been regarded as important factors that affect genetic diversity levels. Outcrossing species usually have considerably higher levels of genetic diversity than selfing species (Hamrick and Godt 1989; Nybom 2004). Previous studies suggested that the mating system of *Rhododendron* may be predominantly outcrossed because of the need for a pollinator (Hirao et al. 2006; Ono et al. 2008; Hirao 2010). In a

field survey, big tree rhododendron is pollinated by insect vectors and birds (S. K. Shen, unpubl. data), which largely promote outcrossing. The seeds of big tree rhododendron are small with wing-like structures that allow them to be dispersed by the wind. Moreover, tree species, even though their populations are declining, usually maintain higher levels of genetic polymorphisms than short-lived herbaceous species (Hamrick and Godt 1989; Nybom 2004). The big tree rhododendron lives for decades to centuries, and this characteristic is highly advantageous to retain genetic variation. The big tree rhododendron is only distributed in two neighbouring populations with small ranges. The two remaining populations may exhibit high genetic diversity derived from the ancestral population; this result is similar to that in a previous study focusing on another critically endangered plant, namely, *Tricyrtis ishiiana* (Setoguchi et al. 2011).

High genetic variation enables species to adapt to changing environments (Zhao et al. 2012b). The presence of moderate to high genetic diversity in the two populations of big tree rhododendron indicated that the current endangered status of this species is not caused by genetic factors (e.g. genetic diversity decline, genetic drift and inbreeding); this result is similar to that in another endangered species, namely, *Tupistra pingbianensis* (Qiao et al. 2010). The main threat to this plant species may be habitat specialization. Although we did not conduct a detailed habitat survey, the limited distribution and few seedlings found in the wild may partly support this hypothesis. However, the factors that lead to the sustainable declining of this population should be further elucidated.

Genetic structure

Genetic structure is affected by several factors, such as breeding systems, genetic drift, population size, seed dispersal, gene flow, evolutionary history and natural selection (Hamrick and Godt 1990). Our analyses of genetic structure showed that the 56 individuals formed a clear separation between CZH and DHT populations except a few admixed individuals; this result indicated weak differentiation (Fig. 2). This conclusion is also supported by the PCoA (Fig. 3). Analysis of molecular variance analysis of two populations showed that 22 % of the genetic variation occurred between CZH and DHT populations, whereas 78 % of the genetic variation occurred within these populations (Table 3). The coefficients of genetic differentiation (G_{st}) and gene flow (N_m) between the two extant populations were 0.110 and 4.065, respectively. Long distance dispersal of pollen and/or seeds results in low genetic differentiation and high gene flow between populations of the same species (Yao et al. 2007; Zhao et al. 2012b). Previous studies found that the seeds of *Rhododendron* species are frequently dispersed by the wind, and the distance of seed dispersal approximately ranges from 30 to 80 m (Ng and Corlett 2000). However, *Rhododendron* pollen can also be transmitted by insect vectors (bees) and birds (Hirao *et al.* 2006; Ono *et al.* 2008; Hirao 2010). Moreover, these pollens can commonly be moved at a distance ranging from 3 to 10 km (Ng and Corlett 2000). The geographical distance between the two extant populations of big tree rhododendron is ~8 km. Thus, it is reasonable to assume that the frequent and continual gene flow between the two populations of big tree rhododendron occurs by pollen dispersal.

Conservation implications

Our study of the genetic structure of big tree rhododendron has important implications for the conservation and management of this narrowly distributed and extremely rare species. Undoubtedly, *in situ* conservation is considered as the most effective method to conserve endangered species (Shen *et al.* 2009). The presence of high genetic diversity in big tree rhododendron indicates that the major factors that threaten the persistence of its population are ecological factors (e.g. habitat specialization) rather than genetic. Considering its habitat specialization and extremely limited distribution, we suggest that management policies should be improved to maintain the appropriate effective population size of big tree rhododendron and to protect its natural habitats. Furthermore, previous studies proposed that mature individuals in populations should be conserved to protect reproductive fitness and evolutionary potential of the species (Cruse-Sanders *et al.* 2005). For instance, adult big tree rhododendrons are critical resources, not only to maintain current genetic diversity but also to provide provenance for its future recovery. Thus, protecting adult trees should be the priority in conservation to ensure ongoing recruitment. In addition, *ex situ* conservation is important to support the recovery of wild populations. The two existing populations have unique genotypes, as detected in the Bayesian clustering analysis. With these findings, we recommend that seeds be collected for germplasm storage and *ex situ* conservation of both the CZH and DHT populations.

Conclusions

Population genetic diversity and structure of extremely small populations of big tree rhododendron were examined in this study using AFLP markers, and we detected moderate to high genetic diversity at the species level, but low genetic differentiation between the two extant populations. These results suggest that some rare and endangered species are able to maintain high levels of genetic diversity even at small population sizes. Our hope is that these results will help design species conservation and management programmes, such as *in situ* and *ex situ* conservation, seed collection for germplasm conservation and reintroduction.

Sources of Funding

This study was financially supported by grants 31360155 and U1302262 from the National Science Foundation of China and grant 2011FB001 from the Application foundation projects of Yunnan Province and Graduate Science Innovation projects of Yunnan University (ynuy201343).

Contributions by the Authors

F.Q.W., S.K.S., Y.H.W., W.B.S. initiated and designed the research; S.K.S. and W.B.S. obtained funding for the study; S.K.S., F.Q.W. and X.J.Z. collected materials and performed the experiments, F.Q.W. and S.K.S. analysed the data and wrote the paper. All authors read, edited and agreed to submit the manuscript.

Acknowledgements

The authors thank Xingchao Zhang, from Administration of Gaoligong Mountain National Nature Reserve, Yunnan Province, for his assistance with field sampling, and Dr Ai-Li Zhang and Dr Guan-Song Yang for their help in the experiments and data analysis.

Literature Cited

Chamberlain DF, Hyam R, Argent G, Fairweather G, Walter KS. 1996. *The genus Rhododendron: its classification and synonymy.* Edinburgh: Royal Botanic Garden Edinburgh.

Chappell M, Robacker C, Jenkins TM. 2008. Genetic diversity of seven deciduous azalea species (*Rhododendron* spp. section *Pentanthera*) native to the eastern United States. *Journal of the American Society for Horticultural Science* **133**:374–382.

Ci XQ, Chen JQ, Li QM, Li J. 2008. AFLP and ISSR analysis reveals high genetic variation and inter-population differentiation in fragmented populations of the endangered *Litsea szemaois* (Lauraceae) from Southwest China. *Plant Systematics and Evolution* **273**:237–246.

Cruse-Sanders JM, Hamrick JL, Ahumada JA. 2005. Consequences of harvesting for genetic diversity in American ginseng (*Panax quinquefolius* L.): a simulation study. *Biodiversity and Conservation* **14**: 493–504.

Doyle JJ, Doyle JL. 1987. A rapid DNA isolation procedure for small quantities of fresh leaf tissue. *Phytochemical Bulletin* **19**:11–15.

Drummond RSM, Keeling DJ, Richardson TE, Gardner RC, Wright SD. 2000. Genetic analysis and conservation of 31 surviving individuals of a rare New Zealand tree, *Metrosideros bartlettii* (Myrtaceae). *Molecular Ecology* **9**:1149–1157.

Earl DA, vonHoldt BM. 2012. STRUCTURE HARVESTER: a website and program for visualizing STRUCTURE output and implementing the Evanno method. *Conservation Genetics Resources* **4**:359–361.

Erfmeier A, Bruelheide H. 2011. Maintenance of high genetic diversity during invasion of *Rhododendron ponticum*. *International Journal of Plant Sciences* **172**:795–806.

Escaravage N, Questiau S, Pornon A, Doche B, Taberlet P. 1998. Clonal diversity in a *Rhododendron ferrugineum* L. (Ericaceae) population inferred from AFLP markers. *Molecular Ecology* **7**: 975–982.

Evanno G, Regnaut S, Goudet J. 2005. Detecting the number of clusters of individuals using the software STRUCTURE: a simulation study. *Molecular Ecology* **14**:2611–2620.

Fang MY, Fang RZ, He MY, Hu LZ, Yang HB, Qin HN, Min TL, Chamberlain DF, Stevens P, Wallace GD, Anderberg A. 2005. *Rhododendron* (Ericaceae). In: Wu ZY, Raven PH, eds. 2005. *Flora of China*, Vol. 14. Beijing and St. Louis: Science Press and Missouri Botanical Garden Press.

Frankham R. 2003. Genetics and conservation biology. *Comptes Rendus Biologies* **326**:22–29.

Frankham R. 2005. Stress and adaptation in conservation genetics. *Journal of Evolutionary Biology* **18**:750–755.

Frankham R, Ballou JD, Briscoe DA. 2002. *Introduction to conservation genetics*. Cambridge: Cambridge University Press.

Fu LG. 1992. *The red list of plants in China*. Beijing: Science Press.

Gordon SP, Sloop CM, Davis HG, Cushman JH. 2012. Population genetic diversity and structure of two rare vernal pool grasses in central California. *Conservation Genetics* **13**:117–130.

Hamrick JL, Godt MJW. 1989. Allozyme diversity in plant species. In: Brown HD, Clegg MT, Kahler AL, eds. *Plant population genetics, breeding and genetic resources*. Sunderland, MA: Sinauer Associates, Inc., 43–46.

Hamrick JL, Godt MJ. 1990. *Plant population genetics, breeding, and genetic resources*. Sunderland, MA: Sinauer, 43–63.

Hirao AS. 2010. Kinship between parents reduces offspring fitness in a natural population of *Rhododendron brachycarpum*. *Annals of Botany* **105**:637–646.

Hirao AS, Kameyama Y, Ohara M, Isagi Y, Kudo G. 2006. Seasonal changes in pollinator activity influence pollen dispersal and seed production of the alpine shrub *Rhododendron aureum* (Ericaceae). *Molecular Ecology* **15**:1165–1173.

Kaljund K, Jaaska V. 2010. No loss of genetic diversity in small and isolated populations of *Medicago sativa* subsp. *falcate*. *Biochemical Systematics and Ecology* **38**:510–520.

Kaneko S, Isagi Y, Nobushima F. 2008. Genetic differentiation among populations of an oceanic island: the case of *Metrosideros boninensis*, an endangered endemic tree species in the Bonin Islands. *Plant Species Biology* **23**:119–128.

Lande R. 1993. Risks of population extinction from demographic and environmental stochasticity and random catastrophes. *The American Naturalist* **142**:911–927.

Li H, Guo HJ, Dao ZL. 2000. *Flora of Gaoligong Mountains*. Beijing: Science Press.

Li Y, Yan HF, Ge XJ. 2012a. Phylogeographic analysis and environmental niche modeling of widespread shrub *Rhododendron simsii* in China reveals multiple glacial refugia during the last glacial maximum. *Journal of Systematics and Evolution* **50**:362–373.

Li YY, Guan SM, Yang SZ, Luo Y, Chen XY. 2012b. Genetic decline and inbreeding depression in an extremely rare tree. *Conservation Genetics* **13**:343–347.

Lopes MS, Mendonça D, Bettencourt SX, Borba AR, Melo C, Baptista C, da Câmara Machado A. 2014. Genetic diversity of an Azorean endemic and endangered plant species inferred from inter-simple sequence repeat markers. *AoB PLANTS* **6**: plu034; doi:10.1093/aobpla/plu034.

Ma YP, Zhao XF, Zhang CQ, Zhao W, Wang TC, Li XY, Sun WB. 2012. Conservation of the giant tree Rhododendron on Gaoligong Mountain, Yunnan, China. *Oryx* **46**:325.

Ma YP, Chen G, Grumbine RE, Dao ZL, Sun WB, Guo HJ. 2013a. Conserving plant species with extremely small populations (PSESP) in China. *Biodiversity and Conservation* **22**:803–809.

Ma YP, Wu ZK, Xue RJ, Tian XL, Gao LM, Sun WB. 2013b. A new species of *Rhododendron* (Ericaceae) from the Gaoligong Mountains, Yunnan, China, supported by morphological and DNA barcoding data. *Phytotaxa* **114**:42–50.

Nei M. 1973. Analysis of gene diversity in subdivided populations. *Proceedings of the National Academy of Sciences of the USA* **70**: 3321–3323.

Ng SC, Corlett RT. 2000. Genetic variation and structure in six *Rhododendron* species (Ericaceae) with contrasting local distribution patterns in Hong Kong, China. *Molecular Ecology* **9**:959–969.

Nybom H. 2004. Comparison of different nuclear DNA markers for estimating intraspecific genetic diversity in plants. *Molecular Ecology* **13**:1143–1155.

Ono A, Dohzono I, Sugawara T. 2008. Bumblebee pollination and reproductive biology of *Rhododendron semibarbatum* (Ericaceae). *Journal of Plant Research* **121**:319–327.

Peakall R, Smouse PE. 2006. GENALEX 6: genetic analysis in Excel. Population genetic software for teaching and research. *Molecular Ecology Notes* **6**:288–295.

Pritchard JK, Stephens M, Donnelly P. 2000. Inference of population structure using multilocus genotype data. *Genetics* **155**: 945–959.

Qiao Q, Zhang CQ, Milne RI. 2010. Population genetics and breeding system of *Tupistra pingbianensis* (Liliaceae), a naturally rare plant endemic to SW China. *Journal of Systematics and Evolution* **48**: 47–57.

Rossetto M, Weaver PK, Dixon KW. 1995. Use of RAPD analysis in devising conservation strategies for the rare and endangered *Grevillea scapigera* (Proteaceae). *Molecular Ecology* **4**:357–364.

Setoguchi H, Mitsui Y, Ikeda H, Nomura N, Tamura A. 2011. Genetic structure of the critically endangered plant *Tricyrtis ishiiana* (Convallariaceae) in relict populations of Japan. *Conservation Genetics* **12**:491–501.

Shen SK, Wang YH, Wang BY, Ma HY, Shen GZ, Han ZW. 2009. Distribution, stand characteristics and habitat of a critically endangered plant *Euryodendron excelsum* H. T. Chang (Theaceae): implications for conservation. *Plant Species Biology* **24**:133–138.

Vos P, Hogers R, Bleeker M, Reijans M, Van de Lee T, Hornes M, Zabeau M. 1995. AFLP: a new technique for DNA fingerprinting. *Nucleic Acids Research* **23**:4407–4414.

Wang Y, Tang SQ, Li XK. 2003. The genetic diversity of the endangered plant *Abies yuanbaoshanensis*. *Chinese Biodiversity* **12**:269–273.

Willi Y, Van Buskirk J, Hoffmann AA. 2006. Limits to the adaptive potential of small populations. *Annual Review of Ecology, Evolution, and Systematics* **37**:433–458.

Yao XH, Ye QG, Kang M, Huang HW. 2007. Microsatellite analysis reveals interpopulation differentiation and gene flow in endangered tree *Changiostyrax dolichocarpa* (Styracaceae) with fragmented distribution in central China. *New Phytologist* **176**: 472–480.

Yeh FC, Yang RC, Boyle T. 1999. POPGENE VERSION 1.31: Microsoft Window-based free Software for Population Genetic Analysis, 1999, ftp://ftp.microsoft.com/Softlib/HPGL.EXE.

Zhao B, Yin ZF, Xu M, Wang QC. 2012*a*. AFLP analysis of genetic variation in wild populations of five *Rhododendron* species in Qinling Mountain in China. *Biochemical Systematics and Ecology* **45**: 198–205.

Zhao XF, Ma YP, Sun WB, Wen X, Milne R. 2012*b*. High genetic diversity and low differentiation of *Michelia coriacea* (Magnoliaceae), a critically endangered endemic in southeast Yunnan, China. *International Journal of Molecular Sciences* **13**:4396–4411.

9

Genomic sequencing and microsatellite marker development for *Boswellia papyrifera*, an economically important but threatened tree native to dry tropical forests

A. B. Addisalem[1,2,3], G. Danny Esselink[1], F. Bongers[2] and M. J. M. Smulders[1*]

[1] Wageningen UR Plant Breeding, Wageningen University and Research Center, PO Box 386, NL-6700 AJ Wageningen, The Netherlands
[2] Center for Ecosystem Studies, Forest Ecology and Forest Management Group, Wageningen University and Research Center, PO Box 47, NL-6700 AA Wageningen, The Netherlands
[3] Wondo Genet College of Forestry and Natural Resources, PO Box 128, Shashemene, Ethiopia

Associate Editor: Kermit Ritland

Abstract. Microsatellite (or simple sequence repeat, SSR) markers are highly informative DNA markers often used in conservation genetic research. Next-generation sequencing enables efficient development of large numbers of SSR markers at lower costs. *Boswellia papyrifera* is an economically important tree species used for frankincense production, an aromatic resinous gum exudate from bark. It grows in dry tropical forests in Africa and is threatened by a lack of rejuvenation. To help guide conservation efforts for this endangered species, we conducted an analysis of its genomic DNA sequences using Illumina paired-end sequencing. The genome size was estimated at 705 Mb per haploid genome. The reads contained one microsatellite repeat per 5.7 kb. Based on a subset of these repeats, we developed 46 polymorphic SSR markers that amplified 2–12 alleles in 10 genotypes. This set included 30 trinucleotide repeat markers, four tetranucleotide repeat markers, six pentanucleotide markers and six hexanucleotide repeat markers. Several markers were cross-transferable to *Boswellia pirrotae* and *B. popoviana*. In addition, retrotransposons were identified, the reads were assembled and several contigs were identified with similarity to genes of the terpene and terpenoid backbone synthesis pathways, which form the major constituents of the bark resin.

Keywords: Conservation genetics; resin; SSR; terpene biosynthesis; terpenoid; tropical dry forest.

Introduction

To implement an effective conservation programme, it is essential to understand the genetic structure of endangered populations and the dynamics of genetic variation over space and time (Karp *et al.* 1997; Burczyk *et al.* 2006; González-Martínez *et al.* 2006; Frankham *et al.* 2010; Nybom *et al.* 2014). Microsatellite or simple sequence repeat (SSR) markers have been widely applied in quantifying the level of genetic variation and its spatial organization, describing the demography and history of populations, and analysing the gene flow and parentage in plants and animals (e.g. Arens *et al.* 2007; Smulders *et al.* 2008; Primmer 2009; Allan and Max 2010). These repeats are abundant in the genome, polymorphic and multiallelic (thus highly informative), have co-dominant inheritance (allowing a direct measurement of heterozygosity),

* Corresponding author's e-mail address: rene.smulders@wur.nl

and markers based on them are frequently transferable across related species (Chase *et al.* 1996; Smulders *et al.* 1997, 2001; Brondani *et al.* 1998; Pastorelli *et al.* 2003; Tuskan *et al.* 2004; Selkoe and Toone 2006; Allan and Max 2010; Fan *et al.* 2013).

Recently, next-generation sequencing technologies have simplified generating large amounts of sequences at affordable cost, thus facilitating the development of molecular markers, including SSRs and single-nucleotide polymorphisms (SNPs) (Edwards *et al.* 2011; Ekblom and Galindo 2011; Castoe *et al.* 2012; Smulders *et al.* 2012; Lance *et al.* 2013; Vukosavljev *et al.* 2015), as well as chloroplast sequences for phylogeographical studies (Van der Merwe *et al.* 2014). The development of markers has thus become feasible also for species for which no prior sequence information exists (Smulders *et al.* 2012), including understudied but economically important crops (Zalapa *et al.* 2012).

Marker development can be based on short-length sequences from genomic DNA sequences or cDNA (RNA-seq). Both sets of reads are useful, but they differ with regard to further data mining. RNA-seq data can be *de novo* assembled into a (partial) transcriptome (Yang and Smith 2013) with some caveats, partly related to the assembler used (Shahin *et al.* 2012). A common denominator appears to be that multiple assemblers need to be compared (Nakasugi *et al.* 2014), but the final result can be compared with the transcriptome of other species. In contrast, it is not straightforward (Vicedomini *et al.* 2013) to assess the quality of a *de novo* assembly of short reads of genomic DNA from a species for which no prior sequence information is available, especially if the genome is large and contains many repeats, and the species is heterozygous or even polyploid. Nevertheless, many studies are based on genomic DNA, as it is easier to extract DNA from dry material of wild species collected in the field (on silica gel) than to try to extract good quality RNA from fresh samples or from samples specifically prepared for RNA extraction. What additional information can be reliably extracted from a single library of short reads of genomic DNA is an open question.

Boswellia papyrifera is currently the number one frankincense-producing tree species in the world (Coppen 2005). Frankincense is an aromatic resinous gum exudate produced from the bark of trees. Its economic value in the world market stems from its use as an ingredient in pharmaceuticals, cosmetics and as a church incense (Groom 1981; Tucker 1986; Lemenih and Teketay 2003). In Ethiopia, besides its value in the national economy, it has a significant contribution in the local livelihoods, providing up to one-third of annual household income, especially in the northern regions of the country (Lemenih *et al.* 2003, 2007; Woldeamanuel 2011).

The population size of *B. papyrifera* is declining in Ethiopia (Abiyu *et al.* 2010; Groenendijk *et al.* 2012; Tolera *et al.* 2013), Eritrea (Ogbazghi *et al.* 2006) and Sudan (Abtew *et al.* 2012). Little or no tree regeneration occurs in its natural range and mortality of adult trees increases. Despite its endangered status and economic importance, very few conservation efforts exist and none are supported by genetic information. The later situation results because genetic markers for the species have not been developed.

In the present study, we applied the Illumina paired-end sequencing technology to sequence genomic DNA of *B. papyrifera* with the goal of identifying microsatellite repeats and developing SSR markers. The reads were also assembled into the first genomic resource for this species, and we present a couple of structural and functional analyses on them.

Methods

Plant material

Boswellia papyrifera is one of the six *Boswellia* species that grow in various parts of Ethiopia. The *B. papyrifera* genotype used for Illumina paired-end sequencing was collected from a natural population at Kafa Humera Wuhdet (14.05265N latitude; 37.13078E longitude) in northwest Ethiopia. Young leaves were collected from growing shoot tips of the plant and preserved in silica gel while in the field and during transportation to the laboratory for DNA extraction. A genomic DNA library for Illumina paired-end sequencing was prepared from 4 µg of DNA following the PCR-based gel-free illumina TruSeq DNA sample prep protocol and sequenced as 2×100 nt paired-end reads on an Illumina HiSeq at Greenomics, Wageningen UR, Wageningen, the Netherlands.

Plant material for SSR marker development

For testing of the SSR loci a set of 12 genotypes were used. Ten of the genotypes represented populations of *B. papyrifera* collected from 10 different regions of Ethiopia. Two genotypes of *Boswellia pirrotae* and *B. popovina* were included for testing the cross-transferability of the markers to closely related species. The *B. pirrotae* sample was from the northwestern part of Ethiopia. *Boswellia popoviana* is endemic to Socotra Island, Yemen, and the dried leaf sample was obtained through the Edinburgh Royal Botanical Garden, UK.

DNA extraction

Total DNA was isolated from silica-dried young leaves following the cetyltrimethylammonium bromide protocol of Fulton *et al.* (1995). As large amounts of phenolic compounds were expected because of the resin content in the leaves, the protocol was modified by the addition of 2 % pvp-40 to the extraction buffer and 1 % mercaptoethanol to the microprep buffer of Fulton *et al.* (1995), added immediately before use. The extraction was followed by purification steps using DNeasy (Qiagen, Venlo, The Netherlands) according to Smulders *et al.* (2010). DNA yield and quality were visually assessed on a 1 % agarose gel.

Sequence filtering

The raw reads were error-corrected using musket (Liu *et al.* 2012*b*). This error-corrected set was used for the repeat assembly. Prinseq-lite 0.20.04 (Schmieder and Edwards 2011) was used for quality control and filtering of reads (minimum read length of 50 nt, minimum average base quality of 25, maximum ambiguous nt (N) of 1) after which the data were used for SSR mining. After low complexity trimming (minimum DUST score of 7 for removal of low complexity reads and removal of duplicate reads, also with Prinseq-lite), paired-end reads with overlapping sequences were connected using connecting overlapped pair-end (Liu *et al.* 2012*a*) in the full mode. Reads were filtered for chloroplast sequences by mapping the reads against the closest chloroplast genome available, which is one of *Citrus sinensis*, using bowtie2 (Langmead and Salzberg 2012, settings -D 20 -R 3 -N 1 -L 20 -i S,1,0.50 -a).

Repeat analysis

Reads from the highly repeated fraction of the genome were extracted and assembled using RepARK (REPetitive motif detection by Assembly of Repetitive k-mers; Koch *et al.* 2014). The motifs present in the repetitive contigs were counted and analysed by blastn (e-value 1e-5) against Repbase v19.08 (database of repetitive DNA elements, Jurka *et al.* 2005).

Assembly and annotation

A *de novo* draft assembly was created from the filtered reads using SOAPdenovo 2.21 (Li *et al.* 2010, settings -K 41 -M 3 -d 4). The gaps emerging during the scaffolding process by SOAPdenovo were closed using GapCloser (vs. 1.12). The contigs >1000 bp of the draft assembly were analysed and functionally annotated using Blast2GO (Conesa *et al.* 2005).

SSR mining and design of primers

Five million of the filtered but not assembled reads were analysed with PAL_FINDER 0.02.03 (Castoe *et al.* 2012) to identify SSRs using slightly adjusted criteria: at least six contiguous repeat units for dinucleotide repeats, four for tri- and tetranucleotide repeats and three for penta- and hexanucleotide repeats (Castoe *et al.* (2012) used six units for trunicleotide repeats). Following Castoe

et al. (2012) the reads with multiple SSR loci were considered a 'compound' repeat if the SSRs had a different repeat motif, but a 'broken' repeat if the SSRs had the same motif. Reverse-complement repeat motifs (e.g. TG and CA) and translated or shifted motifs (e.g. TGG, GTG and GGT) were grouped together, so that there were a total of four unique dinucleotide repeats, 10 unique trinucleotide repeats and so on.

A subset of over 70 000 trinucleotide to hexanucleotide repeat-containing reads was used to further screen potentially amplifiable SSR loci (PALs): loci for which PCR primers could be designed. Primer designing followed the default parameters specified in Primer3 (Rozen and Skaletsky 2000). The reads were then screened for differences in lengths of those sequences that contained these primers (as in Vukosavljev *et al.* 2015). At these loci the sequenced plant may be heterozygous, thus indicating that the locus is polymorphic. These formed the group of potentially polymorphic loci.

SSR loci amplification and analyses of polymorphism

PCRs were performed in a total volume of 10 μL reaction mixture containing 4 μL 2 ng μL^{-1} DNA, 5 μL MP mix from Qiagen kit, 0.8 μL (2 μM) universal fluorescent-labelled primer and 0.2 μL mix of the forward and reverse primers. The fluorescent labelling method described in Schuelke (2000) was adapted to label the primers for analyses of the PCR products with a laser detection system. For this the forward primers were labelled with a universal M13 sequence (AA CAGGTATGACCATGA) at the 5′ end while the reverse primers were tailed with GTTT at their 5′ end according to Brownstein *et al.* (1996) to reduce stutter bands (both tailing sequences are not shown in the sequences in Table 1). A thermal cycling profile was set at 15 min of initial denaturation at 95 °C, followed by 30 cycles of 30 s denaturation at 94 °C, 45 s annealing at 56 °C and 45 s extension at 72 °C. This was followed by additional eight cycles with 53 °C annealing temperature to facilitate the annealing of the fluorescent dye-labelled M13 primer, and a final extension step of 10 min at 72 °C. After amplification 10 μL water was added. Fluorescently labelled amplicons were resolved on a 4200 or 4300 Licor DNA analyser.

Results

Next-generation sequencing

Genomic DNA of one *B. papyrifera* individual was sequenced in order to obtain a library to mine for microsatellite repeats. One lane on an Illumina HiSeq produced 143 458 368 raw reads. Based on k-mer counts, the estimated genome size of *B. papyrifera* was 705 Mb, sequenced at 36× coverage. After error correction and filtering reads

Table 1. Forty-six polymorphic microsatellite markers developed for *B. papyrifera* and their cross-transferability to *B. pirrotae* and *B. popoviana*. [1]A = number of alleles in 10 *B. papyrifera* genotypes. [2]Ho = observed heterozygosity (a tentative figure, as the 10 individuals are from 10 different populations. [3]Amplification was also tested in one individual of *B. pirrotae* (Br) and one of *B. popoviana* (Bv) except where no Bv is indicated. Hom = homozygous and Het = heterozygous, always with products in the same size range as the alleles in *B. papyrifera*, except where noted that they were out of range. No ampl = no amplification.

Name	Primer sequence (5′→3′)	Repeat motif	A[1]	Allele size range (bp)	Quality (Smulders et al. 1997)	Ho[2] based on 10 B. papyrifera genotypes	Other Boswellia species[3]
Bp01	F:TTGTTAAGGCTTTTCTCCTC R:GTTGCTTATCTTTGGCTGAG	(AAG)6	4	119–134	2	0.34	Br = het Bv = hom
Bp02	F:TGAGAAGTTTACCCTTTATGTTT R:TCTCTGCCTCTTCTTCTTATTT	(ATT)13	7	195–219	2	0.78	Br = hom Bv = hom
Bp03	F:ATGGGGAAAGGTTAAAGATC R:CTGCACAACACAAGTTAAGC	(ATC)6	3	123–129	1	0.1	Br = het Bv = het
Bp04	F:TATCAACACTTTTGTTTTGC R:CAATTCGAGTCTCCTCAAC	(TTC)8	2	182–197	3	0.2	Br = het Bv = het
Bp05	F:GGAGCAGGTACCTTGTATGT R:AACAGATCTCTTGGTTTGATT	(AAC)7	5	232–250	1	0.8	Br = hom Bv = hom
Bp06	F: GATCTCCACTTGATCAGGAC R:ACATGGAAAATTGAAAGCAC	(TTC)9	8	263–297	1	0.5	Br = het Bv = het
Bp07	F:GAAACTTTGTGGGTGTTTGT R:TCATCCTCTGACATATCCATT	(ATT)8	3	284–293	1	0.34	Br = hom Bv = hom
Bpo8	F:TTTTCTGTGTTTTGTACGCA R:GCATGCAAGAAATAGGAGAG	(ATT)6	3	207–213	2	0.11	Br = no ampl Bv = no ampl
Bp09	F:TTGATCAATTATTTCGGACA R:AAAATGCAAGTCCTTTGTAA	(ATT)11	7	292–331	1	0.78	Br = no ampl Bv = het
Bp10	F:CTTTGGCAGATTCAAATAGG R:GACACAAGAAAATTGAGGGA	(TTC)6	4	197–213	1	0.11	Br = het Bv = het
Bp11	F:AGAGAATTCCCTAAGGAGAGA R:TCTACAATAGCCCAGCAACT	(TTC)9	6	284–307	1	0.78	Br = hom Bv = het
Bp12	F:ACCCATGATAAAGAGTTCCA R:GAGAACGCCGTTTGAGTT	(ATT)10	7	238–302	2	0.56	Br = het Bv = no ampl
Bp13	F:ATAATTTCCCACCAGGAGAT R:CAACGAACTACAAGTATTGAATG	(ATT)7	3	227–239	1	0.22	Br = hom Bv = hom
Bp14	F:GGCAATTATTTGATCGCTAC R:ATGACATTCATTCGTAACCC	(ATT)15	8	198–253	1	0.44	Br = het Bv = hom
Bp15	F:TATATGCCTTGCTAAGCGTT R:AAACTCCGAGCTGACTACAC	(ATC)10	7	301–337	1	0.78	Br = het Bv = hom
Bp16	F:AAAACTTTGTTTCCTCTCCA R:TCAGAAGGAAGCACTTCAAC	(TCC)11	2	218–221	1	0.33	Br = hom Bv = hom
Bp17	F:AGCAATATTTCCAAAGGACA R:CTGCCCAATAACATAGTTCC	(TTC)11	6	200–215	1	0.4	Br = no ampl Bv = hom
Bp18	F:TTATCTTGTAGTGGGATGGG R:GAGAACTGGTAATCACATGAAA	(TTC)12	6	221–262	2	0.67	Br = hom Bv = no ampl

Continued

Table 1. *Continued*

Name	Primer sequence (5′→3′)	Repeat motif	A[1]	Allele size range (bp)	Quality (Smulders et al. 1997)	Ho[2] based on 10 B. papyrifera genotypes	Other Boswellia species[3]
Bp19	F:GTGCCAGAATTCAGGTATGT R:GGTTGTGAGTCCACCATTAT	(TTC)13	5	287–321	2	0.1	Br = het Bv = hom
Bp20	F:TGCTTTATGACTTTGTTGAGA R:GAACCATCATGCAATTAGTTT	(TTC)15	10	227–266	2	0.5	Br = het Bv = hom
Bp21	F:CAGAGTTAATAATATAAGTAGCAGCA R:CTATGTTCATACTTAGAAAAGTTGG	(TTC)16	12	117–299	1	0.6	Br = hom Bv = hom
Bp22	F:TAAAACCATTTTCAGCAAGG R:AGAACCAGACCTTCAAATCA	(TTC)17	11	237–307	1	0.7	Br = hom Bv = het
Bp23	F:GCGAATTTGCTCTGTAATTC R:TAAGACCCCAAGAAATTGAA	(TTC)20	11	224–266	2	0.8	Br = het Bv = hom
Bp24	F:TATTTGTCAACAGATTGGGG R:CAGTCTAAGTCCACAAACTCC	(CGGGG)3	2	241–251	1	0	Br = hom Bv = hom
Bp25	F:ATCATCATCAGGTGAAGACC R:ATGTCGTTTTCGACTTTCG	(TCTCGC)3	4	261–279	1	0.22	Br = hom Bv = hom
Bp26	F:AAATCATGTTTGGCTAATGG R:TGCAAATGCAAATTAATGG	(TGCC)6	3	235–247	1	0.34	Br = hom Bv = hom
Bp27	F:CTCTAGATGCATAGGGATGG R:AAATATAATCCTAAACCTTGCG	(TCCGGG)3	2	240–246	1	0.25	Br = no ampl Bv = no ampl
Bp28	F:CAAATCCTTGTGATTTCTCC R:AAGTAGCCATAAATAATCATAGGG	(AAGAG)3	4	262–272	1	0.14	Br = het Bv = hom
Bp29	F:ATTTCACAAATCACTTTCGC R:TTAACAAGTAACGCTAACGC	(TC)10(AGCG)5	6	249–264	1	0.43	Br = hom Bv = het
Bp30	F:ATATGCTAGAGACTTGGCCC R:TTTTCAATGCTTGGATGC	(TTGGGC)3	3	200–212	1	0.34	Br = hom Bv = hom
Bp31	F:CAGAACAAAAGTGACAGTTAGC R:GAGGCAAAGAGACTTGACC	(AGAGC)4	4	277–307	2	0.75	Br = hom Bv = no ampl
Bp32	F:TCATAACTTCCAAAATTGAGC R:TTTCTATCTTTGGATCAATGC	(TCTG)4	3	144–156	1	0.11	Br = hom Bv = no ampl
Bp33	F:CGTCTACCTCCTTCTCTTCC R:GTACTAAACCCTCCGTTCG	(TCTCC)3	2	171–181	3	0.33	Br = het Bv = no ampl
Bp34	F:AGAGAACATCCCAAGAATCC R:AGGATGGAGAGCCCTAGC	(ATGGAG)4	4	183–193	1	0.56	Br = het
Bp35	F:GGCTCCTCGCTAACCGACC R:CTCCCAGTCGAGATCGAGCC	(TTGGCG)4	2	224–230	1	0.1	Br = hom
Bp36	F:GGTATAAAGAGAAAGGGATAGAGG R:CACAATTTACTGGCAATGG	(TGTGC)3	4	211–226	2	0.89	Br = hom
Bp37	F:ATCTCGCATTCCTACATCC R:ACGACCTCTTCATCTAACCC	(ATGC)5	2	277–283	1	0.11	Br = hom
Bp38	F:GTTGAGAATGAGAAGAACGG R:CATCAACTTCCTCAAATTCC	(ATC)7,(8)	5	243–273	1	0.22	Br = het

Continued

Table 1. *Continued*

Name	Primer sequence (5′→3′)	Repeat motif	A[1]	Allele size range (bp)	Quality (Smulders et al. 1997)	Ho[2] based on 10 B. papyrifera genotypes	Other Boswellia species[3]
Bp39	F:TCATGGAATAAGAAACCAAA R:TCTTAACATTTCGTCTGCTG	(ATC)8,(9)	8	247–298	2	0.6	Br = het
Bp40	F:AAACAAATATACGTGGCACA R:TCCAAGTGAACATCCAAAAT	(ATT)8,(14)	3	240–255	2	0.3	Br = hom
Bp41	F:TGGGTTTAAAGTATTCTAAAAGG R:CATTAGAAGAGGCAAAATGG	(ATT)8,(9)	4	230–252	2	0.22	Br = hom
Bp42	F:TTATAAGCAGAGCAAATTATAGC R:CTAATTTCGCAATTTAAGGC	(ATT)10,(11)	6	228–264	2	0.4	Br = hom
Bp43	F:CCAAGCCTATACACTTCTTCA R:GATGAATTGGGCTTAGATTG	(TTC)6,(8)	6	272–293	3	0.89	Br = het
Bp44	F:CCATATGGGGATATAGGTCA R:TTGGCCAAGAAGAAACTTAG	(ATT)6,(7)	4	226–235	2	0.25	Br = het (out of range)
Bp45	F:AACAGTTGGTTTAACAACGC R:CTTAAAAGGGAACTGGAAGG	(AACAAG)3,(4)	3	281–293	1	0.67	Br = het
Bp46	F:ATATTCAATTTATCTGTGTGACG R:TTTGATTTCAAAGGAAAACG	(ATATT)3,(4)	2	256–271	2	0.75	Br = hom

for short sequences, sequences with ambiguities (Ns) and low complexity, and excluding redundant sequences, 120 479 203(84 %) paired-end reads and 10 851 777 single-end reads remained.

SSR identification

A search of SSRs in a subset of five million Illumina paired-end reads identified 170 832 reads (3.4 %) containing SSRs. In these reads, a total of 175 607 repeat loci (dinucleotide through hexanucleotide repeats) were identified, which corresponds to one SSR locus per 5.7 kb. Figure 1 shows the frequency of the top-20 repeat motifs. These include all dinucleotide motif repeats (of at least six repeat units long), of which AC and AT repeats were the most abundant. Of the trinucleotide repeats (of at least four repeat units) AAT and AAC were the most frequent, followed by TTC. Excluding the dinucleotide repeats, the remaining 70 415 SSR loci were screened for the presence of sufficient forward and reverse flanking sequences suitable to design primers. This yielded 29 886 (42 %) PALs. Further filtering of these PALs by applying the most stringent criteria aimed at selecting single-copy loci yielded 4071 potentially amplifiable SSR loci.

Polymorphism and amplification of SSR loci

A total of 136 SSR loci (117 randomly picked and 19 loci predicted to be potentially polymorphic as they appeared to have two different alleles in the sequence reads) were tested for amplification and degree of polymorphism in 10 randomly chosen individuals from different populations. Of the 117 randomly picked loci, 82 primer pairs amplified a high-quality PCR product, of which 37 (45 %) were polymorphic with a banding pattern that could be scored clearly (Table 1). Of the 19 primer pairs predicted to be polymorphic, 13 amplified bands of which 9 loci (69 %) were polymorphic, indicating a significantly higher rate of polymorphism (χ^2 test, $P < 0.005$) compared with randomly picked loci. The final set of 46 markers included 30 trinucleotide repeat markers, 4 tetranucleotide repeats, 6 pentanucleotides and 6 hexanucleotide repeats. The number of alleles across the polymorphic loci varied between 2 and 12 with an average value of 4.8 alleles in 10 genotypes. Several of the polymorphic markers with 10–12 alleles were TTC repeats. The heterozygosity per locus ranged widely from 0.10 to 0.89 (average 0.43). It is possible that, when used in larger populations, these markers will show higher estimates of Ho, and additional alleles may be found.

As shown in Table 1, most of the SSRs successfully amplified in *B. pirrotae* and *B. popoviana* (in the latter species the amount of DNA was insufficient to test all markers). Amplification, even if it is in the same size range as the alleles in *B. papyrifera*, is not proof that the marker is

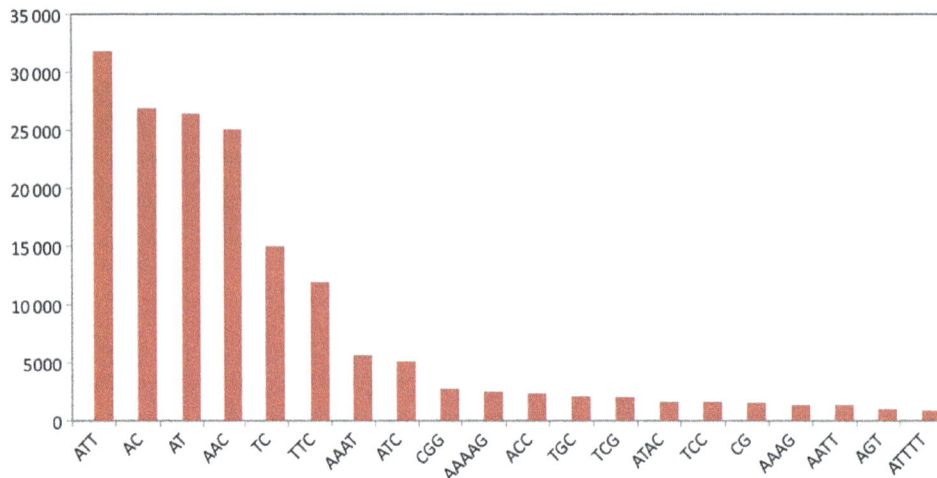

Figure 1. The 20 most frequent SSR motifs obtained, sorted according to frequency.

polymorphic, but heterozygosity (two different alleles in the expected range) is. Based on that criterion at least 19 of the 46 markers are polymorphic in *B. pirrotae* and at least 8 of 33 tested are polymorphic in *B. popoviana*.

Sequence assembly and annotation

The Illumina reads are the first genomic resource generated in the genus *Boswellia*. The repeat fraction was assembled based on k-mer frequency. This produced 49 576 contigs of repeats that were present at least 50× (median length 139 bp, mean length 224 bp, N50 238 bp, maximum length 21 153 bp, total sum = 574 Mbp). Next, 1533 contigs had blastn hits with RepBase, mostly with Copia (639 hits) and Gypsy (523) retrotransposons, alongside EnSpm (114), hAT (72), Satellite (29), TY (23), Harbinger (16), YPrime (14), Helitron (12) and SCTRANSP (3). Intermixed with these elements were hits to the ribosomal RNAs (LSU 56 hits, SSU 41) and also to Caulimoviridae viruses (11).

Using all data in a *de novo* assembly with SOAPdenovo, 444 927 contigs were obtained with a median of 375 bp, a mean contig length of 690 bp, an N50 of 1085 bp and a maximum contig size of 19 236 bp (total sum = 307 Mb genomic DNA sequence). The contigs >1000 bp were blasted against Genbank, and 65 467 were annotated with GO terms (Fig. 2; note that these are overlapping classes).

Terpene biosynthesis genes

Assefa *et al.* (2012) conducted a biophysical and chemical study on resins of *Boswellia* species with special emphasis on *B. papyrifera*. Using the list of identified components, eight contigs of the assembly were found, which represent part of genes of the terpene synthesis pathways, namely pinene synthase, limonene synthase (2×), isoprene synthase (4×) and gamma-terpinene synthase.

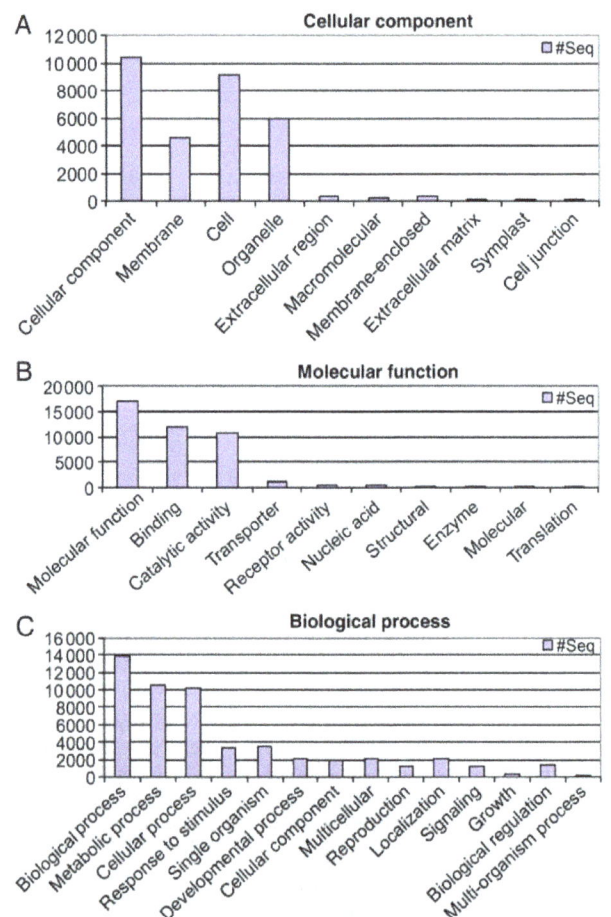

Figure 2. Representation of ontology assignments of the *B. papyrifera* contigs. (A) The 31 086 GO terms of cellular components, (B) the 42 423 GO terms of molecular function and (C) the 54 256 GO terms of biological processes. Note that these are overlapping classes.

We also searched for the enzymes that are involved in terpenoid backbone biosynthesis (according to the Kyoto Encyclopedia of Genes and Genomes pathway

Table 2. MEP/DOXP and mevalonate pathway genes found among the contigs of *B. papyrifera*.

	Name	EC no.
MEP/DOXP pathway		
DXS	1-Deoxy-ᴅ-xylulose-5-phosphate synthase	EC 2.2.1.7
DXR	1-Deoxy-ᴅ-xylulose-5-phosphate reductoisomerase	EC 1.1.1.267
MDS	2-C-methyl-ᴅ-erythritol-2,4-cyclodiphosphate synthase	EC 4.6.1.12
HDS	4-Hydroxy-3-methylbut-2-enyl diphosphate synthase	EC 1.17.7.1
IDI	Isopentenyl diphosphate isomerase	EC 5.3.3.2
GPPS	Geranyl-diphosphate synthase	EC 2.5.1.1
GGPPS	Geranylgeranyl diphosphate synthase	EC 2.5.1.29
CPS	Copalyl diphosphate synthase	EC 5.5.1.12
KS	Kaurene synthase	EC 4.2.3.19
Mevalonate pathway		
AACT	Acetyl-CoA C-acetyltransferase	EC 2.3.1.9
HMGS	Hydroxymethylglutaryl-CoA synthase	EC 2.3.3.10
HMGR	Hydroxymethylglutaryl-CoA reductase	EC 1.1.1.34
MK	Mevalonate kinase	EC 2.7.1.36
PMK	5-Phosphomevalonate kinase	EC 2.7.4.2
MDC	Mevalonate-5-pyrophosphate decarboxylase	EC 4.1.1.33
IDI	Isopentenyl diphosphate isomerase	EC 5.3.3.2
FPPS	Farnesyl diphosphate synthase	EC 2.5.1.10

database). Table 2 lists the enzymes of the mevalonate and non-metavolate (MEP/DOXP) pathways, the two pathways for the synthesis of terpenoid building blocks in plants, which were found among the annotation results. Two of the key enzymes of the MEP pathway, 2-C-methyl-ᴅ-erythritol-4-phosphate cytidylyltransferase (EC 2.7.7.60) and 4-(cytidine 5′-diphospho)-2-C-methyl-ᴅ-erythritol kinase (EC: 2.7.1.148), were not recognized in the set of scaffolds, but reciprocal tBlastx (at 1e-5) against these enzymes identified in *Arabidopsis* did reveal hits with, respectively, 3 and 2 contigs.

Discussion

We have developed the first set of 46 SSR markers for *B. papyrifera*. The markers amplified between 2 and 12 alleles in individuals from 10 different populations across Ethiopia. We based the marker development on DNA sequences from one individual. Most of the markers tested were chosen randomly, but the subset for which we assessed, from the sequence reads, that they probably had two alleles in this individual, gave a significantly higher success rate compared with the randomly chosen ones. This assessment is a technically easy screening step that would improve the efficiency of marker development

in an outbreeding species, even if only sequences from one individual have been generated, as is often the case. It is probably not as efficient as a strategy that generates transcriptome sequences from multiple individuals with the specific aim of testing only those loci on gel for which polymorphisms in repeat length exist among the reads obtained from these individuals (Vukosavljev *et al.* 2015).

The SSR markers were developed based on a set of Illumina paired-end DNA sequence reads from young leaves of a single individual of *B. papyrifera*. The distribution of these reads indicated a genome size of 705 Mb. This is close to the estimate of 682 Mb for *B. serrata*, the only *Boswellia* species listed in the Kew Gardens C-value database.

Mobile elements that are present in multiple copies in the genome were analysed based on sequence homology in k-mers that occurred at high frequency (Koch *et al.* 2014). We have identified a series of retrotransposons, the most common being Copia and Gypsy elements. As these elements are present in large numbers, our Illumina reads probably were a sufficiently good source to determine the presence and relative frequency of various elements.

We also assembled all reads of our paired-end short-read library and obtained 307 Mb of unique sequences. The quality of this assembly is difficult to assess without other independent sources such as libraries of different insert sizes, and we therefore did not compare the results of various assemblers (as, e.g. Shahin *et al.* 2012 did) or merged assemblies (Vicedomini *et al.* 2013). Our resource was searched for genes that are expected to be involved in production of the major compounds of the resin, which in *B. papyrifera* includes diterpenes, triterpenes and nor-triterpenes (Basar 2005; Assefa *et al.* 2012; Bekana *et al.* 2014). The contigs of our assembly gave significant hits for most genes of the core terpene and terpenoid pathways. We have not carried out an in-depth analysis of the sequences in these contigs, as extracting the complete *Boswellia* homologues of these genes would need more bioinformatics steps and independent validation, e.g. by PCR and Sanger sequencing. However, the results indicate that for many genes of interest at least partial sequence information is present.

Genetic information is one of the several tools that facilitate the management of populations and support efforts to conserve threatened species (Moran 2002; Edwards *et al.* 2011). The newly developed SSR markers generated here for *B. papyrifera* can be applied for characterizing the genetic diversity, population structure and processes within populations, such as pollen and seed dispersal distances, information which may assist in identifying conservation units for the species. A study of the population differentiation of *B. papyrifera* across Ethiopia using a subset of these SSR markers is ongoing (Addisalem *et al.*, in prep.). The cross-amplification and polymorphism of the SSR markers in the other two *Boswellia* species, *B. pirrotae* and *B. popoviana,* indicate their potential use for genetic studies of these species and possibly also in other *Boswellia* species. *Boswellia popoviana* is declining and vulnerable in Yemen.

The sequence data generated form the start of a valuable genomic resource for various applications, including estimating the past and present demographic parameters, phylogenetics and phylogeography. With regard to 'conservation genomics', McMahon *et al.* (2014) suggested that genomic sequences are particularly suited to study local adaptation. For most of these applications, genomic sequences need to be generated from several individuals from different populations. This would complement genetic differentiation studies with neutral molecular markers such as SSRs. An exception is the estimation of the the effective population size from SNP density data based on the differences between the alleles at many loci of the heterozygous tree (e.g. Halley *et al.* 2014).

Conclusions

Based on Illumina paired-end sequences, we have developed a set of polymorphic SSR markers for *B. papyrifera*

and two sister species, which will be useful for studying genetic diversity within and differentiation between *Boswellia* populations. We also generated the first genomic resource in *Boswellia.*

Accession Numbers

Accession number in ENA/GenBank for the set of DNA sequences on which the SSR markers were developed: ERS403283.

Sources of Funding

This study was funded by the Netherlands' Fellowship programme (NUFFIC).

Contributions by the Authors

F.B. and M.J.M.S. conceived the study. A.B.A. sampled the plants, carried out the testing and analysed the data. G.D.E. did the bioinformatics analyses. A.B.A., F.B. and M.J.M.S. wrote the paper. All authors have read and approved the submitted manuscript.

Acknowledgements

The authors thank Alan Forrest, the Edinburgh Botanical Garden, for providing the *B. popoviana* sample. Koen Pelgrom and Doret Wouters are thanked for helping in the laboratory and Robert van Loo for help in analysing the genes involved in secondary component synthesis.

Literature Cited

Abiyu A, Bongers F, Eshete A, Gebrehiwot K, Kindu M, Lemenih M, Moges Y, Ogbazghi W, Sterck FJ. 2010. Incense Woodlands in Ethiopia and Eritrea: regeneration problems and restoration possibilities. In: Bongers F, Tenningkeit T, eds. *Degraded forest in eastern Africa: management and restoration.* London, UK: Earthscan, 133–152.

Abtew AA, Pretsch J, Mohamoud TE, Adam YO. 2012. Population status of *Boswellia papyrifera* (Del.) Hochst in the dry woodlands of Nuba Mountains, South Kordofan State, Sudan. *Agriculture and Forestry* **54**:41–50.

Allan GJ, Max TL. 2010. Molecular genetic techniques and markers for ecological research. *Nature Education Knowledge* **3**:2.

Arens P, Van der Sluis Th, Van't Westende WPC, Vosman B, Vos CC, Smulders MJM. 2007. Population differentiation and connectivity among fragmented Moor frog (*Rana arvalis*) populations in the Netherlands. *Landscape Ecology* **22**:1489–1500.

Assefa M, Dekebo H, Kassa H, Habtu A, Fitwi G, Redi-Abshiro M. 2012. Biophysical and chemical investigations of frankincense of *Boswellia papyrifera* from north and northwestern Ethiopia. *Journal of Chemical and Pharmaceutical Research* **4**:1074–1089.

Basar S. 2005. *Phytochemical investigations on Boswellia species. Comparative studies on the essential oils, pyrolysates and boswel-*

lic acids of Boswellia carterii Birdw., Boswellia serrata Roxb., Boswellia frereana Birdw., Boswellia neglecta S. Moore and Boswellia rivae Engl. PhD Thesis, Universität Hamburg, Germany.

Bekana D, Kebede T, Assefa M, Kassa H. 2014. Comparative phytochemical analyses of resins of Boswellia species (B. papyrifera (Del.) Hochst., B. neglecta S. Moore, and B. rivae Engl.) from Northwestern, Southern, and Southeastern Ethiopia. ISRN Analytical Chemistry 2014:374678. http://dx.doi.org/10.1155/2014/374678.

Brondani RPV, Brondani C, Tarchini R, Grattapaglia D. 1998. Development, characterization and mapping of microsatellite markers in Eucalyptus grandis and E. urophylla. Theoretical and Applied Genetics 97:816–827.

Brownstein MJ, Carpten JD, Smith JR. 1996. Modulation of non-templated nucleotide addition by Taq DNA polymerase: primer modifications that facilitate genotyping. BioTechniques 20: 1004–1010.

Burczyk J, Adams WT, Birkes DS, Chybicki IJ. 2006. Using genetic markers to directly estimate gene flow and reproductive success parameters in plants based on naturally regenerated seedlings. Genetics 173:363–372.

Castoe TA, Poole AW, de Koning APJ, Jones KL, Tomback D. 2012. Rapid microsatellite identification from Illumina paired-end genomic sequencing in two birds and a snake. PLoS ONE 7:e30953.

Chase M, Kesseli R, Bawa K. 1996. Microsatellite markers for conservation and population genetics of tropical tree species. American Journal of Botany 83:51–57.

Conesa A, Götz S, García-Gómez JM, Terol J, Talón M, Robles M. 2005. Blast2GO: a universal tool for annotation, visualization and analysis in functional genomics research. Bioinformatics 21:3674–3676.

Coppen JJW. 2005. Overview of international trades and markets. In: Chikamai B, Casadei E, eds. Production and marketing of gum resins: frankincense, myrrh and opoponax. Network for Natural Gums and Resins in Africa (NGARA). Publication Series No. 5. Nairobi, Kenya: NGARA, KEFRI, 5–34.

Edwards CE, Parchman TL, Weekley C. 2011. Assembly, gene annotation and marker development using 454 floral transcriptome sequences in Ziziphus celata (Rhamnaceae), a highly endangered, Florida Endemic Plant. DNA Research 19:1–9.

Ekblom R, Galindo J. 2011. Applications of next generation sequencing in molecular ecology of non-model organisms. Heredity 107: 1–15.

Fan L, Zhang MY, Liu QZ, Li LT, Song Y, Wang LF, Zhang SL, Wu J. 2013. Transferability of newly developed pear SSR markers to other Rosaceae species. Plant Molecular Biology Reporter 31:1271–1282.

Frankham R, Ballou J, Briscoe D. 2010. Introduction to conservation genetics, 2nd edn. UK: Cambridge University Press, pp 644.

Fulton TM, Chunwangse J, Tanksley SD. 1995. Microprep protocol for extraction of DNA from tomato and herbaceous plants. Plant Molecular Biology Reporter 13:207–209.

González-Martínez SC, Krutovsky KV, Neale DB. 2006. Forest-tree population genomics and adaptive evolution. New Phytologist 170:227–238.

Groenendijk P, Eshete A, Sterck FJ, Zuidema P, Bongers F. 2012. Limitations to sustainable frankincense production: blocked regeneration, high adult mortality, and declining population. Journal of Applied Ecology 49:164–173.

Groom N. 1981. Frankincense and myrrh: a study of the Arabian incense trade. London: Longman, 285 p.

Halley YA, Dowd SE, Decker JE, Seabury PM, Bhattarai E, Johnson CD, Rollins D, Tizard IR, Brightsmith DJ, Peterson MJ, Taylor JF, Seabury CM. 2014. A draft de novo genome assembly for the northern Bobwhite (Colinus virginianus) reveals evidence for a rapid decline in effective population size beginning in the late Pleistocene. PLoS ONE 9:e90240.

Jurka J, Kapitonov VV, Pavlicek A, Klonowski P, Kohany O, Walichiewicz J. 2005. Repbase update, a database of eukaryotic repetitive elements. Cytogenetic and Genome Research 110: 462–467.

Karp A, Kresovich S, Bhat KV, Ayad WG, Hodgkin T. 1997. Molecular tools in plant genetic resources conservation: a guide to the technologies. In: IPGRI Technical Bulletin No. 2. Rome, Italy: International Plant Genetic Resources Institute.

Koch P, Platzer M, Downie BR. 2014. RepARK—de novo creation of repeat libraries from whole-genome NGS reads. Nucleic Acids Research 42:e80.

Lance SL, Love CN, Nunziata SO, O'Bryhim JR, Scott DE, Wesley Flynn RW, Jones KL. 2013. 32 species validation of a new Illumina paired-end approach for the development of microsatellites. PLoS ONE 8:e81853.

Langmead B, Salzberg SL. 2012. Fast gapped-read alignment with Bowtie 2. Nature Methods 9:357–359.

Lemenih M, Teketay D. 2003. Frankincense and myrrh resources of Ethiopia: medicinal and industrial uses. SINET Ethiopian Journal of Science 26:161–172.

Lemenih M, Abebe T, Mats O. 2003. Gum and Resin resources from some Acacia, Boswellia, and Commiphora species and their economic contributions in Liban, south-east Ethiopia. Journal of Arid Environments 55:465–482.

Lemenih M, Feleke S, Tadesse W. 2007. Constraints to smallholders production of frankincense in Metema district, North-western Ethiopia. Journal of Arid Environments 71:393–403.

Li R, Zhu H, Ruan J, Qian W, Fang X, Shi Z, Li Y, Li S, Shan G, Kristiansen K, Li S, Yang H, Wang J, Wang J. 2010. De novo assembly of human genomes with massively parallel short read sequencing. Genome Research 20:265–272.

Liu B, Yuan J, Yiu SM, Li Z, Xie Y, Chen Y, Shi Y, Zhang H, Li Y, Lam TW, Luo R. 2012a. COPE: an accurate k-mer-based pair-end reads connection tool to facilitate genome assembly. Bioinformatics 28:2870–2874.

Liu Y, Schröder J, Schmidt B. 2012b. Musket: a multistage k-mer spectrum-based error corrector for Illumina sequence data. Bioinformatics 29:308–315.

McMahon BJ, Teeling EC, Höglund J. 2014. How and why should we implement genomics into conservation? Evolutionary Applications 7:999–1007.

Moran P. 2002. Current conservation genetics: building an ecological approach to the synthesis of molecular and quantitative genetic methods. Ecology of Freshwater Fish 11:30–55.

Nakasugi K, Crowhurst R, Bally J, Waterhouse P. 2014. Combining transcriptome assemblies from multiple de novo assemblers in the allo-tetraploid plant Nicotiana benthamiana. PLoS ONE 9: e91776.

Nybom H, Weising K, Rotter B. 2014. DNA fingerprinting in botany: past, present, future. Investigative Genetics 5:1.

Ogbazghi W, Rijkers T, Wessel M, Bongers F. 2006. The distribution of the francincense tree Boswellia papyrifera in Eritrea: the role of environment and land use. Journal of Biogeography 33:524–535.

Pastorelli R, Smulders MJM, Van't Westende WPC, Vosman B, Giannini R, Vettori C, Vendramin GG. 2003. Characterisation of microsatellite markers in *Fagus sylvatica* L. and *Fagus orientalis* Lipsky. *Molecular Ecology Notes* 3:76–78.

Primmer CR. 2009. From conservation genetics to conservation genomics. *Annals of the New York Academy of Sciences* 1162:357–368.

Rozen S, Skaletsky HJ. 2000. Primer3 on the WWW for general users and for biologist programmers. In: Krawetz S, Misener S, eds. *Bioinformatics methods and protocols: methods in molecular biology*. Totowa, NJ: Humana Press, 365–386.

Schmieder R, Edwards R. 2011. Quality control and preprocessing of metagenomic datasets. *Bioinformatics* 27:863–864.

Schuelke M. 2000. An economic method for the fluorescent labelling of PCR fragments: a poor man's approach to genotyping for research and high-throughput diagnostics. *Nature Biotechnology* 18:233–234.

Selkoe KA, Toone RJ. 2006. Microsatellites for ecologists: a practical guide to using and evaluating microsatellite markers. *Ecology Letters* 9:615–629.

Shahin A, van Gurp T, Peters SA, Visser RGF, van Tuyl JM, Arens P. 2012. SNP markers retrieval for a non-model species: a practical approach. *BMC Research Notes* 5:79.

Smulders MJM, Bredemeijer G, Rus-Kortekaas W, Arens P, Vosman B. 1997. Use of short microsatellites from database sequences to generate polymorphisms among *Lycopersicon esculentum* cultivars and accessions of other *Lycopersicon* species. *Theoretical and Applied Genetics* 94:264–272.

Smulders MJM, Van der Schoot J, Arens P, Vosman B. 2001. Trinucleotide repeat microsatellite markers for black poplar (*Populus nigra* L.). *Molecular Ecology Notes* 1:188–190.

Smulders MJM, Cottrell JE, Lefèvre F, van der Schoot J, Arens P, Vosman B, Tabbener HE, Grassi F, Fossati T, Castiglione S, Krystufek V, Fluch S, Burg K, Vornam B, Pohl A, Gebhardt K, Alba N, Agúndez D, Maestro C, Notivol E, Volosyanchuk R, Pospíšková M, Bordács S, Bovenschen J, van Dam BC, Koelewijn H-P, Halfmaerten D, Ivens B, van Slycken J, Vanden Broeck A, Storme V, Boerjan W. 2008. Structure of the genetic diversity in Black poplar (*Populus nigra* L.) populations across European river systems: consequences for conservation and restoration. *Forest Ecology and Management* 255:1388–1399.

Smulders MJM, Esselink GD, Everaert I, De Riek J, Vosman B. 2010. Characterisation of sugar beet (*Beta vulgaris* L. ssp. vulgaris) varieties using microsatellite markers. *BMC Genetics* 11:41.

Smulders MJM, Vukosavljev M, Shahin A, van de Weg WE, Arens P. 2012. High throughput marker development and application in horticultural crops. *Acta Horticulturae (ISHS)* 961:547–551.

Tolera M, Sass-Klaassen U, Eshete A, Bongers F, Sterck FJ. 2013. Frankincense tree recruitment failed over the past half century. *Forest Ecology and Management* 304:65–72.

Tucker AO. 1986. Frankincense and myrrh. *Economic Botany* 40:425–433.

Tuskan GA, Gunter LE, Yang ZK, Yin Tong M, Sewell MM, DiFazio SP. 2004. Characterization of microsatellites revealed by genomic sequencing of *Populus trichocarpa*. *Canadian Journal of Forestry Research* 34:85–93.

Van der Merwe M, McPherson H, Siow J, Rossetto M. 2014. Next-Gen phylogeography of rainforest trees: exploring landscape-level cpDNA variation from whole-genome sequencing. *Molecular Ecology Resources* 14:199–208.

Vicedomini R, Vezzi F, Scalabrin S, Arvestad L, Policriti A. 2013. GAM-NGS: genomic assemblies merger for next generation sequencing. *BMC Bioinformatics* 14(Suppl. 7):S6.

Vukosavljev M, Esselink GD, Van't Westende WPC, Cox P, Visser RGF, Arens P, Smulders MJM. 2015. Efficient development of highly polymorphic microsatellite markers based on polymorphic repeats in transcriptome sequences of multiple individuals. *Molecular Ecology Resources* 15:17–27.

Woldeamanuel T. 2011. *Dryland resources, livelihoods and institutions: diversity and dynamics in use and management of gum and resin trees in Ethiopia*. PhD Dissertation, Wageningen University, Wageningen, The Netherlands, 169p. ISBN 978-90-8585-962-8.

Yang Y, Smith SA. 2013. Optimizing de novo assembly of short-read RNA-seq data for phylogenomics. *BMC Genomics* 14:328.

Zalapa JE, Cuevas H, Zhu H, Steffan S, Senalik D, Zeldin E, McCown B, Harbut R, Simon P. 2012. Using next-generation sequencing approaches to isolate simple sequence repeat (SSR) loci in the plant sciences. *American Journal of Botany* 99:193–208.

10

Genetic diversity and floral width variation in introduced and native populations of a long-lived woody perennial

Jane C. Stout[1]*, Karl J. Duffy[1,2], Paul A. Egan[1], Maeve Harbourne[1] and Trevor R. Hodkinson[1]

[1] School of Natural Sciences and Trinity Centre for Biodiversity Research, Trinity College Dublin, Dublin 2, Ireland
[2] School of Life Sciences, University of KwaZulu-Natal, Private Bag X01, Scottsville, Pietermaritzburg 3209, South Africa

Associate Editor: Anna Traveset

Abstract. Populations of introduced species in their new environments are expected to differ from native populations, due to processes such as genetic drift, founder effects and local adaptation, which can often result in rapid phenotypic change. Such processes can also lead to changes in the genetic structure of these populations. This study investigated the populations of *Rhododendron ponticum* in its introduced range in Ireland, where it is severely invasive, to determine both genetic and flower width diversity and differentiation. We compared six introduced Irish populations with two populations from *R. ponticum*'s native range in Spain, using amplified fragment length polymorphism and simple sequence repeat genetic markers. We measured flower width, a trait that may affect pollinator visitation, from four Irish and four Spanish populations by measuring both the width at the corolla tip and tube base (nectar holder width). With both genetic markers, populations were differentiated between Ireland and Spain and from each other in both countries. However, populations displayed low genetic diversity (mean Nei's genetic diversity = 0.22), with the largest proportion (76–93 %) of genetic variation contained within, rather than between, populations. Although corolla width was highly variable between individuals within populations, tube width was significantly wider (>0.5 mm) in introduced, compared with native, populations. Our results show that the same species can have genetically distinct populations in both invasive and native regions, and that differences in floral width may occur, possibly in response to ecological sorting processes or local adaptation to pollinator communities.

Keywords: AFLP; corolla tube; floral morphology; invasive plants; microsatellites; population differentiation; SSR.

Introduction

When a species is introduced outside its native range as a result of human activity, genetic diversity and morphological variability within and between populations can vary. This may be a result of the phenotypic plasticity of the species (Chun 2011; Godoy *et al.* 2011), propagule pressure (the number of individuals released and the number of release events; Lockwood *et al.* 2005, but see Nuñez *et al.* 2011) and post-introduction evolutionary processes like inbreeding, drift, hybridization and response to novel selection pressures (Lee 2002; Prentis *et al.* 2008). Introduced populations may be expected to display low genetic diversity due to founder effects from a limited number of initially introduced individuals, with associated negative fitness consequences (Ellstrand and Elam 1993; Young *et al.* 1996), although this is not always the case. Populations of two highly successful introduced plants, *Fallopia japonica* (Japanese Knotweed) and *Eichhornia crassipes* (water hyacinth), have been shown to

* Corresponding author's e-mail address: stoutj@tcd.ie

have very low genetic diversity (Hollingsworth and Bailey 2000; Ren *et al.* 2005).

Post-introduction changes in genetic and morphological variability in plant species may also be related to the life form and breeding system. For example, pollinator limitation, as a result of separation of an introduced plant from its native pollinators, may exert selection pressure on self-compatible populations to evolve from self-incompatible ancestors to ensure seed set (Petanidou *et al.* 2012). This is less likely in perennial species, which have more opportunities for sexual reproduction and the possibility of vegetative spread, and they may retain high genetic diversity post-introduction (Hamrick and Godt 1996). Another possibility is that rather than selection for self-compatibility, introduced plants may be more attractive to resident native pollinators. For example, this could be as a result of the production of larger flowers, which can be more attractive to foraging insects (Eckhart 1993; Conner and Rush 1996). Furthermore, specialized floral characteristics that restrict access to floral rewards, such as narrow or long floral tubes (Suzuki 1994; Stang *et al.* 2006), may be selected against and plants that have more easily accessible rewards may receive more frequent pollinator visitation in their new environment (Armbruster and Baldwin 1998). In self-incompatible *Ipomopsis aggregata*, floral width is under strong selective pressure (Campbell *et al.* 1996) because wider flowers allow increased bill insertion by hummingbird pollinators and a greater proportion of pollen removal. Stang *et al.* (2006) found that in a Spanish floral community there were more visitors to flowers with wide nectar holders. Hence, floral width, particularly nectar holder width, may be an important floral trait determining attractiveness to floral visitors and potential pollinators.

Here we investigate population genetic diversity and floral width variation of *Rhododenron ponticum* (Ericaceae) in expanding introduced populations and compare these with declining native populations. This long-lived woody species was once widely distributed throughout Europe (Cross 1975; Chamberlain *et al.* 1996; Milne and Abbott 2000), but now is primarily found in northern Turkey, the Caucasian states, Lebanon, southern Bulgaria and the Iberian Peninsula (Cross 1975, 1981; Colak *et al.* 1998; Rotherham 2001; Mejías *et al.* 2002). Iberian populations are small and confined to three isolated areas, the largest of which is in the Aljibe Mountains in southern Spain (Castroviejo *et al.* 1993; Mejías *et al.* 2002), where it is classified as endangered (Blanca *et al.* 2000) under IUCN red list criteria (IUCN 1994). In this region it is known from ~20 populations, which, although not undergoing rapid decline (Mejías *et al.* 2007), suffer from very low recruitment, and thus it is considered a vulnerable species in the area (VU: Cabezudo *et al.* 2005). *Rhododenron ponticum*

was introduced as an ornamental plant into Britain and Ireland in the late 18th century (Milne and Abbott 2000; Dehnen-Schmutz and Williamson 2006) and now forms large invasive populations, which are spreading into, and having negative impacts on, native ecosystems on both islands (Cross 1975, 1981; Colak *et al.* 1998). Repeated introductions into many locations over time have created intense propagule pressure (Stephenson *et al.* 2006). Molecular analysis of chloroplast and nuclear ribosomal DNA indicated that British and Irish populations are predominantly derived from Spanish populations, and that hybridisation with North American species (*Rhododendron catawbiense* and *R. maximum*) occurred after *R. ponticum* was introduced in Britain (Milne and Abbott 2000). This was thought to have contributed to the competitive success of populations of *R. ponticum* in Ireland (Erfmeier and Bruelheide 2005; Erfmeier and Bruelheide 2010), where it is particularly successful in the Atlantic climate of the west coast (Cross 1981). However, analyses using amplified fragment length polymorphism (AFLP) markers showed that hybridization is unlikely to have contributed to invasiveness in Irish populations (Erfmeier *et al.* 2011). Relatively low genetic diversity was found in both Irish and Spanish populations, compared with Georgian ones, with weak genetic differentiation among populations within the three countries (Erfmeier and Bruelheide 2011). Studies of growth traits and life history have found that Irish populations had higher rates of annual growth and seedling recruitment (Erfmeier and Bruelheide 2004), and suggest a genetic basis for these traits (Erfmeier and Bruelheide 2005). A previous study comparing the pollination ecology of Irish and Spanish populations has shown that a range of generalist pollinator species visit *R. ponticum* flowers in both ranges, although visitor communities are dominated by different species in Ireland versus Spain, and that a greater volume of nectar is produced in plants from introduced populations (Stout *et al.* 2006). This may indicate that these populations contain individuals with wider flowers that hold greater quantities of nectar. However, no previous studies have investigated floral traits (e.g. nectar holder width) in introduced populations of *R. ponticum*, which may be important given that this species has a mixed mating system, primarily relying on animal-mediated outcrossing to produce seeds for invasion (Mejías *et al.* 2002; Stout 2007a). Selective pressures or ecological sorting in the introduced range could have resulted in larger, more open flowers to enhance attractiveness to resident generalist pollinators.

We examined genetic and floral morphological diversity within and between populations of *R. ponticum* in Ireland, and compared them with native populations in the ancestral range in the Aljibe Mountains in Spain.

The objectives were to: (i) quantify genetic diversity both within and among introduced and native populations; (ii) determine genetic differentiation among introduced and ancestral populations; and (iii) quantify floral width (corolla width and tube width) in the introduced and native range. Specifically, we tested the hypothesis that there is genetic and floral width differentiation among populations in Ireland, and between Irish and Spanish populations.

Methods

Leaf sampling and DNA extraction

Sampling for genetic analysis was carried out in six Irish populations (Table 1), chosen to cover the geographic range of *R. ponticum* within the country, including the west coast (County Galway), the south-west (County Kerry) and the east (County Dublin). Irish populations were relatively large (>100 adult plants, Table 1). In addition, two Spanish populations were sampled within the Parque Natural Los Alcornocales (~5 km inland from the Strait of Gibraltar). These populations were sampled in 2002; they were the largest populations in the Los Alcornacales region, but were still comparatively small (18 and 27 adult plants per population). All of the Spanish populations occur within an ~50 × 30 km area, and are mostly confined to the Aljibe Mountains, where they are restricted to riparian forest habitats (Mejías *et al.* 2007). Nine to 12 individual plants within both introduced and native populations were randomly selected from each population (Table 1). To avoid sampling clones, distinct individuals, separated by >5 m, were selected. We used this sampling procedure as previous work has shown that vast majority of pollinator visits occur within-plant and that the majority of seeds land close to maternal

plants (Stephenson *et al.* 2007; Stout 2007*b*). In addition, this sampling procedure ensured that replicate samples were taken in the native range to compare with invasive populations.

Leaf material was collected and stored in silica gel (Chase and Hills 1991). DNA was extracted from ~0.1 g of dried material using a modified 2× hexadecyltryltrimethyl-ammonium bromide procedure (Doyle and Doyle 1987; Hodkinson *et al.* 2007), and was purified with JetQuick columns (GENOMED Gmbh) according to the manufacturer's protocol. Two polymerase chain reaction (PCR)-based methods were employed to assess genetic diversity: AFLPs (Vos *et al.* 1995) and nuclear microsatellites—simple-sequence repeats (SSRs).

AFLP protocol

Sampled DNA was restricted with the endonucleases EcoRI and MseI and ligated to appropriate double-stranded adapters according to the manufacturer's protocols. Amplified fragment length polymorphism analysis was performed according to the AFLP plant mapping protocol of Applied Biosystems, Inc. Two steps of amplification followed: a pre-selective amplification using primer pairs with one selective base was followed by a selective amplification to further reduce the number of fragments. For the second amplification, the following three selective primer pairs were selected sequentially: EcoRI-ACA/MseI-CAG, EcoRI-AAG/MseI-CTC and EcoRI-AGC/MseI-CAG. The products were sized using an Applied Biosystems 310 Genetic Analyzer with GeneScan version 3.1 and Genotyper version 3.7 software. Amplified fragment length polymorphism profiles were manually scored with the presence of each peak recorded as '1' and the absence of a peak as '0'. Only peaks ranging from 50 to 500 bp were scored. A peak was scored as present if it was separated by at

Table 1. *Rhododendron ponticum* populations used for genetic analysis and genetic diversity estimates within populations using (i) AFLP markers and (ii) SSR markers. Size, approximate number of mature, flowering plants in a population; N, number of individuals analysed; Tb, total number of bands; Pb, number of private bands; P, percentage of polymorphic loci at the 5 % level; H_j, Nei's genetic diversity; N_a, observed allele number; N_e, effective allele number; H_O, observed heterozygosity; H_E, expected heterozygosity; H, average heterozygosity.

Region	Population	Position	Size	(i) AFLP					(ii) SSR					
				N	Tb	Pb	P	H_j	N	N_a	N_e	H_O	H_E	H
Ireland	Howth	53.377N 6.07W	~150	10	277	7	68.9	0.234	10	3.25	2.64	1.000	0.591	0.48
	Glencullen	53.23N 6.272W	~150	9	268	13	66.7	0.226	10	2.50	2.30	1.000	0.547	0.48
	Gortderraree	51.988N 9.558W	>1000	10	254	6	63.2	0.218	10	3.00	2.86	0.750	0.465	0.49
	Gortracussane	52.006N 9.54W	>1000	10	239	2	59.5	0.211	11	3.75	2.92	0.750	0.494	0.48
	Recess	54.467N 0.739W	~100	10	274	5	68.2	0.227	10	3.25	2.89	1.000	0.672	0.48
	Kylemore	53.561N 9.866W	>1000	10	264	4	65.7	0.204	10	3.00	2.57	0.750	0.450	0.49
Spain	El Palancar	36.082N 5.543W	18	10	274	14	68.2	0.225	10	2.00	2.00	0.500	0.250	0.53
	Las Corzas	36.111N 5.528W	27	10	278	14	69.2	0.225	12	3.50	2.44	0.500	0.289	0.53

least 1 bp and has a relatively high peak height threshold (Meudt and Clarke 2007). In order to reduce genotyping error, AFLP profiles were scored at least twice by individuals with no knowledge of the origin of plant material.

SSR protocol

No SSR markers have been published for *R. ponticum*, and so nuclear SSR amplification of seven polymorphic loci isolated from *R. metternichii* var. *hondoense* was screened according to the methods described in Naito *et al.* (1998), of which four were informative for *R. ponticum* (RM3D2, RM2D2, RM9D6 and RM2D5). Polymerase chain reaction amplification followed (Naito *et al.* 1998), and the amplicons were sized on an Applied Biosystems 310 Genetic Analyzer with GeneScan version 3.1 and Genotyper version 3.7 software.

Floral width

In addition to quantifying genetic variation in native and invasive populations of *R. ponticum*, in 2011 we quantified floral width in representative plants in both regions to test whether nectar holder width varied between populations and the two regions. To estimate floral width, two measurements were made on each flower in the field using dial callipers (Moore and Wright, CDP150M), with a precision of 0.01 mm: (i) the width of the corolla at the widest point between the upper wing petals and (ii) the width of the corolla tube at the base. These traits were measured as they represent the extent to which *R. ponticum* flowers are open to insect pollinators in order to access nectar

rewards. Due to logistical constraints, and the fact that these data were collected separately from the leaf material for the population genetic study, only relatively few measurements were taken per population and in only one of the populations (El Palancar, Spain) sampled for genetic analysis. Measurements were made in four Irish and four Spanish populations (Table 2). From each population, five completely open flowers (third floral phase, i.e. with corolla wide open, stigma receptive and protruding beyond anthers; Mejías *et al.* 2002) from each of five individual plants were randomly selected for measurement.

Data analysis

Population genetic diversity. For the AFLP data set, genetic diversity estimates were calculated with AFLPsurv V.1.0 (Vekemans *et al.* 2002). To estimate allelic frequencies, the Bayesian method with a non-uniform prior distribution of allele frequencies (Zhivotovsky 1999) was used. Due to the mixed mating system of the species (Stout 2007a), we assumed some deviation ($F_{IS} = 0.1$) from the Hardy–Weinberg equilibrium. Statistics of gene diversity were calculated according to Lynch and Milligan (1994). For each population, we calculated the proportion of polymorphic loci (P) and Nei's gene diversity (H_j). For the SSR data set, GenAlEx 6.2 (Peakall and Smouse 2006) was used to test for departures from the Hardy–Weinberg equilibrium. Observed heterozygosity (H_O) and Nei's expected heterozygosity (H_E) were calculated with GenAlEx 6.2, and the average heterozygosity was calculated with PopGene 1.32 (Yeh *et al.* 2000).

Table 2. *Rhododendron ponticum* populations used for flower morphology measurements.

Region	Population	Position	Size	Elevation (m)	Habitat type
Ireland	Crossover	52.894N 6.400W	75	165	Riparian woodland, *Quercus petraea* dominant, with *Betula pendula*
	Dunran	53.060N 6.102W	125	156	Mixed forest plantation, mainly *Pinus contorta* with *Q. petraea* in patches
	Tropperstown	53.017N 6.274W	50	185	Open forest, *Q. petraea*, *Fraxinus excelsior* and *B. pendula* dominant
	Shankhill	53.192N 6.427W	225	281	Mixed forest plantation, *Fagus sylvatica* and *F. excelsior*, some *P. contorta*
Spain	El Palancar	36.081N 5.543W	50	495	Stream gulley, patchy *Q. suber* forest on the edge of grazed grassland
	Llanos del Juncal	36.105N 5.540W	125	747	Cloud forest, *Q. canariensis* with *Crataegus monogyna* and *Ilex perado*
	Garganta de Puerto Oscuro	36.518N 5.632W	100	605	Stream valley, mixed forest cover of *Q. canariensis* with *Q. suber* patches
	Garganta del Aljibe	36.538N 5.635W	75	469	Stream valley, *Q. canariensis* dominant, *Arbutus unedo* common

Population genetic structure. Euclidean pairwise genetic distances were calculated in GenAlEx 6.2, which allows a common pathway for subsequent statistical analysis for both dominant AFLP markers and codominant SSR markers (Maguire et al. 2002). For both data sets, genetic distances were calculated using Eq. (1),

$$E = n[1 - (2nxy/2n)] \qquad (1)$$

where n is the total number of polymorphic bands and $2nxy$ is the number of markers shared by two individuals (Peakall et al. 1995; Maguire et al. 2002). Total genetic diversity was partitioned among groups of populations, among populations within groups and within populations using a hierarchical analysis of molecular variance (AMOVA) in GenAlEx 6.2. Genetic structure was tested with AMOVA on the genetic distance matrix (9999 permutations) produced for both sets of markers (Weir 1996). Analysis of molecular variance output nomenclature follows that of Excoffier et al. (1992) in that variation was summarized both as the proportion of the total variance and as φ-statistics (F_{ST} analogues). Pairwise genetic distances among populations and their level of significance for both the AFLP and SSR markers were also obtained from the AMOVA (9999 permutations). In addition, a non-hierarchical AMOVA was performed to test population differentiation in Ireland and Spain separately.

Unweighted pair group method with arithmetic mean cluster analysis (UPGMA) was performed in PopGene 1.32 using Nei's genetic distance (Nei 1972) to analyse the patterns of population-level genetic distances across all populations for both the AFLP and SSR data sets. A Mantel test was used to compare pairwise genetic differences from the AFLP and SSR data.

Floral width. Corolla width and tube width were compared between regions (Ireland and Spain), among populations within regions and among plants within populations, using hierarchical (nested) ANOVA (with 'region', Ireland or Spain, as a fixed factor, 'population' nested within the region as a random factor and 'plant' nested within the population as a random factor; $n = 5$). Analyses were conducted using GMAV5 for Windows (University of Sydney, Australia). Data were tested for heterogeneity of variances using Cochran's test prior to analysis ($P = 0.0976$ and 0.0978 for corolla and tube width data, respectively) and were not transformed. Post-hoc Student–Newman–Keuls (SNK) tests were used to determine which means differed from each other (using the standard threshold of significance $\alpha = 0.05$).

Results

AFLP markers

A total of 402 reliable peaks were produced from the three AFLP primer combinations: 132 EcoRI-ACA/MseI-CAG, 144 EcoRI-AGC/MseI-CAG and 126 EcoRI-AAG/MseI-CTC among the 79 R. ponticum individuals surveyed. Overall, 95.3 % of the loci were polymorphic. Genetic diversity (H_j) was similar within both Irish and Spanish populations (range 0.204–0.234 over all populations) and more than half (59–69 %) of loci within populations were found to be polymorphic (Table 1). Analysis of molecular variance revealed that 93 % of the variance was found among individuals within populations (Table 3a). Significant ($P < 0.001$) but low genetic differentiation was recorded among populations relative to the total ($\varphi_{PT} = 0.070$), among populations within regions (Ireland and Spain) ($\varphi_{PR} = 0.044$, $P < 0.001$) and among regions ($\varphi_{RT} = 0.028$, $P = 0.005$) (Table 3a). Pairwise φ_{PT} values between populations were variable, ranging from <0.001 to 0.133 (Table 4a). A non-hierarchical AMOVA (not shown) revealed less, although significant, differentiation among Irish populations ($\varphi_{PT} = 0.037$; $P = 0.001$) than among Spanish populations ($\varphi_{PT} = 0.073$; $P = 0.01$). When individuals were grouped into populations, the UPGMA separated the Spanish population, El Palancar from other populations, and grouped the Las Corzas population with the Irish populations (Fig. 1A).

SSR markers

A total of 29 alleles were detected within the 83 R. ponticum individuals and across the four nuclear SSR loci examined (RM3D2 = 6, RM2D2 = 9, RM9D6 = 8, RM2D5 = 6). There were no significant departures from the Hardy–Weinberg equilibrium for any of the markers across populations. All populations had similar percentages of heterozygosity and the effective allele number ranged from 2.00 to 2.92 (Table 1). Analysis of molecular variance revealed that 76 % of the variance was partitioned among individuals within populations (Table 3b). There was, however, significant ($P < 0.001$) and relatively high genetic differentiation among all populations ($\varphi_{PT} = 0.243$), among populations within regions ($\varphi_{PR} = 0.133$, $P < 0.001$) and between regions ($\varphi_{RT} = 0.127$, $P < 0.001$) (Table 3b). Pairwise φ_{PT} values between populations were variable, ranging from <0.001 to 0.427 (Table 4b). Non-hierarchical AMOVAs (not shown) revealed significant differentiation between Irish populations ($\varphi_{PT} = 0.126$; $P < 0.001$) and no differentiation between Spanish populations ($\varphi_{PT} = 0.104$; $P = 0.065$). In contrast to the AFLP results, the UPGMA grouped the two Spanish populations with the Irish Glencullen population. Within this group, the Glencullen population grouped most closely with

Table 3. Analysis of molecular variance based on: (a) 402 AFLP loci and (b) four nuclear SSR loci. φ_{RT} is the correlation of individuals from the same region (Spain or Ireland) relative to that of the total; φ_{PR} is the correlation between individuals within a population relative to that of individuals within the same region; φ_{PT} is the correlation between individuals within a population relative to that of individuals of the total (Peakall et al. 1995).

(a)

Source	Df	SS	MS	Variation	%
Between regions	1	100.333	100.333	1.279	3
Among populations	6	371.499	61.916	1.947	4
Within populations	71	3033.128	42.720	42.720	93
	φ value	P value			
φ_{RT}	0.028	0.004			
φ_{PR}	0.044	<0.001			
φ_{PT}	0.070	<0.001			

(b)

Source	Df	SS	MS	Variation	%
Between regions	1	16.066	16.066	0.333	13
Among populations	6	30.702	5.117	0.304	12
Within populations	75	148.823	1.984	1.984	76
	φ value	P value			
φ_{RT}	0.127	<0.001			
φ_{PR}	0.133	<0.001			
φ_{PT}	0.243	<0.001			

the El Palancar population (Fig. 1B). No relationship was found between linear AFLP and SSR individual pairwise genetic distance matrices ($r_{xy} = 0.360$, $P = 0.076$).

Floral width

There were no significant differences in corolla width between regions (mean $\pm$ SE: Ireland 45.72 $\pm$ 0.95, Spain 43.09 $\pm$ 1.18 mm) or between populations within regions (Table 5), but SNK tests revealed that there were significant differences among plants within populations in the Irish population at Shankill and in the Spanish populations ($P < 0.05$). There were significant differences in tube width between regions, among populations within regions and within populations (Table 5). Corolla tubes were significantly wider in Ireland compared with Spain (mean $\pm$ SE: Ireland 3.24 $\pm$ 0.16, Spain 2.73 $\pm$ 0.12 mm; Fig. 2). Student–Newman–Keuls post-hoc tests revealed that floral tubes were significantly wider in Crossover compared with the other populations in Ireland, and varied significantly between plants within populations in Ireland, but not in Spain ($P < 0.05$, Fig. 2B).

Discussion

This study shows that both genetic diversity and floral width vary between populations of R. ponticum, but that native, rare populations are genetically and morphologically distinct from invasive populations. We found similar levels of genetic diversity in both declining native (Spanish) and invasive introduced (Irish) populations of R. ponticum. However, populations and regions were also genetically differentiated, a trend detected with both AFLP and SSR markers. Such genetic differentiation is expected when genetic drift and inbreeding occur in geographically isolated populations (Oakley and Winn 2012). Our findings of low genetic diversity and genetic similarity between Spanish and Irish populations support the findings of Erfmeier and Bruelheide (2011) using AFLP markers on native and invasive populations of R. ponticum. Although the level of diversity detected with AFLP and SSR markers was low, it was within the range for a species with a mixed mating system (Nybom and Bartish 2000). Rhododenron ponticum primarily relies on insect-mediated pollination for sexual reproduction (Mejías et al. 2002; Stout 2007a); however, clones may be prominent within populations

Table 4. Genetic distances (Nei 1972) based on the (a) AFLP data set and (b) SSR data set. Values and levels of significance are given in the lower left and upper right of triangle, respectively. Significances are based on random permutations (9999). $*P < 0.05$, $**P < 0.01$, $***P < 0.001$.

(a)

Spain		Ireland						
El Palancar	Las Corzas	Howth	Glencullen	Gortderraree	Gortracussane	Recess	Kylemore	*Spain*
	**	*	***	**	***	**	***	El Palancar
0.073		ns	ns	*	**	ns	**	Las Corzas
								Ireland
0.039	0.036		ns	ns	ns	ns	*	Howth
0.093	0.041	0.028		ns	*	ns	*	Glencullen
0.072	0.071	<0.001	0.022		*	ns	**	Gortderraree
0.104	0.090	0.026	0.041	0.039		ns	*	Gortracussane
0.092	0.034	0.028	0.041	0.040	0.040		ns	Recess
0.133	0.087	0.051	0.049	0.086	0.053	0.020		Kylemore

(b)

Spain		Ireland						
El Palancar	Las Corzas	Howth	Glencullen	Gortderraree	Gortracussane	Recess	Kylemore	*Spain*
	ns	***	**	***	**	***	***	El Palancar
0.104		***	**	**	**	***	***	Las Corzas
								Ireland
0.376	0.292		*	**	ns	ns	***	Howth
0.176	0.196	0.133		**	ns	**	**	Glencullen
0.212	0.147	0.236	0.134		ns	**	*	Gortderraree
0.184	0.140	0.049	0.069	0.030		ns	*	Gortracussane
0.387	0.305	0.062	0.174	0.131	<0.001		**	Recess
0.427	0.323	0.298	0.252	0.094	0.110	0.125		Kylemore

resulting from vegetative spread (Mejías *et al.* 2002). We found 28 SSR genotypes in the Irish samples, of which 12 occurred more than once, and so may be clonal, with one to eight unique genotypes per population **[see Supporting Information]**. Therefore, pollen transfer between neighbouring plants may result in bi-parental inbreeding (where both parents are closely related), rather than outcrossing. In fact, observations of the behaviour of pollinators suggest that the levels of geitonogamy (pollen transfer among flowers of the same plant) may be high, since the main pollinators of *R. ponticum* in Ireland (*Bombus* spp.) tend to move between flowers on the same plant far more frequently than between flowers on different plants (Stout 2007*b*).

The native Spanish *R. ponticum* populations had more private AFLP bands (each population had 14 private bands) than invasive Irish ones (which had a mean of six private bands). Spanish populations contained fewer than 30 individual plants possibly due to range contraction and lack of sexual regeneration (Mejías *et al.* 2002). The introduced Irish populations were generally larger than the Spanish ones and were expanding as a result of both sexual and vegetative reproduction (Stout *et al.* 2006), but they may have derived from a small number of founding individuals. Thus Irish populations contained lower genetic variation and a lower number of private bands. Examination of polymorphic chloroplast DNA would be useful to explore founder effects further, as if there were a small number of founding individuals, we might expect low plastid diversity.

Our data show that native Spanish populations have probably experienced a recent genetic bottleneck as they have both low overall expected heterozygosity (H_j), as estimated by AFLPs, and low allelic diversity, as estimated by SSRs. In addition, Irish Glencullen population groups with Spanish populations in both the AFLP and SSR analyses (and Irish Howth, Glencullen and Recess are not significantly differentiated from the Spanish Las Corzas population in AFLP pairwise φ_{PT} comparisons), which reveals the similarity of Irish and Spanish populations. However, the grouping of one of the Irish populations (Glencullen) with the Spanish populations could be due to homoplasy (a similar genetic structure due to convergence): this population also has more private alleles (alleles that are unique to a particular population from many populations sampled), and other Irish populations may have grouped together as they have fewer private alleles.

The population differentiation found, even between pairs of populations in each geographical location in

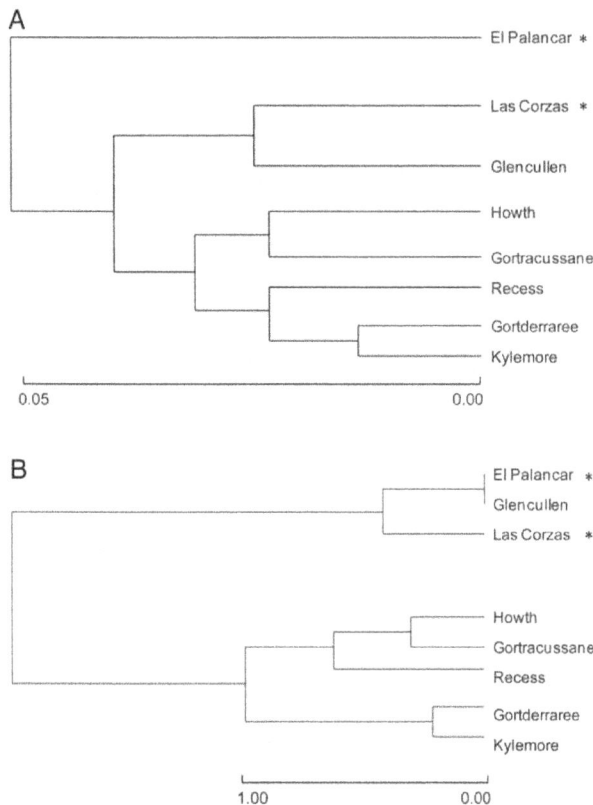

Figure 1. Rooted UPGMA tree depicting relationships between the populations investigated based on: (A) the AFLP data set and (B) the SSR data set using Nei's genetic distance (Nei 1972). Asterisks indicate native Spanish populations and scale bars represent genetic distance.

Ireland (counties Galway, Kerry and Dublin), suggests that the gene flow is limited. This supports findings of Stephenson et al. (2007) who examined seed dispersal in this species and concluded that a very small proportion (0.02 %) of seeds moved >50 m, and Stout (2007b) who found that pollen dispersal was also likely to be limited (with 98 % of bee moves between flowers <1 m apart). Thus, with limited gene flow via both pollen and seeds, spread is likely to be the result of populations spreading in the form of an 'invasion front' and/or repeated introductions. Hence, management should focus on containing existing populations and preventing new introductions.

Corolla width varied much more within populations than between them, particularly in Spain. This suggests that this trait is not under strong selection pressure or is naturally highly variable, and our measurements are consistent with other descriptions of corolla width (e.g. Mejías et al. (2002) describe flowers as having a corolla of up to 6 cm in diameter). Tube width is clearly a highly variable trait in R. ponticum, varying between individuals within populations, among populations and between regions, similar to the patterns found for genetic diversity. No previous published studies have described tube width in R. ponticum. However, corolla tubes were, on average, >0.5 mm wider in Irish populations than in Spanish ones. Wider corolla tubes may be associated with increased pollinator visitation rates, because a greater range of pollinating insects can access the nectar from more open flowers with wider tubes. Wider corolla tubes in the introduced range may be an advantage, given that introduced species have to rely on native generalist insects for pollination. Indeed, a range of generalist visitor species were recorded visiting R. ponticum in Ireland, including solitary bees, bumblebees and hoverflies (Stout et al. 2006). The most common pollinators in Ireland are bumblebees, which have relatively long tongues (compared with hoverflies and solitary bees) and the ability to rob nectar if corolla tubes are too narrow for them to probe (Stout et al. 2006). However, the most effective pollinators are large queen bumblebees (Stout

Table 5. Nested ANOVA results comparing corolla widths and tube widths of flowers between regions (Ireland and Spain), among populations (four per country, nested within regions) and within population (among five sampled plants nested within populations) ($n = 5$).

	Corolla width				Tube width			
	MS	F	Df	P	MS	F	Df	P
Between regions	346.37	3.02	1,6	0.133	13.00	6.76	1,6	0.041
Among populations	114.77	1.32	6,32	0.277	1.92	3.47	6,32	0.010
Within populations	86.94	4.93	32,160	<0.001	0.55	2.98	32,160	<0.001
Error	17.64				0.19			

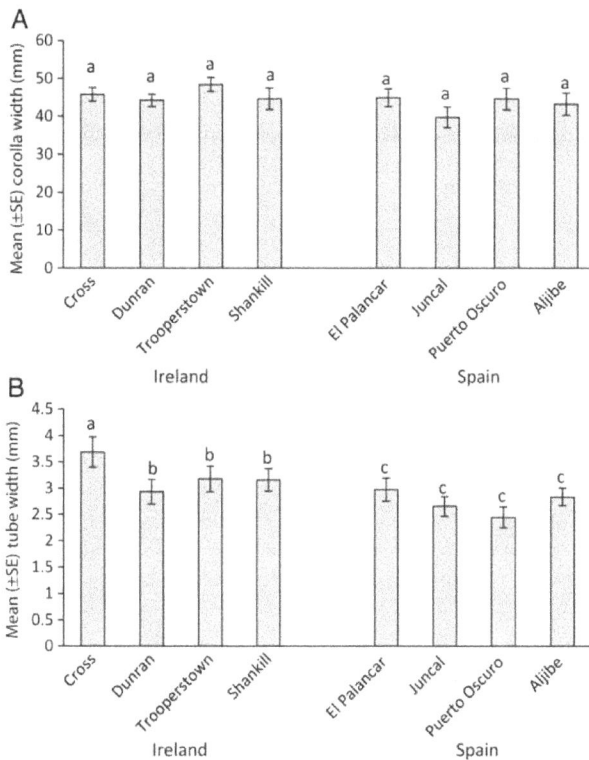

Figure 2. Mean ($\pm$ SE) (A) corolla width in millimetre and (B) tube width in millimetre of flowers from each population within each region (Ireland and Spain). Means were calculated from five flowers from each of five plants within each population, and then for the population. Letters above scale bars correspond to the results of multiple comparison tests (SNK, $P < 0.05$).

2007*b*), which may be able to visit and collect nectar more efficiently if corolla tubes are wider (Suzuki 1994). Further comparisons of pollinator behaviour in the invasive and native range would be needed to test this hypothesis. Given that populations have diverged at a genetic level, we can also expect divergence at a morphological level (although the same populations were not compared for molecular and morphological analyses in this study). It may be non-adaptive processes, such as ecological sorting, different introduction times of *R. ponticum* to Ireland historically and phenotypic plasticity (i.e. changed phenotype in response to environmental conditions, e.g. Herrera 1992), that have a strong effect on the results observed. Indeed genetic drift and founder effects may explain the relatively low, though significant, levels of genetic differentiation observed. More work is needed to clarify whether such non-adaptive processes alone, or in combination with adaptive evolution in novel habitats, drive population differentiation in non-native populations. Although *R. ponticum* is very long-lived, it was introduced 200–250 years ago, and there has been time for post-introduction evolutionary change. Indeed, studies have found evidence for

post-introduction selection affecting vegetative growth in *R. ponticum* (Erfmeier and Bruelheide 2005). Alternatively, initial introductions into Ireland may have been of wider tubed individuals; historical herbarium flower specimens could be used to investigate this, as long as the corolla tube is visible and intact. However, this is relatively unlikely given that there have been repeated introductions (Stephenson *et al.* 2006). In addition, there is little evidence for this floral difference to be a result of post-introduction hybridization with other *Rhododendron* species (Erfmeier *et al.* 2011).

Conclusions

Using both the AFLP and SSR markers, we have shown that invasive *R. ponticum* in Ireland has low genetic diversity and populations are closely related to ancestral Spanish ones, but are also differentiated from one another, with limited between-population gene flow. Introduced individuals produce flowers with wider corolla tubes, which may attract more floral visitors due to increased nectar availability. The results of this study show that native Spanish populations are distinct and should be the focus of continuing conservation attention due to their restricted distribution and small size.

Sources of Funding

This work was funded by an Enterprise Ireland Postdoctoral Fellowship (PD/2001/050), a Trinity Trust Award, a British Ecological Society Small Ecological Project Grant, a grant from the Percy Sladen Memorial Fund (Linnean Society) and a Science Foundation Ireland Research Frontiers Programme grant (10/RFP/EOB2842). K.J.D. is funded by the University of KwaZulu-Natal Postdoctoral fellowship.

Contributions by the Authors

The lead author designed the study, obtained the funding, carried out fieldwork, laboratory work, data analysis and writing; all other authors contributed to study design, fieldwork, laboratory work, data analysis and writing.

Acknowledgements

The authors thank Dr Naomi Kingston for assistance and advice in the laboratory and Dr Juan Arroyo for showing them the location of populations in Spain and to Parque Natural 'Los Alcornocales' for permission to sample there. They are grateful to three anonymous reviewers for comments that greatly improved this paper.

Literature Cited

Armbruster WS, Baldwin BG. 1998. Switch from specialized to generalized pollination. *Nature* **394**:632.

Blanca G, Cabezudo B, Hernández-Bermejo JE, Herrera CM, Molero Mesa J, Muñoz J, Valdés B. 2000. *Libro rojo de la flora silvestre amenazada de Andalucía, tomo I: Especies en peligro de extinción.* Sevilla, Spain: Junta de Andalucía.

Cabezudo B, Talavera S, Blanca G, Salazar C, Cueto M, Valdés B, Hernández Bermejo JE, Herrera CM, Rodríguez Hiraldo C, Navas D. 2005. Lista Roja de la flora vascular de Andalucía. Junta de Andalucía.

Campbell DR, Waser NM, Price MV. 1996. Mechanisms of hummingbird-mediated selection for flower width in *Ipomopsis aggregata*. *Ecology* **77**:1463–1472.

Castroviejo S, Aedo C, Gómez Campo C, Lainz M, Montserrat P, Morales R, Mufioz Garmendia F, Nieto Feliner G, Rico E, Talavera S, Villar L. 1993. *Flora Ibérica*, Vol. IV, Madrid, Spain: Real Jardín Botánico (C.S.I.C.).

Chamberlain D, Hyam R, Argent G, Fairweather G, Walker KS. 1996. *The genus Rhododendron: its classification and synonymy.* Edinburgh: Royal Botanic Garden Edinburgh.

Chase MW, Hills HG. 1991. Silica gel: an ideal material for field preservation of leaf samples. *Taxon* **40**:215–220.

Chun Y. 2011. Phenotypic plasticity of introduced versus native purple loosestrife: univariate and multivariate reaction norm approaches. *Biological Invasions* **13**:819–829.

Colak AH, Cross JR, Rotherham ID. 1998. *Rhododendron ponticum* in native and exotic environments, with particular reference to Turkey and the British Isles. *Journal of Practical Ecology and Conservation* **2**:34–41.

Conner JK, Rush S. 1996. Effects of flower size and number on pollinator visitation to wild radish, *Raphanus raphanistrum*. *Oecologia* **105**:509–516.

Cross JR. 1975. Biological flora of the British Isles *Rhododendron ponticum*. *Journal of Ecology* **63**:345–365.

Cross JR. 1981. The establishment of *Rhododendron ponticum* in the Killarney oakwoods, S.W. Ireland. *Journal of Ecology* **69**:807–824.

Dehnen-Schmutz K, Williamson M. 2006. *Rhododendron ponticum* in Britain and Ireland: social, economic and ecological factors in its successful invasion. *Environment and History* **12**:325–350.

Doyle JJ, Doyle JL. 1987. A rapid DNA isolation procedure for small quantities of fresh leaf material. *Phytochemical Bulletin* **19**:11–15.

Eckhart VM. 1993. Do hermaphrodites of gynodioecious *Phacelia linearis* (Hydrophyllaceae) trade off seed-production to attract pollinators? *Biological Journal of the Linnean Society* **50**:47–63.

Ellstrand NC, Elam DR. 1993. Population genetic consequences of small population-size—implications for plant conservation. *Annual Review of Ecology and Systematics* **24PG**:289–296.

Erfmeier A, Bruelheide H. 2004. Comparisons of native and invasive *Rhododendron ponticum* populations: growth, reproduction and morphology under field conditions. *Flora* **199**:120–132.

Erfmeier A, Bruelheide H. 2005. Invasive and native *Rhododendron ponticum* populations: is there evidence for genotypic differences in germination and growth? *Ecography* **28**:417–428.

Erfmeier A, Bruelheide H. 2010. Invasibility or invasiveness? Effects of habitat, genotype, and their interaction on invasive *Rhododendron ponticum* populations. *Biological Invasions* **12**:657–676.

Erfmeier A, Bruelheide H. 2011. Maintenance of high genetic diversity during invasion of *Rhododendron ponticum*. *International Journal of Plant Sciences* **172**:795–806.

Erfmeier A, Tsaliki M, Ross C, Bruelheide H. 2011. Genetic and phenotypic differentiation between invasive and native *Rhododendron* (Ericaeae) taxa and the role of hybridization. *Ecology and Evolution* **1**:392–407.

Excoffier L, Smouse PE, Quattro JM. 1992. Analysis of molecular variance inferred from metric distances among DNA haplotypes—application to human mitochondrial DNA restriction data. *Genetics* **131**:479–491.

Godoy O, Valladares F, Castro-Díez P. 2011. Multispecies comparison reveals that invasive and native plants differ in their traits but not in their plasticity. *Functional Ecology* **25**:1248–1259.

Hamrick JL, Godt MJW. 1996. Effects of life history traits on genetic diversity in plant species. *Philosophical Transactions of the Royal Society of London. Series B: Biological Sciences* **351**:1291–1298.

Herrera CM. 1992. Historical effects and sorting processes as explanations for contemporary ecological patterns: character syndromes in Mediterranean woody plants. *American Naturalist* **140**:421–446.

Hodkinson TR, Waldren S, Parnell JAN, Kelleher CT, Salamin K, Salamin N. 2007. DNA banking for plant breeding, biotechnology and biodiversity evaluation. *Journal of Plant Research* **120**:17–29.

Hollingsworth ML, Bailey JP. 2000. Evidence for massive clonal growth in the invasive weed *Fallopia japonica* (Japanese Knotweed). *Botanical Journal of the Linnean Society* **133**:463–472.

IUCN. 1994. IUCN Red List Categories and Criteria version 2.3. IUCN Gland, Switzerland and Cambridge, UK.

Lee CE. 2002. Evolutionary genetics of invasive species. *Trends in Ecology and Evolution* **17**:386–391.

Lockwood JL, Cassey P, Blackburn T. 2005. The role of propagule pressure in explaining species invasions. *Trends in Ecology and Evolution* **20**:211–269.

Lynch M, Milligan BG. 1994. Analysis of population genetic structure with RAPD markers. *Molecular Ecology* **3**:91–99.

Maguire TL, Peakall R, Saenger P. 2002. Comparative analysis of genetic diversity in the mangrove species *Avicennia marina* (Forsk.) Vierh. (Avicenniaceae) detected by AFLPs and SSRs. *Theoretical and Applied Genetics* **104**:388–398.

Mejías JA, Arroyo J, Ojeda F. 2002. Reproductive ecology of *Rhododendron ponticum* (Ericaceae) in relict Mediterranean populations. *Botanical Journal of the Linnean Society* **140**:297–311.

Mejías JA, Arroyo J, Marañón T. 2007. Ecology and biogeography of plant communities associated with the post Plio-Pleistocene relict *Rhododendron ponticum* subsp. *baeticum* in southern Spain. *Journal of Biogeography* **34**:456–472.

Meudt HM, Clarke AC. 2007. Almost forgotten or latest practice? AFLP applications, analyses and advances. *Trends in Plant Science* **12**:106–117.

Milne RI, Abbott RJ. 2000. Origin and evolution of invasive naturalized material of *Rhododendron ponticum* L. in the British Isles. *Molecular Ecology* **9**:541–556.

Naito K, Isagi Y, Nakagoshi N. 1998. Isolation and characterization of microsatellites of *Rhododendron metternichii* Sieb et Zucc. var. *hondoense* Nakai. *Molecular Ecology* **7**:925–931.

Nei M. 1972. Genetic distance between populations. *American Naturalist* **106**:283–292.

Nuñez MA, Moretti A, Simberloff D. 2011. Propagule pressure hypothesis not supported by an 80-year experiment on woody species invasion. *Oikos* **120**:1311–1316.

Nybom H, Bartish IV. 2000. Effects of life history traits and sampling strategies on genetic diversity estimates obtained with RAPD markers in plants. *Perspectives in Plant Ecology, Evolution and Systematics* **3**:93–114.

Oakley CG, Winn AA. 2012. Effects of population size and isolation on heterosis, mean fitness, and inbreeding depression in a perennial plant. *New Phytologist* **196**:261–270.

Peakall R, Smouse PE. 2006. GenAlEx 6: genetic analysis in excel. Population genetic software for teaching and research. *Molecular Ecology Notes* **6**:288–295.

Peakall R, Smouse PE, Huff DR. 1995. Evolutionary implications of allozyme and RAPD variation in diploid populations of dioecious Buffalograss, *Buchloe dactyloides*. *Molecular Ecology* **4**:135–147.

Petanidou T, Godfree RC, Song DS, Kantsa A, Dupont YL, Waser NM. 2012. Self-compatibility and plant invasiveness: comparing species in native and invasive ranges. *Perspectives in Plant Ecology, Evolution and Systematics* **14**:3–12.

Prentis PJ, Wilson JRU, Dormontt EE, Richardson DM, Lowe AJ. 2008. Adaptive evolution in invasive species. *Trends in Plant Science* **13**:288–294.

Ren M-X, Zhang Q-G, Zhang D-Y. 2005. Random amplified polymorphic DNA markers reveal low genetic variation and a single dominant genotype in *Eichhornia crassipes* populations throughout China. *Weed Research* **45**:236–244.

Rotherham ID. 2001. *Rhododendron* gone wild: conservation impications of *Rhododendron ponticum* in Britain. *Biologist* **48**:7–11.

Stang M, Klinkhamer PGL, van der Meijden E. 2006. Size constraints and flower abundance determine the number of interactions in a plant–flower visitor web. *Oikos* **112**:111–121.

Stephenson CM, MacKenzie ML, Edwards C, Travis JMJ. 2006. Modelling establishment probabilities of an exotic plant, *Rhododendron ponticum*, invading a heterogeneous, woodland landscape using logistic regression with spatial autocorrelation. *Ecological Modelling* **193**:747–758.

Stephenson CM, Kohn DD, Park KJ, Atkinson R, Edwards C, Travis JM. 2007. Testing mechanistic models of seed dispersal for the invasive *Rhododendron ponticum* (L.). *Perspectives in Plant Ecology, Evolution and Systematics* **9**:15–28.

Stout JC. 2007a. Reproductive biology of the invasive exotic shrub, *Rhododendron ponticum* L. (Ericaceae). *Botanical Journal of the Linnean Society* **155**:373–381.

Stout JC. 2007b. Pollination of invasive *Rhododendron ponticum* (Ericaceae) in Ireland. *Apidologie* **38**:198–206.

Stout JC, Parnell JAN, Arroyo J, Crowe TP. 2006. Pollination ecology and seed production of *Rhododendron ponticum* in native and exotic habitats. *Biodiversity and Conservation* **15**:755–777.

Suzuki K. 1994. Pollinator restriction in the narrow-tube flower type of *Mertensia ciliata* (James) G Don (Boraginaceae). *Plant Species Biology* **9**:69–73.

Vekemans X, Beauwens T, Lemaire M, Roldán-Ruiz I. 2002. Data from amplified fragment length polymorphism (AFLP) markers show indication of size homoplasy and of a relationship between degree of homoplasy and fragment size. *Molecular Ecology* **11**:139–151.

Vos P, Hogers R, Bleeker M, Reijans M, Vandelee T, Hornes M, Frijters A, Pot J, Peleman J, Kuiper M, Zabeau M. 1995. AFLP—a new technique for DNA fingerprinting. *Nucleic Acids Research* **23**:4407–4414.

Weir BS. 1996. *Genetic data analysis II: methods for discrete population genetic data*. Sunderland, Massachusetts, USA: Sinauer Associates, Inc.

Yeh FC, Yang R-C, Boyle TBJ, Ye Z-H, Mao JX. 2000. POPGENE, Microsoft windows based software for population genetic analysis. Version 1.32. Alberta, Canada: Molecular Biology and Biotechnology Centre, University of Alberta.

Young A, Boyle T, Brown T. 1996. The population genetic consequences of habitat fragmentation for plants. *Trends in Ecology and Evolution* **11**:413–418.

Zhivotovsky LA. 1999. Estimating population structure in diploids with multilocus dominant DNA markers. *Molecular Ecology* **8**:907–913.

11

Analysis of population genetic structure and gene flow in an annual plant before and after a rapid evolutionary response to drought

Rachel S. Welt[1,3]*, Amy Litt[2,4] and Steven J. Franks[1,2]

[1] Department of Biological Sciences, Fordham University, Bronx, New York, NY 10458, USA
[2] The New York Botanical Garden, Bronx, New York, NY 10458, USA
[3] Present address: Department of Herpetology, American Museum of Natural History, New York, NY 10024, USA
[4] Present address: Botany and Plant Sciences, UC Riverside, Riverside, CA 92521, USA

Associate Editor: F. Xavier Picó

Abstract. The impact of environmental change on population structure is not well understood. This study aimed to examine the effect of a climate change event on gene flow over space and time in two populations of *Brassica rapa* that evolved more synchronous flowering times over 5 years of drought in southern California. Using plants grown from seeds collected before and after the drought, we estimated genetic parameters within and between populations and across generations. We expected that with greater temporal opportunity to cross-pollinate, due to reduced phenological isolation, these populations would exhibit an increase in gene flow following the drought. We found low but significant F_{ST}, but no change in F_{ST} or Nm across the drought, in contrast to predictions. Bayesian analysis of these data indicates minor differentiation between the two populations but no noticeable change in structure before and after the shift in flowering times. However, we found high and significant levels of F_{IS}, indicating that inbreeding likely occurred in these populations despite self-incompatibility in *B. rapa*. In this system, we did not find an impact of climate change on gene flow or population structuring. The contribution of gene flow to adaptive evolution may vary by system, however, and is thus an important parameter to consider in further studies of natural responses to environmental change.

Keywords: *Brassica rapa*; climate change; drought; gene flow; phenological isolation; population structure; resurrection study.

Introduction

Climate change is a major environmental concern and one of the greatest threats to biodiversity (Williams *et al.* 2008; Bellard *et al.* 2012). Climate can influence natural systems from a genetic to an ecosystem level, affecting individuals' fitness and species' survival, and is a major factor in determining community composition and ecosystem function (Bellard *et al.* 2012). The complex relationships between biotic and abiotic factors, combined with the intricate responses of species to a changing climate and other threats to biodiversity, make understanding the current and future impacts of global climate change difficult as well as of great importance (Parmesan and Yohe 2003; Pearson and Dawson 2003; Walther 2010).

* Corresponding author's e-mail address: rwelt@amnh.org

Climate change can act as a strong selective pressure in natural populations (Jump and Peñuelas 2005; Gienapp *et al.* 2008; Hoffmann and Sgró 2011). Several recent studies have illustrated the potential for rapid evolution in response to changes in species' surroundings (Huey *et al.* 2000; Stockwell *et al.* 2003; Franks *et al.* 2007). Adaptive responses in both plants and animals frequently involve the evolution of phenological traits (Fitter and Fitter 2002; Peñuelas *et al.* 2002; Menzel *et al.* 2006; Hoffmann *et al.* 2010), which often depend on seasonal cues, such as temperature and/or precipitation conditions (Franks *et al.* 2007; Jarrad *et al.* 2008). Consequently, changes in climate patterns can have major impacts on phenology, and evolutionary changes in phenology may further affect adaptive ability by altering patterns of gene flow between populations (Garant *et al.* 2007; Franks and Weis 2009). For example, in populations that are differentially affected by changes in environmental conditions, phenology (i.e. timing of reproduction) may become more or less synchronized, impacting the opportunity for reproduction across populations.

Generally, gene flow is expected to reduce local adaptation by homogenizing populations found in differing environments, or by spreading detrimental alleles across populations (Ellstrand *et al.* 1989; Holt and Gomulkiewicz 1997). However, gene flow may also serve to introduce potentially adaptive alleles to populations, and increase genetic diversity, which natural selection can act upon to provide an evolutionary response (Ellstrand *et al.* 1989; Räsänen and Hendry 2008; North *et al.* 2011; Sexton *et al.* 2011). An improved understanding of both the effects of climate on evolutionary changes in phenology, and the consequences of these phenological changes for gene flow and local adaptation, will be valuable in predicting how species may respond to changes in climate (Garant *et al.* 2007).

Adaptation can provide a response to changes in climate in some natural populations (Franks *et al.* 2014). Franks *et al.* (2007) found that an extended drought in southern California led to an adaptive response in the flowering times of two populations of *Brassica rapa* (field mustard). The drought, lasting from 2000 to 2004, corresponded to a shift to earlier mean flowering time in both populations, causing an abbreviated growing season and resulting in more synchronous flowering times between the populations following the drought (Franks *et al.* 2007; Franks and Weis 2009).

Seeds were collected from both sites in May 1997 (prior to the drought) and June 2004 (immediately following the drought), at the end of the growing season for both populations in these years. A parent-offspring analysis from a greenhouse study demonstrated an additive genetic basis for this evolutionary change in flowering time (Franks *et al.* 2007). Additionally, further analysis has shown a decrease in phenological isolation of the two sites between 1997 (Isolation Index = 33 %) and 2004 (Isolation Index = 14 %) as a result of the change in flowering time (Franks and Weis 2009).

The current study focuses on the impacts of the change in flowering times on the genetics of these natural populations. We analysed the population genetics of this system before and after the drought to examine genetic structuring and gene flow across this phenological change. While previous studies have used population genetic data to infer historic levels of gene flow and genetic structure, this study is unique in using the 'resurrection approach' (Franks *et al.* 2008) of comparing population genetic patterns in ancestors and descendants using stored seeds.

Based on previous analyses of this system (Franks *et al.* 2007; Franks and Weis 2009), we expected to find an increase in gene flow between the populations following the drought, corresponding to decreased phenological, and potentially reproductive, isolation. Given the long-distance (>5 km) pollinator-mediated pollen dispersal found in related species (Devaux *et al.* 2005, 2007), we expected that our sites (3 km apart) would be connected by some degree of gene flow. Regardless, our approach allowed us to determine genetic changes that occurred both within and between populations over time. Changes in population structure would not only elucidate how an adaptive evolutionary change influenced genetic patterns, but may also have implications for the ability of these populations to further evolve in response to continued changes in climate. The results of this study can contribute to the growing understanding of natural responses to climate change events, which will be necessary in aiding in the conservation of species as the environment continues to change.

Methods

Study system

Brassica rapa is a weedy plant, native to Eurasia but currently distributed worldwide with cultivars and naturalized populations (Suwabe *et al.* 2002). The populations of interest in this study are found in southern California, where *B. rapa* was introduced ~300 years ago and has become naturalized (Franks *et al.* 2007). In these populations, *B. rapa* is an annual plant, produces perfect flowers and is self-incompatible. Once an individual plant has begun to flower, many new flowers may open each day, each flower lasting ~3 days, for the remainder of the growing season. In southern California, *B. rapa* germinates in the winter, flowers in the spring and continues to produce flowers until the end of the rainy season (late spring to summer). Flowers are pollinated by many species of vagile insects, such as bees and flies, allowing for the dispersal of pollen across these insects' foraging range (Davis *et al.* 1994; Warwick *et al.* 2003).

The populations were located in upper Newport Harbor Back Bay (BB) and near the San Joaquin Marsh Arboretum (Arb), in Irvine, CA, USA (Fig. 1). The two locations are ~3 km apart and are characterized by differing soil conditions. Back Bay soil is sandy and dry in comparison to Arb, where the soil has a higher water table and clay content, aiding in moisture retention (Franke *et al.* 2006). These two populations also differed in their flowering time ranges, with BB flowering earlier than Arb prior to the drought (Franks *et al.* 2007). Both populations comprised hundreds to many thousands of individuals both before and after the drought.

Samples

Seeds collected from individuals in the Arb and BB populations in 1997 and 2004 were stored at 4 °C. Over 10 000 seeds were collected from individuals across both populations to obtain representative samples of the gene pools. Seeds from this stock were grown for tissue collection on light carts in the Franks laboratory at Fordham University. Plants were watered daily and fertilized weekly yielding

259 total samples (56 Arb 1997, 39 BB 1997, 89 Arb 2004 and 75 BB 2004) (Table 1).

DNA extraction

Leaf tissue was collected from individuals following the development of true leaves. Approximately 50 mg of tissue per individual was removed, dried and stored in silica gel until extraction. Dried tissue was homogenized using Qiagen's TissueLyser system (Qiagen, Valencia, CA, USA). DNA was extracted using a Qiagen DNeasy plant kit following manufacturer's DNeasy 96 Protocol. This extraction process yielded 200 μL of eluted DNA, which was stored at −20 °C.

Microsatellite analysis

Ten microsatellite loci were chosen based on repeatable amplification success and high degree of polymorphism: BN12A (Szewc-McFadden *et al.* 1996), Na10-A08, Na10-D09, Na10-G10, Ni4-A03, Ol10-D08, Ra2-E04, Ra2-E12 (Lowe *et al.* 2002, 2004), BRMS-040 and BRMS-037 (Suwabe *et al.* 2002). PCR was carried out in 12.5 μL reactions using TC-5000 thermal cyclers (Techne, Bibby Scientific Ltd, Staffordshire, UK) to isolate and amplify microsatellite loci. These reactions included 0.5 μL of unlabelled forward and reverse primers, 0.5 μL of

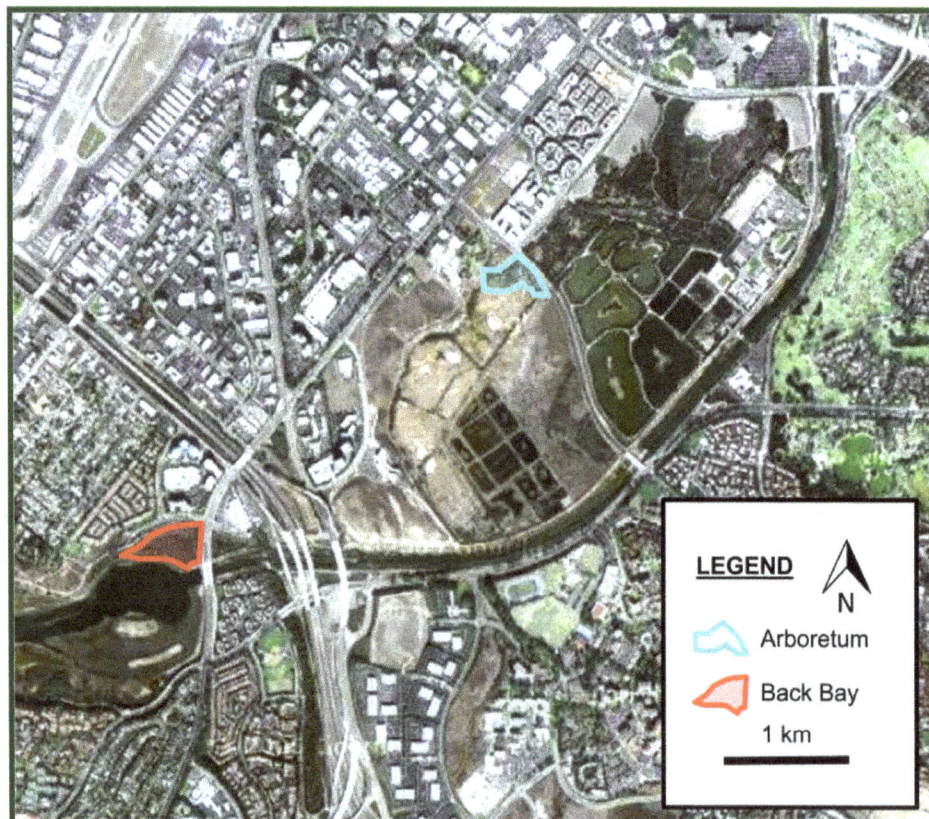

Figure 1. Map of Irvine, California showing the location of the two populations: Arboretum and Back Bay.

Table 1. Genetic diversity characteristics by population and year: population size (N), per cent polymorphic loci (P), number of PA, number of alleles per locus (A), expected heterozygosity (H_e), observed heterozygosity (H_o) and inbreeding coefficient (F_{IS}). Standard error is provided in parentheses. An asterisk indicates significance from zero at $P < 0.05$.

Population	Year	N	P (%)	PA	A	H_e	H_o	F_{IS}
BB	1997	39	90	0	3.700 (0.633)	0.517 (0.071)	0.332 (0.057)	0.369*
BB	2004	75	100	8	5.000 (0.715)	0.503 (0.066)	0.349 (0.065)	0.309*
Arb	1997	56	100	4	4.200 (0.629)	0.490 (0.065)	0.332 (0.065)	0.330*
Arb	2004	89	90	3	4.100 (0.737)	0.453 (0.068)	0.301 (0.066)	0.340*

template DNA, RNase-free water and AmpliTaq Gold 360 Master Mix (Life Technologies Inc., Grand Island, NY, USA). Cycling protocol began with an initial denaturation at 95 °C for 5 min, followed by 30 cycles of amplification at 95 °C for 30 s, optimized annealing temperatures [**see Supporting Information**] for 30 s, and 72 °C for 30 s. This concluded with a final extension at 72 °C for 7 min.

A second PCR was run to incorporate a fluorescent label (WellRED Dye D2, D3 or D4) (Sigma-Aldrich Co., St. Louis, MO, USA) using 0.5 μL reverse primer and 0.5 μL fluorescently labelled primer (M13F(-20) [D2], M13F(-40) [D3] or M13R [D4]) corresponding to the tag sequence added to the 5′ end of the initial forward primer. This was included in a reaction with 0.5 μL of 1 : 100 diluted PCR product from the first PCR, along with RNase-free water and AmpliTaq Gold 360 Master Mix. This second PCR followed the same cycling protocol as the first. PCR products were run on 1 % agarose gel to confirm the presence of target microsatellite fragments. Fragment length was then analysed with a Beckman Coulter CEQ 8800 using Genetic Analysis System v.9.0 (Beckman Coulter, Fullerton, CA, USA). Samples were run with 0.5 μL of Beckman Coulter 400 bp Size Standard and PCR product at volumes according to their fluorescent label: 2.5 μL of M13F(-20), 2.5 μL of M13F(-40) and 1.5 μL of M13R. Alleles were scored using GENEMARKER v.1.90 (SoftGenetics, State College, PA, USA).

Statistical analysis

GENALEX v.6.41 (Peakall and Smouse 2006) was used to determine measures of microsatellite diversity, including number of alleles (A), expected heterozygosity (H_e), observed heterozygosity (H_o) and inbreeding coefficient (F_{IS}) for each locus. ARLEQUIN v.3.5.1.3 (Excoffier and Lischer 2010) was then used to calculate Hardy–Weinberg Equilibrium (HWE) and to test for linkage disequilibrium (LD) across all loci. Departure from HWE was determined for each locus using the Markov chain method with a length of 1.0×10^5 steps (Guo and Thompson 1992). Linkage disequilibrium was determined using an Expectation Maximization algorithm with 1.0×10^4 permutations (Lewontin and Kojima 1960; Slatkin 1994; Slatkin

and Excoffier 1996). The frequency of null alleles for a locus was estimated using an Expectation Maximization algorithm (Dempster et al. 1977) in GENEPOP v 4.0.10 (Raymond and Rousset 1995; Rousset 2008). These parameters describe the loci and can indicate the influence that each has in defining measures of structure in these populations.

GENALEX was also used to determine measures of allelic diversity within these populations including polymorphism (P) across all loci, the number of private alleles (PA), A, H_e and H_o. F_{IS} was calculated using ARLEQUIN for each population. F-statistics, such as F_{IS} (inbreeding) and F_{ST} (fixation), are used to describe the fixation of alleles for individuals in a population, and for populations in a metapopulation, respectively (Wright 1951; Weir and Cockerham 1984). Several analogues of F_{ST} have been developed to account for the mutational characteristics of microsatellites (Slatkin 1995; Michalakis and Excoffier 1996; Hedrick 2005; Jost 2008). In addition to F_{ST}, we also calculated ϕ_{ST} (Michalakis and Excoffier 1996), which takes into account the step-wise mutation model believed to be characteristic of many microsatellites. Because F_{ST} and gene flow are inversely related (Wright 1951), we used F_{ST} to test the hypothesis that gene flow increased following the evolutionary shift to greater flowering synchrony between the populations.

Pairwise comparisons of F_{ST} and ϕ_{ST}, as well as AMOVA were performed in GENALEX. To determine whether levels of fixation and gene flow changed across this period, we compared F_{ST} values across these years. Significance was established using 95 % confidence intervals (CI) as determined through 1000 bootstrapping replicates to estimate F_{ST} using GDA v.1.1 (Lewis and Zaykin 2002).

Bayesian estimates of genetic clustering probabilistically assigns individuals to populations defined by allele frequencies at multiple loci, and were determined using STRUCTURE v.2.3.2.1 (Pritchard et al. 2000) for 5.0×10^4 burn-in repetitions and 1.0×10^6 MCMC simulations at four iterations. This allowed for an estimate of the number of genetic units, K, following Evanno et al. (2005) using STRUCTURE HARVESTER web v.0.6.92 (Earl and vonHoldt 2012). Outputs of four iterations at $K = 3$ were run in

CLUMPP v.1.1.2 (Jakobsson and Rosenberg 2007) to align these clustering results, and were then visualized using DISTRUCT v.1.1 (Rosenberg 2004).

Data were reanalysed using only loci in HWE in each population **[see Supporting Information]**, as the analyses assume that these microsatellites are acting as neutral markers and a deviation from HWE may be an indication of an effect of selection. The same procedures were used to estimate F_{IS} within each population, F_{ST} and ϕ_{ST} between Arb and BB in 1997 and 2004 and to compare F_{ST} between the 1997 and 2004 generations. Results for these analyses are noted along with the results of analyses run with all 10 loci.

Results

Allelic characteristics

Across the populations, the loci analysed showed a mean polymorphism of 95 % per population (Table 1). Between two and nine alleles were found per locus (mean = 5.8 alleles per locus) **[see Supporting Information]** and between zero and eight PA were found in each population (Table 1). Of the 38 polymorphic loci (for 10 loci across four populations), 26 exhibited departure from HWE ($P < 0.05$). Out of 180 total locus pairs, 22 pairs across all populations showed LD. However, these locus pairs were not linked in each population. The estimated frequency of null alleles ranged from $< 1.0 \times 10^{-4}$ to 0.3558 for all loci. The loci in HWE were found to have a lower frequency of null alleles. Therefore, supplementary analyses performed using only loci in HWE, also minimized the influence of allelic dropout.

Intra-population characteristics

We found high genetic diversity, and some changes in the number of PA, but no significant change in overall genetic diversity over time **[see Supporting Information]**. In the BB population, the number of PA increased from zero to eight between the sampling years (Table 1). The mean number of alleles per locus estimated for these samples did not change significantly (3.7 ± 0.633 (± SE) in 1997 to 5.0 ± 0.715 in 2004) (Table 1) **[see Supporting Information]**. However, in Arb, the number of PA decreased

from four to three, with no significant change in the number of alleles per locus (4.2 ± 0.629 in 1997 and 4.1 ± 0.737 in 2004) (Table 1) **[see Supporting Information]**.

All populations showed fewer than expected heterozygotes and high levels of inbreeding (F_{IS}) (Table 1). F_{IS} was significantly greater than zero for all populations and generations (Table 1). Levels of F_{IS} did not change significantly over time **[see Supporting Information]**. Measures of F_{IS} using only loci in HWE showed lower levels of inbreeding (between 0.065 for Arb97 and 0.085 for BB04) **[see Supporting Information]** than the analyses performed using all 10 loci (Table 1), which is expected as random mating is an assumption of HWE.

Genetic structure and gene flow

We estimated fixation between all pairs of populations (BB 97, BB 04, Arb 97 and Arb 04) as F_{ST} and ϕ_{ST}. Fixation between populations was low but significant for all pairs except BB 97 and BB 04. F_{ST} ranged from 0.027 (between Arb 97 and Arb 04) to 0.039 (BB 97 and Arb 04) (Table 2). Pairwise comparisons of ϕ_{ST} ranged from 0.067 (Arb 97 and Arb 04) to 0.108 (Arb 04 and BB 04).

In addition to pairwise comparisons, we used AMOVA to measure F_{ST} and ϕ_{ST} in both 1997 and 2004. We found that estimates of F_{ST} and ϕ_{ST} were low but significant and increased slightly from 1997 to 2004 (Table 3) **[see Supporting Information]**. AMOVA calculations of loci in HWE found low estimates of F_{ST} and ϕ_{ST} between populations in both years with no change in variance within populations across this time (92 % of allelic variance was found within populations for F_{ST} estimates in 1997 and 2004, and 86 % for ϕ_{ST} estimates in 1997 and 2004).

To compare fixation between the two populations before and after the drought, we permuted F_{ST} (Fig. 2) using 1000 bootstrap replications. We estimated F_{ST} to be 0.053 (0.019 – 0.082; 95 % CI) between BB and Arb in 1997 and 0.076 (0.024 – 0.136) in 2004. As these estimates of F_{ST} do not overlap with zero, this suggests significant levels of differentiation between the populations in both years. However, the estimates of F_{ST} were not significantly different between these 2 years (Fig. 2), indicating a lack of change in fixation or gene flow between the generations. Analysis using loci in HWE similarly shows no

Table 2. Pairwise estimates of fixation (F_{ST}, ϕ_{ST}) and gene flow (Nm) between populations and generations. F statistics (F_{ST} in (a) and ϕ_{ST} in (b)) given below the diagonal and Nm given above. F statistics significantly different from zero ($P < 0.05$) are indicated with an asterisk.

(a) F_{ST}	Arb 04	Arb 97	BB 04	BB 97	(b) ϕ_{ST}	Arb 04	Arb 97	BB 04	BB 97
Arb 04		9.173	6.245	6.201	Arb 04		3.488	2.057	2.404
Arb 97	0.027*		7.400	8.292	Arb 97	0.067*		2.265	3.001
BB 04	0.038*	0.033*		28.961	BB 04	0.108*	0.099*		38.347
BB 97	0.039*	0.029*	0.009		BB 97	0.094*	0.077*	0.006	

Table 3. Hierarchical AMOVA tables. The proportion of genetic variation is partitioned among populations, within populations and within individuals. Overall fixation, F_{ST}, and corresponding Nm provided for both years (1997 in (a); 2004 in (b)). F_{ST} significantly different from zero ($P < 0.05$) are indicated with an asterisk.

	Source	df	SS	MS	Est. var.	%
(a) 1997						
	Among populations	1	16.602	16.602	0.143	5
	Among individuals	93	319.635	3.437	0.890	33
	Within individuals	95	157.500	1.658	1.658	62
	Total	189	493.737		2.691	100
	F_{ST}	0.053*				
	Nm	4.448				
(b) 2004						
	Among populations	1	34.977	34.977	0.195	7
	Among individuals	162	520.151	3.211	0.801	31
	Within individuals	164	264.000	1.610	1.610	62
	Total	327	819.128		2.605	100
	F_{ST}	0.075*				
	Nm	3.088				

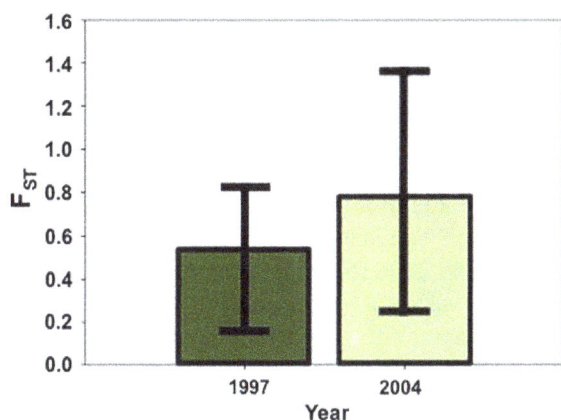

Figure 2. Population fixation (F_{ST}) between 1997 and 2004. F_{ST} estimated with 1000 bootstrap replications. Error bars display 95 % CI.

significant change in F_{ST} between Arb and BB during this time.

Bayesian analysis of this data suggests the existence of three distinct clusters ($K = 3$) in this system **[see Supporting Information]**. The greater proportion of individuals assigned to the genetic cluster represented by yellow in BB as compared with Arb indicates some genetic distinctiveness between these locations in both years. However, every individual sampled is estimated, in some proportion, to be assigned to each of these three clusters (yellow, orange and blue) (Fig. 3) suggesting low levels of structure, and consequently high levels of gene flow or genetic similarity in 1997 and 2004.

Discussion

The results of this study show relatively high levels of inbreeding within the Arb and BB populations, and low fixation between these locations in 1997 and 2004. Despite a previous finding of reduced phenological, and potentially reproductive isolation between these populations following a 5-year drought, we did not find a significant change in gene flow or in genetic structuring across this time.

Genetic diversity

Microsatellite data from these populations revealed high allelic diversity. Across all loci, we found 58 alleles, compared with 47 described in the literature **[see Supporting Information]** (Szewc-McFadden et al. 1996; Lowe et al. 2002, 2004; Suwabe et al. 2002). We found no significant shift in within population genetic parameters between 1997 and 2004 **[see Supporting Information]**. The presence of eight new PA in BB in 2004, as compared with 1997, may indicate an introduction of new alleles by mutation (Selkoe and Toonen 2006), as no nearby populations are known to exist to introduce new alleles via gene flow. Mutation rate is dependent on the effective population size and can vary due to environmental stresses, allowing for different trends in allelic diversity at each site (Hoffmann and Hercus 2000). However, the increase in PA in BB may also be due, fully or in part, to the relatively small sample size of BB 97 compared with BB 04 and a

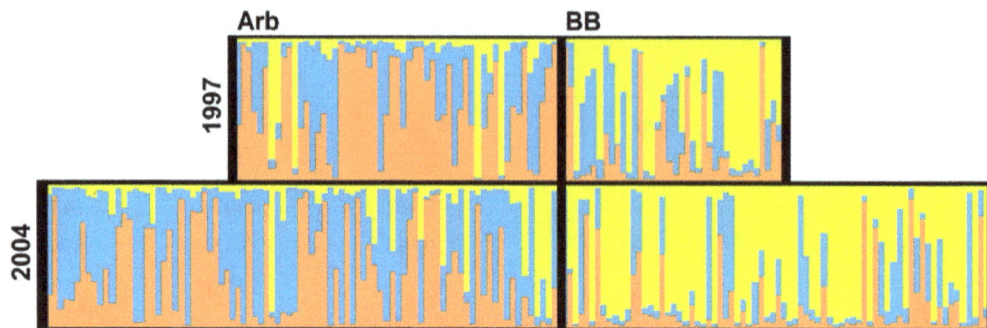

Figure 3. Genetic clustering assignment. Bayesian cluster analysis using allelic data from 10 microsatellite loci calculates three ($K = 3$) distinct genetic units (yellow, blue and orange). Assignment of these units is applied to each individual sampled in all four populations.

failure to detect low frequency alleles with our sampling effort. Thus, these populations appear to have high diversity at these loci, but there is little evidence to suggest changes in diversity over time.

Intra-population genetic structuring

We found evidence for high levels of inbreeding in our populations despite the genetic self-incompatibility of *B. rapa* (Sobotka *et al.* 2000). This high level of inbreeding could be due to local seed and pollen dispersal causing mating between relatives, or to assortative mating. Some level of assortative mating was likely in this system due to the range of flowering times, as individuals may only cross with other individuals contemporaneously in flower (Weis and Kossler 2004; Franks and Weis 2009), and individuals with similar flowering times may be more likely related to one another than expected by chance. Levels of inbreeding did not change significantly between these years despite changes in the flowering phenology of each population (Franks and Weis 2009). As the drought resulted in an abbreviated growing season in 2004, this could allow for a greater proportion of individuals to be in flower at the same time thereby reducing the likelihood of potential crosses between closely related individuals. However, genetic similarity among all individuals may have allowed these populations to maintain high F_{IS} across these generations. Ultimately, this shift in phenology did not appear to influence the genetic structuring within each population.

Fixation and gene flow over time

We found low levels of fixation, and correspondingly high levels of gene flow or shared common ancestry, between the two populations both before and after the drought. The lowest values of fixation were found between years within the same population, which is likely due to the 2004 generation maintaining much of the genetic character of its ancestral 1997 populations. However, we found very little deviation from these F_{ST} values between

all other population comparisons, indicating maintenance of, and no large change in, population connectivity over these generations.

We calculated F_{ST} and ϕ_{ST} in both of these generations to investigate whether a change in gene flow occurred with the decreased phenological isolation between the populations. Although these parameters have been critiqued as estimates of genetic differentiation, particularly for microsatellite data (Whitlock and McCauley 1999; Jost 2008), the focus of our analysis is to determine if the populations became more genetically similar over time, as would occur if gene flow increased. Thus, examining these parameters and their change over time is appropriate for our goals. Both F_{ST} and ϕ_{ST} indicate similar patterns of fixation between the BB and Arb populations with lower, significant fixation occurring prior to the drought in 1997 and higher, significant fixation following the drought in 2004 (Table 3) **[see Supporting Information]**. These values of fixation are very low and represent minimal structure between the populations. The statistical significance of these measures suggests a deviation from no fixation, however the low estimates of F_{ST} may not represent biologically meaningful levels of structure. The 95 % bootstrapped CI indicate no significant change in F_{ST} or gene flow over time (Fig. 2). This result was not expected given the increased phenological opportunity for gene flow. However, in some systems, a decrease in phenological isolation may not cause changes in gene flow, particularly when minimal structure is present prior to the phenological change.

The data were also analysed using Bayesian statistics, which showed little genetic distinctiveness between these populations and generations. Three genetic clusters were recovered for this dataset, and all individuals in each population were assigned to these clusters to varying degrees (Fig. 3). This indicates a high degree of genetic relatedness of the BB and Arb sites across these years. Despite some variation across the two sites in terms of the relative proportions of each cluster, the assignment of all individuals sampled to each of the three clusters suggests that these populations had a recent shared ancestry. Furthermore, no major shift in proportion of these genetic units is evident across time (Fig. 3). Overall, this assessment of the data supports the results of F_{ST} analyses. The minimal observed change in structure and gene flow may have been due to an insignificant change in migration rate over time, or these results may have instead been impacted by the populations' genetic similarity or large effective population sizes (Neigel 2002).

Evolutionary, environmental and community considerations

The lack of an expected change in gene flow over time could be due to several factors including the evolutionary history of these populations, seed banks, a low initial degree of genetic structure between the populations or a direct effect of the drought on the movement of pollen between populations. While research has been done to examine the phenological traits of these populations, little is known about their evolutionary history preceding the initial Franks et al. (2007) study, and pollination dynamics are not well understood. The low levels of fixation maintained between the two populations across seven generations (Table 3) [see Supporting Information] could have resulted if they had been very genetically similar across the data collection period. For example, if both populations had been established by individuals from the same source population, one population had been established from the other, or they were effectively a single population in the recent past, they would exhibit low levels of fixation. In any of these cases, fixation could have increased over time, given the 3 km separating these populations, allowing for divergence through local effects of genetic drift and mutation at neutral loci. However, it is possible that neither of these processes had an effect on fixation in this time, or that high levels of gene flow diminished a trend toward fixation.

Seed banks present in these populations might influence phenology and genetics. Both populations likely have persistent seed banks, as seed viability remains high over time, especially under the dry conditions that these populations experience outside of the growing season, and as persistent seed banks were found in the related species, Brassica napus (Simard et al. 2002). The presence of a seed bank would serve to hinder the evolutionary response to selection across this climatic shift, as genotypes adapted to previous conditions could emerge after conditions changed. However, evolutionary change in phenology was previously documented in these populations across these years (Franks et al. 2007) despite the likely presence of a seed bank. Nevertheless, a seed bank

could have been a contributing factor to the lack of change in gene flow seen in these populations.

Additionally, given that little information exists regarding the pollinators of this system in these years, it is unknown whether they were able to easily traverse the 3 km distance separating the populations. Some sources define 3 km as the maximum distance that insects will forage (Rieger et al. 2002), while others believe that flowering plant populations within a range of 10–20 km can still exhibit substantial levels of gene flow, depending on the type of insect pollinator (dePamphilis and Wyatt 1989; Ellstrand 1992; Cresswell et al. 2000). Therefore, understanding whether the distance between these populations is a reasonable foraging range for this system's pollinators is necessary to properly interpret these results. The observed levels of fixation suggest very high levels of gene flow but this may instead be a consequence of historic patterns of population connectivity (Heywood 1991; Slatkin 1995; Whitlock and McCauley 1999).

The landscape surrounding these two populations is highly urbanized and developed (Fig. 1), but contains managed conservation areas surrounding the populations, including the UCI Arboretum grounds and the Upper Newport Bay Nature Preserve and Ecological Reserve. Although weedy populations of other wild and landscaped flowering plants are present in the area and could potentially serve as stepping stones in pollinators moving between the two study populations, the extent to which this occurs, and the effects of urbanization in this area on pollinator movement, is not known.

Another potential explanation for the absence of a significant change in gene flow with decreased phenological isolation is the effect of the drought, or other unknown changes, on pollinators. As B. rapa is primarily pollinated by insects, gene flow would require the foraging activity of these pollinators. As climate change is negatively impacting many systems across the globe (Hannah et al. 2002; Parmesan and Yohe 2003), the effect of this drought may have altered the population sizes of insects in the area or caused some modification of their behaviour that impacted their ability to transfer pollen across this range (Ollerton 1996; Rafferty and Ives 2011). Therefore, if the potential for increased gene flow existed in these populations over this drought period, negative impacts of the drought on pollinators could have had effects throughout the community, which may have minimized the level of gene flow achieved through this change (Ollerton 1996; Williams et al. 2008; Walther 2010; Rafferty and Ives 2011). As little is known about the evolutionary history of this system and about the community at the time of this drought, any variation in gene flow or genetic structuring cannot be directly attributed to a change in climate. However, an effect of the drought on the pollinators in this area would likely have consequences for gene flow (Cresswell *et al.* 2000).

Conclusions

Despite the lack of a significant change in genetic structuring of these populations in response to a climate change event, this case highlights the importance of understanding the effects of environmental changes on gene flow between natural populations. Further research examining the genetic responses of natural populations to climate change across multiple systems will be useful in increasing our understanding of the evolutionary consequences of climate change (Franks and Hoffmann 2012) and establishing a baseline understanding with which to advise management decisions.

Sources of Funding

Funding was provided by a Sigma Xi Grant-in-Aid-of-Research (G20101015154375) (USA) to R.S.W., a Faculty Research Grant from Fordham and a grant (DEB-1142784) from the National Science Foundation (USA) to S.J.F. and by the New York Botanical Garden.

Contributions by the Authors

All authors took part in designing the experiment, discussing the results and composing and editing the manuscript. R.S.W. carried out the research, analysed the data and drafted the manuscript.

Acknowledgements

We thank Evon Hekkala, Jim Lewis, Anthony Caragiulo, Anna Nowogrodski, Vinson Doyle and Rebecca Lalchan for assistance with this research, as well as two anonymous reviewers for helpful comments which improved this manuscript. Seeds were made available by Art Weis and Emily Austen.

Supporting Information

The following additional information is available in the online version of this article –

Table S1. 'HWE estimates of F_{IS}' provides estimates of F_{IS}, and corresponding 95 % CI, from 1000 bootstrap replicates using only loci found to be in HWE. Loci used are listed for each population and generation.

Table S2. 'HWE estimates of F_{ST}' provides estimates of F_{ST}, and corresponding 95 % CI (when available), from 1000 bootstrap replicates using only loci found to be in HWE. Loci used are listed for each generation.

Table S3. 'Hierarchical AMOVA tables using loci in HWE' (F_{ST}) provides estimates of F_{ST} and the distribution of mo-

lecular variance for both generations, using only loci in HWE.

Table S4. 'Hierarchical AMOVA tables using loci in HWE' (ϕ_{ST}) provides estimates of ϕ_{ST} and the distribution of molecular variance for both generations, using only loci in HWE.

Table S5. 'Loci in HWE' provides a list of the loci estimated to be in HWE for each population and generation.

Table S6. 'Microsatellite loci characteristics' lists allelic parameters calculated for each of the 10 loci used in this study.

Table S7. 'Hierarchical AMOVA tables' provides estimates of ϕ_{ST} and the distribution of molecular variance for both generations, using all 10 loci.

Table S8. 'Within population parameters' provides estimates of standard deviation for genetic parameters of each population and generation.

Table S9. 'STRUCTURE likelihood outputs' provides likelihood scores used to determine the optimal number of clusters comprising these populations.

Table S10. 'Within population parameters over time' provides the results of Student's *t*-test comparing within population genetic parameters across generations.

Table S11. 'CI for within population fixation' provides the 95 % CI corresponding to estimates of F_{IS} from 1000 bootstrap replicates, for each population and generation.

Table S12. 'Estimates of F_{IS} across populations' provides global estimates of F_{IS} for each generation, and across generations.

Literature Cited

Bellard C, Bertelsmeier C, Leadley P, Thuiller W, Courchamp F. 2012. Impacts of climate change on the future of biodiversity. *Ecology Letters* 15:365–377.

Cresswell JE, Osborne JL, Goulson D. 2000. An economic model of the limits to foraging range in central place foragers with numerical solutions for bumblebees. *Ecological Entomology* 25:249–255.

Davis AR, Sawhney VK, Fowke LC, Low NH. 1994. Floral nectar secretion and ploidy in *Brassica rapa* and *B napus* (Brassicaceae). I. Nectary size and nectar carbohydrate production and composition. *Apidologie* 25:602–614.

Dempster AP, Laird NM, Rubin DB. 1977. Maximum likelihood from incomplete data via the EM algorithm. *Journal of the Royal Statistical Society: Series B (Statistical Methodology)* 39:1–38.

dePamphilis CW, Wyatt R. 1989. Hybridization and introgression in buckeyes (*Aesculus*: Hippocastanaceae): a review of the evidence and a hypothesis to explain long-distance gene flow. *Systematic Botany* 14:593–611.

Devaux C, Lavigne C, Falentin-Guyomarc'h H, Vautrin S, Lecomte J, Klein EK. 2005. High diversity of oilseed rape pollen clouds over an agro-ecosystem indicates long-distance dispersal. *Molecular Ecology* 14:2269–2280.

Devaux C, Lavigne C, Austerlitz F, Klein EK. 2007. Modelling and estimating pollen movement in oilseed rape (*Brassica napus*) at the landscape scale using genetic markers. *Molecular Ecology* **16**: 487–499.

Earl DA, vonHoldt BM. 2012. STRUCTURE HARVESTER: a website and program for visualizing STRUCTURE output and implementing the Evanno method. *Conservation Genetics Resources* **4**:359–361.

Ellstrand NC. 1992. Gene flow by pollen: implications for plant conservation genetics. *Oikos* **63**:77–86.

Ellstrand NC, Devlin B, Marshall DL. 1989. Gene flow by pollen into small populations: data from experimental and natural stands of wild radish. *Proceedings of the National Academy of Sciences of the USA* **86**:9044–9047.

Evanno G, Regnaut S, Goudet J. 2005. Detecting the number of clusters of individuals using the software STRUCTURE: a simulation study. *Molecular Ecology* **14**:2611–2620.

Excoffier L, Lischer HEL. 2010. Arlequin suite ver 3.5: a new series of programs to perform population genetics analyses under Linux and Windows. *Molecular Ecology Resources* **10**:564–567.

Fitter AH, Fitter RSR. 2002. Rapid changes in flowering time in British plants. *Science* **296**:1689–1691.

Franke DM, Ellis AG, Dharjwa M, Freshwater M, Fujikawa M, Padron A, Weis AE. 2006. A steep cline in flowering time for *Brassica rapa* in southern California: population-level variation in the field and the greenhouse. *International Journal of Plant Sciences* **167**:83–92.

Franks SJ, Hoffmann AA. 2012. Genetics of climate change adaptation. *Annual Review of Genetics* **46**:185–208.

Franks SJ, Weis AE. 2009. Climate change alters reproductive isolation and potential gene flow in an annual plant. *Evolutionary Applications* **2**:481–488.

Franks SJ, Sim S, Weis AE. 2007. Rapid evolution of flowering time by an annual plant in response to a climate fluctuation. *Proceedings of the National Academy of Sciences of the USA* **104**:1278–1282.

Franks SJ, Avise JC, Bradshaw WE, Conner JK, Etterson JR, Mazer SJ, Shaw RG, Weis AE. 2008. The resurrection initiative: storing ancestral genotypes to capture evolution in action. *BioScience* **58**: 870–873.

Franks SJ, Weber JJ, Aitken SN. 2014. Evolutionary and plastic responses to climate change in terrestrial plant populations. *Evolutionary Applications* **7**:123–139.

Garant D, Forde SE, Hendry AP. 2007. The multifarious effects of dispersal and gene flow on contemporary adaptation. *Functional Ecology* **21**:434–443.

Gienapp P, Teplitsky C, Alho JS, Mills JA, Merilä J. 2008. Climate change and evolution: disentangling environmental and genetic responses. *Molecular Ecology* **17**:167–178.

Guo SW, Thompson EA. 1992. Performing the exact test of Hardy–Weinberg proportion for multiple alleles. *Biometrics* **48**:361–372.

Hannah L, Midgley GF, Lovejoy T, Bond WJ, Bush M, Lovett JC, Scott D, Woodward FI. 2002. Conservation of biodiversity in a changing climate. *Conservation Biology* **16**:264–268.

Hedrick PW. 2005. A standardized genetic differentiation measure. *Evolution* **59**:1633–1638.

Heywood JS. 1991. Spatial analysis of genetic variation in plant populations. *Annual Review of Ecology, Evolution, and Systematics* **22**:335–355.

Hoffmann AA, Hercus MJ. 2000. Environmental stress as an evolutionary force. *BioScience* **50**:217–226.

Hoffmann AA, Sgró CM. 2011. Climate change and evolutionary adaptation. *Nature* **470**:479–485.

Hoffmann AA, Camac JS, Williams RJ, Papst W, Jarrad FC, Wahren C-H. 2010. Phenological changes in six Australian subalpine plants in response to experimental warming and year-to-year variation. *Journal of Ecology* **98**:927–937.

Holt RD, Gomulkiewicz R. 1997. How does immigration influence local adaptation? A reexamination of a familiar paradigm. *The American Naturalist* **149**:563–572.

Huey RB, Gilchrist GW, Carlson ML, Berrigan D, Serra L. 2000. Rapid evolution of a geographic cline in size in an introduced fly. *Science* **287**:308–309.

Jakobsson M, Rosenberg NA. 2007. CLUMPP: a cluster matching and permutation program for dealing with label switching and multimodality in analysis of population structure. *Bioinformatics* **23**: 1801–1806.

Jarrad FC, Wahren C-H, Williams RJ, Burgman MA. 2008. Impacts of experimental warming and fire on phenology of subalpine open-heath species. *Australian Journal of Botany* **56**:617–629.

Jost L. 2008. G_{ST} and its relatives do not measure differentiation. *Molecular Ecology* **17**:4015–4026.

Jump AS, Peñuelas J. 2005. Running to stand still: adaptation and the response of plants to rapid climate change. *Ecology Letters* **8**: 1010–1020.

Lewis PO, Zaykin D. 2002. Genetic Data Analysis (GDA): computer program for the analysis of allelic data, version 1.1 [online]. http://hydrodictyon.eeb.uconn.edu/people/plewis/software.php.

Lewontin RC, Kojima K. 1960. The evolutionary dynamics of complex polymorphisms. *Evolution* **14**:450–472.

Lowe AJ, Jones AE, Raybould AF, Trick M, Moule CL, Edwards KJ. 2002. Transferability and genome specificity of a new set of microsatellite primers among *Brassica* species of the U triangle. *Molecular Ecology Notes* **2**:7–11.

Lowe AJ, Moule C, Trick M, Edwards KJ. 2004. Efficient large-scale development of microsatellites for marker and mapping applications in *Brassica* crop species. *Theoretical and Applied Genetics* **108**:1103–1112.

Menzel A, Sparks TH, Estrella N, Koch E, Aasa A, Ahas R, Alm-Kübler K, Bissolli P, Braslavská O, Briede A, Chmielewski FM, Crepinsek Z, Curnel Y, Dahl Å, Defila C, Donnelly A, Filella Y, Jatczak K, Måge F, Mestre A, Nordli Ø, Peñuelas J, Pirinen P, Remišová V, Scheifinger H, Striz M, Susnik A, Van Vliet AJH, Wielgolaski F-E, Zach S, Zust A. 2006. European phenological response to climate change matches the warming pattern. *Global Change Biology* **12**: 1969–1976.

Michalakis Y, Excoffier L. 1996. A generic estimation of population subdivision using distances between alleles with special reference for microsatellite loci. *Genetics* **142**:1061–1064.

Neigel JE. 2002. Is F_{ST} obsolete? *Conservation Genetics* **3**:167–173.

North A, Pennanen J, Ovaskainen O, Laine A-L. 2011. Local adaptation in a changing world: the roles of gene-flow, mutation, and sexual reproduction. *Evolution* **65**:79–89.

Ollerton J. 1996. Reconciling ecological processes with phylogenetic patterns: the apparent paradox of plant-pollinator systems. *The Journal of Ecology* **84**:767–769.

Parmesan C, Yohe G. 2003. A globally coherent fingerprint of climate change impacts across natural systems. *Nature* **421**:37–42.

Peakall R, Smouse PE. 2006. GENELEX 6: genetic analysis in Excel. Population genetic software for teaching and research. *Molecular Ecology Notes* **6**:288–295.

Pearson RG, Dawson TP. 2003. Predicting the impacts of climate change on the distribution of species: are bioclimate envelope models useful? *Global Ecology and Biogeography* **12**:361–371.

Peñuelas J, Filella I, Comas P. 2002. Changed plant and animal life cycles from 1952 to 2000 in the Mediterranean region. *Global Change Biology* **8**:531–544.

Pritchard JK, Stephens M, Donnelly P. 2000. Inference of population structure using multilocus genotype data. *Genetics* **155**:945–959.

Rafferty NE, Ives AR. 2011. Effects of experimental shifts in flowering phenology on plant-pollinator interactions. *Ecology Letters* **14**: 69–74.

Räsänen K, Hendry AP. 2008. Disentangling interactions between adaptive divergence and gene flow when ecology drives diversification. *Ecology Letters* **11**:624–636.

Raymond M, Rousset F. 1995. GENEPOP (version 1.2): population genetics software for exact tests and ecumenicism. *Journal of Heredity* **86**:248–249.

Rieger MA, Lamond M, Preston C, Powles SB, Roush RT. 2002. Pollen-mediated movement of herbicide resistance between commercial canola fields. *Science* **296**:2386–2388.

Rosenberg NA. 2004. DISTRUCT: a program for the graphical display of population structure. *Molecular Ecology Notes* **4**: 137–138.

Rousset F. 2008. Genepop'007: a complete re-implementation of the Genepop software for Windows and Linux. *Molecular Ecology Resources* **8**:103–106.

Selkoe KA, Toonen RJ. 2006. Microsatellites for ecologists: a practical guide to using and evaluating microsatellite markers. *Ecology Letters* **9**:615–629.

Sexton JP, Strauss SY, Rice KJ. 2011. Gene flow increases fitness at the warm edge of a species' range. *Proceedings of the National Academy of Sciences of the USA* **108**:11704–11709.

Simard M-J, Légère A, Pageau D, Lajeunesse J, Warwick S. 2002. The frequency and persistence of volunteer canola (*Brassica napus*) in Québec cropping systems. *Weed Technology* **16**:433–439.

Slatkin M. 1994. Linkage disequilibrium in growing and stable populations. *Genetics* **137**:331–336.

Slatkin M. 1995. A measure of population subdivision based on microsatellite allele frequencies. *Genetics* **139**:457–462.

Slatkin M, Excoffier L. 1996. Testing for linkage disequilibrium in genotypic data using the Expectation-Maximization algorithm. *Heredity* **76**:377–383.

Sobotka R, Sáková L, Curn V. 2000. Molecular mechanisms of self-incompatibility in Brassica. *Current Issues in Molecular Biology* **2**:103–112.

Stockwell CA, Hendry AP, Kinnison MT. 2003. Contemporary evolution meets conservation biology. *Trends in Ecology and Evolution* **18**: 94–101.

Suwabe K, Iketani H, Nunome T, Kage T, Hirai M. 2002. Isolation and characterization of microsatellites in *Brassica rapa* L. *Theoretical and Applied Genetics* **104**:1092–1098.

Szewc-McFadden AK, Kresovich S, Bliek SM, Mitchell SE, McFerson JR. 1996. Identification of polymorphic, conserved simple sequence repeats (SSRs) in cultivated *Brassica* species. *Theoretical and Applied Genetics* **93**:534–538.

Walther G-R. 2010. Community and ecosystem responses to recent climate change. *Philosophical Transactions of the Royal Society B: Biological Sciences* **365**:2019–2024.

Warwick SI, Simard MJ, Légère A, Beckie HJ, Braun L, Zhu B, Mason P, Séguin-Swartz G, Stewart CN. 2003. Hybridization between transgenic *Brassica napus* L. and its wild relatives: *Brassica rapa* L., *Raphanus raphanistrum* L., *Sinapis arvensis* L., and *Erucastrum gallicum* (Willd.) O.E. Schulz. *Theoretical and Applied Genetics* **107**:528–539.

Weir BS, Cockerham CC. 1984. Estimating F-statistics for the analysis of population structure. *Evolution* **38**:1358–1370.

Weis AE, Kossler TM. 2004. Genetic variation in flowering time induces phenological assortative mating: quantitative genetic methods applied to *Brassica rapa*. *American Journal of Botany* **91**:825–836.

Whitlock MC, McCauley DE. 1999. Indirect measures of gene flow and migration: $F_{ST} \neq 1/(4Nm + 1)$. *Heredity* **82**:117–125.

Williams SE, Shoo LP, Isaac JL, Hoffmann AA, Langham G. 2008. Towards an integrated framework for assessing the vulnerability of species to climate change. *PLoS Biology* **6**:2621–2626.

Wright S. 1951. The genetical structure of populations. *Annals of Human Genetics* **15**:323–354.

12

Population structure and genetic diversity of the perennial medicinal shrub *Plumbago*

Sayantan Panda[1†], Dhiraj Naik[2†] and Avinash Kamble[1*]

[1] Department of Botany, Savitribai Phule Pune University, Ganeshkhind, Pune 411007, India
[2] Department of Environmental Sciences, Indian Institute of Advanced Research, Koba Institutional Area, Gandhinagar 382007, India

Associate Editor: Kermit Ritland

Abstract. Knowledge of the natural genetic variation and structure in a species is important for developing appropriate conservation strategies. As genetic diversity analysis among and within populations of *Plumbago zeylanica* remains unknown, we aimed (i) to examine the patterns and levels of morphological and genetic variability within/among populations and ascertain whether these variations are dependent on geographical conditions; and (ii) to evaluate genetic differentiation and population structure within the species. A total of 130 individuals from 13 populations of *P. zeylanica* were collected, covering the entire distribution area of species across India. The genetic structure and variation within and among populations were evaluated using inter-simple sequence repeat (ISSR) and randomly amplified DNA polymorphism (RAPD) markers. High levels of genetic diversity and significantly high genetic differentiation were revealed by both the markers among all studied populations. High values of among-population genetic diversity were found, which accounted for 60 % of the total genetic variance. The estimators of genetic diversity were higher in northern and eastern populations than in southern and western populations indicating the possible loss of genetic diversity during the spread of this species to Southern India. Bayesian analysis, unweighted pair group method with arithmetic average cluster analysis and principal coordinates analysis all showed similar results. A significant isolation-by-distance pattern was revealed in *P. zeylanica* by ISSR ($r = 0.413$, $P = 0.05$) and RAPD ($r = 0.279$, $P = 0.05$) analysis. The results obtained suggest an urgent need for conservation of existing natural populations along with extensive domestication of this species for commercial purpose.

Keywords: Genetic diversity; molecular markers; *Plumbago zeylanica*; population structure.

Introduction

An understanding of the patterns of genetic variation within and among populations of medicinal plants is essential for devising optimum genetic resource management strategies for their conservation, sustainable utilization and genetic improvements. Natural populations of medicinal plant species are extensively exploited due to their heavy demands. In such cases, long-term survival as well as semi-domesticated nature of many medicinal plants depends on the maintenance of sufficient genetic variability within and among populations to accommodate new selection pressures exerted by continuous environmental changes (Barrett and Kohn 1991). Genetic diversity maintained in a plant species would be

* Corresponding author's e-mail address: kambleavinash2000@gmail.com
† Equal contribution.

influenced by many processes, such as the long-term evolutionary history and the characteristics of the species, including genetic drift, gene flow, and reproductive mode and mating system (Hamrick and Godt 1996). Thus, an accurate estimate of genetic diversity of medicinal plant species is influenced by many processes such as the long-term evolutionary process as well as information useful for developing conservation plans to preserve genetic diversity (Falk and Holsinger 1991).

Medicinal plants in India are gaining much attention and are being cultivated widely by the farmers but many of them are still in semi-domesticated nature. Several studies have examined the effects of cultivation on the genetic diversity of crop plant species and forest tree species populations in India (Bahulikar et al. 2004; Shaanker et al. 2004; Bodare et al. 2013; Harish et al. 2014). Unfortunately, a handful of studies have examined genetic variation of *Plumbago* and other Plumbaginaceae family members in India inspite of their high economical benefit (Britto et al. 2009; Ding et al. 2012; Haji et al. 2014), and none of them have examined wide-scale genetic structure of *P. zeylanica*. Moreover, very little is known about the impact of environmental conditions such as latitude, longitude and other meteorological variables on genetic structure of this species (Britto et al. 2009; Haji et al. 2014).

Plumbago zeylanica, a perennial shrub of family Plumbaginaceae, is widely dispersed in wild throughout India and has also been introduced as a plantation species. It is native to warmer tropical and sub-tropical regions of world, grows naturally in India, Sri Lanka and in South West Asia (Pant et al. 2012). In the recent decades, Plumbago is widely spread in tropical and sub-tropical regions of Australia, Asia and Africa (Tilak et al. 2004; Jamal et al. 2014) including Ethiopia (Tilak et al. 2004). It occurs in deciduous woodland, savannas and scrub forest with an elevation of 300–2000 m. Plant consists of slender stems with thin, glabrous and ovate leaves. The flowers of *P. zeylanica* are dioecious and the pollination is primarily carried out by insect and wind. They are characterized by having a tube-shaped calyx with glandular trichomes secreting sticky mucilage. This plant exhibits both sexual reproduction and clonal growth by rooted shoots (Pant et al. 2012). As a traditional Indian medicinal shrub, *P. zeylanica* has a variety of important biological functions, such as inhibiting tumour cell growth, anti-ulceration, anti-deyspepsic and enhancing immunity (Tilak et al. 2004; Jamal et al. 2014), and has been extensively used to treat chronic diseases. The propagation of *P. zeylanica* seems to be unpredictable due to poor seed viability, improper seed germination and lower seedling recruitment in field conditions are also reported (Pant et al. 2012). With the growth of commercial demand in recent years, excessive exploitation has shrunk the

natural resource of this species to a narrow distribution, and its survival has been seriously threatened. Previous studies have mainly focused on the resource distribution, its morphological characteristics, dynamics and pharmacological properties (Tilak et al. 2004; Hafeez et al. 2012, 2013; Jamal et al. 2014). Therefore, to formulate conservation strategies for existing natural populations, we aimed to assess the genetic diversity and differentiation between and among populations of *Plumbago* by using randomly amplified polymorphic DNA (RAPD) and inter-simple sequence repeats (ISSR) markers which are widely used because of low cost, easy access and high polymorphic nature. Unlike SSRs these markers do not require prior knowledge of genome sequence. Despite limitations regarding reproducibility of RAPD, combination of ISSR and RAPD markers has been used for understanding population genetic diversity and structure in a number of species (Naik et al. 2009; Ding et al. 2013; Harish et al. 2014).

We focused on large-scale population genetic analysis of *P. zeylanica* using RAPD and ISSR markers to (i) evaluate the wide-range genetic structure of 13 populations selected to cover its distribution across India, (ii) infer relationship between latitude and the components of genetic variation in *P. zeylanica* populations, (iii) compare the population genetic structure in *P. zeylanica* populations using two dominant markers and (iv) provide necessary information for developing conservation strategies for this endangered medicinal shrub.

Methods

Ethics statement

No national permissions were required for this study as it did not involve critically endangered or protected species. No specific permissions were required to access the study sites; the collections were made on public lands.

Study sites and plant sample collection

Healthy seedlings of *P. zeylanica* were randomly selected from each population site covering an area of 50 km and collected during the month of October–December 2010. Data on the coordinates and altitudes of all the population sites are presented [see Supporting Information— Table S1]. All the plants collected from different regions were established in the poly house. Authentication of all the plant specimens was done at Botanical Survey of India (BSI), Pune, Maharashtra, India. A total of 13 natural populations of *P. zeylanica* were sampled across four different regions of India, which represented a wide geographic distribution in a range from 97 to 801 m in elevation and 30–394 cm in annual rainfall [see Supporting Information—Table S1 and Fig. S1]. To examine the latitudinal pattern of genetic variation within the species, populations were grouped into northern, southern, western and eastern sectors. Ten quantitative and three qualitative morphological traits were measured and examined in 130 individuals that were genotyped **[see Supporting Information—Table S2]**. Ten quantitative morphological descriptors were selected from the International Plant Genetic Resources Institute (IPGRI) descriptors (IPGRI 1993) in the studied populations.

Genomic DNA extraction

Genomic DNA extraction was carried out using CTAB procedure (de la Cruz *et al.* 1997) with minor modifications. DNA concentration was determined by comparing the intensity of the ethidium bromide stained bands with that of similarly stained bands of known amount of Lambda DNA (Fermentas, USA). The concentration of each DNA sample was made to 10 ng μL^{-1}.

RAPD and ISSR genotyping

Thirty-three ISSR primers from primer set no. 9 (University of British Columbia, Canada) and 50 RAPD primers (Operon, Eurofins Genomics, India) were selected for this study based on the presence of clear, repeatable and polymorphic amplified bands. The amplification was carried out in 20 μL reaction volume and consisted of 0.1 mM of each dNTP, 1 U *Taq* polymerase, $1\times$ of *Taq* polymerase buffer, 1.6 mM $MgCl_2$ (Fermentas, USA) and 20 ng genomic DNA. DNA amplification was performed in a thermocycler (Corbett Research, Australia) programmed for an initial denaturation at 94 °C for 5 min, 44 cycles of denaturation at 94 °C for 1 min, annealing at 50 °C/37 °C (for ISSR and RAPD primers, respectively) for 45 s and extension at 72 °C for 1 min and a final extension at 72 °C for 10 min. The amplified products were separated on 2 % agarose gel and stained with ethidium bromide (2 $\mu g \ mL^{-1}$). The reproducibility of DNA amplification profiles was tested by repeating the polymerase chain reactions (PCRs) twice with 20 of the 33 selected ISSR primers and 20 of the 50 selected RAPD primers.

Data analysis

Morphological structure of Plumbago populations.
To describe the structure of individual morphological diversity, a dissimilarity matrix was computed based on the 13 morphological traits. Data for each quantitative trait was scored from 10 randomly chosen plants. From these measurements, the mean, standard deviation (SD) and coefficient of variation (CV) were calculated for each morphological character. The morphological similarity among individuals was then assessed by a principal coordinates analysis (PCoA) using the R package *ade4*.

Allelic loci scoring.
Reproducible and well-defined bands obtained after PCR amplification using each RAPD and ISSR primers were scored as 1 or 0 for the presence or absence of bands and a binary matrix was generated for RAPD and ISSR markers. Based on the binary data matrix, we estimated the total number of polymorphic loci and percentage polymorphic loci per primer combination.

Genetic diversity analysis using RAPD and ISSR markers.
The genetic diversity parameters were calculated for each population and for each marker using POPGENE version 1.32 (Yeh *et al.* 2000). The percentage of polymorphic bands (PPB), Nei's gene diversity (*H*) (Nei 1973), Shannon's index, Nei's unbiased genetic distance, Nei's genetic differentiation index among populations (G_{ST}) and gene flow (N_m) was estimated using POPGENE. The obtained genetic distance matrix was then used to construct the dendrogram using the unweighted pair group method with arithmetic average (UPGMA) algorithm MEGA version 6.0.5 (Tamura *et al.* 2013). To assess percent distribution of genetic variation among and within populations, a hierarchical analysis of molecular variance (AMOVA) was performed using GenAlEx 6.2 software (Excoffier *et al.* 1992; Peakall and Smouse 2012). Genetic distance was tested against geographic distance by Mantel test with 999 random permutations using GenAlEx 6.2 software (Excoffier *et al.* 1992). The effect of latitude on genetic diversity was analysed by He, PPB for each population for both the markers. In addition, AMOVA was conducted to estimate genetic variation among latitudinal sectors and a linear regression was tested against latitude using Sigma-Plot, version 10.0, considering He and PPB as dependent variables.

Population structure analysis.
The Bayesian clustering method was implemented to deduce population structure using STRUCTURE 2.2.3 software (Pritchard *et al.* 2000; Falush *et al.* 2003). STRUCTURE performs Bayesian assignments of individuals to a given number of genetically homogenous clusters (*K*) of populations. Twenty replications of each proposed *K* value (from *K* = 1 through *K* = 20) were investigated under no-admixture ancestry and the correlated allele frequencies model by running 100 000 iterations of each *K*, with a burn-in length of 100 000 iterations. To assist the determination of optimal *K*, ΔK was estimated as described (Evanno *et al.* 2005). The probability distribution [ln *P(D)*] and ΔK were retrieved from the STRUCTURE HARVESTOR software (Earl and von-Holdt 2012). Bar chart for the proportion of the member coefficient of each individual for each *K* was summarized

using CLUMPP (Jakobsson and Rosenberg 2007) and visualized in DISTRUCT (Rosenberg 2004).

Results

Morphological variability

In multidimensional analysis of morphological data matrix containing quantitative and qualitative characters, sampled population of *P. zeylanica* was significantly distinguishable and was quite variable representing high levels of inter- and intra-population variation [see Supporting Information—Fig. S2]. The principal component analysis (PCA) represented that the first two components, which had eigenvalue higher than 1, denotes the total of 77.6 % of whole phenotypic variability, contributing to all the variables to the morphological diversity of sampled populations [see Supporting Information—Fig. S3 and Table S2]. The most discriminative quantitative characters were length of the inflorescence axis, number of inflorescence per vine, number of flower per inflorescence, distance between two adjacent flowers and length of petal based on correlation of these characters with PC1. Principal component analysis indicated that the habit type and trichome colour on the inflorescence axis were the two best qualitative diagnostic characters. Both traits were highly correlated with first PCA axis, and showed a semi-overlapping pattern of variation among the sampled populations.

ISSR and RAPD polymorphism

A total of 130 individuals belonging to 13 populations of *P. zeylanica* were surveyed. Among and within studied populations generated a total of 229 fragments by using 20 selected ISSR primers, of which 169 (73.8 %) were polymorphic [see Supporting Information—Table S3]. Each primer amplified 10–19 bands with an average of 14.6. The size of the amplified fragments ranged from 200 to 2000 bp. In general, ISSR variation within populations was very low and varied erratically across localities (Table 1).

The RAPD analysis yielded a total of 232 loci for 130 individuals generated by using 20 selected primers, of which 78.9 % (183 fragments) were polymorphic between individuals [see Supporting Information—Table S4]. In comparison to ISSR profiling, RAPD variation within populations was higher in *P. zeylanica* and varied across the populations (Table 2).

Genetic diversity

Based on ISSR profiling, genetic diversity of the species across all the populations was with an average $H = 0.034$ (SE = 0.001) and Ne, PPB and I were on average 1.04 (ranged from 1.01 to 1.06), 11 (ranged from 6.55 to 15.28) and 0.04 (ranged from 0.02 to 0.06), respectively (Table 1). As a general pattern, genetic diversity and percent polymorphic bands of *P. zeylanica* populations decreased with increasing latitude [see Supporting Information—Fig. S4]. Overall

Table 1. Genetic diversity within populations of *P. zeylanica* using ISSR markers. PPB, percentage of polymorphic bands.

Populations	Region	Effective number of alleles (n_e)	PPB (%)	Nei's genetic diversity (h)	Shannon's information index (I)
Solan, Himachal Pradesh	North	1.03	08.30	0.021	0.033
Panipat, Haryana	North	1.04	10.48	0.028	0.045
Ananthagiri, Vikarabad, Andhra Pradesh	South	1.04	11.35	0.028	0.047
Tirupati, Chittor, Andhra Pradesh	South	1.04	10.92	0.026	0.044
Kolli, Salem, Tamilnadu	South	1.02	09.17	0.018	0.032
Coimbatore, Tamilnadu	South	1.06	15.72	0.042	0.069
RFRI Campus, Jorhat, Assam	East	1.01	06.55	0.012	0.022
NBU Campus, Siliguri, West Bengal	East	1.02	06.99	0.016	0.027
Kadma, Bankura, West Bengal	East	1.03	11.79	0.024	0.042
Ajra-Amboli, Kolhapur, Maharashtra	East	1.03	09.61	0.022	0.037
JNVU Campus, Jodhpur, Rajasthan	West	1.05	13.10	0.035	0.057
Ellora, Aurangabad, Maharashtra	West	1.05	13.54	0.036	0.058
Shendi, Bhandardara, Maharashtra	West	1.06	15.28	0.038	0.063
Average		1.04	11.0	0.03	0.04
Species-level		1.44 (0.347)	73.8	0.26 (0.173)	0.41 (0.230)

Table 2. Genetic diversity within populations of *P. zeylanica* using RAPD markers. PPB, percentage of polymorphic bands.

Populations	Region	Effective number of alleles (n_e)	PPB (%)	Nei's genetic diversity (h)	Shannon's information index (I)
Solan, Himachal Pradesh	North	1.07	09.05	0.040	0.057
Panipat, Haryana	North	1.09	11.21	0.049	0.070
Ananthagiri, Vikarabad, Andhra Pradesh	South	1.09	11.64	0.051	0.073
Tirupati, Chittor, Andhra Pradesh	South	1.08	10.34	0.045	0.065
Kolli, Salem, Tamilnadu	South	1.06	08.62	0.037	0.054
Coimbatore, Tamilnadu	South	1.11	14.22	0.062	0.089
RFRI Campus, Jorhat, Assam	East	1.05	06.47	0.028	0.040
NBU Campus, Siliguri, West Bengal	East	1.06	07.76	0.034	0.048
Kadma, Bankura, West Bengal	East	1.09	12.50	0.054	0.078
Ajra-Amboli, Kolhapur, Maharashtra	East	1.07	09.05	0.039	0.056
JNVU Campus, Jodhpur, Rajasthan	West	1.11	13.79	0.060	0.087
Ellora, Aurangabad, Maharashtra	West	1.09	12.50	0.054	0.078
Shendi, Bhandardara, Maharashtra	West	1.12	15.95	0.069	0.100
Average		1.08	11.0	0.05	0.07
Species-level		1.52 (0.329)	78.9	0.31 (0.155)	0.47 (0.197)

Table 3. Analysis of molecular variance for 130 individuals of *P. zeylanica* using ISSR markers, significance tests after 1000 random permutations. df, degrees of freedom; SSD, sum of squares; TVP, total variance component.

Source of variation	df	SSD	Variance components	TVP (%)	P-value
Variance among region	3	790.71	0.75	2	0.007
Variance among populations (variance within region)	9	2157.27	22.62	61	0.001
Variance within populations	117	1574.50	13.45	37	0.001
Total	129	4522.48	36.83	100	

genetic diversity indices showed that the city Coimbatore from southern region of India has the highest diversity and Jorhat from eastern region of India has the lowest. Genetic diversity of western region populations was double compared with eastern region populations (Table 1). Southern and northern region populations showed intermediate genetic diversity when compared with western region populations (Table 1).

Similar to ISSR analysis, RAPD analysis showed decreased genetic diversity and PPBs of *P. zeylanica* populations with increasing latitude **[see Supporting Information—Fig. S4]**. Assuming a Hardy–Weinberg equilibrium, *H* was with an average of 0.051 ± 0.002 whereas Ne, PPB and I were on average 1.08 (ranged from 1.05 to 1.12), 11 (ranged from 6.47 to 15.95) and 0.07 (ranged from 0.04 to 0.10), respectively (Table 2). All genetic diversity indices showed

that Jodhpur from western region has the highest diversity and Jorhat towards the east has the lowest. Genetic diversity of western region populations was double compared with eastern region populations (Table 2). Southern and northern region populations showed intermediate genetic diversity when compared with western region populations (Table 2).

Genetic differentiation and gene flow

Distribution of total genetic variation by nested AMOVA for ISSR dataset revealed that most of the total variance is attributable to genetic variation among populations (61 %) (Table 3). However, 37 and 2 % of the variance was partitioned within populations and among regions, respectively. Analysis of molecular variance revealed a low level of genetic differentiation, but this was highly significant ($P < 0.001$) between within and among

Table 4. Analysis of molecular variance for 130 individuals of *P. zeylanica* using RAPD markers, significance tests after 1000 random permutations. df, degrees of freedom; SSD, sum of squares; TVP, total variance component.

Source of variation	df	SSD	Variance components	TVP (%)	P-value
Variance among region	3	706.36	0.00	0	0.656
Variance among populations (variance within region)	9	2154.14	22.27	57	0.001
Variance within populations	117	1946.20	16.63	43	0.001
Total	129	4806.70	38.90	100	

populations (Table 3). Pronounced level of genetic differentiation among populations ($\phi_{ST} = 0.627$, $P < 0.001$) and limited estimated gene flow between populations ($N_m = 0.06$) was observed. In RAPD dataset for the same populations, nested AMOVA showed that vast majority of variance (57 %) was among populations, while 43 and 0 % of the variance was partitioned within populations and among regions, respectively (Table 4). A significant level of genetic differentiations was observed within ($\phi_{ST} = 0.572$, $P < 0.001$) and among ($\phi_{ST} = 0.571$, $P < 0.001$) populations, while low among-population gene flow ($N_m = 0.09$) was observed, using RAPD dataset.

Genetic relationship and structuring

POPGENE analysis revealed that Nei's unbiased genetic distance ranged from 0.151 (Coimbatore vs. Aurangabad) to 0.633 (Aurangabad vs. Jorhat) in ISSR dataset, while with RAPD dataset, Nei's unbiased genetic distance ranged from 0.584 (Kadma vs. Jorhat) to 0.913 (Ananthagiri to Aurangabad). The UPGMA tree based on Nei's unbiased genetic distance resolved them into three clusters **[see Supporting Information—Figs S5 and S6]** mainly in correspondence with eastern and western region. The Siliguri population formed a sole group, and the remaining populations formed the other group, which can be subdivided into two subgroups. Ananthagiri and Aurangabad populations, which showed highest similarity, were grouped together. This cluster further grouped with Coimbatore and Ajra. These four populations, which were from Deccan Platue, formed one major cluster. Populations close to Himalaya formed a separate cluster, which showed bootstrap values of 56 %. Bootstrap values ranged from 44 to 95 % for each population cluster. These values indicated that each population showed high confidence limits for clustering. Bootstrap values at the nodes joining two different populations ranged from 53 to 89 %. While Ananthagiri and Aurangabad populations showed the bootstrap value of 95 %, hierarchical clustering of these with Coimbatore and Ajra populations showed bootstrap values of 86 and 84 %, respectively. Similarly, the bootstrap value for the cluster of Solan and Panipat populations was 56 %. Clustering of

these with Siliguri showed moderately low bootstrap values of 44 %.

The two-dimensional PCoA for 230 individuals in 13 populations based on ISSR dataset accounted for 28.15 % (axis 1) and 18.43 % (axis 2) of total variance, respectively **[see Supporting Information—Fig. S7]**. As expected, Silguri, Jorhat, Solan, Kadma, Panipat and Bhandarada populations occupied similar position along the axis 1. The other populations such as Ajra, Aurangabad, Coimbatore and Ananthagiri have similar genetic similarity.

In the STRUCTURE analysis using ISSR dataset, the real *K* value with the highest value of ΔK for the 130 individuals was 2 (Fig. 1) followed by levelling off and accompanied by an increase in variance. A diagnostic for true value of *K* is a decrease in slope and increase in variance of ln *P*(*D*) (Fig. 1). The proportion of each individual in each population assigned into two sub-clusters (Clusters I and II) (Fig. 2), resulted in agreement with UPGMA dendrogram **[see Supporting Information—Fig. S5]**. Several single populations were assigned to specific clusters for higher values of *K* (Fig. 2). Although several nodes were poorly supported by bootstraps, the results of the UPGMA tree showed a similar pattern as that of the STRUCTURE analysis **[see Supporting Information—Fig. S5]** (Fig. 2).

The population structure of *P. zeylanica* using RAPD dataset inferred using the method of Pritchard *et al.* (2000) showed a spatial pattern of genetic distances among the populations which was similar to the results of UPGMA dendrogram. On the basis of the method of Evanno *et al.* (2005), all the analysed genotypes were split into $K = 2$ groups (Fig. 3). Several single populations were assigned to specific clusters for higher values of *K* (Fig. 3). However, the clustering patterns for values $K > 4$ showed complicated multimodality, which could be because of assignment of individuals to clusters in inconsistent between runs. This indicates some of the models were difficult to fit into the data. However, some populations were not fully supported with any of the two clusters; rather, it appears to be admixed **[see Supporting Information—Fig. S4]**. Although several nodes were poorly supported

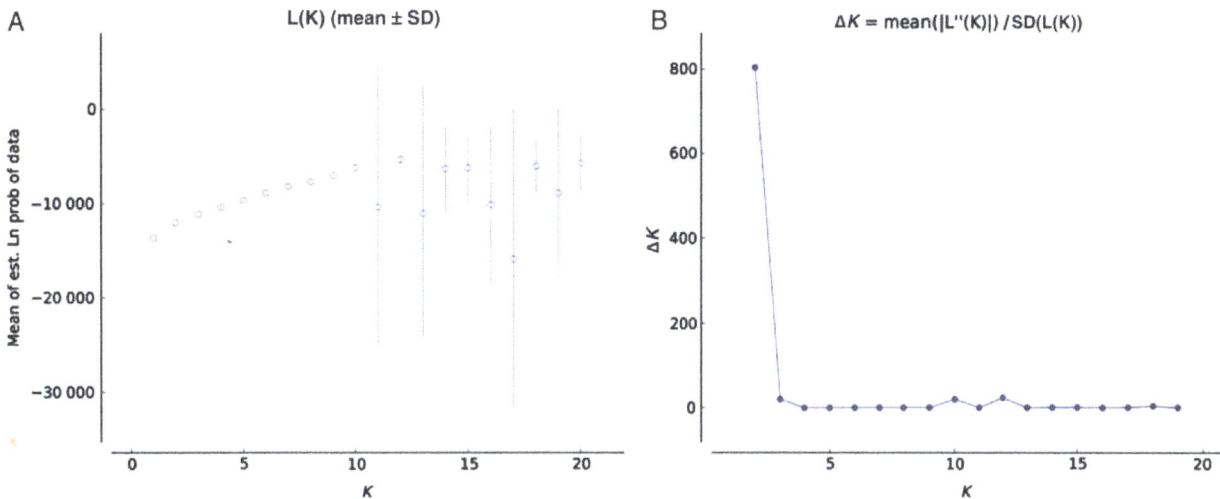

Figure 1. Line graphs from the STRUCTURE model of ln P(D) (a measure of the natural logarithm of the posterior probability, P of the data, D) and ΔK for sampled P. zeylanica populations using ISSR marker (P), where K is the hypothesized number of populations. (A) The mean values of ln P(D) and SD from 10 runs for each value of K = 1–20. (B) The distributions of ΔK over K = 1–20.

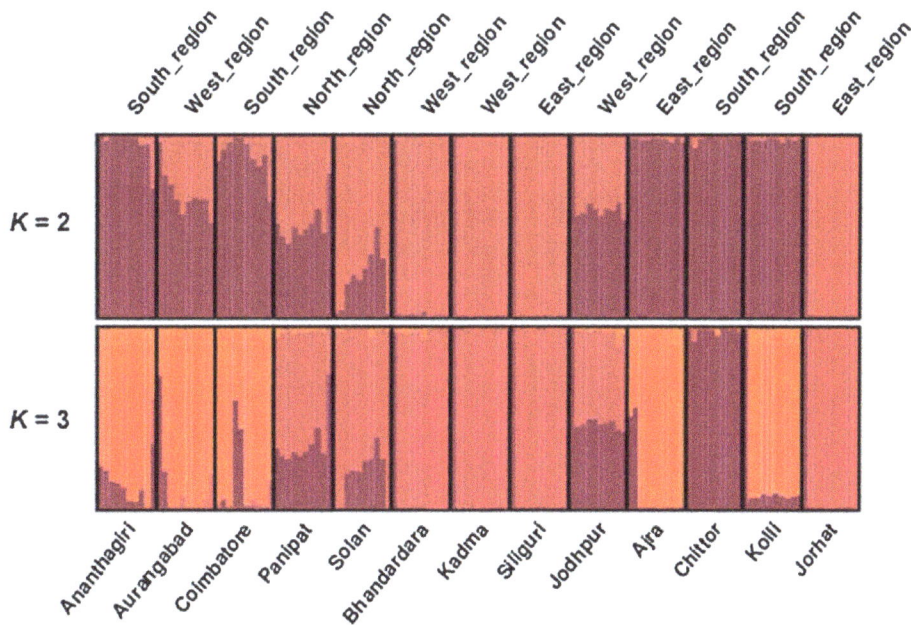

Figure 2. Genetic structure of *P. zeylanica* populations inferred from Bayesian clustering method (Pritchard *et al.* 2000) STRUCTURE plot of 130 wild *P. zeylanica* individuals using ISSR markers. The y-axis shows the proportion membership into the various clusters. Each coloured vertical bar represents a single individual and the 10 individuals from each of the 13 sampled populations are grouped together. Vertical black bars have been included as visual separators between the populations.

by bootstraps, the results of the UPGMA tree showed a similar pattern as that of the STRUCTURE analysis **[see Supporting Information—Fig. S6]** (Fig. 4). The Mantel's test result showed a positive correlation between geographic and genetic distance with significance detected using ISSRs (r = 0.413, P = 0.05, 999 permutations), and by RAPDs (r = 0.279, P = 0.05, 999 permutations) (data not shown).

Discussion

This study provides a first report of broad survey of genetic variation in *P. zeylanica* along the latitudinal gradient in India. There is a substantial variation in environmental conditions along the geographic range which encompasses the edaphic conditions supporting natural vegetation of *P. zeylanica* and climatic conditions. Under changing environmental conditions, genetic diversity is

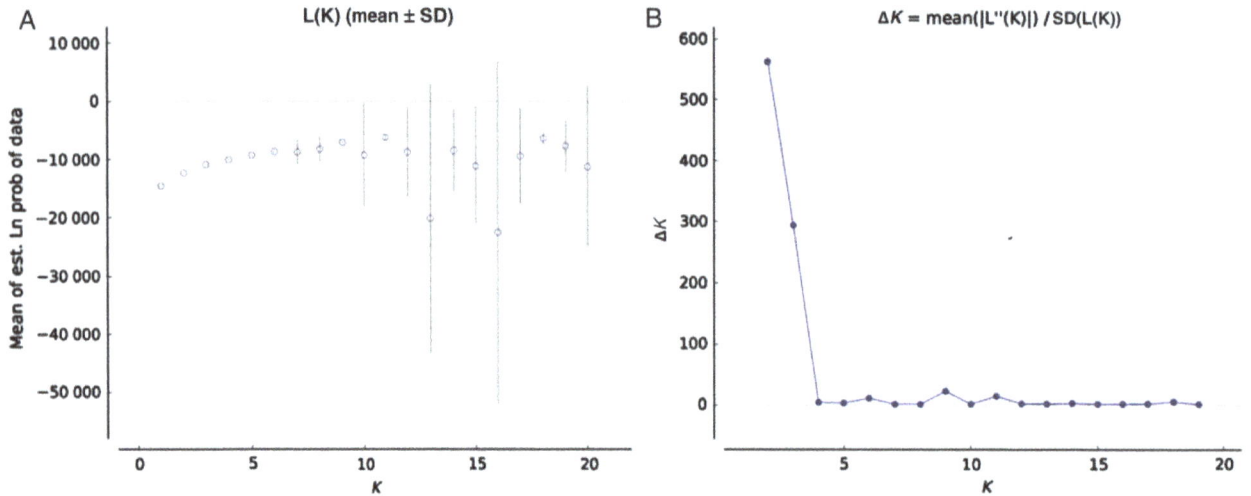

Figure 3. Line graphs from the STRUCTURE model of ln $P(D)$ (a measure of the natural logarithm of the posterior probability, P of the data, D) and ΔK for sampled *P. zeylanica* populations using RAPD marker (P), where K is the hypothesized number of populations. (A) The mean values of ln $P(D)$ and SD from 10 runs for each value of $K = 1$–20. (B) The distributions of ΔK over $K = 1$–20.

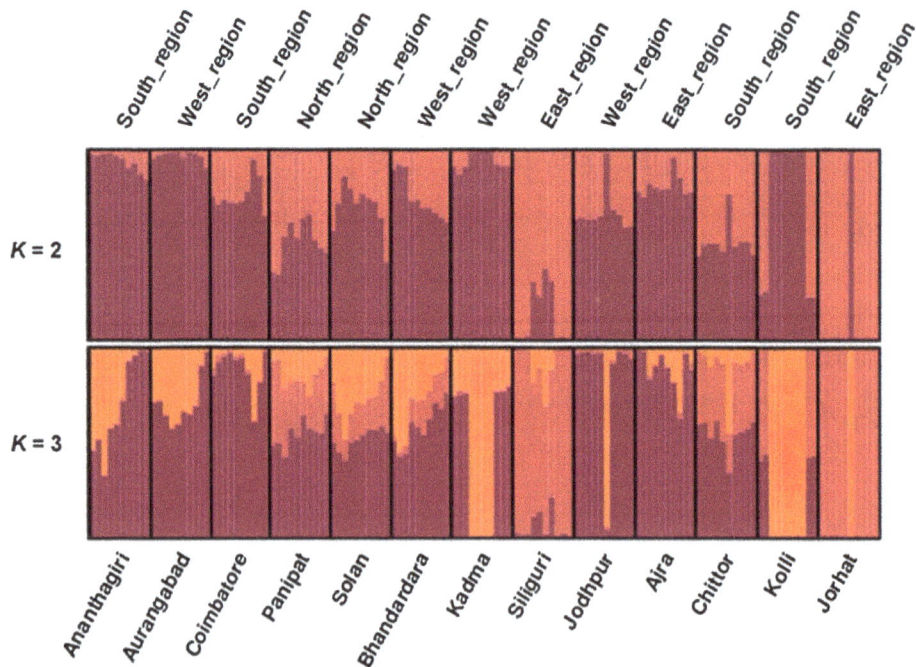

Figure 4. Genetic structure of *P. zeylanica* populations inferred from Bayesian clustering method (Pritchard *et al.* 2000) STRUCTURE plot of 130 wild *P. zeylanica* individuals using RAPD markers. The *y*-axis shows the proportion membership into the various clusters. Each coloured vertical bar represents a single individual and the 10 individuals from each of the 13 sampled populations are grouped together. Vertical black bars have been included as visual separators between the populations.

crucial for effective management and developing conservation strategies for valuable, endemic and medicinally important species *P. zeylanica*. To understand the extent of genetic diversity, genetic structure and differentiation among *P. zeylanica* populations occurring in different geographic regions of India, two PCR-based molecular markers, namely RAPD and ISSR, were selected based on their applicability in other medicinal plant taxa that were used for both inter- and intra-population analysis (Pither *et al.* 2003; Juchum *et al.* 2007; Naik *et al.* 2009).

Based on morphometric data, it was observed that flower-oriented traits showed distinct variability among the populations of *P. zeylanica* [**see Supporting Information— Table S2 and Fig. S8**]. The UPGMA analysis based on average taxonomic distance among the populations produced two distinct clusters. Examination of character coefficient revealed that many of the original variables were strongly and positively correlated with PC1 including flower size and shape, colour of trichomes on sepal, leaf area, stem height and plant habits. Principal component analysis showed that populations from different geographic locations with higher mean annual precipitation tended to be large in size in all floral and vegetative traits [see Supporting Information—Fig. S1]. However, other traits of supposedly taxonomic importance such as flower shape and leaf area exerted only minor influence on PC1 [see Supporting Information—Fig. S8]. Two distinctive colours of glandular trichomes were noted, namely semi-transparent and purple within the populations studied, which might have an adaptive response against phytophagous insect herbivory (Jaime et al. 2013). Plumbaginaceae is considered to be related to Droceraceae (Jaime et al. 2013) and it was reported that glandular trichomes in the genus *Plumbago* is capable of secreting proteases in response to chemical stimulation but not in a way that would be typical of a true carnivore (Jaime et al. 2013). Beside the genus *Plumbago*, the genus from same family *Limonium* shows many leaf and gland characteristics common to the Plumbago and the Caryophyllales carnivores that might be expected in the last common ancestor with the carnivores (Meimberg et al. 2000). The observed variation in herbivore pressure among taxa likely caused by habitat differentiation might have played a role in trait differentiation through divergent selection or may be due to interaction with some specific insect. The change in colour of trichomes might be to attract insects (Healy et al. 2009), as purple colour is more attractive over semi-transparent and this might be in reference to nutrient levels in soil (Pawar et al. 2008). In present study, populations with purple trichomes were mainly from Deccan platue region namely Coimbatore, Ananthagiri, Ajra and Aurangabad [see Supporting Information—Fig. S2]. These regions are known to have nitrogen poor soil as it is formed from igneous rocks (Pawar et al. 2008). To overcome this nitrogen deficiency, the populations spread to Deccan platue might have been evolving towards insectivorous habit and the change in trichome colour from semi-transparent to purple might be an adaptive trait in attracting more insects.

To correlate phenotypic variation with molecular phylogeny, ISSR and RAPD markers were used because they can detect very low levels of genetic variation, making them powerful genetic markers which have been used in genetic diversity and population genetic studies of wild plants (Pither et al. 2003; Bahulikar et al. 2004; Juchum et al. 2007; Naik et al. 2009; Bodare et al. 2013). In the present study, we have shown that these

markers revealed a significant genetic variation among geographically separated sub-populations of *P. zeylanica* in Indian regions. Inter-simple sequence repeats and RAPDs also revealed diversity within each population. The obtained results based on heterozygosity data for both the markers in accordance with various studies in other wild plant species suggest the line of evidence that inter-population genetic diversity is higher than the intra-populations, suggesting that a significant genetic variation is maintained due to outcrossing events in this species.

In 13 populations of *P. zeylanica* studied, we found that Shannon's index of genetic diversity was estimated to be an average 0.04 (by ISSRs) and 0.07 (by RAPDs) (Tables 1 and 2). These values can be compared with other plant species with similar life histories. Our survey of 13 populations of *P. zeylanica* revealed high genetic diversity [ISSR and RAPD analysis revealed band polymorphism = 73.8 and 78.9 %, Nei's gene diversity $(H) = 0.26$ and 0.31, Shannon's index $(I) = 0.41$ and 0.47, respectively] at the species level (Tables 1 and 2). Similar results were observed in genetic diversity studies of southern India populations of *P. zeylanica* (Britto et al. 2009; Haji et al. 2014). Limited sample size was the major limitation of the previous study; moreover, Haji et al. (2014) did not consider the geographical distribution of the species. Genetic diversity at the population level was observed to be comparatively lower [from ISSR and RAPD assay band polymorphism = 11 % for both, Nei's gene diversity $(H) = 0.03$ and 0.05, Shannon's index $(I) = 0.04$ and 0.07).

High level of among-population genetic differentiation was revealed in *P. zeylanica* population using both ISSR and RAPD markers ($G_{ST} = 0.90$ and 0.84 from ISSR and RAPD analysis, respectively), indicating that the populations were subjected to genetic isolation. Similar results were also reported in many other medicinal and endemic species such as *Torreya jackii* (Li and Jin 2007), *Rhodiola alsia* (Xia et al. 2005), and *Rheum palmatuma* and *R. tanguticum* (Wang et al. 2012). The higher genetic differentiation of population within a species is driven by various factors such as genetic isolation or genetic drift, pollination and breeding system and geographic distribution range (Young et al. 1996; Hogbin and Peakall 1999). The geographic distribution and topographical barriers can lead to difficulties in seed dispersal resulting in limited gene flow among populations (Hamrick and Godt 1996). Mantel test showed a significant isolation-by-distance pattern in *P. zeylanica*, indicating genetic isolation has a significant effect on genetic variation and structure in this species. We also observed that the gene flow between the populations was low ($N_m = 0.06$ and 0.09). The sampling sites chosen in this work were geographically widely separated (Solan and Coimbatore = ~2800 km; Jodhpur and Jorhat = ~2600 km) and this may account for the high genetic diversity and gene differentiation observed in this species. However, some of the sampling sites (Kolli and Coimbatore) were separated by ~150 km, and the probable causes underlying genetic diversity of these populations could not be attributed to geographical separation alone. From the present study it is revealed that the populations of P. zeylanica from different eco-geographical regions of India were grouped into two major clusters, one with purple trichomes and other with semi-transparent trichomes on sepal, this is also supported by both ISSR and RAPD data (Figs 2 and 4).

Habitat fragmentation is an important cause of alteration in the population structure of plants (Young et al. 1996). In our study, the fragmented populations of P. zeylanica represented by Jorhat, Siliguri and Kolli populations showed lower diversity (H = 0.12, 0.16 and 0.18, respectively, from ISSR analysis and 0.028, 0.034 and 0.037, respectively, from RAPD analysis) when compared with the natural forest represented by Bhandardara population (ISSR and RAPD assay revealed H = 0.038 and 0.069, respectively). This could be attributed to a reduction in the gene pool and increased gene differentiation in fragmented forests, which arises due to the presence of fewer individuals in the population and induced inbreeding (Young et al. 1996). High levels of gene differentiation were also reported in other species existing in the fragmented habitats. Some of these were also endangered medicinally important species like Sinopodophyllum hexandrum (G_{ST} = 0.62) (Xiao et al. 2006), Podophyllum hexandrum (G_{ST} = 0.75) (Alam et al. 2009), Saruma henryi (G_{ST} = 0.69) (Zhou et al. 2010), Magnolia officinalis (G_{ST} = 0.67) (Yu et al. 2011), indicating that fragmentation, besides bringing about a high level of differentiation in populations, also leads to a reduction in their numbers. A critical number of individuals are required to sustain a population, reduction in the number of individuals below this level affects the sustainability of the species and leads to their elimination (Rajora and Mosseler 2001). Plant species that are not yet endangered, but whose populations have been fragmented due to anthropogenic activities, also show high levels of differentiation such as Dactylorhiza hatagirea (G_{ST} = 0.25) (Warghat et al. 2012) or Curcuma zedoaria (Islam et al. 2006). Plumbago zeylanica belongs to the latter category of plants, as it is largely pollinated by bees and other insects preferentially due to extensive network of sticky glands present on the floral surface. The seed dispersal in P. zeylanica populations is due to herbivores, which limits their spread. The limited seed spread and seedling establishment may contribute to a reduced gene flow in this species (Lowe et al. 2005). Between- and within-population genetic diversity of the species also depends on the type of pollination and subsequent breeding system. Dioecious species like Eurya nitida (HS = 0.13) (Bahulikar et al. 2004), as well as other outcrossing species like Taxus fauna (HS = 0.12) (Shah et al. 2008), and Changium smyrnioides (HS = 0.11) (Qiu et al. 2004), showed high genetic diversity within populations.

Genetic diversity within a species is shaped over long periods of time through evolutionary genetic processes acting in combination on that species (Rajora and Mosseler 2001). The evolutionary history of P. zeylanica is not known, but the species is widely distributed in the different eco-climatic zones of India, which is suggestive of a broad genetic base (Pant et al. 2012; Haji et al. 2014). Reduction and fragmentation in wild medicinal plants due to over-exploitation in the forest cover could be one of the main causes that led to an increase in genetic differentiation and reduced gene flow between populations.

Conclusions

The results of the present study suggest that P. zeylanica has higher genetic diversity at species and population level as assessed by two different molecular markers however, extraordinarily high among-population genetic differentiation existed in P. zeylanica. A larger proportion of genetic variation was observed among populations. A greater effort should be made to preserve all the extant populations and their habitats in the field, especially with those populations with higher genetic diversity. Considering higher demands for raw tissue material and heavy exploitations, these wild Plumbago resources have long been subjected to excessive collections. It would be sustainable if plantation of new populations can be established to meet the market demand. This way it can alleviate the excessive collection of natural resources of P. zeylanica.

To conclude, the present study generated information useful for developing appropriate conservation strategies which would ensure that there is less anthropogenic destruction of existing habitats, increase in the natural population size, optimization and improvement of cultivation practices ensuring constant supply of plant material without exploiting the natural populations.

Sources of Funding

Our work was funded by University Grant Commission, New Delhi, India under DRS-SAP III programme.

Contributions by the Authors

Conceived and designed the experiments: S.P. and A.K. Performed the experiments: S.P. Analysed the data: S.P.,

D.N., and A.K. Contributed reagents/materials/analysis tools: S.P., D.N. and A.K. Wrote the paper: D.N., A.K. and S.P.

Acknowledgements

The authors thank the Government Forest Department of India for assisting in field work. The authors are grateful to Botanical Survey of India (BSI), Western Circle, Pune, India for authenticating the plant species. D.N. acknowledges the facilities provided by IIAR, Gandhinagar, India.

Supporting Information

The following additional information is available in the online version of this article –

Table S1. Sampling details of *P. zeylanica* populations.

Table S2. Descriptive statistics of the 13 traits measured on populations collected.

Table S3. ISSR primers used for ISSR analysis.

Table S4. RAPD primers used for RAPD analysis.

Figure S1. Location of 13 sampled *P. zeylanica* populations in India.

Figure S2. Types of glandular trichomes on sepal of *P. zeylanica*.

Figure S3. Principal component analysis (PCoA) of growth and morphological variables.

Figure S4. Regression of genetic diversity (A: He and B: PPB) on latitude (oS) for ISSR (a and b) and RAPD (c and d) marker data and (C: He and D: PPB) of *P. zeylanica* populations.

Figure S5. Unrooted dendrogram from the neighbor-joining analysis of ISSR markers.

Figure S6. Unrooted dendrogram from the neighbor-joining analysis of RAPD markers.

Figure S7. Principal component analysis (PCoA) using ISSR and RAPD data.

Figure S8. Types of growth habit observed in *P. zeylanica* populations.

Literature Cited

Alam MA, Gulati P, Gulati AK, Mishra GP, Naik PK. 2009. Assessment of genetic diversity among *Podophyllum hexandrum* genotypes of the North-Western Himalayan region for podophyllotoxin production. *Indian Journal of Biotechnology* 8:391–399.

Bahulikar RA, Lagu MD, Kulkarni BG, Pandit SS, Suresh HS, Rao MKV, Ranjekar PK, Gupta VS. 2004. Genetic diversity among spatially isolated populations of *Eurya nitida* Korth. (Theaceae) based on inter-simple sequence repeats. *Current Science* 86:824–831.

Barrett SRH, Kohn J. 1991. Genetic and evolutionary consequences of small population size in plants: implications for conservation. In: Falk DA, Holsinger KE, eds. *Genetics and conservation of rare plants*. New York: Oxford University Press, 3–30.

Bodare S, Tsuda Y, Ravikanth G, Uma Shaanker R, Lascoux M. 2013. Genetic structure and demographic history of the endangered tree species *Dysoxylum malabaricum* (Meliaceae) in Western Ghats, India: implications for conservation in a biodiversity hotspot. *Ecology and Evolution* 3:3233–3248.

Britto AJD, Mahesh R, Sujin RM, Dharmar K. 2009. Detection of DNA polymorphism by RAPD-PCR fingerprint in *Plumbago zeylanica* L. from Western ghats. *Madras Agricultural Journal* 96:291–292.

de la Cruz M, Ramirez F, Hernandez H. 1997. DNA isolation and amplification from cacti. *Plant Molecular Biology Reporter* 15:319–325.

Ding G, Zhang D, Yu Y, Zhao L, Zhang B. 2012. Phylogenetic relationship among related genera of Plumbaginaceae and preliminary genetic diversity of *Limonium sinense* in China. *Gene* 506:400–403.

Ding G, Zhang D, Yu Y, Zhao L, Zhang B. 2013. Analysis of genetic variability and population structure of the endemic medicinal *Limonium sinense* using molecular markers. *Gene* 520:189–193.

Earl DA, vonHoldt BM. 2012. STRUCTURE HARVESTER: a website and program for visualizing STRUCTURE output and implementing the Evanno method. *Conservation Genetics Resources* 4:359–361.

Evanno G, Regnaut S, Goudet J. 2005. Detecting the number of clusters of individuals using the software STRUCTURE: a simulation study. *Molecular Ecology* 14:2611–2620.

Excoffier L, Smouse PE, Quattro JM. 1992. Analysis of molecular variance inferred from metric distances among DNA haplotypes: application to human mitochondrial DNA restriction data. *Genetics* 131:479–491.

Falk D, Holsinger K. 1991. *Genetics and conservation of rare plants*. New York: Oxford University Press.

Falush D, Stephens M, Pritchard JK. 2003. Inference of population structure using multilocus genotype data: linked loci and correlated allele frequencies. *Genetics* 164:1567–1587.

Hafeez BB, Jamal MS, Fischer JW, Mustafa A, Verma AK. 2012. Plumbagin, a plant derived natural agent inhibits the growth of pancreatic cancer cells in *in vitro* and *in vivo* via targeting EGFR, Stat3 and NF-κB signaling pathways. *International Journal of Cancer* 131:2175–2186.

Hafeez BB, Zhong W, Fischer JW, Mustafa A, Shi X, Meske L, Hong H, Cai W, Havighurst T, Kim K, Verma AK. 2013. Plumbagin, a medicinal plant (*Plumbago zeylanica*)-derived 1,4-naphthoquinone, inhibits growth and metastasis of human prostate cancer PC-3M-luciferase cells in an orthotopic xenograft mouse model. *Molecular Oncology* 7:428–439.

Haji RFA, Bhargava M, Akhoon BA, Kumar A, Brindavanam NB, Verma V. 2014. Correlation and functional differentiation between different markers to study the genetic diversity analysis in medicinally important plant *Plumbago zeylanica*. *Industrial Crops and Products* 55:75–82.

Hamrick JL, Godt MJW. 1996. Effects of life history traits on genetic diversity in plant species. *Philosophical Transactions of the Royal Society B: Biological Sciences* 351:1291–1298.

Harish, Gupta AK, Phulwaria M, Rai MK, Shekhawat NS. 2014. Conservation genetics of endangered medicinal plant *Commiphora wightii* in Indian Thar Desert. *Gene* 535:266–272.

Healy RA, Palmer RG, Horner HT. 2009. Multicellular secretory trichome development on soybean and related *Glycine* gynoecia. *International Journal of Plant Sciences* 170:444–456.

Hogbin PM, Peakall R. 1999. Evaluation of the contribution of genetic research to the management of the endangered plant *Zieria prostrata*. *Conservation Biology* 13:514–522.

International Plant Genetic Resources Institute (IPGRI). 1993. *Diversity for development*. Rome: IBPGR/ICRISAT, 69 p.

Islam MA, Kloppstech K, Esch E. 2006. Population genetic diversity of *Curcuma zedoaria* (Christm.) Roscoe—a conservation prioritised medicinal plant in Bangladesh. *Conservation Genetics* 6: 1027–1033.

Jaime R, Rey PJ, Alcántara JM, Bastida JM. 2013. Glandular trichomes as an inflorescence defence mechanism against insect herbivores in *Iberian columbines*. *Oecologia* 172:1051–1060.

Jakobsson M, Rosenberg NA. 2007. CLUMPP: a cluster matching and permutation program for dealing with label switching and multi-modality in analysis of population structure. *Bioinformatics* 23: 1801–1806.

Jamal MS, Parveen S, Beg MA, Suhail M, Chaudhary AGA, Damanhouri GA, Abuzenadah AM, Rehan M. 2014. Anticancer compound plumbagin and its molecular targets: a structural insight into the inhibitory mechanisms using computational approaches. *PLoS ONE* 9:e87309.

Juchum FS, Leal JB, Santos LM, Almeida MP, Ahnert D, Corrêa RX. 2007. Evaluation of genetic diversity in a natural rose-wood population (*Dalbergia nigra* Vell. Allemão ex Benth.) using RAPD markers. *Genetics and Molecular Research* 6: 543–553.

Li J, Jin Z. 2007. Genetic variation and differentiation in *Torreya jackii* Chun, an endangered plant endemic to China. *Plant Science* 172: 1048–1053.

Lowe AJ, Boshier D, Ward M, Bacles CFE, Navarro C. 2005. Genetic resource impacts of habitat loss and degradation; reconciling empirical evidence and predicted theory for neotropical trees. *Heredity* 95:255–273.

Meimberg H, Dittrich P, Bringmann G, Schlauer J, Heubl G. 2000. Molecular phylogeny of Caryophyllidae s.l. based on *matK* sequences with special emphasis on carnivorous taxa. *Plant Biology* 2:218–228.

Naik D, Singh D, Vartak V, Paranjpe S, Bhargava S. 2009. Assessment of morphological and genetic diversity in *Gmelina arborea* Roxb. *New Forests* 38:99–115.

Nei M. 1973. Analysis of gene diversity in subdivided populations. *Proceedings of the National Academy of Sciences of the USA* 70: 3321–3323.

Pant M, Lal A, Rana S, Rani A. 2012. *Plumbago zeylanica* L.: a mini review. *International Journal of Pharmaceutical Applications* 3: 399–405.

Pawar NJ, Pawar JB, Kumar S, Supekar A. 2008. Geochemical eccentricity of ground water allied to weathering of basalts from the Deccan volcanic province, India: insinuation on CO_2 consumption. *Aquatic Geochemistry* 14:41–71.

Peakall R, Smouse PE. 2012. GenAlEx 6.5: genetic analysis in Excel. Population genetic software for teaching and research—an update. *Bioinformatics* 28:2537–2539.

Pither R, Shore JS, Kellman M. 2003. Genetic diversity of the tropical tree *Terminalia amazonia* (Combretaceae) in naturally fragmented populations. *Heredity* 91:307–313.

Pritchard JK, Stephens M, Donnelly P. 2000. Inference of population structure using multilocus genotype data. *Genetics* 155:945–959.

Qiu YX, Hong DY, Fu CX, Cameron KM. 2004. Genetic variation in the endangered and endemic species *Changium smyrnioides* (Apiaceae). *Biochemical Systematics and Ecology* 32:583–596.

Rajora OP, Mosseler A. 2001. Challenges and opportunities for conservation of forest genetic resources. *Euphytica* 118:197–212.

Rosenberg NA. 2004. Distruct: a program for the graphical display of population structure. *Molecular Ecology Notes* 4:137–138.

Shaanker RU, Ganeshaiah KN, Rao MN, Aravind NA. 2004. Ecological consequences of forest use: from genes to ecosystem—a case study in the Biligiri Rangaswamy temple wildlife sanctuary, South India. *Conservation and Society* 2:347–363.

Shah A, Li DZ, Gao LM, Li HT, Möller M. 2008. Genetic diversity within and among populations of the endangered species *Taxus fuana* (Taxaceae) from Pakistan and implications for its conservation. *Biochemical Systematics and Ecology* 36:183–193.

Tamura K, Stecher G, Peterson D, Filipski A, Kumar S. 2013. MEGA6: molecular evolutionary genetics analysis version 6.0. *Molecular Biology and Evolution* 30:2725–2729.

Tilak JC, Adhikari S, Devasagayam TPA. 2004. Antioxidant properties of *Plumbago zeylanica*, an Indian medicinal plant and its active ingredient, plumbagin. *Redox Report* 9:219–227.

Wang X, Yang R, Feng S, Hou X, Zhang Y, Li Y, Ren Y. 2012. Genetic variation in *Rheum palmatum* and *Rheum tanguticum* (Polygonaceae), two medicinally and endemic species in China using ISSR markers. *PLoS ONE* 7:e51667.

Warghat A, Bajpai PK, Murkute A, Sood H, Chaurasia OP, Srivastava RB. 2012. Genetic diversity and population structure of *Dactylorhiza hatagirea* (Orchidaceae) in cold desert Ladakh region of India. *Journal of Medicinal Plant Research* 6:2388–2395.

Xia T, Chen S, Chen S, Ge X. 2005. Genetic variation within and among populations of *Rhodiola alsia* (Crassulaceae) native to the Tibetan Plateau as detected by ISSR markers. *Biochemical Genetics* 43: 87–101.

Xiao M, Li Q, Wang L, Guo L, Li J, Tang L, Chen F. 2006. ISSR analysis of the genetic diversity of the endangered species *Sinopodophyllum hexandrum* (Royle) Ying from Western Sichuan Province, China. *Journal of Integrative Plant Biology* 48:1140–1146.

Yeh F, Yang RC, Boyle T. 2000. *POPGENE—for the analysis of genetic variation among and within populations using co-dominant and dominant markers*. Version 1.32.

Young A, Boyle T, Brown T. 1996. The population genetic consequences of habitat fragmentation for plants. *Trends in Ecology and Evolution* 11:413–418.

Yu HH, Yang ZL, Sun B, Liu R. 2011. Genetic diversity and relationship of endangered plant *Magnolia officinalis* (Magnoliaceae) assessed with ISSR polymorphisms. *Biochemical Systematics and Ecology* 39:71–78.

Zhou TH, Qian ZQ, Li S, Guo ZG, Huang ZH, Liu ZL, Zhao GF. 2010. Genetic diversity of the endangered Chinese endemic herb *Saruma henryi* Oliv. (Aristolochiaceae) and its implications for conservation. *Population Ecology* 52:223–231.

13

Genetic diversity and population structure of an important wild berry crop

Laura Zoratti[1][*][†], Luisa Palmieri[2][†], Laura Jaakola[3,4] and Hely Häggman[1]

[1] Department of Genetics and Physiology, University of Oulu, PO Box 3000, FI-90014 Oulu, Finland
[2] Department of Food Quality and Nutrition, Research and Innovation Center, Fondazione Edmund Mach, Via E. Mach, 1-38010 San Michele all'Adige (TN), Italy
[3] Department of Arctic and Marine Biology, UiT The Arctic University of Norway, Climate Laboratory, 9037 Tromsø, Norway
[4] Norwegian Institute of Bioeconomy Research, NIBIO Holt, PO Box 115, 1431 Ås, Norway

Associate Editor: Kermit Ritland

Abstract. The success of plant breeding in the coming years will be associated with access to new sources of variation, which will include landraces and wild relatives of crop species. In order to access the reservoir of favourable alleles within wild germplasm, knowledge about the genetic diversity and the population structure of wild species is needed. Bilberry (*Vaccinium myrtillus*) is one of the most important wild crops growing in the forests of Northern European countries, noted for its nutritional properties and its beneficial effects on human health. Assessment of the genetic diversity of wild bilberry germplasm is needed for efforts such as *in situ* conservation, on-farm management and development of plant breeding programmes. However, to date, only a few local (small-scale) genetic studies of this species have been performed. We therefore conducted a study of genetic variability within 32 individual samples collected from different locations in Iceland, Norway, Sweden, Finland and Germany, and analysed genetic diversity among geographic groups. Four selected inter-simple sequence repeat primers allowed the amplification of 127 polymorphic loci which, based on analysis of variance, made it possible to identify 85 % of the genetic diversity within studied bilberry populations, being in agreement with the mixed-mating system of bilberry. Significant correlations were obtained between geographic and genetic distances for the entire set of samples. The analyses also highlighted the presence of a north–south genetic gradient, which is in accordance with recent findings on phenotypic traits of bilberry.

Keywords: Bilberry; genetic diversity; germplasm; ISSR; population structure; *Vaccinium myrtillus*.

Introduction

The success of plant breeding over the past century has been associated with a narrowing of the available genetic diversity within elite germplasm of species. New sources of variation include landraces and wild relatives of crop species, and although exploiting wild relatives as a source of novel alleles is challenging, it has provided notable successes in crop improvement (Tester and Langridge 2010). Most crop geneticists agree that the enrichment of the cultivated gene pool will be necessary to meet

* Corresponding author's e-mail address: zoratti.laura@gmail.com
† These authors contributed equally to the realization of the present work.

the challenges that lie ahead associated with global environmental changes (Feuillet *et al.* 2008). However, many advances are still needed to access the extensive reservoir of favourable alleles within wild germplasm. These include increasing our understanding of the molecular basis for key traits and expanding existing phenotyping and genotyping of germplasm collections (Feuillet *et al.* 2008). Therefore, knowledge of the genetic diversity and the population structure of wild species is crucial for their management as well as conservation (Burdon and Wilcox 2007; Zhao *et al.* 2014).

Vaccinium is a genus of ~450 plant species in the family Ericaceae that are widely distributed in the Northern Hemisphere and also in the mountains of tropical Asia and Central and South America (Song and Hancock 2011). The species within this genus present different levels of ploidy (2*x*, 4*x* and 6*x*; *x* = 12), which results in evident morphological differences. Regarding domestication and commercial fruit crop production, the most important species are *V. corymbosum* (highbush blueberry), *V. virgatum* (rabbit-eye blueberry), *V. angustifolium* (lowbush blueberry), *V. macrocarpon* (cranberry) and *V. vitis-idaea* (lingonberry). The genus also contains the wild *V. myrtillus* (bilberry) and a number of other currently non-cultivated *Vaccinium* species that show great potential as new berry crops (Song and Hancock 2011). Bilberry belongs to the section *Myrtillus*, and it is a diploid species (2*n* = 2*x* = 24; Song and Hancock 2011). The plant is a deciduous woody dwarf shrub, and it grows typically in pine and spruce heath forests and old peat bogs in Europe, North America, Greenland and northern parts of Asia, including Japan and Greenland (Nestby *et al.* 2011). Bilberry reproduces clonally through rhizomes and also sexually, with an outcrossing rate ranging from 0.66 to 0.75, and it is therefore considered to belong to the group of mixed-mating species (Jacquemart 1993).

Bilberry is an important wild fruit crop, especially in Northern and Eastern European countries, where the berries are picked from the wild and are either sold on the fresh market or frozen for use in food industries to make jams, juices and flavourings. The fruits are also important to the pharmaceutical industry, as they are naturally rich in polyphenols and other antioxidant compounds, which have potential beneficial effects on human health. These berries contain great amounts of flavonoids, in particular anthocyanins, which can reach up to 500 mg/100 g fresh weight (Lätti *et al.* 2008); they also produce carotenoids (Bunea *et al.* 2011) and lower amounts of ascorbic acid (Cocetta *et al.* 2012).

There is an increasing demand for these berries due to their high nutritional value (Martinussen *et al.* 2009), although they are still poorly exploited from a commercial point of view. Despite the fact that the average wild berry

yield in Scandinavia has been estimated to be approximately 1 billion kg year^{-1}, only ~5–10 % of the annual crop is utilized for private or commercial consumption (Paassilta *et al.* 2009). Nestby *et al.* (2011) underlined the need for developing an improved production system, in which high yields of good-quality bilberries are produced at manageable costs. So far, cultivation of the species has been very limited and the berries used for commercial purposes are mainly harvested from forests. Therefore, development of forest management systems is considered a good option to achieve this purpose (Nestby *et al.* 2011). Forest management systems will initially require efforts to identify areas in which plants produce high yields and high-quality bilberries. Plants with high-quality characteristics can be identified by phytochemical content or phenotypic traits of interest (e.g. plant productivity, fruit antioxidant content and fruit shelf life) and by genotypic-based methods, whereby the detected molecular polymorphisms are correlated to phenotypic traits. The genotypic-based methods are generally effective, they only need a small amount of DNA (Tanya *et al.* 2011) and they are not affected by environmental factors or developmental stages of the plants. Inter-simple sequence repeats (ISSRs) have shown to be good markers for assessing the genetic diversity of wild *Vaccinium* species from wide geographical areas of collection, in particular, lingonberry (*V. vitis-idaea*; Debnath 2007) and lowbush blueberry (*V. angustifolium*; Debnath 2009). Moreover, ISSR markers were able to detect more polymorphisms than random amplified polimorphic DNA in the same species (Debnath 2009). Therefore, ISSR markers were chosen for our study where the aims were to (i) test the applicability of ISSR markers on bilberry; (ii) determine genetic relationships and diversity among bilberry populations derived from biomes in Northern Europe and (iii) find markers to be used in conservation and management of bilberry in forests of Northern Europe.

Methods

Study sites and sampling

Thirty-two individual bilberry samples derived from different seeds collected at different latitudes in several Nordic countries were included in this study (Table 1, Fig. 1): Iceland (IS1, IS2), Norway (N2, N4, N7), Sweden (R), Finland (S, P, M, L) and Germany (K). The plants were established in 2003 from bilberry seeds harvested from a pool of ripe berries collected in an area of 10 × 10 m and micropropagated *in vitro* (Jaakola *et al.* 2001) at the Botanical Gardens of the University of Oulu (Finland). Plantlets were grown in growth rooms under controlled conditions (+22 °C under 16 h photoperiod, white fluorescent Osram 18 W, 1.8 W m^{-2}).

Table 1. Provenances of bilberry genotypes and genetic diversity parameters based on ISSR markers. Number of samples analysed (N), number of different alleles (Na), number of effective alleles (Ne), number of private bands (Np), percentage of polymorphic loci ($P \%$), expected heterozygosity (He) and Shannon's Information index (I).

Provenance	ID	Country	Latitude (°N)	Longitude (°E)	Altitude (m above sea level)	Genotype ID	N	Na	Ne	Np	P %	He	I
Kleifarveugr	IS1	Iceland	66°07'	−18°38'	178	IS1_a, IS1_3, IS1_4	3.000	0.646	1.158	2	24.40	0.092	0.137
Strandavegur	IS2	Iceland	65°47'	−21°22'	10	IS2_a, IS2_1, IS2_5	3.000	0.803	1.161	0	29.92	0.102	0.156
Storfjord	N2	Norway	69°23'	20°16'	3	N2_2, N2_5, N2_6	3.000	0.661	1.142	0	23.62	0.086	0.129
Trondelag	N4	Norway	63°32'	10°53'	420	N4, N4_3, N4_5	3.000	0.614	1.111	0	22.05	0.072	0.112
Storgata	N7	Norway	60°54'	10°44'	173	N7. N7_5, N7_6	3.000	0.772	1.193	0	31.50	0.115	0.173
Kvikkjokk	R	Sweden	66°57'	17°43'	327	R1, R2, R3	3.000	0.693	1.182	2	25.98	0.103	0.151
Sodankylä	S	Finland	67°25'	26°35'	189	S1, S3	2.000	0.630	1.128	2	18.11	0.075	0.110
Muhos	M	Finland	64°48'	25°59'	39	M, M1, M5	3.000	0.835	1.198	2	34.65	0.122	0.184
Parkano	P	Finland	62°02'	23°02'	117	P, P_1, P_10	3.000	0.961	1.251	6	39.37	0.148	0.220
Lapinjärvi	L	Finland	60°37'	26°11'	21	L2, L3, L6	3.000	0.803	1.199	2	33.86	0.121	0.182
Kiel	K	Germany	54°20'	10°08'	14	K2, K6, K10	3.000	0.945	1.276	2	43.31	0.162	0.242

Genotyping

Genomic DNA was isolated from shoot tips of actively growing *in vitro*-cultured bilberry shoot tips, using the EZNA™ SP Plant DNA Mini Kit (Omega Bio-tek Inc., Norcross, GA, USA) following the manufacturer's instructions. The concentration of DNA was estimated with the NanoDrop N-1000 spectrophotometer (NanoDrop Technologies, Thermo Scientific, Wilmington, DE, USA) at 260 nm. Fifteen primers representing di-, tri-, tetra- and pentamer repeats, previously used to characterize other *Vaccinium* species (Debnath 2007), were considered for the study. Of these, UBC-825, UBC-857, UBC-873 and UBC-881, which gave clear banding patterns, were used for the final study (Table 2). Polymerase chain reaction (PCR) was performed in a final volume of 25 μL including 10 ng of DNA template, $1 \times$ Optimized DyNAzyme™ buffer (10 mM Tris–HCl pH 8.3, 1.5 mM $MgCl_2$, 50 mM KCl, 0.1 % Triton X-100; Finnzyme, Espoo, Finland), 0.3 μM of each primer, 200 μM dNTPs, 0.8 U of DyNAzyme™ II DNA Polymerase (Finnzyme). The thermal profile consisted of 10 min at 94 °C, followed by 31 cycles of 1 min at 94 °C, 1 min at 46.5 °C and 2 min at 72 °C and a final extension at 72 °C for 10 min. The PCR reaction was purified using sodium acetate–ethanol DNA precipitation. One microlitre of the purified sample was analysed on a capillary electrophoresis system, Agilent 2100 Bioanalyzer with DNA7500 kit (Agilent Technologies, Santa Clara, CA, USA) according to the manufacturer's instructions. Each primer–clone sample combination was repeated at least two times and only replicated bands were included in the analyses. Fragments of similar size across individuals were assumed to be homologous.

Data analysis

Genetic diversity. The amplification product sizes were scored using 2100 Expert Software (Agilent Technologies). The results were transformed into a binomial matrix as present (1) or absent (0) for each marker. Since the ISSR marker is dominant, we assumed that each band represented the single bi-allelic locus (Debnath 2007). Band patterns were analysed in order to determine the level of polymorphism (total number of bands), number of polymorphic bands, proportion of polymorphic bands and the resolving power (Rp) detected for each primer. Resolving power was calculated according to Prevost and Wilkinson (1999); this measure is based on the distribution of alleles among the genotypes and it estimates the discrimination capacity of each primer. Thus, the resolving power of a primer is defined as Rp = $\sum$Ib, where Ib (band informativeness) takes the value of $1 - [2 \times |0.5 - p|]$ and p is the ratio of genotypes sharing the band. Moreover, the binomial matrix was

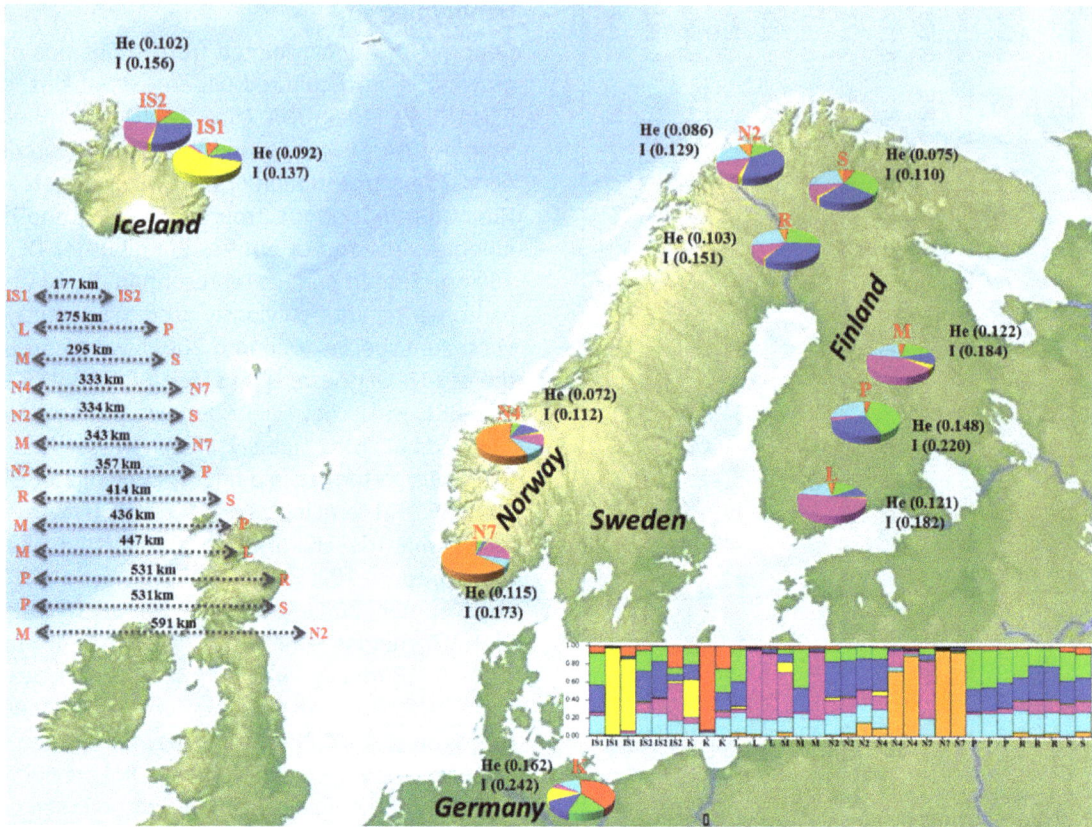

Figure 1. Map of sampling sites in Northern Europe, including ID (according to Table 1). The pie chart represents rather average coefficients of membership resulting from the genetic structure analysis (best fit model, $K = 7$). Each colour represents a different gene pool. The barplot represents each accession as a single vertical bar broken into K colour segments, with lengths proportional to the estimate probability of membership in each inferred cluster. Spatial autocorrelation analysis results, and geographical distances of correlated populations, are reported on the left of the figure (ID, grey arrows and geographical distance in kilometres).

Table 2. Molecular ISSR primers used for bilberry genotypes' discrimination. Y = (C or T) in ISSR primer sequences; repeat motif and the data on DNA profile and polymorphism generated in 32 bilberry samples; total number of bands (NB), number of polymorphic bands (NPB), proportion of polymorphic bands (PPB), rank of molecular weights (RW); resolving power (Rp).

Primer name	Sequence	NB	NPB	PPB (%)	RW (bp)	Rp
UBC-825	$(AC)_8T$	33	31	93.9	310–2100	8.625
UBC-857	$(AC)_8YG$	37	37	100	80–6600	13.87
UBC-873	$(GACA)_4$	25	25	100	80–2600	9.18
UBC-881	$(GGGTG)_3$	32	32	100	60–1400	14.31

used to produce the input matrix following GenAlex version 6.1 software manual instructions (Peakall and Smouse 2006) and analysed using the same software to estimate genetic diversity parameters, i.e. number of different bands, number of different bands with frequency $\geq 5\%$, number of private bands, number of locally common bands frequency $\geq 5\%$ found in ≤ 25 and $\leq 50\%$ of populations, mean of expected heterozygosity (He) and the Shannon's Information Index (I) calculated as $I = -1 \times (p \times \ln(p) + q \times$

$\ln(q))$, where p and q are the estimated allele frequencies.

Analysis of molecular variance (AMOVA) was used to partition the total genetic variance into 'within-populations' or 'among-populations' levels. The software GenAlex was used to generate a matrix of pairwise genetic distances between individuals and to calculate the following variance components: degrees of freedom, sum of squares, mean sum of squares, estimated variance and conversion of estimated variances to percentage of total variance. The number of permutations for significance testing was set at 9999. Canonical correspondence analysis (CCA) was done using Past software v. 2.17c (Hammer *et al.* 2001), to determine the relative importance of geographical factors in the spatial organization of genetic diversity among genotypes. This analysis, originally designed for relating species composition to different predictive variables (Ter Braak 1986), has been successfully used to describe the relationship between environmental variables and genetic composition (Angers *et al.* 1999; Girard and Angers 2006; Dell'Acqua *et al.* 2014). The analysis was performed using a geographical variables/genetic data matrix where longitude, latitude and altitude were used as geographic factors. Here, we consider individuals as sites, and alleles at outlier loci as objects. The first three input file columns contain geographical variables, following the CCA Past software v. 2.17c instructions.

Spatial genetic analysis. Spatial analysis was conducted using the genetic spatial autocorrelation (SA) (Smouse and Peakall 1999; Peakall *et al.* 2003) option in GenAlex version 6. The pairwise geographical distance matrices were calculated considering them as the crow flies distances in kilometres and were used with the previously obtained genetic distance matrices to generate an autocorrelation coefficient r for each distance class using two different options. The autocorrelation coefficient (r) is similar to Moran's I (Moran 1950) and ranged from -1 to 1. The significance level was tested by constructing a two-tailed 95 % confidence interval around the null hypothesis of no spatial genetic structure, which is $r = 0$. The autocorrelation analysis was performed using a multiple distance class simulation. Since the distance classes have to be small enough to capture the spatial pattern of interest, while large enough to include an adequate number of pairwise comparisons for statistical testing, we characterized the spatial genetic relationship kilometre intervals from 0 up to 2400. This allowed us to determine the strength of autocorrelation and to what extent the autocorrelation decays with increasing distance. To test the null hypothesis of no spatial structure, confidence limits were calculated using permutation and bootstrapping (999 interactions).

Directional autocorrelation analysis was carried out by testing for the direction of maximum genetic correlation using the bearing procedure implemented in PASSAGE version 2 (Rosenberg and Anderson 2011). The bearing method analysed the correlation coefficient (r) between geographical distance and genetic relatedness under fixed bearing angles (degrees north of due east). For each sector, r is calculated from each distance pair weighted by the cosine of its angle with respect to the angle of the centre arc of the respective sector. Geographical locations were imported with associated genetic data. PASSAGE calculates a distance matrix from genetic data and geographical co-ordinates and an angular matrix from geographical co-ordinates. The correlation was calculated for each 10° sector, from 0° to 180°, and the significance was estimated using 999 randomizations.

Grouping of bilberry individuals by STRUCTURE and Cluster analyses. The software STRUCTURE 2.3.3 (Pritchard *et al.* 2000; Falush *et al.* 2003), which, by means of iterative algorithms, identifies clusters of related individuals from multi-locus genotypes, was used to examine the genetic structure of populations. Ten independent runs of STRUCTURE were performed for each K value from 1 to 11. Each run consisted of a burn-in period of 100 000 steps, followed by 1 000 000 Markov Chain Monte Carlo replicates, assuming an admixture model and correlated allele frequencies. No prior information was used to define the clusters. The most likely K was chosen comparing the average estimates of the likelihood of the data, $\ln(\mathrm{Pr}(X|K))$, for each value of K (Pritchard *et al.* 2000), as well as calculating the *ad hoc* statistics ΔK, based on the rate of change in the ln probability of data between successive K values (Evanno *et al.* 2005). Furthermore, the Past 2.17c (Hammer *et al.* 2001) software was used to generate a matrix using the Dice similarity index. This matrix was used to construct a Ward's dendrogram tree.

Results

Genetic diversity

Polymerase chain reaction assays using four primers selected in the initial tests allowed 127 ISSR loci to be amplified from the DNA samples derived from 32 bilberry individuals. The detected loci ranged between 60 and 6600 bp (within the limits of the Agilent DNA7500 kit that allowed the detection of fragments between 50 and 7500 bp). The average number of loci per primer was 31.75, with the highest number of loci ($n = 37$) detected by the UBC-857 primer and the smallest number ($n = 25$) by the UBC-873 primer (Table 2). Of the 127 amplified loci, 126 (99.24 %) were polymorphic. The greatest discrimination power among samples was obtained with primer UBC-881 yielding an Rp value of 14.31, while the lowest Rp value of 8.62 was yielded with primer UBC-825. Except for the primer pair of UBC-825, other primers produced 100 % polymorphic bands (Table 2). The number of bands within populations ranged between 50 and 72, with a number of private bands that ranged between 0 and 6 and a mean expected heterozygosity value that ranged between 0.072 and 0.162. The highest percentage of polymorphic loci (P %) was established in Kiel individuals (43.31 %), while the lowest was observed in S individuals (18.11 %). Moreover, the mean number of different alleles over all loci (Na) for each population ranged between 0.614 and 0.961, the mean number of effective alleles (Ne) ranged between 1.111 and 1.276, and the Shannon's Information Index ranged between 0.110 and 0.242 (Table 1). Analysis of molecular variance indicated that 15 % of the total genetic variance was attributable to among-populations diversity and the rest (85 %) to within-populations diversity. The value of ΦPT (0.190 with a maximum of 0.795) indicated a great level of genetic differentiation among populations. We used a CCA to investigate the possible aggregation or differentiation of analysed genotypes. Results of the CCA (Fig. 2) revealed the strong influence of the geographical position on genotype aggregation (Axis 1 = 63.3 % of variance; Axis 2 = 35.7 %). The correlation biplot, which considers both the direction and the relative length of the vectors, underlines clustering for most of the individuals having the same origins. For instance, the clustering of Finnish individuals (P, L, M and one S sample, Fig. 1) occurred according to longitude, and the clustering of the same Finnish samples with Norwegian individuals derived from the closer latitudes (Fig. 1) was evident. The individuals from Norway (N4 and N7) overlapped and were in close proximity to Swedish individuals probably due to the same influence of all geographical variables on all

these samples. IS1, IS2 and K individuals are clustering alone as previously shown, but IS1 and IS2 seem to be strongly influenced by negative longitude values while the K individuals seem to be most influenced by lower values of latitude.

Spatial genetic structure

In the correlogram that resulted from SA analysis, the y-axis average SA coefficient, r, has a function of distance class on the x-axis. The maximum number of distance classes obtained using the even sample classes option was 52. It is apparent in Fig. 3 that there is a highly significant positive SA (the r value falls above or at the 95 % confidence interval, close to a correlation of zero) at distance classes 0–100 evidencing non-random spatial genetic structure within populations. Moreover, the positive correlation decreases until 600 km ($r = 0.028$). Figure 1 reports the IDs and the geographical distances of populations showing positive r values. The bearing analysis from PASSAGE indicated the strongest correlation occurred along a north–south axis ($r = 0.856$, $P = 0.001$), while the weakest occurred in the east–west direction ($r = -0.188$, $P = 0$).

Clustering of bilberry individuals by STRUCTURE analysis

STRUCTURE analysis assigned genotypes into respective groups on the basis of their allele frequencies. According to the user-defined settings, the programme assumed

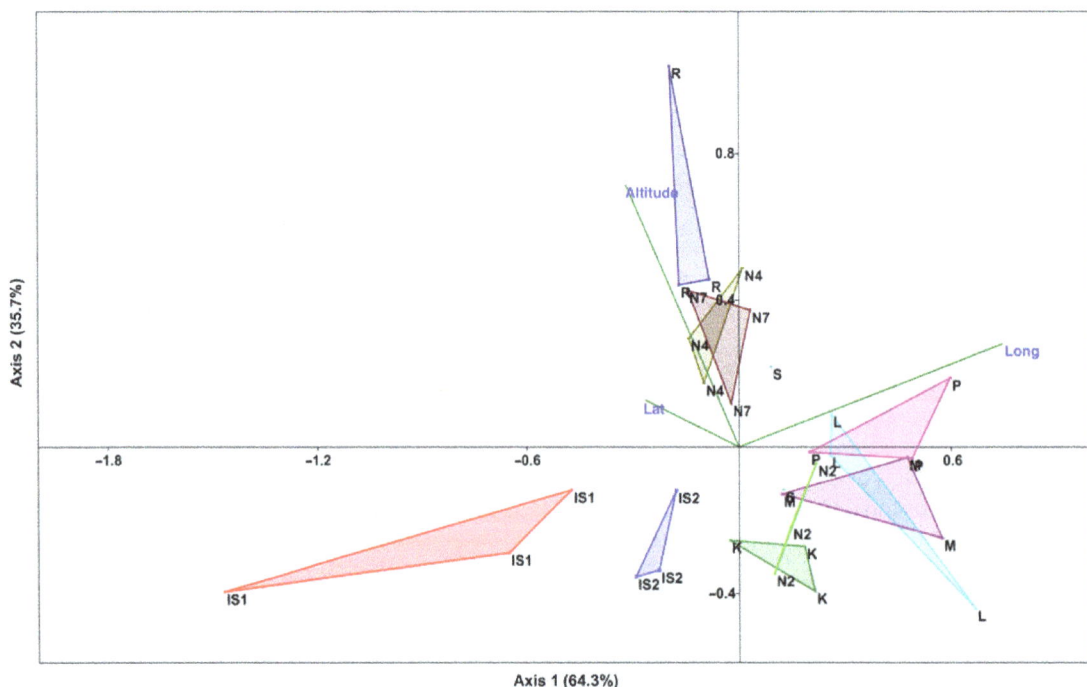

Figure 2. Canonical correspondence analysis ordination biplot representing genotype aggregation and geographical variables (solid arrows). The CCA explained 64.3 and 35.7 % of the variation on the first two axes.

that genotypes are admixed and the allele frequencies are correlated as a consequence of shared ancestry and/or migration. Bayesian cluster, based on an admixture model, presumes that each individual has inherited some proportion of its ancestry from each of the K genotypes (Pritchard et al. 2000). According to the Evanno's method, the STRUCTURE analysis indicated $K = 7$ as the most likely number of gene pools. These pools represented most geographical groups but with a substantially different proportion of membership (q) of each group in each gene pool. Considering the mean value for each geographical group of genotypes (Q), six different patterns of genetic makeup were evidenced. The first type of pattern included two Norwegian groups (N4 and N7) out of three with the highest Q in gene pool 7 (0.572 and 0.636, respectively). Finnish populations represented two distinct patterns with the highest Q value in gene pool 5 for L and M groups (0.508 and 0.420, respectively) and in gene pool 2 for P and S groups (0.411 and 0.295, respectively). The IS1 population had the maximum Q value in gene pool 4 (0.59) and the K group in gene pool 1. Finally, individuals from Iceland (IS2), Norway (N2) and Sweden (R) had similar Q values in gene pools 3, 6 and 7 (Fig. 1).

The Dice distances between pairs of populations were calculated based on the 127 analysed bands. The dendrogram was built using Ward's method (Fig. 4) and showed clusters comparable with clusters evidenced from the STRUCTURE analysis, except for P and R genotypes, which clustered alone as previously assessed.

Discussion

Plant genetic resources are essential for sustainable agriculture and food security. One of the best ways to preserve them is *in situ* management, which becomes particularly important in cases of wild crop species such as bilberry. Molecular markers can provide important information regarding genetic polymorphism and essential knowledge for development and improvement of plant populations; however, to date only a few studies have been carried out on the assessment of bilberry germplasm genetic diversity (Raspé et al. 2004; Albert et al. 2008).

In our study, ISSR markers were used to evaluate genetic variation among 32 bilberry individuals located in Fennoscandia and Germany. Four primers detected significant genetic variation among the genotypes thanks to their high polymorphism level (99 %). A similar high level of diversity in the same genus (80.4 %) was reported in highbush and rabbit-eye blueberry (Garriga et al. 2013). The ability to discriminate bilberry samples varied between different ISSR primers, in agreement with the findings in lingonberry (Debnath 2007) and lowbush blueberry (Debnath 2009).

The high percentage value of genetic diversity within-population obtained from AMOVA can be explained as a natural selection mechanism to reduce fitness costs due to geitonogamous self-pollination in bilberry as previously reported by Albert et al. (2008). The same results

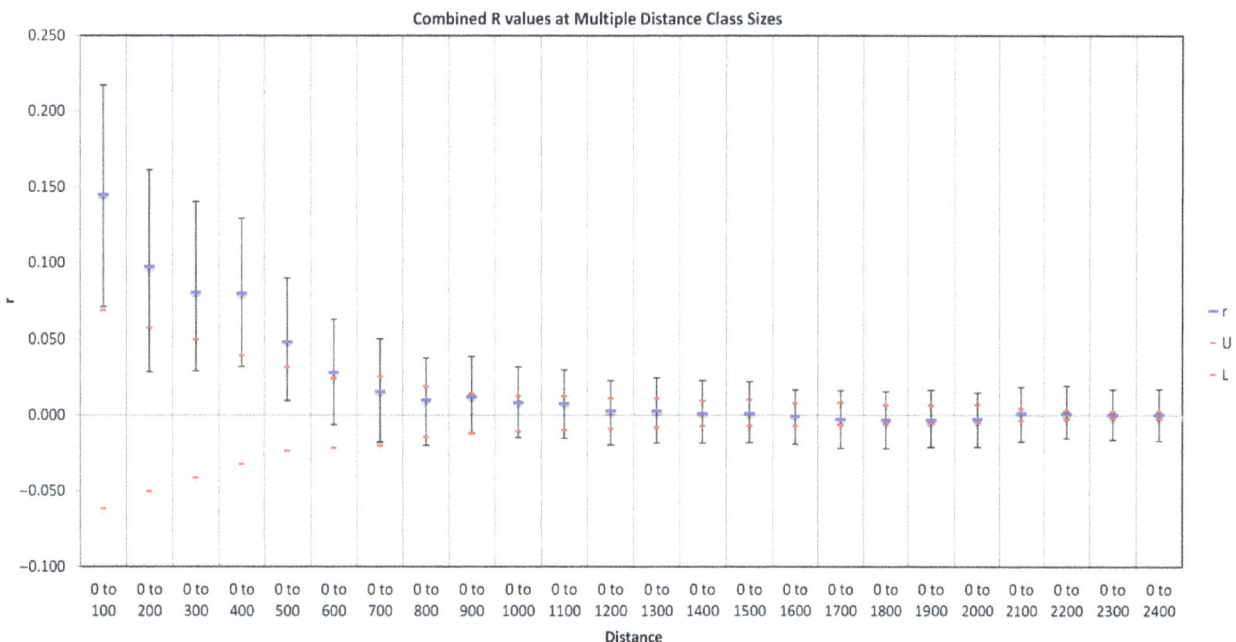

Figure 3. Results of multiple SA analyses for increasing distance class sizes to determine SA. Confidence limits for the *r* values are indicated and were estimated by permutation (999 interactions). Upper (U) and lower (L) confidence limits were generated for the null hypothesis of no SA (*r* = 0) by bootstrap (999 interactions).

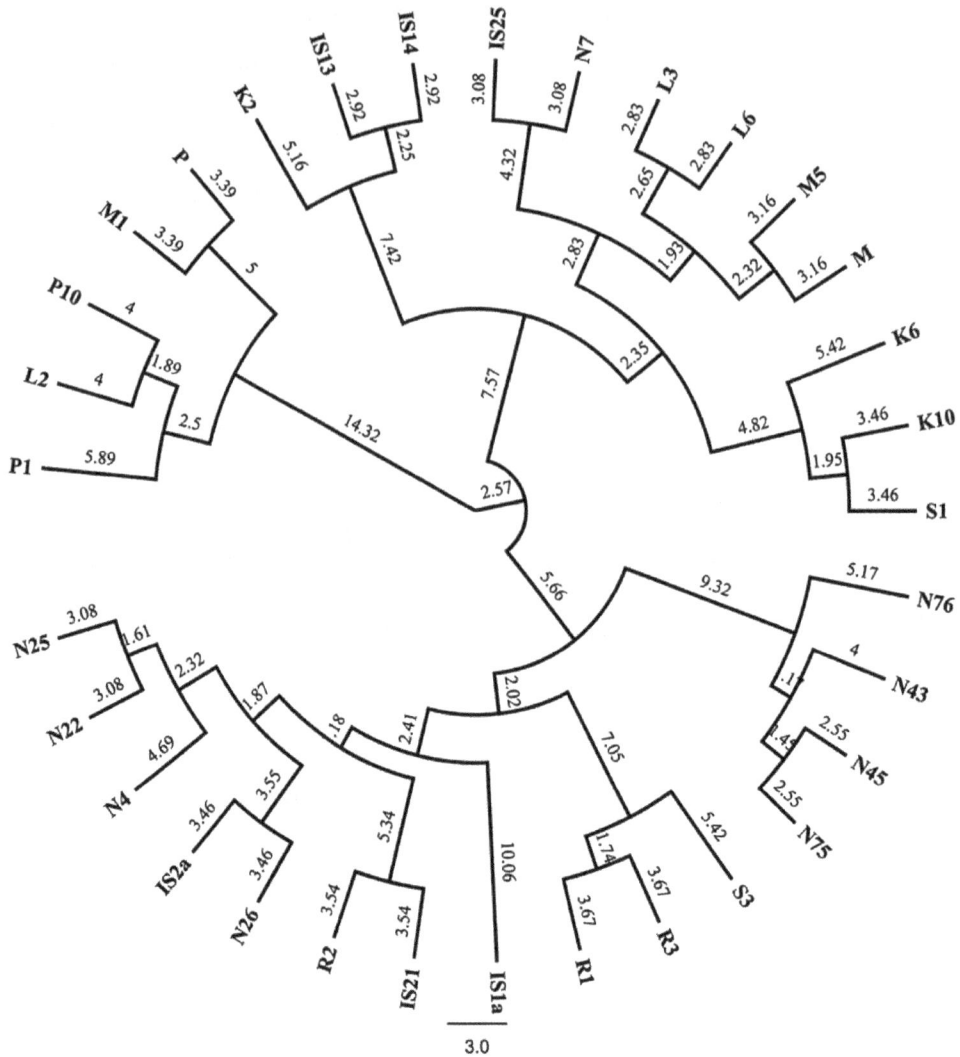

Figure 4. Dendogram of bilberry individuals using Ward's method. Numbers above branches indicate bootstrap values based on 10 000 replicates.

were also found in other wild *Vaccinium* species such as lingonberry (Persson and Gustavsson 2001; Garkava-Gustavsson *et al.* 2005; Debnath 2007), bog bilberry (*V. uliginosum*; Eidesen *et al.* 2007) and wild lowbush blueberry (Debnath 2009). Moreover, the great genetic differentiation among geographically distant populations, determined from the positive value of ΦPT, is in accordance with the theory that the level of genetic heterogeneity among individuals is greater in species with geographically disjunctive populations than in species with more continuous distributions (Hamrick and Godt 1996; Premoli *et al.* 2001). Therefore, in the present study, we focussed on the genetic diversity of bilberry plants separated by long distances, clones of which are discontinuous and isolated from one another by mountains or seas. The effect of these natural barriers together with common results of different clusters and spatial and

population structure analyses provided evidence of a great association between the Finnish and the Norwegian N2 populations on one side, the Norwegian populations N4 and N7 on the other side, and split IS and K populations from all the others. Moreover, the presence of a north–south genetic gradient is in accordance with recent findings on bilberry phenotypic traits (Lätti *et al.* 2008; Åkerström *et al.* 2010; Uleberg *et al.* 2012). These results are also supported by other studies carried out on different species that show how intra-specific genetic variation increases or decreases in relation to the physical distance separating the individuals, and is also showing the influence of the geographic structure of natural populations (Brooks *et al.* 2015). To date, most conservation activities have focussed on the species level; however, also genetic variation at an intra-specific level needs to be considered to avoid loss of diversity derived from severe inbreeding, resulting in lowered fitness and increasing risk of extinction. Moreover, the determination and the conservation of the within-population genetic diversity of one species could increase evolutionary resilience when different and geographically separated environments are connected. When populations are interconnected along climatic and geographical gradients, there is the potential for *in situ* adaptive evolution (Sgrò *et al.* 2011). Furthermore, plant species conserved in key biodiversity areas are an essential genetic source to develop new varieties for future breeding work and to avoid the diversity loss derived from severe inbreeding. Finally, genetic diversity conservation might become crucial in a biotic or abiotic crisis where only very rare genotypes may be resistant to a new disease, pathogen strain or environmental condition. This was further supported by the highest values for He, Shannon Index and number of polymorphic bands.

Further studies with a higher number of markers and samples from the distribution area of bilberry are needed to identify the key biodiversity areas of the species.

Conclusions

Advances in genotyping techniques combined with more sophisticated statistical methods provide the means by which among- and within-population genetic diversity can be estimated in the absence of any prior specific information. This valuation is necessary to conserve the biodiversity of specific areas. In this perspective, we assessed the intra- and inter-population genetic diversity of 32 individuals collected from different North European countries. We found the presence of significant correlations between geographic and genetic distances, which placed Norwegian, Finnish, Icelandic and German genotypes in separate groups. The present results indicate how key biodiversity areas of the wild *V. myrtillus* species could be individualized as a useful source of biodiversity for future ecological studies and breeding purposes.

Sources of Funding

The study was funded by the Thule Institute (2013–16) (to H.H.) and by Nordic Innovation Centre—New Nordic Food project 'Bilberry: Towards functional food markets' (2007–10) (to H.H. and L.J.).

Contributions by the Authors

L.Z. performed the DNA analyses and the ISSR data scoring and was involved in writing and editing; L.P. performed all the statistical analyses and was involved in writing and editing; L.J. and H.H. provided contribution to the concept and the design of the work. All authors read and approved the final manuscript.

Acknowledgements

The partners of the New Nordic Food project are acknowledged for the bilberry seed samples from Iceland, Norway and Sweden. We thank Aaron Bergdahl for English revision of the manuscript.

Literature Cited

Åkerström A, Jaakola L, Bång U, Jäderlund A. 2010. Effects of latitude-related factors and geographical origin on anthocyanidin concentrations in fruits of *Vaccinium myrtillus* L. (bilberries). *Journal of Agricultural and Food Chemistry* **58**:11939–11945.

Albert T, Raspé O, Jacquemart A-L. 2008. Influence of clonal growth on selfing rate in *Vaccinium myrtillus* L. *Plant Biology* **10**:643–649.

Angers B, Magnan P, Plante M, Bernatchez L. 1999. Canonical correspondence analysis for estimating spatial and environmental effects on microsatellite gene diversity in brook charr (*Salvelinus fontinalis*). *Molecular Ecology* **8**:1043–1053.

Brooks TM, Cuttelod A, Faith DP, Garcia-Moreno J, Langhammer P, Pérez-Espona S. 2015. Why and how might genetic and phylogenetic diversity be reflected in the identification of key biodiversity areas? *Philosophical Transactions of the Royal Society B: Biological Sciences* **370**:20140019.

Bunea A, Rugină DO, Pintea AM, Bunea CI, Socaciu C. 2011. Comparative polyphenolic content and antioxidant activities of some wild and cultivated blueberries from Romania. *Notulae Botanicae Horti Agrobotanici Cluj-Napoca* **39**:70–76.

Burdon RD, Wilcox PL. 2007. Population management: potential impacts of advances in genomics. *New Forests* **34**:187–206.

Cocetta G, Karppinen K, Suokas M, Hohtola A, Häggman H, Spinardi A, Mignani I, Jaakola L. 2012. Ascorbic acid metabolism during bilberry (*Vaccinium myrtillus* L.) fruit development. *Journal of Plant Physiology* **169**:1059–1065.

Debnath SC. 2007. Inter simple sequence repeat (ISSR) to assess genetic diversity within a collection of wild lingonberry (*Vaccinium vitis-idaea* L.) clones. *Canadian Journal of Plant Science* **87**:337–344.

Debnath SC. 2009. Development of ISSR markers for genetic diversity studies in *Vaccinium angustifolium*. *Nordic Journal of Botany* **27**: 141–148.

Dell'acqua M, Fricano A, Gomarasca S, Caccianiga M, Piffanelli P, Bocchi S, Gianfranceschi L. 2014. Genome scan of Kenyan *Themeda triandra* populations by AFLP markers reveals a complex genetic structure and hints for ongoing environmental selection. *South African Journal of Botany* **92**:28–38.

Eidesen PB, Alsos IG, Popp M, Stensrud Ø, Suda J, Brochmann C. 2007. Nuclear vs. plastid data: complex pleistocene history of a circumpolar key species. *Molecular Ecology* **16**:3902–3925.

Evanno G, Regnaut S, Goudet J. 2005. Detecting the number of clusters of individuals using the software STRUCTURE: a simulation study. *Molecular Ecology* **14**:2611–2620.

Falush D, Stephens M, Pritchard JK. 2003. Inference of population structure using multilocus genotype data: linked loci and correlated allele frequencies. *Genetics* **164**:1567–1587.

Feuillet C, Langridge P, Waugh R. 2008. Cereal breeding takes a walk on the wild side. *Trends in Genetics* **24**:24–32.

Garkava-Gustavsson L, Persson HA, Nybom H, Rumpunen K, Gustavsson BA, Bartish I. 2005. RAPD-based analysis of genetic diversity and selection of lingonberry (*Vaccinium vitis-idaea* L.) material for ex situ conservation. *Genetic Resources and Crop Evolution* **52**:723–735.

Garriga M, Parra PA, Caligari PDS, Retamales JB, Carrasco BA, Lobos GA, García-Gonzáles R. 2013. Application of inter-simple sequence repeats relative to simple sequence repeats as a molecular marker system for indexing blueberry cultivars. *Canadian Journal of Plant Science* **93**:913–921.

Girard P, Angers B. 2006. The impact of postglacial marine invasions on the genetic diversity of an obligate freshwater fish, the longnose dace (*Rhinichthys cataractae*), on the Quebec peninsula. *Canadian Journal of Fisheries and Aquatic Sciences* **63**: 1429–1438.

Hammer Ø, Harper DAT, Ryan PD. 2001. Past: paleontological statistics software package for education and data analysis. *Palaeontologia Electronica* **4**:1–9.

Hamrick JL, Godt MJW. 1996. Effects of life history traits on genetic diversity in plant species. *Philosophical Transactions of the Royal Society B: Biological Sciences* **351**:1291–1298.

Jaakola L, Tolvanen A, Laine K, Hohtola A. 2001. Effect of N^6-isopentenyladenine concentration on growth initiation in vitro and rooting of bilberry and lingonberry microshoots. *Plant Cell, Tissue and Organ Culture* **66**:73–77.

Jacquemart AL. 1993. Floral visitors of *Vaccinium* species in the high Ardennes, Belgium. *Flora* **188**:263–273.

Lätti AK, Riihinen KR, Kainulainen PS. 2008. Analysis of anthocyanin variation in wild populations of bilberry (*Vaccinium myrtillus* L.) in Finland. *Journal of Agricultural and Food Chemistry* **56**:190–196.

Martinussen I, Rohloff J, Uleberg E, Junttila O, Hohtola A, Jaakola L, Häggman H. 2009. Climatic effects on the production and quality of bilberries (*Vaccinium myrtillus*). *Latvian Journal of Agronomy* **12**:71–74.

Moran PAP. 1950. Notes on continuous stochastic phenomena. *Biometrika* **37**:17–23.

Nestby R, Percival D, Martinussen I, Opstad N, Rohloff J. 2011. The European blueberry (*Vaccinium myrtillus* L.) and the potential for cultivation. *European Journal of Plant Science and Biotechnology* **5**:5–16.

Paassilta M, Moisio S, Jaakola L, Häggman H. 2009. Voice of the Nordic wild berry industry. Oulu, Finland: University of Oulu Press.

Peakall R, Smouse PE. 2006. GENALEX 6: genetic analysis in Excel. Population genetic software for teaching and research. *Molecular Ecology Notes* **6**:288–295.

Peakall R, Ruibal M, Lindenmayer DB. 2003. Spatial autocorrelation analysis offers new insights into gene flow in the Australian bush rat, *Rattus fuscipes*. *Evolution* **57**:1182–1195.

Persson HA, Gustavsson BA. 2001. The extent of clonality and genetic diversity in lingonberry (*Vaccinium vitis-idaea* L.) revealed by RAPDs and leaf-shape analysis. *Molecular Ecology* **10**: 1385–1397.

Premoli AC, Souto CP, Allnutt TR, Newton AC. 2001. Effects of population disjunction on isozyme variation in the widespread *Pilgerodendron uviferum*. *Heredity* **87**:337–343.

Prevost A, Wilkinson MJ. 1999. A new system of comparing PCR primers applied to ISSR fingerprinting of potato cultivars. *Theoretical and Applied Genetics* **98**:107–112.

Pritchard JK, Stephens M, Donnelly P. 2000. Inference of population structure using multilocus genotype data. *Genetics* **155**: 945–959.

Raspé O, Guillaume P, Jacquemart AL. 2004. Inbreeding depression and biased paternity after mixed-pollination in *Vaccinium myrtillus* L. (Ericaceae). *International Journal of Plant Sciences* **165**:765–771.

Rosenberg MS, Anderson CD. 2011. PASSaGE: pattern analysis, spatial statistics and geographic exegesis. Version 2. *Methods in Ecology and Evolution* **2**:229–232.

Sgrò CM, Lowe AJ, Hoffmann AA. 2011. Building evolutionary resilience for conserving biodiversity under climate change. *Evolutionary Applications* **4**:326–337.

Smouse PE, Peakall R. 1999. Spatial autocorrelation analysis of individual multiallele and multilocus genetic structure. *Heredity* **82**: 561–573.

Song GQ, Hancock F. 2011. *Vaccinium*. In: Kole C, ed. *Wild crop relatives: genomic breeding resource. Temperate fruits*. New York: Springer, 210–213.

Tanya P, Taeprayoon P, Hadkam Y, Srinives P. 2011. Genetic diversity among *Jatropha* and *Jatropha*-related species based on ISSR markers. *Plant Molecular Biology Reporter* **29**:252–264.

Ter Braak CJF. 1986. Canonical correspondence analysis: a new eigenvector technique for multivariate direct gradient analysis. *Ecology* **67**:1167–1179.

Tester M, Langridge P. 2010. Breeding technologies to increase crop production in a changing world. *Science* **327**:818–822.

Uleberg E, Rohloff J, Jaakola L, Trôst K, Junttila O, Häggman H, Martinussen I. 2012. Effects of temperature and photoperiod on yield and chemical composition of Northern and Southern clones of bilberry (*Vaccinium myrtillus* L.). *Journal of Agricultural and Food Chemistry* **60**:10406–10414.

Zhao D, Yang J, Yang S, Kato K, Luo J. 2014. Genetic diversity and domestication origin of tea plant *Camellia taliensis* (Theaceae) as revealed by microsatellite markers. *BMC Plant Biology* **14**:14.

14

Mode of inheritance for biochemical traits in genetically engineered cotton under water stress

Muhammad Ali Abid[1,2,†], Waqas Malik[1,2*†], Azra Yasmeen[3], Abdul Qayyum[1], Rui Zhang[2], Chengzhen Liang[2], Sandui Guo[2] and Javaria Ashraf[1]

[1] Genomics Lab, Department of Plant Breeding and Genetics, Faculty of Agricultural Sciences and Technology, Bahauddin Zakariya University, Multan 60000, Pakistan
[2] Biotechnology Research Institute, Chinese Academy of Agricultural Sciences, 100081 Beijing, China
[3] Department of Agronomy, Faculty of Agricultural Sciences and Technology, Bahauddin Zakariya University, Multan 60000, Pakistan

Associate Editor: Kermit Ritland

Abstract. Drought is an abiotic environmental stress that can significantly reduce crop productivity. We examined the mode of inheritance for different biochemical traits including total soluble proteins, chlorophyll *a*, chlorophyll *b*, total chlorophyll, carotenoids, total phenolic contents and enzymatic antioxidants (superoxide dismutase, peroxidase and catalase), and their relationship with *Bacillus thuringiensis* (*Bt*) toxin under control and drought conditions. Eight genetically diverse cotton genotypes were selfed for two generations to ensure homozygosity. Fifteen F_1 hybrids were developed by crossing five non-*Bt* female lines with three *Bt* male testers. The F_1 hybrids and eight parents were finally evaluated under control (100 % field capacity (FC)) and drought (50 % FC) conditions in 2013. The biochemical traits appeared to be controlled by non-additive gene action with low narrow sense heritability estimates. The estimates of general combining ability and specific combining ability for all biochemical traits were significant under control and drought conditions. The genotype-by-trait biplot analysis showed the better performance of *Bt* cotton hybrids when compared with their parental genotypes for various biochemical traits under control and drought conditions. The biplot and path coefficient analyses revealed the prevalence of different relationships between Cry1Ac toxin and biochemical traits in the control and drought conditions. In conclusion, biochemical traits could serve as potential biochemical markers for breeding *Bt* cotton genotypes without compromising the optimal level of *Bt* toxin.

Keywords: Biochemical markers; carotenoids; Cry1Ac toxin; enzymatic antioxidants; non-additive gene action.

Introduction

Plants are more vulnerable to unfavourable environmental conditions during growth, development and reproduction due to their sessile nature (Trewavas 2002). Drought is one of the major factors limiting crop production and commonly leads to substantial losses in yield. Plants have evolved a variety of different mechanisms at morphological, physiological, cellular and biochemical levels to overcome water stress conditions (Fang and Xiong 2015). In addition to naturally occurring mechanisms, more than 80 years of breeding activities have led to an increase in crop yield under drought conditions. Although

* Corresponding author's e-mail address: waqasmalik@bzu.edu.pk
† These authors contributed equally to this work.

fundamental research has provided considerable gains in our understanding of the responses of plants to water deficits, there is still a large gap between yields of crops in stress and non-stress environments. Minimizing the 'yield gap' and increasing yield stability under different water-deficient conditions are of strategic importance for plant scientists (Cattivelli et al. 2008).

Water stress leads to the production of reactive oxygen species (ROS), and their accumulation causes toxicity, peroxidation of cellular membranes, oxidation of carbo-hydrates, proteins, lipids and even DNA (Apel and Hirt 2004; Shah et al. 2011). The balance between ROS gener-ation and scavenging is maintained by various enzymatic antioxidants like superoxide dismutase (SOD), catalase (CAT), peroxidases (PODs) and non-enzymatic antioxi-dants including phenolics, ascorbate, carotenoids and tocopherols (Gill and Tuteja 2010). Superoxide dismutase, POD and CAT comprise the main enzymatic antioxidant system that catabolize free radicals and limit the poten-tial for oxidative damage (Apel and Hirt 2004). Superoxide dismutase catalyses the superoxide (O_2^-) to O_2 and H_2O_2, then POD and CAT catalyse the conversion of H_2O_2 to H_2O and O_2 (Mittler et al. 2004). In addition to this, leaf water potential influences the photosynthetic process by reduc-tion in CO_2 fixation due to stomatal closure (Flexas et al. 2004), disturbing photosynthetic pigments like chloro-phyll and carotenoids and damaging the photosynthetic apparatus (Wahid and Rasul 2005; Parida et al. 2007). The concentration of Bacillus thuringiensis (Bt) protein in dif-ferent transgenic crops including cotton is significantly influenced by water deficit conditions (Wang et al. 2001; Luo et al. 2008). Reduction in efficacy of Bt crystalline endotoxins in genetically engineered crop due to abiotic stresses would result in poor control over targeted pests and may increase resistance against Bt proteins (Tabashnik et al. 2008).

The interaction of Bt toxin production and water stress could be particularly important to determine whether transgenic crops will continue to be effective against tar-get insects/pests in the future. Therefore, information regarding the inheritance of stress-related traits could be helpful for plant breeders to devise a breeding strat-egy. The line × tester analysis provides better estimates for genetics components, detection of suitable parents and superior crosses needed for selection procedures in further generations (Ganapathy et al. 2005; Ahuja and Dhayal 2007). Selection for the improvement of specific trait in an earlier segregating population would be effect-ive if the trait of interest is controlled by additive effects. However, in cases of greater proportion of non-additive (epistatic and dominant) gene effects, the selection should be carried out in later generations (Jagtap 1986). Genes with additive effects would result in increased

general combining ability (gca), and non-additive gene action is responsible for specific combining ability (sca). Further, the selection procedure could be more effective by investigating the nature of association among traits, often leading to decisive results about breeding of plants for a specific purpose (Cakmakci et al. 1998). In this regard, biplot and path coefficient analyses are reliable biometrical techniques (Dewey and Lu 1959; Yan and Kang 2003).

Transgenic cotton expressing Bt toxin has been the most rapidly adopted genetically engineered crop world-wide (James 2002; Barwale et al. 2004; Dong et al. 2005). This insect-resistant cotton is effective in controlling lepidopteran insects and benefits farmers and the envir-onment by reducing the synthetic insecticidal sprays and preserving the population of beneficial arthropods (Gianessi and Carpenter 1999; Tabashnik et al. 2002). The sustainability of this technology depends largely upon adequate concentration of Bt protein during the entire growth period of plants. However, expression of the Bt transgene is affected by water stress environment (Traore et al. 2000). Keeping in mind the increasing short-age of water in the world and its impact on cotton pro-duction, it is imperative to investigate the genetic pattern of various biochemical traits for drought toler-ance in cotton. The objective of this study was to investi-gate the mode of inheritance and nature of association among various biochemical traits in interspecific and intraspecific crosses of cotton under normal and drought conditions. These findings would pave the way for cotton breeders to develop drought-tolerant Bt cotton varieties.

Methods

Plant materials

The plant materials comprised eight genetically diverse cotton genotypes. Five genotypes belonging to Gossy-pium hirsutum (SA-1357, MNH-814, VH-303, MNH-886 and FH-142) and one belonging to G. barbadense (GIZA-7) have white fibre colour and were collected from the Cotton Research Station in Multan, Pakistan. Two genotypes belonging to G. hirsutum (BZUG1 and BZUB) have green and brown fibre colour, respectively, and were collected from the Department of Plant Breed-ing and Genetics, Bahauddin Zakariya University, Multan, Pakistan.

Development of breeding materials

The genotypes were selfed for two generations during 2011–12 to ensure their purity. Four seeds of each selfed genotype were sown in earthen pots (40 cm diameter and 75 cm deep) containing 16 kg of sandy loam soil dur-ing November 2012, in a glasshouse having automatic temperature controls. Temperatures of $30\,°C \pm 5$ and $20\,°C \pm 5$ were maintained during day and night, respectively. At the 15th day of emergence, plants were thinned, maintaining only two healthy seedlings per pot. The recommended cultural practices were adopted during the conduct of experiment. Lines × Testers (5 × 3) crosses were made at the flowering stage. All the non-*Bt* genotypes, i.e. SA-1357, MNH-814, BZUG1, BZUB and GIZA-7, were used as female parent (lines), while three *Bt* genotypes MNH-886, FH-142 and VH-303 served as male parent (testers). Crossed bolls were hand-picked and ginned using a single-ruler ginning machine to derive F_0 seed.

Evaluation of breeding materials

In May 2013, four F_0 seeds from each of the 15 crosses along with their parents were sown in plastic pipes (90 cm depth and 3 cm diameter) in two sets (i.e. control and drought) in a glasshouse. Temperatures of $30\,°C \pm 5$ day and $20\,°C \pm 5$ night were maintained using the automatic cooling and heating systems of the glasshouse. Clay loam and farmyard manure in a ratio of $3:1$ were used as media in pipes to facilitate plant growth. The experiment was laid out in a completely randomized design with three replications and each replication comprised five pipes. The same amount of water was given to the both sets of plants, and thinning was carried out at the 15th day of emergence to have only one plant per pipe. After the 15th day of plant emergence, two different levels, 100 % field capacity (control) and 50 % field capacity (drought), were maintained on a gravimetric basis (Nachabe 1998). These field capacity levels were maintained up to harvesting.

Sample collection

The fully expanded leaf samples from plants of both control and drought treatments were collected at 90 days after emergence because at this stage, the plant has a maximum number of developing bolls and an optimum amount of *Bt* toxin is very necessary, along with other biochemical traits, to avoid boll worm attack and drought stress. The collected leaf samples were stored immediately at $-80\,°C$ for different biochemical analyses. All the spectrophotometric analyses of biochemical traits were conducted using Implen-Nanophotometer (Germany) in the Genomics Lab at the Department of Plant Breeding and Genetics, Bahauddin Zakariya University, Multan, Pakistan.

Determination of total soluble proteins

For the extraction of total soluble proteins (TSP, mg g^{-1}), 0.5 g of leaf sample was ground in 1 mL of 50 mM phosphate buffer with pH 7.2. The ground material was centrifuged at 12 000 r.p.m. for 5 min and supernatant was transferred to another 1.5 mL centrifuged tube. Bradford assay (Bradford 1976) was used to quantify the TSP by constructing a standard curve (10, 20, 30, 40 and 50 µg mL^{-1}) for reaction mixture of bovine serum albumin, dye stock (Coomassie Brilliant Blue G-250 dye) and distilled water. The absorbance of reaction mixture for the standard curve and that of the sample was recorded at 595 nm.

Determination of leaf chlorophyll and carotenoid contents

Leaf carotenoids, chlorophyll (*a* and *b*) and total chlorophyll contents were analysed by grinding 0.5 g of the leaf sample in 80 % acetone solution followed by filtration through Whatman #1 paper. The absorbance of filtrate was recorded at 663, 644 and 452.5 nm. The contents of chlorophyll *a*, chlorophyll *b*, total chlorophyll and carotenoids were calculated (µg mL^{-1}) according to formulae given by Metzner et al. (1965).

$$\text{Chlorophyll } a = (10.3 \times E663) - (0.98 \times E644)$$

$$\text{Chlorophyll } b = (19.7 \times E644) - (3.87 \times E663)$$

$$\text{Total chlorophyll} = \text{chlorophyll } a + \text{chlorophyll } b$$

$$\text{Carotenoids} = 4.2 \times E452.5 - \{(0.0264 \times \text{chlorophyll } a) + (0.426 \times \text{chlorophyll } b)\}$$

where E is the absorbance at that specific wavelength.

Total phenolic contents

The total phenolic contents (TPC, mg GAE g^{-1}) of leaf samples were quantified according to Ainsworth and Gillespie (2007). Gallic acid solutions of different concentrations (500, 250, 150 and 100 mg L^{-1}) were prepared to plot the calibration curve by determining absorbance at 760 nm. For preparation of the sample, 0.5 g of cotton leaf was ground in 80 % acetone solution followed by filtration through Whatman #1 paper. The volume of filtrate was increased to 10 mL by adding acetone solution. For preparation of the reaction mixture, a 20 µL sample or standard was added in 100 µL of Follin–Ciocalteu reagent, 1.58 mL of distilled water within 8 min and mixed with 300 µL of 20 % (w/v) sodium carbonate solution. The prepared reaction mixture was kept in darkness for 2 h. Total phenolic contents of samples were determined at 760 nm.

Determination of enzymatic antioxidants

For preparation of the enzyme extract, a 0.5 g leaf sample was ground in 5 mL of 50 mM phosphate buffer, pH 7.8. The extract was centrifuged at 15 000 r.p.m. for 20 min and supernatant was transferred to separate 1.5 mL tube and kept in darkness.

Superoxide dismutase EC number (1.15.1.1). Superoxide dismutase activity was determined following Giannopolitis and Ries (1977) using its ability to inhibit the photochemical reduction of nitrobluetetrazolium (NBT) at 560 nm. The reaction mixture was primed by mixing 50 μL of enzyme extract, 1 mL of 50 μM NBT, 500 μL of 75 mM ethylenediaminetetraacetic acid, 950 μL of 50 mM phosphate buffer, 1 mL of 1.3 μM riboflavin and 500 μL of 13 mM methionine. Test tubes containing the reaction mixture were incubated under 30 W fluorescent lamp illuminations for 5 min. The reaction was stopped when the fluorescent lamp was switched off and covered with aluminium foil. Test tubes containing the same reaction mixture without enzyme extract served as blank. Blue formazan was developed due to photoreduction of NBT, which was measured using absorbance at 560 nm. Superoxide dismutase activity was expressed as IU min^{-1} mg^{-1} of protein.

Peroxidase EC number (1.11.1.7). Peroxidase activity (mmol min^{-1} mg^{-1} protein) was determined according to the method described by Chance and Maehly (1955). Peroxidase activity was determined by guaiacol oxidation, and the unit of POD was defined as 0.01 absorbance change min^{-1} mg^{-1} protein. The 3 mL reaction mixture was prepared by mixing 100 μL of enzyme extract, 2 mL of 50 mM phosphate buffer, 500 μL of 40 mM H_2O_2 and 400 μL of 20 mM guaiacol. The change in absorbance was recorded at 470 nm for every 20 s up to 5 min.

Catalase EC number (1.11.1.6). Catalase activity was estimated according to Chance and Maehly (1955), which involved the initial decomposition of H_2O_2. The 3 mL reaction mixture for the determination of CAT contained 2 mL of 50 mM phosphate buffer, 900 μL of 5.9 mM H_2O_2 and 100 μL of enzyme extract. Absorbance was observed for every 30 s to 5 min at 240 nm. The unit of CAT activity was defined as decomposition of μmol of H_2O_2 min^{-1} mg^{-1} protein.

Cry1Ac protein concentration assay

The concentration of Cry 1Ac (in $\mu g\ g^{-1}$) in cotton leaf extracts was determined through enzyme-linked immunosorbent assay following Shan *et al.* (2007). An ice-cold 1× sample extraction buffer (500 μL) was used to homogenize the lyophilized tissue. The lyophilized tissue was macerated through mortar-driven pestle at 3000 r.p.m. for 30 s,

then chilled on ice for 30 s and macerated for 30 s again, centrifuged at 8000 r.p.m. for 15 min. Then the supernatant was collected for the determination of Cry1Ac protein. The antibody, buffer blank, standards and controls (negative and positive) were added to each well and incubated at 37 °C. After 45 min, the buffered enzyme was added and incubated for 30 min at room temperature. Finally, the absorbance was recorded at 405 nm.

Statistical analysis

The data for all biochemical traits, i.e. TSP, chlorophyll *a*, chlorophyll *b*, carotenoids, total chlorophyll, TPC and enzymatic antioxidants (SOD, POD and CAT) under both control and drought conditions, were analysed following the line × tester analysis (Singh and Chaudhary 1999). The sum of square for genotypes was subdivided into variation among parents, among parents vs. crosses and among crosses. The sum of square for parents was also subdivided into variation among lines, among testers and among line × testers.

Estimation of variance components and heritabilities. The estimates of variance for combining abilities, genetic components and heritabilities were calculated using the mean square values. The variances due to gca and sca were tested against their respective error variances, derived from the analysis of variance of the different traits as follows:

I. Covariance of half sib line = Cov.H.S. (line)
$$= \frac{Ms_l - Ms_{l \times t}}{rt}$$

II. Covariance of half sib tester = Cov.H.S. (tester)
$$= \frac{Ms_t - Ms_{l \times t}}{rl}$$

III. Covariance of half sib (average)
$$= \frac{1}{r(2lt - l - t)} \left\{ \frac{(l-1)(Ms_l) + (t-1)(Ms_t)}{l + t - 2} - Ms_{l \times t} \right\}$$

IV. Covariance of full sib
$$= \frac{(Ms_l - Ms_e) + (Ms_t - Ms_e) + (Ms_{l \times t} - Ms_e)}{3r}$$
$$+ \frac{6r Cov.H.S. - r(l+t) Cov.H.S.}{3r}$$

where *l*, *t*, *r*, MS_l, MS_t, $MS_{l \times t}$ and MS_e are number of lines, number of testers, number of replications, mean square of lines, mean square of testers, mean square of line × tester and error mean square, respectively.

General combining ability variance and sca variance were calculated following the formulae

$$\text{V. } \sigma^2_{gca} = \text{Cov.H.S.} = \left(\frac{1+F}{4}\right)\sigma^2_A = \frac{1}{2\sigma^2_A}$$

So, $\sigma^2_A = 2\sigma^2_{gca}$

$$\text{VI. } \sigma^2_{sca} = \left(\frac{1+F}{2}\right)^2 \sigma^2_D = \sigma^2_D$$

So, $\sigma^2_D = \sigma^2_{sca}$

Additive and dominance genetic variances were calculated by taking inbreeding coefficient as one ($F = 1$). Narrow sense heritability (h^2) was calculated using the formula

$$\text{VII. } h^2 = \frac{\sigma^2_A}{\sigma^2_P}$$

Per cent contribution of lines, testers and lines × testers.

I. Per cent contribution of lines

$$= \frac{\text{Sum square of lines}}{\text{Sum square of crosses}} \times 100$$

II. Per cent contribution of tester

$$= \frac{\text{Sum square of testers}}{\text{Sum square of crosses}} \times 100$$

III. Per cent contribution of lines × testers

$$= \frac{\text{Sum square of lines × testers}}{\text{Sum square of crosses}} \times 100$$

Estimation of combining ability effects. General combining ability and sca were calculated from the two-way table of lines vs. testers in which each value was total over replications (Singh and Chaudhary 1999)

$$\text{I. gca effects of } i\text{th line} = gi = \frac{xi..}{tr} - \frac{x...}{ltr}$$

$$\text{II. gca effects of } j\text{th tester} = gj = \frac{x.j.}{lr} - \frac{x...}{ltr}$$

$$\text{III. sca effects of } i\text{th cross} = sij = \frac{xij}{r} - \frac{xi..}{tr} - \frac{x.j.}{lr} - \frac{x...}{ltr}$$

where l is the number of lines, t the number of testers, r the number of replications, $xi..$ the sum of ith line over all testers and replications, $x...$ the sum of means of all

crosses of lines and testers over replications, $x.j.$ the sum of jth tester over lines and replications, xij the sum of mean ijth hybrid combination over replications.

Estimation of standard error for combining ability effects.

$$\text{I. SE}(gi) \text{ lines} = \left(\frac{\text{MSe}}{tr}\right)^{1/2}$$

$$\text{II. SE}(gi - gj) \text{ lines} = \left(\frac{2\text{MSe}}{tr}\right)^{1/2}$$

$$\text{III. SE}(gij) \text{ crosses} = \left(\frac{\text{MSe}}{r}\right)^{1/2}$$

$$\text{IV. SE}(gj) \text{ testers} = \left(\frac{\text{MSe}}{lr}\right)^{1/2}$$

$$\text{V. SE}(gi - gj) \text{ tester} = \left(\frac{2\text{MSe}}{lr}\right)^{1/2}$$

$$\text{VI. SE}(sij - ski) \text{ crosses} = \left(\frac{2\text{MSe}}{r}\right)^{1/2}$$

Test of significance for gca and sca effects.

$$\text{I. } Ti\text{(cal) for gca of lines} = \left(\frac{gi - 0}{\text{SE}(gi)}\right)$$

$$\text{II. } Tj\text{(cal) for gca of testers} = \left(\frac{gj - 0}{\text{SE}(gj)}\right)$$

$$\text{III. } Tij\text{(cal) for sca of crosses} = \left(\frac{sij - 0}{\text{SE}(sij)}\right)$$

The gca effects of lines and testers and sca effects of crosses were marked significant (*$P < 0.05$) and highly significant (**$P < 0.01$) when values of Ti, Tj and Tij were $\geq$ 't' tabulated values at infinity (∞) error degree of freedom.

Biplot and path coefficient analysis. The genotype-by-trait (GT) biplot analysis was computed by following (Yan and Kang 2003):

$$\text{I. } \frac{\alpha_{ij} - \beta_j}{\sigma_j} = \sum_{n=1}^{2} \lambda_n \xi_{in} \eta_{jn} + \varepsilon_{ij} = \sum_{n=1}^{2} \xi_{in} \eta_{jn} + \varepsilon_{ij}$$

where α_{ij} is the mean value of genotype i for trait j, β_j the mean value of all genotypes for trait j, σ_j the standard deviation of trait j among genotype means, λ_n the singular value for principal component (PCn), ξ_{in} the PCn score for genotype i, η_{jn} the PCn score for trait j and ε_{ij} is the residual associated with genotype i in trait j.

The path coefficient was performed following Dewey and Lu (1959). This technique involves partitioning of correlation coefficients to direct and indirect effects through alternate pathways of casual variables over resultant variables. *Bacillus thuringiensis* protein (Cry1Ac) was considered as resultant variable, while other studied traits were casual variables. The figures of path analysis were generated using PAST statistical packages (Hammer et al. 2001).

Results

Genetic effects and heritability estimates

There was significant variation among parents, parent vs. crosses, crosses and line × tester interaction for the biochemical traits (Table 1). The variance of gca was lower than the variance of sca for TSP, chlorophyll *a*, chlorophyll *b*, carotenoids, total chlorophyll, TPC and enzymatic antioxidants (SOD, POD and CAT) under control and drought conditions. The degree of dominance $(\sigma_D^2/\sigma_A^2)^{1/2}$ and $\sigma_{sca}^2/\sigma_{gca}^2$ ratio was greater than unity for all biochemical traits under both treatments. The amount of narrow sense heritabilities was low for all traits under both conditions, i.e. control and drought. Further, the narrow sense heritabilities under control were inconsistent with the narrow sense heritabilities under drought for all studied traits. The maximum amount of heritability was observed for carotenoids (20.28) and total chlorophyll (14.20) under control and drought conditions, respectively. The maternal genotypes (lines) were found superior for chlorophyll *a*, chlorophyll *b*, total chlorophyll and carotenoids, while the contribution of line × tester interaction was greater for TSP, TPC, SOD, POD and CAT under control treatment. However, contributions of paternal genotypes (tester) were lower for all biochemical traits. Under drought condition, a greater contribution of tester was recorded for carotenoids, and the contribution of lines was greater for chlorophyll *a*, chlorophyll *b*, total chlorophyll, TPC, SOD and POD, but for line × tester interaction, the contribution of TSP and CAT was higher (Table 2).

The estimates of gca effects of parents varied significantly for TSP, chlorophyll *a*, chlorophyll *b*, total chlorophyll, carotenoids, TPC and enzymatic antioxidants (SOD, POD and CAT) under control and drought conditions. Among lines, MNH-814 had maximum gca effects for TSP and SOD. The Egyptian cotton Giza-7 was found to have highest gca effects for chlorophyll *a*, chlorophyll *b*, total

chlorophyll, carotenoids, TPC and CAT. The tester MNH-886 had the highest significant gca effects for carotenoids, TPC and POD. The line VH-303 showed maximum gca effects for TSP, SOD and CAT. The tester FH-142 was found better for chlorophyll contents under control condition, while under drought conditions, SA-1357 had maximum gca effects for carotenoids; MNH-814 had high gca effect for chlorophyll *a*, *b*, total chlorophyll contents and CAT; BZUG1 had high gca effect for SOD; BZUB had high gca effect for TSP and POD; and Giza-7 showed maximum gca effects for TPC. Among testers, MNH-886 was found better with maximum gca effects for TPC, SOD and CAT, and VH-303 showed maximum gca effects for TSP, total chlorophyll, chlorophyll *b* and POD, while FH-142 had highest gca effects for chlorophyll *a* and carotenoids (Table 3).

The intraspecific hybrid BZUB × VH-303 had highly significant positive sca effects for most of the studied traits, i.e. TSP, chlorophyll contents, carotenoids, SOD and POD. Similarly, the interspecific hybrid Giza-7 × FH-142 surpass all intraspecific hybrids with highest sca effects for TSP, SOD and POD under control condition, while the estimates of sca effects illustrated that intraspecific cotton hybrid BZUB × MNH-886 had significant sca effects for chlorophyll contents, carotenoids, TPC, POD and CAT. Similarly, cotton hybrids MNH-814 × FH-142 had significant sca effects for TSP, TPC, SOD, POD and CAT under drought condition (Table 4).

Biplot analysis

The angle between trait vectors of *Bt* toxin (Cry1Ac) and other biochemical traits except SOD, POD, CAT and TSP was <90 ° under control (Fig. 1A). However, the trait vector of *Bt* toxin Cry1Ac had a >90 ° angle with TSP, TPC and POD under drought condition (Fig. 1B).

The identification and evaluation of elite cotton genotypes for different biochemical markers under both stressed and non-stressed conditions was done by GT biplot analysis. Genotype by trait biplot analysis depicted that three genotypes VH-303, BZUB and SA-1357; interspecific hybrids (Giza-7 × MNH-886 and Giza-7 × FH-142) and two intraspecific hybrids BZUG1 × VH-303 and MNH-814 × VH-303 were at the vertex of the polygon under control condition (Fig. 1A). Among these hybrids, BZUG1 × VH-303 and MNH-814 × VH-303 were found near the trait vectors of TSP, SOD, POD and CAT, and a hybrid SA-1357 × VH-303 was near to the origin of biplot. However, two interspecific hybrids Giza-7 × MNH-886 and Giza-7 × FH-142 were found farthest from the origin but at or near the trait vectors of total chlorophyll contents, TPC, carotenoids and Cry1Ac under control condition (Fig. 1A). Similarly, the GT biplot analysis revealed that intraspecific hybrids (BZUG1 × MNH-886, BZUG1 ×

Table 1. Analysis of variance for biochemical traits under control and drought conditions in *Bt* cotton. SOV, source of variation; df, degrees of freedom; Tr, treatments; C, control; D, drought; MS, mean square; F, F ratio; TSP, total soluble protein; TPC, total phenolic contents; SOD, superoxide dismutase; POD, peroxidase; CAT, catalase.

SOV	df	Tr	TSP		Chlorophyll a		Chlorophyll b		Carotenoids		Total chlorophyll		TPC		SOD		POD		CAT	
			MS	F	MS	F	MS	F	MS	F	MS	F	MS	F	MS	F	MS	F	MS	F
Replications	2	C	2.14	0.31	0.38	2.00	0.16	3.13	0.04	3.71	0.20	0.39	0.05	1.65	464.71	1.08	794.22	11.66	23.35	2.81
		D	46.18	10.65	0.00	0.03	0.01	0.52	0.00	0.08	0.04	0.73	0.01	1.36	230.45	0.74	16.26	1.49	110.45	26.77
Genotypes	22	C	124.84	18.07	3.88	20.42	3.53	67.27	0.15	14.42	14.20	27.56	0.77	24.33	3 82 156.75	885.35	11 797.99	173.19	1272.24	153.38
		D	105.78	24.39	1.15	26.58	2.71	163.08	0.41	28.61	6.74	119.24	4.66	1317.31	198 754.82	634.76	22 817.12	2093.78	604.26	146.45
Parents	7	C	45.17	6.54	1.15	6.05	0.76	14.43	0.16	15.51	3.23	6.26	1.24	39.40	258 268.48	598.33	3115.49	45.73	534.01	64.38
		D	99.47	22.93	1.27	29.40	3.12	187.76	0.69	47.41	8.11	143.48	2.04	577.40	355 542.64	1135.50	32 115.24	2947.01	562.93	136.43
Parents vs. crosses	1	C	596.07	86.26	44.99	237.06	37.52	714.11	0.74	72.26	164.64	319.43	0.34	10.77	264 607.01	613.02	33 866.41	497.13	3389.14	408.60
		D	99.14	22.86	4.96	114.72	7.22	434.90	1.17	80.63	24.14	427.05	1.97	558.01	33 738.86	107.75	132 233.48	12 134.23	1603.05	388.53
Crosses	14	C	131.01	18.96	2.30	12.13	2.49	47.48	0.10	9.75	8.95	17.36	0.56	17.76	452 497.30	1048.31	14 562.94	213.77	1490.15	179.66
		D	109.41	25.22	0.82	18.87	2.18	131.33	0.22	15.49	4.81	85.13	6.16	1741.51	132 147.76	422.04	10 352.61	949.99	553.59	134.17
Lines	4	C	187.40	1.67	5.27	4.60	5.57	3.79	0.22	6.41	20.17	4.13	0.69	1.52	110 546.33	0.30	22 197.73	1.57	1109.78	0.76
		D	163.63	1.60	1.42	3.38	4.19	2.97	0.08	0.50	9.94	3.75	12.43	2.73	266 502.36	3.14	19 387.15	2.58	914.57	1.96
Testers	2	C	94.66	0.85	0.99	0.87	0.44	0.30	0.12	12.01	2.74	0.56	0.75	1.66	1 49 617.84	3.87	896.78	0.06	2350.83	1.60
		D	30.18	0.30	1.18	2.81	1.22	0.86	0.73	4.30	3.18	1.20	0.05	0.01	52 018.82	0.61	3581.16	0.48	177.27	0.38
Lines × testers	8	C	111.90	16.19	1.15	6.03	1.47	27.99	0.03	3.34	4.88	9.47	0.45	14.29	374 192.65	866.90	14 162.08	207.89	1465.16	176.64
		D	102.11	23.54	0.42	9.73	1.41	85.16	0.17	11.66	2.65	46.95	4.55	1287.40	85 002.68	271.47	7528.20	690.82	467.19	113.23
Error	44	C	6.91	–	0.19	–	0.05	–	0.01	–	0.52	–	0.03	–	431.65	–	68.12	–	8.30	–
		D	4.34	–	0.04	–	0.02	–	0.01	–	0.06	–	0.00	–	313.12	–	10.90	–	4.13	–

Table 2. Estimates of genetic components, heritabilities and per cent contribution of lines, testers and line × tester to the total variation for biochemical traits under control and drought conditions in Bt cotton. Tr, treatments; C, control; D, drought; σ^2_{gca}, variance of gca; σ^2_{sca}, variance of sca; σ^2_A, additive genetic variance; σ^2_D, dominant genetic variance; $(\sigma^2_D/\sigma^2_A)^{1/2}$, degree of dominance; h^2(n.s), narrow sense heritability; df, degrees of freedom; TSP, total soluble protein; TPC, total phenolic contents; SOD, superoxide dismutase; POD, peroxidase; CAT, catalase.

Genetic components	Tr	TSP	Chlorophyll a	Chlorophyll b	Carotenoids	Total chlorophyll	TPC	SOD	POD	CAT
σ^2_{gca}	C	0.68	0.04	0.04	0.00	0.14	0.00	2768.35	14.17	0.88
	D	0.26	0.01	0.03	0.00	0.08	0.06	1666.75	99.85	3.06
σ^2_{sca}	C	34.99	0.32	0.47	0.01	1.46	0.14	124 587.00	4697.99	485.62
	D	32.59	0.13	0.47	0.05	0.87	1.52	28 229.86	2505.77	154.35
$\sigma^2_{gca}/\sigma^2_{sca}$	C	0.02	0.13	0.08	0.29	0.10	0.03	0.02	0.00	0.00
	D	0.01	0.11	0.06	0.04	0.09	0.04	0.06	0.04	0.02
$\sigma^2_{sca}/\sigma^2_{gca}$	C	51.81	7.78	13.06	3.45	10.13	36.03	45.00	331.51	549.77
	D	126.26	9.00	17.19	26.23	11.35	26.71	16.94	25.10	50.53
σ^2_A	C	1.35	0.08	0.07	0.01	0.29	0.01	5536.69	28.34	1.77
	D	0.52	0.03	0.05	0.00	0.15	0.11	3333.49	199.71	6.11
σ^2_D	C	34.99	0.32	0.47	0.01	1.46	0.14	124 587.00	4697.99	485.62
	D	32.59	0.13	0.47	0.05	0.87	1.52	28 229.86	2505.77	154.35
$(\sigma^2_D/\sigma^2_A)^{1/2}$	C	5.10	1.97	2.56	1.31	2.25	4.24	4.74	12.87	16.58
	D	7.95	2.12	2.93	3.62	2.38	3.65	2.91	3.54	5.03
h^2(n.s)	C	3.12	13.87	12.12	20.28	12.72	4.33	4.24	0.59	0.36
	D	1.38	14.20	10.10	5.62	14.20	6.95	10.46	7.35	3.71
Contribution of lines	C	40.87	65.43	63.81	62.80	64.43	35.03	6.98	43.55	21.28
	D	42.73	49.82	54.95	10.73	59.04	57.65	57.62	53.51	47.20
Testers	C	10.32	6.15	2.50	17.61	4.38	19.02	45.77	0.88	22.54
	D	3.94	20.72	7.99	46.26	9.45	0.11	5.62	4.94	4.58
Lines × testers	C	48.81	28.42	33.69	19.59	31.19	45.96	47.25	55.57	56.19
	D	53.33	29.46	37.05	43.01	31.52	42.24	36.76	41.55	48.22

Table 3. General combining ability effects (gca) indicating the breeding value of lines and testers for biochemical traits under control and drought conditions in *Bt* cotton. *Significant at 5 % level of probability. **Significant at 1 % level of probability. Tr, treatments; C, control; D, drought; SE(gi), standard error (gca effects for lines); SE($gi–gj$) line, standard error (between gca effects of two lines); SE(gj), standard error (gca effects for testers); SE($gi–gj$) tester, standard error (between gca effects of two testers); TSP, total soluble protein; TPC, total phenolic contents; SOD, superoxide dismutase; POD, peroxidase; CAT, catalase.

	Tr	TSP	Chlorophyll a	Chlorophyll b	Carotenoids	Total chlorophyll	TPC	SOD	POD	CAT
Lines										
SA-1357	C	−3.56**	−0.26*	−0.75**	−1.01**	0.01	−0.05	44.94**	−31.49**	−15.22**
	D	−1.05	0.18*	−0.25**	−0.07	0.15**	−0.74**	7.49	−5.10**	−11.62**
MNH-814	C	5.58**	−0.96**	−0.72**	−1.68**	−0.18**	−0.28**	143.60**	−37.57**	0.22
	D	−0.79	0.52**	1.09**	1.62**	−0.11**	0.76**	−163.73**	−21.13**	9.49**
BZUG1	C	4.36**	0.29*	0.83**	1.13**	0.02	−0.24**	−15.40*	66.51**	1.34
	D	2.35**	−0.09	0.02	−0.07	0.01	−0.88**	185.49**	−5.17**	7.79**
BZUB	C	−3.26**	−0.19	−0.19*	−0.38	−0.09*	0.21*	−11.52*	−37.68**	−2.30
	D	5.46**	−0.56**	−0.77**	−1.33**	−0.05	−0.85**	153.82**	77.44**	4.41**
GIZA-7	C	−3.12**	1.11**	0.82**	1.93**	0.24**	0.35**	−161.62**	40.23**	15.96**
	D	−5.97**	−0.06	−0.09*	−0.14*	0.01	1.70**	−183.07**	−46.04**	−10.07**
SE(gi)	C	0.88	0.15	0.08	0.24	0.03	0.06	6.93	2.75	0.96
	D	0.69	0.07	0.04	0.08	0.04	0.02	5.90	1.10	0.68
SE($gi–gj$) lines	C	1.24	0.21	0.11	0.34	0.05	0.08	9.79	3.89	1.36
	D	0.98	0.10	0.06	0.11	0.06	0.03	8.34	1.56	0.96
Testers										
MNH-886	C	−0.03	−0.25*	−0.17**	−0.42*	0.06*	0.25**	−202.47**	7.10**	−0.52
	D	0.47	−0.31**	−0.18**	−0.50**	−0.09**	0.06**	61.42**	−6.80**	2.14**
VH-303	C	2.53**	−0.01	−0.01	−0.02	−0.10**	−0.08*	357.94**	1.14	12.77**
	D	1.13*	0.08	0.33**	0.41**	−0.16**	−0.01	−55.98**	17.69**	−3.97**
FH-142	C	−2.50**	0.26*	0.17**	0.44*	0.04*	−0.17**	−155.46**	−8.24**	−12.25**
	D	−1.59**	0.23**	−0.15**	0.08	0.25**	−0.05**	−5.44	−10.88**	1.82**
SE(gj)	C	0.68	0.11	0.06	0.19	0.03	0.05	5.36	2.13	0.74
	D	0.54	0.05	0.03	0.06	0.03	0.02	4.57	0.85	0.52
SE($gi–gj$) tester	C	0.96	0.16	0.08	0.26	0.04	0.06	7.59	3.01	1.05
	D	0.76	0.08	0.05	0.09	0.04	0.02	6.46	1.21	0.74

Table 4. Specific combining ability effects indicating genetic value of crosses due to interaction of their parents for biochemical traits in Bt cotton. *Significant at 5 % level of probability. **Significant at 1 % level of probability. SE(ij), standard error (sca effects for crosses); SE(sij–skl), standard error (between sca effects of two crosses); Tr, treatments; C, control; D, drought; TSP, total soluble protein; TPC, total phenolic contents; SOD, superoxide dismutase; POD, peroxidase; CAT, catalase.

Crosses	Tr	TSP	Chlorophyll a	Chlorophyll b	Carotenoids	Total chlorophyll	TPC	SOD	POD	CAT
SA-1357 × MNH-886	C	5.22**	−0.81**	−0.68**	−1.49**	−0.03	−0.18*	403.47**	61.78**	11.90**
	D	−1.89	0.23*	0.18*	0.41**	0.07	0.26**	1.58	−42.88**	−1.19
SA-1357 × VH-303	C	−3.75*	0.73**	0.79**	1.53**	−0.04	0.41**	−197.94**	−26.70**	−16.10**
	D	3.88**	−0.22*	−0.46**	−0.68**	0.04	0.82**	33.64**	51.76**	3.62*
SA-1357 × FH-142	C	−1.47	0.08	−0.11	−0.03	0.07	−0.22*	−205.54**	−35.09**	4.19*
	D	−1.99*	−0.01	0.27*	0.26*	−0.12*	−1.08**	−35.22**	−8.88**	−2.43*
MNH-814 × MNH-886	C	5.68**	0.71**	0.60**	1.31**	−0.03	0.04	−142.19**	−10.15*	2.76*
	D	−7.40**	−0.01	−0.14*	−0.15	0.01	0.21**	−129.53**	−45.78**	−17.74**
MNH-814 × VH-303	C	−3.77*	−0.29	−0.28*	−0.58	0.05	−0.02	405.40**	3.09	−2.64
	D	−2.02*	0.18	0.79**	0.98**	−0.08	−1.08**	−97.47**	−26.31**	−4.91**
MNH-814 × FH-142	C	−1.91	−0.42*	−0.32*	−0.73*	−0.03	−0.01	−263.20**	7.06	−0.12
	D	9.42**	−0.18	−0.65**	−0.83**	0.07	0.87**	227.00**	72.09**	22.65**
BZUG1 × MNH-886	C	0.93	0.1	0.27*	0.38	0.05	0.56**	−99.19**	41.82**	20.14**
	D	6.87**	−0.05	0.39**	0.34*	−0.28**	0.75**	−67.42**	21.29**	11.59**
BZUG1 × VH-303	C	2.98*	−0.55*	−1.08**	−1.63**	0.03	−0.15	−38.60**	43.46**	−14.52**
	D	−1.60	−0.28*	−0.32*	−0.60**	−0.01	−0.22*	171.98**	20.34**	1.40
BZUG1 × FH-142	C	−3.91*	0.45*	0.81**	1.26**	−0.05	−0.41**	137.80**	−85.29**	−5.62**
	D	−5.27**	0.33**	−0.07	0.26*	0.29**	−0.53**	−104.56**	−41.63**	−12.99**
BZUB × MNH-886	C	−1.95	−0.11	−0.28*	−0.39	−0.06	−0.36**	−185.45**	−9.79*	3.38*
	D	0.29	0.33**	0.50**	0.84**	0.20**	0.33**	−51.09**	25.22**	5.31**
BZUB × VH-303	C	2.81*	0.47*	0.48**	0.95**	0.15*	−0.14	265.52**	9.85*	−3.40*
	D	2.40*	0.15	−0.68**	−0.53**	0.18*	−1.17**	−19.36*	−1.87	0.51
BZUB × FH-142	C	−0.86	−0.36	−0.20	−0.57	−0.09	0.51**	−80.08**	−0.06	0.02
	D	−2.68*	−0.49**	0.18*	−0.31*	−0.38**	0.84**	70.44**	−23.35**	−5.81**
GIZA-7 × MNH-886	C	−9.89*	0.11	0.09	0.19	0.06	−0.06	23.36*	−83.66**	−38.18**
	D	2.12*	−0.50**	−0.93**	−1.43**	0.03	−1.56**	246.47**	42.15**	2.02
GIZA-7 × VH-303	C	1.73	−0.36	0.09	−0.27	−0.16**	−0.08	−434.38**	−29.71**	36.65**
	D	−2.65*	0.16**	0.66**	0.82**	−0.13*	1.66**	−88.80**	−43.92**	−0.61
GIZA-7 × FH-142	C	8.15*	0.25	−0.17	0.08	0.09*	0.14	411.02**	113.38**	1.53
	D	0.53	0.34**	0.27**	0.61**	0.14**	−0.10**	−157.67**	1.77	−1.41
SE(ij)	C	1.52	0.25	0.13	0.41	0.06	0.1	12.00	4.77	1.66
	D	1.20	0.12	0.07	0.14	0.07	0.03	10.22	1.91	1.17
SE(sij–skl)	C	2.15	0.36	0.19	0.59	0.08	0.14	16.96	6.74	2.35
	D	1.70	0.17	0.11	0.19	0.10	0.05	14.45	2.70	1.66

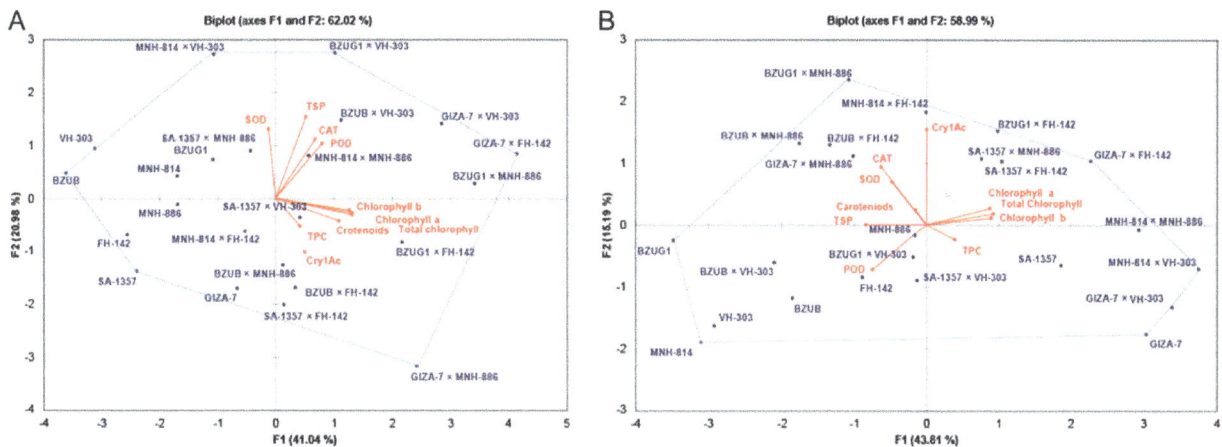

Figure 1. Biplot analysis for biochemical traits in *Bt* cotton. (A) Control and (B) drought conditions. TSP, total soluble protein; TPC, total phenolic contents; SOD, superoxide dismutase; POD, peroxidase; CAT, catalase.

FH-142 and MNH-814 × VH-303) and two interspecific hybrids Giza-7 × FH-142, and Giza-7 × VH-303 along with three genotypes, i.e. Giza-7, MNH-814 and BZUG1, were at the vertex of the polygon under drought condition. Among these cotton hybrids, BZUG1 × MNH-886 and BZUG1 × FH-142 were observed near the trait vectors of Cry1Ac, SOD and CAT. Parental genotypes, viz. MNH-814 and BZUG1, were found close to the trait vectors of TSP and POD (Fig. 1B).

Path coefficient analysis

The total chlorophyll contents, carotenoids and TSP had positive direct effects on Cry1Ac toxin under control conditions, while the TPC, SOD, POD and CAT had negative direct effects on Cry1Ac toxin. Results also depicted that the TSP had a positive direct effect on chlorophyll *a*, chlorophyll *b*, SOD, POD and CAT but had a negative direct effect on TPC. In addition, TSP had an indirect negative effect on POD via SOD. The positive indirect effect of chlorophyll *a* and *b* on carotenoids and Cry1Ac toxin was observed under control condition (Fig. 2A). The TSP, total chlorophyll contents and chlorophyll *a* and *b* had negative direct and indirect effects on Cry1Ac toxin under drought condition. However, TSP had direct positive effect on SOD, POD and CAT and negative effect on TPC and chlorophyll *a* and *b*, respectively (Fig. 2B).

Discussion

Bacillus thuringiensis cotton is the product of modern agricultural research that is continuously replacing the cultivation of non-*Bt* cotton cultivars. Different *Bt* cotton genotypes have different abilities to tolerate drought by varying the expression of various biochemical traits according to their genetic potential. It has also been reported that drought is known to affect the efficacy

of transgenes in genetically modified crops (Bruns and Abel 2003). Therefore, exploring the mode of inheritance for the various biochemical traits analysed in this study will enable plant breeders to develop *Bt* cotton genotypes having a desired amount of *Bt* toxin and antioxidant activity under water stressed and non-stressed conditions.

Non-additive gene action was involved in the expression of all biochemical traits including TSP, chlorophyll *a*, chlorophyll *b*, carotenoids, total chlorophyll, TPC and enzymatic antioxidants (SOD, POD and CAT) under control and drought conditions. These findings are confirmed by the reports of Mehndiratta and Phul (1983), Greish *et al.* (2005) and Immanuel *et al.* (2006) in pearl millet, tomato and maize, respectively. Our findings also suggested that heterosis breeding would be more fruitful for the development of drought-tolerant *Bt* cotton hybrids. However, Song *et al.* (2014) reported the preponderance of additive genetic effects for SOD and POD activity in cotton. The difference in gene action could be attributed to environmental factors and different genetic make-up of breeding material.

Heritability is a good index for transmission of traits from parents to offspring, and the scope of trait improvement through selection breeding depends upon the magnitude of heritability (Neelima and Chenga 2008). In our findings, low amount of narrow sense heritability coupled with higher degree of dominance $(\sigma_D^2/\sigma_A^2)^{1/2}$ and ratio $\sigma_{sca}^2/\sigma_{gca}^2$ for all biochemical traits under both conditions further confirmed the prevalence of non-additive genetic effects. These findings demonstrated that selection based on all these biochemical traits in early segregating generations would be less efficient (Saleem *et al.* 2009; Rad *et al.* 2012). The low amount of heritabilities for different biochemical traits also suggested that direct selection based on these traits will not yield encouraging

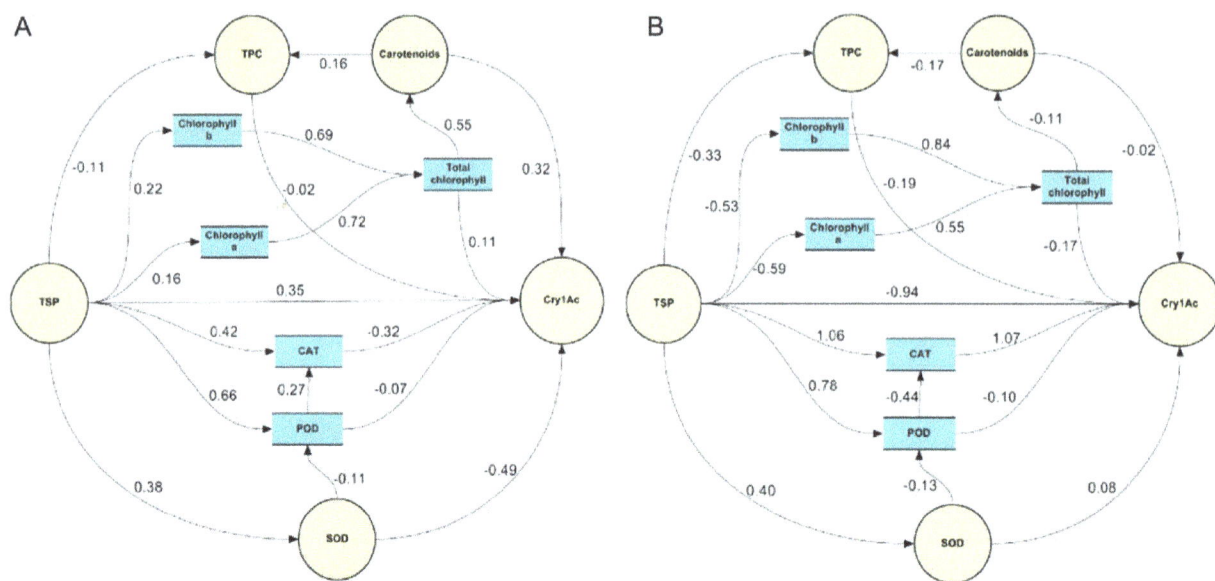

Figure 2. Path coefficient analysis for biochemical traits in *Bt* cotton. (A) Control and (B) drought conditions. TSP, total soluble protein; TPC, total phenolic contents; SOD, superoxide dismutase; POD, peroxidase; CAT, catalase.

results as they are influenced not only by genotype but also by environment and genotype × environment interaction (Rosielle and Hamblin 1981; Ribaut *et al.* 1997).

Estimates of combining ability demonstrated the breeding value of parental lines towards the development of a desired genotype. General combining ability is due to additive-type gene action, whereas sca is administrated by the non-additive nature of genes (Wu *et al.* 2006; Jenkins *et al.* 2007). In the present study, parental lines, viz. MNH-814, GIZA-7, MNH-886 and FH-142, with highly significant gca effects for biochemical traits, i.e. TSP, chlorophyll *a*, chlorophyll *b*, total chlorophyll, carotenoids, TPC and enzymatic antioxidants (SOD, POD and CAT), suggested that these genotypes had more tendency to pass these biochemical traits to their progenies under control conditions (Rad *et al.* 2012), whereas parental genotypes, i.e. MNH-814, BZUG1, BZUB, MNH-886 and VH-303, with more gca effects were found superior combiners for chlorophyll *a*, *b*, total chlorophyll contents, TSP and enzymatic antioxidants under drought. Our findings also suggested that parents with high gca effects for specific traits might have additive gene action and these parents could be used for the development of genotypes having greater amounts of these biochemical traits through hybridization followed by selection breeding (Saleem *et al.* 2009). Specific estimates of combining abilities showed that intraspecific hybrid (BZUB × VH-303) and interspecific hybrid (GIZA-7 × FH-142) were found best combiners for TSP, chlorophyll contents, carotenoids, SOD, POD and CAT under control condition, whereas BZUB × MNH-886 and MNH-886 × FH-142 were best specific combiners for chlorophyll contents, carotenoids,

TPC, TSP, SOD, POD and CAT under drought conditions. The results suggested that these hybrids represent an ample scope for hybrid development of *Bt* cotton with more tolerance to drought (Ashokkumar *et al.* 2010). In our study, the hybrids with negative value of sca for different biochemical traits under both conditions, i.e. normal and drought, indicated the existence of different genes with minor effects in each line or prevalence of epistasis (Abd *et al.* 2009).

The correlation coefficient among various traits can be estimated using biplot analysis. In this biometrical technique, a trait vector is drawn from the origin to each marker of the traits, and cosine of the angle among trait vectors determine the nature of association among them (Yan and Rajcan 2002), whereas for traits having a multidirectional relationship with each other, path analysis could be a useful biometrical tool in predicting correlation responses to directional selection and in identifying traits that may not be important from a breeding point of view but that can serve as precursors for the important ones (Rasheed *et al.* 2009). The hybrids at the vertex and near any trait vector could be attributed best or poor for that particular trait (Yan *et al.* 2007). In biplot analysis, those hybrids that were found to be best for different biochemical traits under control and drought conditions suggested that better performance of these hybrids when compared with their parents might be due to accumulation of favourable alleles and increase in heterozygosity at different loci, because heterozygosity is always superior to homozygosity (Springer and Stupar 2007). In our study, the positive association coupled with positive direct effects among various biochemical traits suggested that the prevalence of pleiotropic gene effects or linkage and selection based on these traits will lead to simultaneous improvement in these biochemical traits (Iqbal *et al.* 2003). However, negative association and negative direct effect among different traits can be broken by random mating leading to rigorous selection breeding (Miller and Rawlings 1967).

The biochemical basis of negative associations between Cry1Ac toxin with TSP under control and drought conditions and negative association coupled with negative direct effect of enzymatic antioxidant (SOD, POD and CAT) on Cry1Ac protein only under control conditions suggested that all available nitrogen might be routed for enzymes and protein synthesis associated with growth and survival therefore cannot be allocated to toxin (Cry1Ac) synthesis (Coviella *et al.* 2002). However, the positive association and positive direct effects of SOD and CAT over Cry1Ac toxin under drought conditions signify the capability of eliminating the free radicals of reactive oxygen by the *Bt* cotton. In our study, negative association and negative direct effect of TPC with TSP and *Bt* toxin could be supported by the 'Protein Competition Model of phenolic allocation'. According to this model, the metabolic pathways of plants allocated to either phenolics or soluble proteins compete for a common precursor phenylalanine, which acts as limiting factor (Jones and Hartley 1999). In this study, our finding regarding negative association of chlorophyll contents with soluble protein under control and drought conditions might be attributed to the concept that plants are able to shift allocation between carbon-based and nitrogen-based defensive compounds depending upon the availability of carbon and nitrogen nutrients (Coviella *et al.* 2002). Similarly, positive direct effects and positive association of chlorophyll *a*, chlorophyll *b*, total chlorophyll and carotenoids with Cry1Ac toxin under control illustrated that amount of these biochemical traits corresponds to the amount of Cry1Ac protein (Hosagoudar *et al.* 2008; Poongothai *et al.* 2010).

Conclusions

The mode of inheritance for various biochemical traits suggest that there is room for the improvement of these traits under water stressed and non-stressed conditions. The biplot and path coefficient analyses proved to be effective biometrical tools in revealing that biochemical traits behaved differentially with Cry1Ac toxin under control and drought conditions. The differential behaviour of these biochemicals with Cry1Ac suggested that these traits can serve as biochemical markers while breeding *Bt* cotton. Further, critical understanding about inheritance and the association between different biochemical traits

are likely to pave the way for breeding cotton genotypes having a desired level of Cry1Ac toxin with ample tendency to withstand drought conditions.

Sources of Funding

This study was supported by the research grant for Department of Plant Breeding and Genetics, Bahauddin Zakariya University, Multan.

Contributions by the Authors

M.A.A., W.M. and A.Y. conceived and design the study. M.A.A. and W.M. conducted the experiments. All authors contributed in data analysis and manuscript preparation.

Acknowledgements

We are thankful to Dr Asif Ali Khan, Department of Plant Breeding and Genetics, University of Agriculture, Faisalabad, for his valued input and technical support. We are also thankful to Dr Saghir Ahmad Cotton Research Station, Multan, for providing seed of some cotton genotypes.

Literature Cited

Abd AAA, Mohamed AAA, Gaballah MM. 2009. Genetic studies of some physiological and shoot characters in relation to drought tolerance in rice. *Journal of Agricultural Research* 4:964–990.

Ahuja SL, Dhayal LS. 2007. Combining ability estimates for yield and fiber quality traits in 4 × 13 line × tester crosses of *Gossypium hirsutum*. *Euphytica* **153**:87–98.

Ainsworth EA, Gillespie KM. 2007. Estimation of total phenolic content and other oxidation substrates in plant tissues using Follin-Ciocalteu reagent. *Nature Protocols* 2:875–877.

Apel K, Hirt H. 2004. Reactive oxygen species: metabolism, oxidative stress, and signal transduction. *Annual Review of Plant Biology* **55**:373–399.

Ashokkumar K, Ravikesavan R, Prince SJ. 2010. Combining ability estimates for yield and fiber quality traits in line x tester crosses of upland cotton (*Gossypium hirsutum*). *International Journal of Biology* **2**:179–189.

Barwale RB, Gadwa VR, Zehr U, Zehr B. 2004. Prospects for *Bt* cotton technology in India. *Agbioforum* 7:23–26.

Bradford MM. 1976. A rapid and sensitive method for the quantitation of microgram quantities of protein utilizing the principle of protein-dye binding. *Annals of Biochemistry* **72**:248–254.

Bruns HA, Abel CA. 2003. Nitrogen fertility effects on *Bt*-endotoxin and nitrogen concentrations of maize during early growth. *Agronomy Journal* **95**:207–211.

Cakmakci S, Unay A, Acikgoz E. 1998. An investigation on determination of characters regarding to seed and straw yield using different methods in common vetch. *Turkish Journal of Agriculture and Forestry* **22**:161–166.

Cattivelli L, Rizza F, Badeck F-W, Mazzucotelli E, Mastrangelo AM, Francia E, Mare C, Tondelli A, Stanca AM. 2008. Drought tolerance improvement in crop plants: an integrated view from breeding to genomics. *Field Crops Research* **105**:1-14.

Chance B, Maehly AC. 1955. Assay of catalase and peroxidases. *Methods in Enzymology* **2**:764-775.

Coviella CE, Stipanovic RD, Trumble JT. 2002. Plant allocation to defensive compounds: interactions between elevated CO_2 and nitrogen in transgenic cotton plants. *Journal of Experimental Botany* **53**:323-331.

Dewey DR, Lu KH. 1959. A correlation and path coefficient analysis of components of crested wheat grass and seed production. *Agronomy Journal* **51**:515-518.

Dong HZ, Li WJ, Tang W, Li ZH, Zhang DM. 2005. Increased yield and revenue with a seedling transplanting system for hybrid seed production in *Bt* cotton. *Journal of Agronomy and Crop Science* **191**:116-124.

Fang Y, Xiong L. 2015. General mechanisms of drought response and their application in drought resistance improvement in plants. *Cellular and Molecular Life Sciences* **72**:673-689.

Flexas J, Bota J, Loreto F, Cornic G, Sharkey TD. 2004. Diffusive and metabolic limitations to photosynthesis under drought and salinity in C_3 plants. *Plant Biology* **6**:269-279.

Ganapathy S, Nadarajan N, Saravanan S, Shanmugathan M. 2005. Heterosis for seed cotton yield and fiber characters in cotton (*Gossypium hirsutum* L). *Crop Research Hisar* **30**:451-454.

Gianessi LP, Carpenter JE. 1999. *Agricultural biotechnology insect control benefits*. Washington, DC: National Center for Food and Agricultural Policy.

Giannopolitis CN, Ries SK. 1977. Superoxide dismutases II. Purification and quantitative relationship with water-soluble protein in seedlings. *Plant Physiology* **59**:315-318.

Gill SS, Tuteja N. 2010. Reactive oxygen species and antioxidant machinery in abiotic stress tolerance in crop plants. *Plant Physiology and Biochemistry* **48**:909-930.

Greish SMA, Swidan SA, Fouly AHME, Guirgis AA, Raheem AAAE. 2005. Evaluation of performance and gene action of quantitative characters in some local and exotic tomato genotypes. I. Morphological and physiological traits. *Zagazig Journal of Agriculture* **32**:93-107.

Hammer O, Harper DAT, Ryan PD. 2001. PAST: paleontological statistics software package for education and data analysis. *Palaeontologia Electronica* **4**:4-9.

Hosagoudar GN, Chattannavar SN, Kulkarni S. 2008. Biochemical studies in *Bt* and non-*Bt* cotton genotypes against alternaria blight disease (*Alternaria macrospora* Zimm.). *Karnataka Journal of Agricultural Science* **21**:70-73.

Immanuel SC, Nagarajan P, Vijendra DLD. 2006. Heterotic expression and combining ability analysis for qualitative and quantitative traits in inbreds of maize (*Zea mays* L.). *Crop Research* **32**:77-85.

Iqbal M, Chang MA, Iqbal MZ, Hassan M, Nasir A, Islam N. 2003. Correlation and path co-efficient analysis of earliness and agronomic characters of upland cotton in Multan. *Pakistan Journal of Agronomy* **2**:160-168.

Jagtap DR. 1986. Combining ability in upland cotton. *Indian Journal of Agricultural Sciences* **12**:833-840.

James C. 2002. *Global review of commercialized transgenic crops: 2001 Feature: Btcotton. ISAAA Briefs No. 26*. Ithaca, NY: ISAAA.

Jenkins JN, Mccarty JC Jr, Wu J, Saha S, Gutierrez O, Hayes R, Stelly DM. 2007. Genetic effects of thirteen *Gossypium barbadense* L. chromosome substitution lines in top crosses with upland cotton cultivars: II. Fiber quality traits. *Crop Science* **47**: 561-570.

Jones CG, Hartley SE. 1999. A protein competition model of phenolic allocation. *Oikos* **86**:27-44.

Luo Z, Dong H, Li W, Ming Z, Zhu Y. 2008. Individual and combined effects of salinity and waterlogging on *Cry1Ac* expression and insecticidal efficacy of *Bt* cotton. *Crop Protection* **27**:1485-1490.

Mehndiratta PD, Phul PS. 1983. Genetic and cytoplasmic effects on chlorophyll content in pearl millet. *Theoretical and Applied Genetics* **65**:339-342.

Metzner H, Rau H, Senger H. 1965. Untersuchungen zur Synchronisierbarkeit einzelner Pigmentmangel-Mutanten von *Chlorella*. *Planta* **65**:186-194.

Miller PA, Rawlings JO. 1967. Breakup of initial linkage blocks through intermating in a cotton breeding population. *Crop Science* **7**: 199-204.

Mittler R, Vanderauwera S, Gollery M, Van Breusegem F. 2004. Reactive oxygen gene network of plants. *Trends in Plant Science* **9**: 490-498.

Nachabe MH. 1998. Refining the definition of field capacity in the literature. *Journal of Irrigation and Drainage Engineering* **124**: 230-232.

Neelima S, Chenga RV. 2008. Genetic parameters of yield and fiber quality traits in American cotton (*Gossypium hirsutum* L.). *Indian Journal of Agricultural Research* **42**:67-70.

Parida AK, Dagaonkar VS, Phalak MS, Umalkar GV, Aurangabadkar LP. 2007. Alterations in photosynthetic pigments, protein and osmotic components in cotton genotypes subjected to short-term drought stress followed by recovery. *Plant Biotechnology Report* **1**:37-48.

Poongothai S, Ilavarasan R, Karrunakaran CM. 2010. *Cry1Ac* levels and biochemical variations in *Bt* cotton as influenced by tissue maturity and senescence. *Journal of Plant Breeding and Crop Science* **2**:96-103.

Rad MRN, Kadir MA, Yusop MR. 2012. Genetic behaviour for plant capacity to produce chlorophyll in wheat (*Triticum aestivum*) under drought stress. *Australian Journal of Crop Science* **6**:415-420.

Rasheed A, Malik W, Khan AA, Murtaza N, Qayyum A, Noor E. 2009. Genetic evaluation of fiber yield and yield components in fifteen cotton (*Gossypium hirsutum*) genotypes. *International Journal of Agriculture & Biology* **11**:581-585.

Ribaut J-M, Hu X, Hoisington D, Gonzales-De-Leon D. 1997. Use of STSs and SSRs as rapid and reliable preselection tools in a marker assisted selection backcross scheme. *Plant Molecular Biology Reporter* **15**:154-162.

Rosielle AA, Hamblin J. 1981. Theoretical aspects of selection for yield in stress and non-stress environment. *Crop Science* **21**: 943-946.

Saleem MY, Asghar M, Haq MA, Rafique T, Kamran A, Khan AA. 2009. Genetic analysis to identify suitable parents for hybrid seed production in tomato (*Lycopersicon esculentum* Mill.). *Pakistan Journal of Botany* **41**:1107-1116.

Shah AR, Khan TM, Sadaqat HA, Chatha AA. 2011. Alterations in leaf pigments in cotton (*Gossypium hirsutum*) genotypes subjected to drought stress conditions. *International Journal of Agriculture & Biology* **13**:902-908.

Shan G, Embrey SK, Schaffer BW. 2007. A highly specific enzyme linked immunosorbent assay for the detection of *Cry1Ac* insecticidal crystal protein in transgenic Wide Strike cotton. *Journal of Agricultural and Food Chemistry* **55**:5974-5979.

Singh RK, Chaudhary BD. 1999. *Biometrical methods in quantitative genetics analysis*, 1st edn. New Delhi: Kalyani Publisher.

Song M, Fan S, Pang C, Wei H, Yu S. 2014. Genetic analysis of the anti-oxidant enzymes, methane dicarboxylic aldehyde (MDA) and chlorophyll content in leaves of the short season cotton (*Gossypium hirsutum* L.). *Euphytica* **198**:153–162.

Springer NM, Stupar RM. 2007. Allelic variation and heterosis in maize: how do two halves make more than a whole? *Genome Research* **17**:264–275.

Tabashnik BE, Dennehy TG, Sims MA, Larkin K, Head GP, Moar WJ, Carriere Y. 2002. Control of resistant pink bollworm (*Pectinophora gossypiella*) by transgenic cotton that produces *Bacillus thuringiensis* toxin Cry2Ab. *Applied and Environmental Microbiology* **68**:3790–3794.

Tabashnik BE, Gassmann AJ, Crowder DW, Carriére Y. 2008. Insect resistance to *Bt* crops: evidence versus theory. *Nature Biotechnology* **26**:199–202.

Traore SB, Carlson RE, Pilcher CD, Rice ME. 2000. *Bt* and non *Bt* maize growth and development as affected by temperature and drought stress. *Agronomy Journal* **92**:1027–1035.

Trewavas A. 2002. Plant cell signal transduction: the emerging phenotype. *The Plant Cell* **14**(Suppl):S3–S4.

Wahid A, Rasul E. 2005. Photosynthesis in leaf, stem, flower and fruit. In: Pessarakli M, ed. *Handbook of photosynthesis*. 2nd edn. Florida: CRC Press, 479–497.

Wang LM, Wang JB, Sheng FF, Zhang XK, Liu RZ. 2001. Influences of water logging and drought on different transgenic *Bt* cotton cultivars. *Cotton Science* **2**:87–90.

Wu J, Jenkins JN, Mccarty JC, Wu D. 2006. Variance component estimation using the additive, dominance, and additive and additive model when genotypes vary across environments. *Crop Science* **46**:174–179.

Yan W, Kang MS. 2003. *GGE biplot analysis: a graphical tool for breeders, genetics, and agronomists*. Florida: CRC Press.

Yan W, Rajcan I. 2002. Biplot analysis of test sites and trait relations of soybean in Ontario. *Crop Science* **42**:11–20.

Yan W, Kang MS, Ma B, Woods S, Cornelius PL. 2007. GGE biplot vs. AMMI analysis of genotype-by-environment data. *Crop Science* **47**:643–655.

15

Insight into infrageneric circumscription through complete chloroplast genome sequences of two *Trillium* species

Sang-Chul Kim, Jung Sung Kim and Joo-Hwan Kim*

Department of Life Science, Gachon University, Seongnamdaero 1342, Seongnam-si, Gyeonggi-do 461-701, Korea

Associate Editor: Chelsea D. Specht

Abstract. Genomic events including gene loss, duplication, pseudogenization and rearrangement in plant genomes are valuable sources for exploring and understanding the process of evolution in angiosperms. The family Melanthiaceae is distributed in temperate regions of the Northern Hemisphere and divided into five tribes (Heloniadeae, Chionographideae, Xerophylleae, Melanthieae and Parideae) based on the molecular phylogenetic analyses. At present, complete chloroplast genomes of the Melanthiaceae have been reported from three species. In the previous genomic study of Liliales, a *trnI*-CAU gene duplication event was reported from *Paris verticillata*, a member of Parideae. To clarify the significant genomic events of the tribe Parideae, we analysed the complete chloroplast genome sequences of two *Trillium* species representing two subgenera: *Trillium* and *Phyllantherum*. In *Trillium tschonoskii* (subgenus *Trillium*), the circular double-stranded cpDNA sequence of 156 852 bp consists of two inverted repeat (IR) regions of 26 501 bp each, a large single-copy (LSC) region of 83 981 bp and a small single-copy (SSC) region of 19 869 bp. The chloroplast genome sequence of *T. maculatum* (subgenus *Phyllantherum*) is 157 359 bp in length, consisting of two IRs (25 535 bp), one SSC (19 949 bp) and one LSC (86 340 bp), and is longer than that of *T. tschonoskii*. The results showed that the cpDNAs of Parideae are highly conserved across genome structure, gene order and contents. However, the chloroplast genome of *T. maculatum* contained a 3.4-kb inverted sequence between *ndhC* and *rbcL* in the LSC region, and it was a unique feature for subgenera *Phyllantherum*. In addition, we found three different types of gene duplication in the intergenic spacer between *rpl23* and *ycf2* containing *trnI*-CAU, which were in agreement with the circumscription of subgenera and sections in Parideae excluding *T. govanianum*. These genomic features provide informative molecular markers for identifying the infrageneric taxa of *Trillium* and improve our understanding of the evolution patterns of Parideae in Melanthiaceae.

Keywords: Chloroplast genome; comparative genomics; gene duplication; single inversion; *Trillium maculatum*; *Trillium tschonoskii*, *trnI*-CAU.

Introduction

The chloroplast that characterizes all green plants (Viridiplantae) originated from an endosymbiotic event between independent living cyanobacteria and a non-photosynthetic host (Dyall *et al.* 2004). Chloroplast genomes of flowering plants are typically circular double-stranded DNA molecules, and usually contain two inverted repeat (IR) regions (IRA and IRB) separated by

* Corresponding author's e-mail address: kimjh2009@gachon.ac.kr

a large single-copy and a small single-copy (LSC and SSC, respectively) regions (Ravi *et al.* 2008). The plastid genome is mostly stable in structure, gene content and gene order across land plant lineages (Jansen *et al.* 2005). Due to this stability, it demonstrated great utility for developing phylogenetic hypotheses across the plant tree of life (Jansen *et al.* 2007; Zhang *et al.* 2011; Li *et al.* 2013). Within seed plants, plastid genomes usually contain 101–118 unique genes with the majority of those 66–82 coding for proteins involved in photosynthesis and gene expression, 29–32 of these genes code for transfer RNAs and 4 code for the ribosomal RNA genes (Jansen and Ruhlman 2012). The advance of next-generation sequencing has facilitated rapid growth of complete chloroplast genomes due to time-saving and low-cost advantages (Shendure and Ji 2008). To date, ~500 complete chloroplast DNA genome sequence data have been released in GenBank's Organelle Genome Resources (http://www.ncbi.nlm.nih.gov/genome).

Melanthiaceae, a member of Liliales, comprises 17 genera and ~178 species of perennial herbs that are mostly distributed in the temperate regions of the Northern Hemisphere (Zomlefer *et al.* 2001). Species of this family are characterized by their extrorse anthers and carpels bearing three distinct styles (Rudall *et al.* 2000). The family has been divided into five tribes: Heloniadeae, Chionographideae, Xerophylleae, Melanthieae and Parideae (The Angiosperm Phylogeny Group 2009). Prior to any molecular systematic analyses, Melanthiaceae were divided into several taxonomically independent families by Takhtajan (1997) due to their unique autapomorphies. Trilliaceae, which is now recognized as tribe Parideae (Trillieae), are unique in having solitary flowers, berries, membranous nectary and large chromosomes with five chromosomes as the base number. The phylogeny of species within the Trilliaceae (now Parideae) was highly debated by many researchers using molecular and morphological data (Kato *et al.* 1995; Osaloo *et al.* 1999; Osaloo and Kawano 1999; Farmer and Schilling 2002; Farmer 2006). Tribe Parideae includes three genera: *Paris*, *Trillium* and *Pseudotrillium*. *Paris* has 4–15 leaves in a whorl, flowers 4-merous or more and inner perianth segments that are much narrower than outer ones, while *Trillium* has only 3 leaves in a whorl, flowers 3-merous and inner perianth segments that are a little narrower than the outer ones. *Pseudotrillium* has thick, tough, heart-shaped leaves, spotted petals and flower stalks that extend until the ripe fruit touches the ground. *Trillium* has been divided into two subgenera differing in the presence of pedicel: subgenus *Trillium* (with pedicels) and *Phyllantherum* (without pedicels) (Freeman 1969, 1975). The monophyly of subgenus *Phyllantherum* was strongly supported in many previous studies (Osaloo *et al.* 1999;

Osaloo and Kawano 1999; Farmer and Schilling 2002; Farmer 2006). On the other hand, subgenus *Trillium* is rendered a paraphyletic group by the inclusion of *Phyllantherum*.

Currently, complete chloroplast genomes of the Melanthiaceae have been reported from *Paris verticillata* (KJ433485; Do *et al.* 2014), *Veratrum patulum* (KF437397; Do *et al.* 2013) and *Chionographis japonica* (KF951065; Bodin *et al.* 2013), which represent three tribes of Parideae, Melanthieae and Chionographideae, respectively. In this study, we analysed complete chloroplast genome sequences of subgenera *Trillium* and *Phyllanthrum* of *Trillium* to better understand the evolution of the chloroplast genomes in tribe Parideae and across the Melanthiaceae. We analysed the sequence variation between two subgenera and proposed novel molecular markers for phylogenetic studies by comparing the two newly generated genome sequences. In addition, we characterized the *trnI*-CAU duplication event in Parideae, detected in *P. verticillata* chloroplast genome (KJ433485), to determine the origin of the repeating unit. Consequently, these results provide additional knowledge about the patterns of the chloroplast genome evolution within tribe Parideae.

Methods

DNA extraction, sequencing and annotation

We collected *Trillium tschonoskii* from Ulleung Island, South Korea. The voucher specimen and plant materials were deposited at the herbarium (GCU) and Medicinal Plant Resources Bank (MPRB) of Gachon University. *Trillium maculatum* was obtained from the Abraham Baldwin Agricultural College, USA (voucher No. Susan Farmer 19990006). We used silica gel-dried leaves from each species to extract total genomic DNA using the DNeasy Plant Mini Kit (Qiagen, Seoul, South Korea).

The Hiseq 2000 system was employed to sequence chloroplast genomes of *T. tschonoskii* and *T. maculatum*. Raw data were assembled using Geneious ver. 7. 1 (Biomatters Ltd, New Zealand) with default settings. After trimming the sequences, we mapped pair-end reads to the reference sequence of *P. verticillata* (KJ433485). Aligned contigs were ordered according to the reference genome and the gaps were filled via direct sequencing of polymerase chain reaction (PCR) products with newly designed primers. In addition, the ambiguous sequences including low assembly coverage regions and the borders of the four junctions between LSC, SSC and IR regions were confirmed using the Sanger method.

Complete chloroplast genomes of both species were annotated by Geneious ver. 7. 1 (Biomatters Ltd), with manual corrections for putative start and stop codons. The exon positions of protein-coding genes and intron were determined using released Liliales chloroplast genome sequences as references. All tRNA sequences were confirmed utilizing the web-based online tool of tRNAScan-SE (Schattner *et al.* 2005) with default settings to corroborate tRNA boundaries identified by Geneious. The genome maps were generated using OGDraw (OrganellarGenomeDRAW; Lohse *et al.* 2007) followed by manual modification.

Comparison of the chloroplast genome sequences of two subgenera

The simple sequence repeats (SSRs) were analysed using Phobos Version 3.3.12 (Kraemer *et al.* 2009), with thresholds of eight repeat units for mononucleotide SSRs, four repeat units for dinucleotide, trinucleotide SSRs and three repeat units for tetranucleotide, pentanucleotide and hexanucleotide SSRs. All the detected repeats were manually verified, and the redundant results were removed. We aligned the plastid genome sequences of two *Trillium* using MAFFT (Katoh *et al.* 2002). The identified insertion/deletion mutations (indels) from the results were confirmed by reassembling the whole reads generated by HiSeq 2000. The single nucleotide polymorphisms (SNPs) were analysed using Geneious 7.1 (Kearse *et al.* 2012), and each indel and SNP were separated based on the position excluding one of IR regions. Since we are comparing only two genomes, we quantified the sequence divergence as the ratio of aligned nucleotide sites within specifically different regions (*p*-distance). Sanger sequences and assembled genomes were calculated using mean *p*-distance in MEGA 6.0 (Tamura *et al.* 2013).

Twenty-nine species, representing the two subgenera of *Trillium* in Parideae, were selected for comparative sequencing of inversion. The PCR amplification primers were designed based on the sequence comparisons among three chloroplast genome sequences of two *Trillium* species (in this study), and *P. verticillata* (KJ433485). Presence and absence PCR amplifications were carried out using various combinations of the three primers (I1F: 5′-CCC TAG GTT TTT TTC TTC AAG-3′, I1R: 5′-TTA TGT AGC TTA TCC TTT AGA CC-3′ and I2R: 5′-AGA AGG TCT ACG GTT CGA G-3′).

trnI-CAU duplication pattern in the tribe Parideae

To clarify the *trnI*-CAU duplication pattern in the tribe Parideae, we designed two primers (Primer 1: 5′-GAA GAG TTC GAC CCA ATG CT-3′, Primer 2: 5′-TTA TGA AAC TCT TTG ACC CC-3′) for amplifying the intergenic spacer (IGS) region of *rpl23-ycf2* based on the identical sequence among the three species (*P. verticillata, T. maculatum* and *T. tschonoskii*). The PCR condition for IGS region of *rpl23-ycf2* was at initial denaturation at 94 °C for 5 min, followed by 30 cycles of denaturation at 94 °C for 1 min, annealing at 50 °C for 1 min and extension at 72 °C for 2 min, with a final extension at 72 °C for 5 min. We obtained variously sized PCR products ranging from 500 to 1200 bp, and compared the sequences of this region from 33 species covering the infrageneric classification of the tribe. Sequence editing and assembly were performed using Sequencher (ver. 5.1). The sequence alignment was initially performed using MAFFT (Katoh *et al.* 2002) and was adjusted manually.

Results

Comparison of the complete chloroplast genomes of subgenera *Trillium* and *Phyllantherum*

We sequenced the complete chloroplast genome sequence of two *Trillium* species, *T. tschonoskii* (subgenera *Trillium*; GenBank accession number KR780076) and *T. maculatum* (subgenera *Phyllantherum*; GenBank accession number KR780075) (Fig. 1). In total, 4 292 702 (*T. tschonoskii*) and 18 348 134 (*T. maculatum*) paired-end reads were generated. Out of those, 60 805 and 246 240 reads were identified as the chloroplast genome sequences for *T. tschonoskii* and *T. maculatum*, respectively. The chloroplast genome of *T. tschonoskii* was composed of 156 852 bp in length (AT content 62.5 %), and it comprised a LSC region (83 981 bp), a SSC region (19 869 bp) and two IR regions (26 501 bp), while *T. maculatum* was 157 359 bp in length (AT content 62.5 %, 86 340 bp of LSC, 19 949 bp of SSC and 25 535 bp of IRs).

The gene content and order were slightly different between both species because of the *rpl22* position in the IR-LSC boundary and *trnI*-CAU duplication in IR. While the *rpl22* gene remained in the LSC region of the *T. maculatum* plastid genome, this gene was present in the IR region of *T. tschonoskii* plastid genome (Fig. 2). In total, 116 genes of *T. maculatum* were identified and consisted of 78 coding genes, 4 rRNA genes, 31 tRNA genes and 3 pseudogenes, while those of *T. tschonoskii* were 115 genes without tRNA gene duplication [**see Supporting Information—Table S1**]. In addition, *T. tschonoskii* has 7 coding genes, 4 rRNA genes, 9 tRNA genes, 2 pseudogenes, whereas *T. maculatum* has 8 coding genes, 4 rRNA genes, 8 tRNA genes, 2 pseudogenes, duplicated in the IR region, making a total of 138 genes and 137 genes presented in the *T. tschonoskii* and *T. maculatum* chloroplast genome, respectively. Among these genes, 22 intron-containing genes were found including

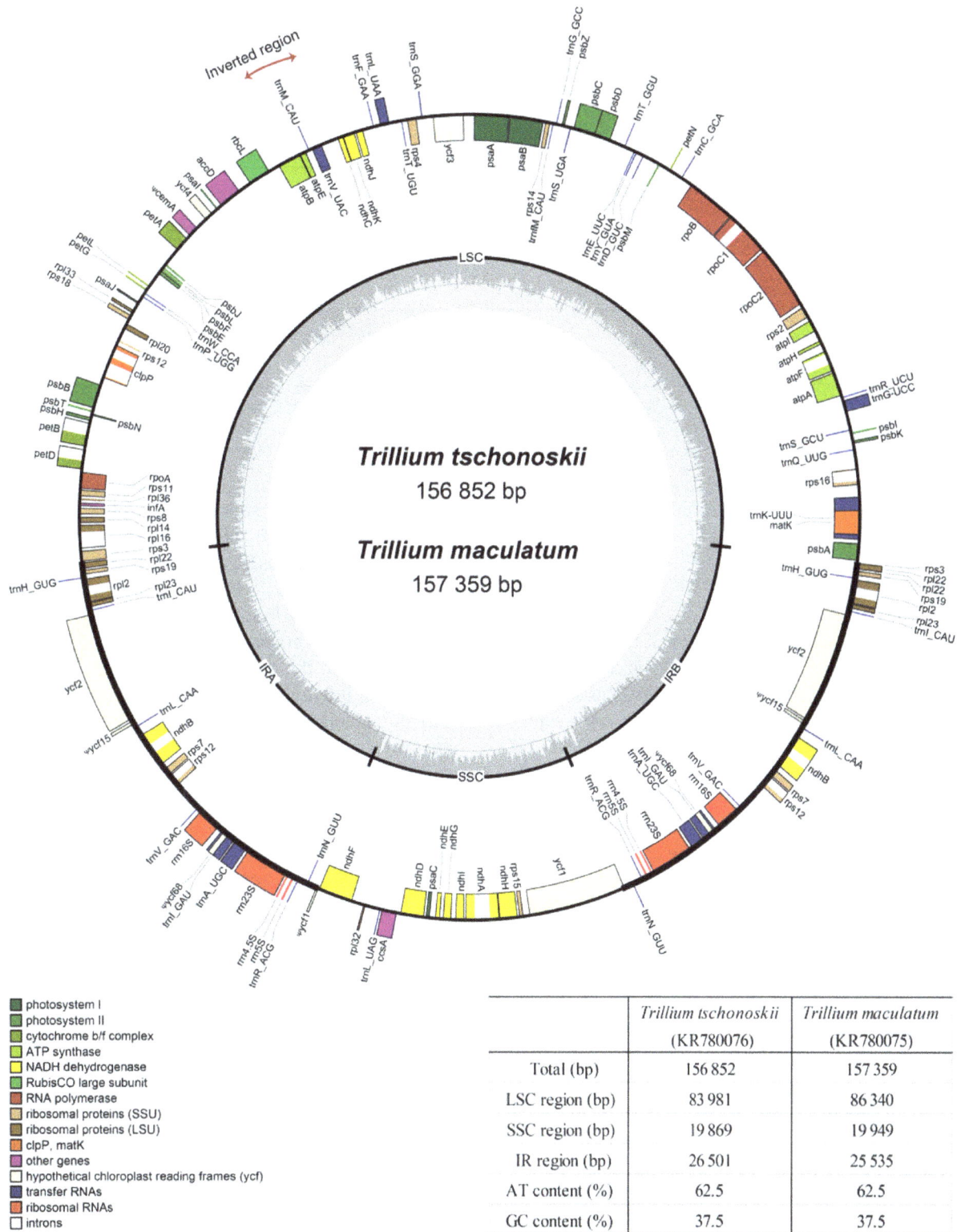

Figure 1. Gene maps and summary of the *T. tschonoskii* Maxim. and *T. maculatum* Raf. chloroplast genomes. IR, inverted repeat; LCS, large single-copy region; SSC, small single-copy region.

15 protein-coding genes and 7 tRNA genes. Among them, *ycf3* and *clpP* gene contained two introns. The *trnK*-UUU has the largest intron (*T. tschonoskii*: 2614 bp,

T. maculatum: 2640 bp) including the *matK* gene. *Ycf15* and *ycf68* in the IR region were pseudogenized because of the presence of several internal stop codons.

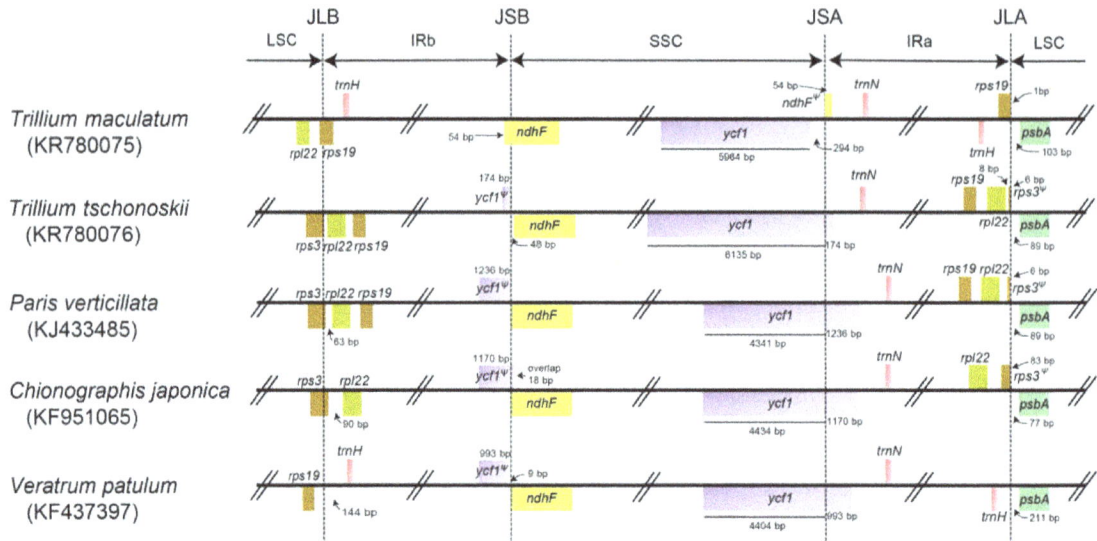

Figure 2. Comparison of the IR boundaries among five species within Melanthiaceae.

Furthermore, the *cemA* gene located in the LSC of both genomes was also pseudogenized.

Characterization of single inversion in subgenus *Phyllantherum*

Based on comparison of *T. maculatum*, *T. tschonoskii* and *P. verticillata*, a single inversion of 3.4 kb is characterized in the chloroplast genome of *T. maculatum*. This inversion is located between the *ndhC* and *rbcL* genes. We designed three different primers including I1F (5′-CCC TAG GTT TTT TTC TTC AAG-3′), I1R (5′-TTA TGT AGC TTA TCC TTT AGA CC-3′) and I2R (5′-AGA AGG TCT ACG GTT CGA G-3′) to confirm and clarify the distribution of this inversion throughout the genus *Trillium*. Specifically, the primer pairs of I1F and I1R worked only in the normal type, while I2R and I1R primer pairs were utilized for the recognized inversion type among examined species. The results showed that the inversion occurred in all examined species of the subgenera *Phyllantherum* (Fig. 3A and B).

Indels, SNPs and SSR between two subgenera of *Trillium*

A total of 402 indels were detected between *T. maculatum* and *T. tschonoskii*, and most indels were located in the IGS regions (78.2 %). 66.2 % of the total number of indels were found in the LSC, while 22.1 and 11.7 % were present in the SSC and IR regions, respectively [Table 1, **see Supporting Information—Table S2**]. The average length of indels was 74.8 bp, and the largest indel was located in *ycf1* and *ycf2*. The frequency of 1 bp indels was 10.6 %, while 79.3 % of all indels were over 20 bp in length. In rRNA sequences, one indel of 3 bp and four indels of 5 bp were found in 16S rRNA and 23S rRNA. In addition, indel events were identified in 20

coding genes of both species (*accD*, *atpB*, *ccsA*, *cemA*, *clpP*, *infA*, *matK*, *ndhF*, *rpl2*, *rpl20*, *rpl22*, *rpl32*, *rpoC1*, *rpoC2*, *rps11*, *rps15*, *rps18*, *rps19*, *ycf1* and *ycf2*).

A total of 2861 SNPs were detected between *T. maculatum* and *T. tschonoskii* (Table 2), and 1620 SNPs were transversions. In total, 1707 (59.7 %) SNPs were located in the coding regions, and 1154 (40.3 %) were within IGS regions or within introns.

In our result of SNPs, *p*-distance values in coding regions range from 0.002 to 0.23 and the average value was 0.02. On the other hand, the average *p*-distance value in non-coding regions was 0.034. Figure 4 shows the average *p*-distance for five classes of genomic regions: protein-coding genes, tRNAs, rRNAs, IGSs and introns. The IGS divergence is almost double that of the next highest class (genes). Introns hold the lowest sequence divergence, at an average of 0.011%.

We detected SSRs longer than 8 bp in *T. maculatum*, *T. tschonoskii* and *P. verticillata* chloroplast genomes by the method of Qian *et al.* (2013). According to Qian *et al.*, the threshold was set because 8 bp or longer SSRs are prone to slipstrand mispairing, which is thought to be the primary mutational mechanism causing their high level of polymorphism. In this analysis, the total number of SSRs was 204 in *P. verticillata*, 205 in *T. maculatulatum* and 213 in *T. tschonoskii* (Table 3). The most abundant type of SSR in Parideae was a mononucleotide, with 138 in *P. verticillata*, 121 in *T. maculatulatum* and 133 in *T. tschonoskii*. In addition to mononucleotide SSRs, there are 52 dinucleotide SSRs in *P. verticillata*, 57 in *T. maculatulatum* and 53 in *T. tschonoskii*. Trinucleotide SSRs were less frequent with 6, 14 and 7 in *P. verticillata*, *T. maculatum* and *T. tschonoskii*, respectively. The hexanucleotide SSRs were found only in *Trillium* species.

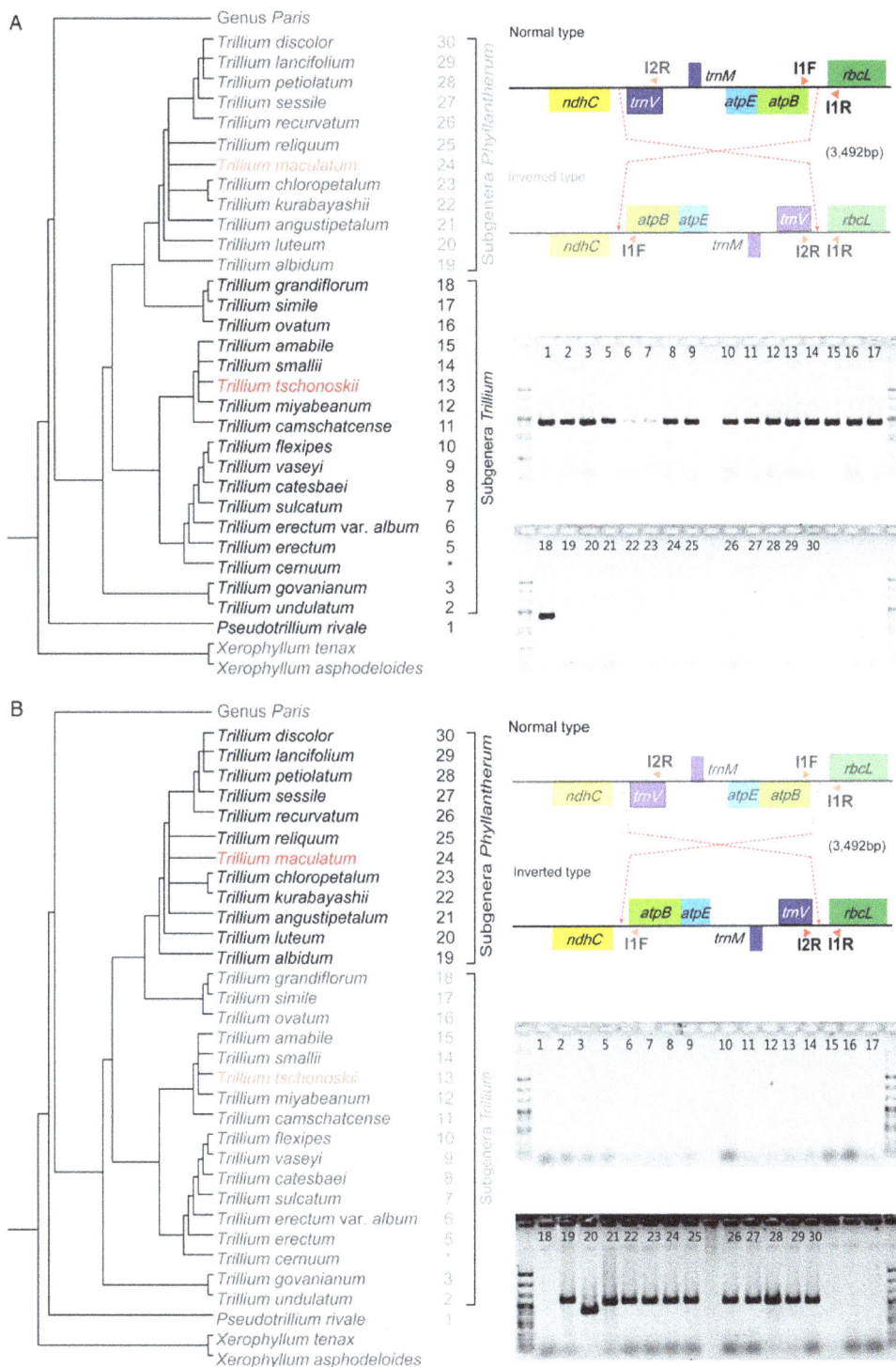

Figure 3. Confirmation of inversion (3492 bp) between *ndhC* and *rbcL* in the genus *Trillium*. (A) Design of primer to amplify junction regions between *atpB* and *rbcL* regions. The positions of *atpB* and *rbcL* genes in LSC regions are drawn based on the sequence assembly results of *T. tschonoskii*, *T. maculatum* in this study (red text). *The data downloaded from the NCBI. The forward primer I1F contains the sequence in *atpB* region. The sequence of the reverse primer (I1R) is located in the *rbcL* gene. Polymerase chain reaction amplification of IGS between *atpB* and *rbcL*. Relationships of Parideae lineages followed the phylogenetic trees of S. C. Kim, J. S. Kim, W. C. Mark, F. F. Michael and J. H. Kim (unpublished data). (B) Primers were designed to amplify junction regions between *trnV*-UAC and *rbcL* regions. The positions of *trnV*-UAC and *rbcL* genes in LSC regions are drawn based on the sequence assembly results of *T. tschonoskii*, *T. maculatum* in this study (red text). *The data downloaded from the NCBI. The forward primer I2R contains the sequence in *trnV*-UAC region. The sequence of the reverse primer I1R is located in the *rbcL* gene. Polymerase chain reaction amplification of IGS between *trnV*-UAC and *rbcL*. Relationships of Parideae lineages followed the phylogenetic trees of S. C. Kim, J. S. Kim, W. C. Mark, F. F. Michael and J. H. Kim (unpublished data).

Table 1. The number and total length of insertion–deletion mutations between the chloroplast genomes of *T. tschonoskii* and *T. maculatum* in Parideae.

Region	Number of indels	Total length of indels
IGS	262	5139
Intron	51	528
Coding gene	86	3005

The majority of mononucleotide repeats were A-T rich (Table 3).

Type of *trnI*-CAU of Parideae

We compared the sequences of the IGS region between *rpl23* and *ycf2* using 33 species including *Xerophyllum* to understand the evolutionary implication of *trnI*-CAU duplication, which was reported from the *Paris* chloroplast genome (Do *et al.* 2014). Based on the results, we found that this region is of highly variable length among the species, and we distinguished three major types based on the number of copies of *trnI*-CAU (Fig. 5). Type A was composed of a single *trnI*-CAU and was found in *Xerophyllum*, *Pseudotrillium rivale* and *T. undulatum*. It was also identified in several *Trillium* and *Paris* species, but with variable lengths: in subgenus *Trillium* species, the sequences ranged from 207 to 445 bp, in which there are two tandem repeats of 'CAG GTA TTA TCA TAC TGA AA' (20 bp) and 'CAT ATT ATC ATA CTG AAA' (18 bp). Similarly, in subgenus *Daiswa* of *Paris*, there were 24 bp random tandem repeats of TAT AAC TTA ACA GGA ATC ATC GTA. Type B contained two copies of *trnI*-CAU. This type is found in subgenus *Phyllantherum* of *Trillium* and section *Kinugasa* of subgenus *Paris*. The lengths of tandem repeat sequences were 180 bp (subgenera *Phyllantherum*) and 155 bp (section *Kinugasa* of subgenus *Paris*), which included 74 bp of *trnI*-CAU. Remarkably, section *Kinugasa* (*Paris japonica*) has the longest length of IGS between *rpl23* and *ycf2* among the tribe Parideae. Type C, possessing three copies of *trnI*-CAU genes in the sequenced region, was detected in *T. govanianum* and section *Paris* of subgenus *Paris*. They included three fully repeated units including *trnI*-CAU, and the lengths were 155 and 139 bp, respectively.

Discussion

Comparison of complete plastid genomes of subgenera *Trillium* and *Phyllantherum*

The plastid genome structure of the two *Trillium* species, *T. maculatum* and *T. tschonoskii*, have a typical form found in most angiosperms (Zhang *et al.* 2011; Kim and Kim 2013; Li *et al.* 2013; Qian *et al.* 2013). The

T. tschonoskii chloroplast genome was 507 bp shorter than *T. maculatum*, and we confirmed that the length variation among Parideae chloroplast genomes including *Paris verticillata* occurred by gene deletion and duplication as well as its IR expansion.

Although chloroplast genomes are considered highly conserved among land plants, sequence polymorphisms were often observed among closely related species. From the *T. tschonoskii* and *T. maculatum* chloroplast genome sequences, we confirmed that 402 indels and 2861 SNPs were present between the two species.

In addition, we found that SSRs (i.e. microsatellites), composed of 1–6 bp in length per unit, are distributed throughout both genomes. The SSRs have been accepted as one of the major molecular markers for genome variation between species or within populations due to their high polymorphism within the species and have been widely practiced for analysing plant population structure, diversity, differentiation and maternity analysis (Liu *et al.* 2013). Simple sequence repeats have successfully been applied to the study of Poaceae, Brassicaceae and Solanaceae (Provan *et al.* 1997, 1999; Bryan *et al.* 1999; Flannery *et al.* 2006). Simple sequence repeats detected in the present study will provide basic information for the further analysis of genetic diversity in Parideae.

Based on our results, the IR/LSC boundary and the IR/SSC boundary differed between the two subgenera of *Trillium*. Inverted repeat/large single-copy junction was expanded to a part of *rps3* in *T. tschonoskii*, whereas that of *T. maculatum* was found at *rps19*. The *ycf1* was completely located in SSC of *T. tschonoskii*, but a part of the *ycf1* gene was duplicated in IR of *T. maculatum*. Within the Parideae, the IR boundary pattern of *T. tschonoskii* was more similar to *P. verticillata* than *T. maculatum* (Fig. 2).

Inversion events in Melanthiaceae

Inversions caused by the recombination between repeated sequences are considered to be a main mechanism for changes in gene order among plastid genomes (Jansen and Ruhlman 2012). Most of the reported inversions in plastid genomes are in the LSC region (Kim *et al.* 2005). In subtribe *Phaseolinae* of Fabaceae, there is a 78 kb inversion between *trnH/rpl14* and *rps19/rps8* in the chloroplast genome (Tangphatsornruang *et al.* 2010). Additionally, Kim *et al.* (2005) reported that the inversion occurred in the spacer between tRNA^{Gly} and tRNA^{Ser} genes of *Lactuca sativa*. Also, they defined two inversions that characterize Asteraceae. The two inversions were identical across all members of Asteraceae, suggesting that the inversion events are likely to occur simultaneously or within a short period of time following the origin of the family. In Campanulaceae, >50 large

Table 2. Single nucleotide polymorphisms found between the plastid genomes of *T. tschonoskii* and *T. maculatum*. (A) Single nucleotide polymorphisms in coding gene. (B) Single nucleotide polymorphisms in intron. (C) Single nucleotide polymorphisms in IGS regions. Bold values represent *p*-distance >0.08.

(A)

Gene	Aligned length (bp)	No. SNP	p-Distance	Gene	Aligned length (bp)	No. SNP	p-Distance	Gene	Aligned length (bp)	No. SNP	p-Distance
psbD	1062	2	0.002	ycf4	555	4	0.007	rpl2	828	14	0.017
psaB	2205	5	0.002	cemA	695	5	0.007	rrn5S	121	2	0.017
psaA	2253	4	0.002	rrn23S	2814	19	0.007	rpl36	114	2	0.018
atpE	405	1	0.002	ndhD	1503	10	0.007	rrn4.5S	103	2	0.019
psbB	1527	3	0.002	ndhE	306	2	0.007	rrn16S	1494	30	0.02
petB	648	1	0.002	ndhI	543	4	0.007	rpl14	369	9	0.024
psbA	1062	3	0.003	rps15	276	3	0.007	ycf2	7209	146	0.024
ndhK	768	2	0.003	rpoB	3213	25	0.008	rps7	468	11	0.024
ndhC	363	1	0.003	psbF	120	1	0.008	rpl20	387	9	0.025
rbcL	1434	5	0.003	rps8	399	3	0.008	rps18	363	8	0.026
petA	963	3	0.003	psbN	132	1	0.008	rps3	657	18	0.027
ndhB	1533	5	0.003	ycf15	234	2	0.009	trnT-UGU	73	2	0.027
ccsA	969	3	0.003	pebT	108	1	0.009	trnE-UUC	73	2	0.027
rps16	252	1	0.004	rps14	303	3	0.01	trnC-GCA	71	2	0.028
atpA	1524	6	0.004	rps4	606	6	0.01	trnQ-UUG	72	2	0.028
atpH	246	1	0.004	rpoA	1023	10	0.01	rps12	372	11	0.03
psbC	1422	6	0.004	infA	243	2	0.01	rpl33	201	7	0.035
psbE	252	1	0.004	rpoC1	2097	22	0.011	rpl22	387	13	0.035
petD	483	2	0.004	trnS-GCU	88	1	0.011	rpl32	156	6	0.04
atpB	1521	7	0.005	rpl23	282	3	0.011	trnP-UGG	74	3	0.041
ycf68	376	2	0.005	ndhF	2232	26	0.012	trnI-CAU	74	3	0.041
ndhG	531	3	0.006	matK	1554	22	0.014	rps11	405	19	0.048
ndhA	1083	7	0.006	trnH-GUG	74	1	0.014	**rps19**	351	24	0.084
ndhH	1182	7	0.006	trnW-CCA	74	1	0.014	**clpP**	639	65	0.111
atpF	555	4	0.007	rpl16	411	6	0.015	**ycf1**	6778	664	0.121
rpoC2	4140	27	0.007	rps2	711	12	0.017	**accD**	1566	323	0.23

(B)

Gene	Aligned length (bp)	No. SNP	p-Distance
atpF intron	812	14	0.018
clpP intron 1	709	12	0.034
clpP intron 2	983	29	0.012
ndhA intron	1077	6	0.006
ndhB intron	695	1	0.001
petB intron	823	4	0.005
petD intron	747	5	0.007
rpl16 intron	1075	26	0.026
rpl2 intron	664	3	0.005
rpoC1 intron	714	9	0.013
rps16 intron	783	11	0.015
trnI-GAU intron	936	2	0.002
trnK-UUU intron	1109	34	0.012
trnL-UAA intron	538	2	0.004
trnV-UAC	595	4	0.007
ycf3 intron1	737	7	0.006
ycf3 intron2	738	8	0.01

(C)

IGS	Aligned length (bp)	No. SNP	p-Distance
trnL-CAA_ndhB	578	1	0.002
ndhB_rps7	323	1	0.003
atpI_rps2	242	1	0.004
rps12_trnV-GAC	1905	8	0.004
psaI_ycf4	376	2	0.005
petB_petD	205	1	0.005
psbE_petL	952	6	0.006
psbB_psbT	168	1	0.006
ycf4_cemA	785	5	0.007
rrn16S_trnI-GAU	296	2	0.007
trnA-UGC_rrn23S	144	1	0.007
trnR-ACG_trnN-GUU	572	4	0.007
rps14_psaB	132	1	0.008
psbN_psbH	124	1	0.008
infA_rps8	296	2	0.008
rrn23S_rrn4.5S	102	2	0.02
rpl32_trnL-UAG	938	16	0.02
ndhH_rps15	110	2	0.02
rpoB_trnC-GCA	859	16	0.022
rps19_trnH-GUG	147	3	0.022
atpA_atpF	92	2	0.023
trnG-GCC_trnfM-CAU	132	3	0.023
petA_psbJ	1139	26	0.023
petG_trnW-CCA	138	3	0.023
rps8_rpl14	179	4	0.023
ndhE_ndhG	301	4	0.023
rps15_ycf1	429	9	0.023
trnK-UUU_rps16	783	23	0.024
rpl20_clpP	1208	25	0.024
trnL-UAG_ccsA	82	2	0.025

Continued

Table 2. *Continued*

IGS	Aligned length (bp)	No. SNP	p-Distance	IGS	Aligned length (bp)	No. SNP	p-Distance
trnC-GCA_petN	831	8	0.01	ndhD_psaC	119	3	0.025
trnD-GUC_trnY-GUA	412	4	0.01	cemA_petA	242	6	0.026
trnS-GGA_rps4	307	3	0.01	trnQ-UUG_psbK	360	9	0.027
ycf15_trnL-CAA	674	2	0.01	trnG-UCC_trnR-UCU	150	4	0.027
psbM_trnD-GUC	1031	11	0.011	trnT-UGU_trnL-UAA	739	21	0.029
trnT-GGU_psbD	1016	12	0.011	rpoA_rps11	68	2	0.029
ndhJ_ndhK	89	1	0.011	atpH_atpI	655	19	0.031
psaJ_rpl33	482	3	0.011	rps16_trnQ-UUG	1204	37	0.037
petD_rpoA	179	2	0.011	trnW-CCA_trnP-UGG	167	6	0.037
ndhG_ndhI	283	3	0.011	ndhF_rpl32	778	28	0.038
trnF-GAA_ndhJ	686	8	0.012	psbK_psbI	396	15	0.039
rps2_rpoC2	246	3	0.013	psaC_ndhE	380	14	0.04
ycf3_trnS-GGA	759	8	0.013	psbH_petB	134	5	0.042
rps4_trnT-UGU	320	4	0.013	trnE-UUC_trnT-GGU	724	26	0.044
trnL-UAA_trnF-GAA	386	5	0.013	rpl33_rps18	200	8	0.049
atpF_atpH	474	7	0.015	clpP_psbB	507	23	0.049
psbZ_trnG-GCC	296	4	0.015	rps11_rpl36	151	7	0.051
trnM_CAU- atpE	206	3	0.015	trnS-GCU_trnG-UCC	1178	57	0.052
psaA_ycf3	642	10	0.016	psbI_trnS-GCU	124	6	0.054
petL_petG	183	3	0.016	psbC_trnS-UGA	140	8	0.057
rpl36_infA	154	2	0.016	rpl23_trnI-CAU	210	11	0.065
rpl14_rpl16	126	2	0.016	rpoC1_rpoB	37	2	0.077
psbA_trnK-UUU	243	10	0.017	**trnN-GUU_ndhF**	782	31	0.086
petN_psbM	712	12	0.017	**rpl22_rps19**	115	4	0.091
trnY-GUA_trnE-UUC	59	1	0.017	**rps3_rpl22**	80	5	0.098
ccsA_ndhD	242	4	0.018	**accD_psaI**	285	20	0.099
rpoC2_rpoC1	154	3	0.019	**psbT_psbN**	65	10	0.154
trnP-UGG_psaJ	388	7	0.019	**trnH-GUG_rpl2**	44	8	0.186
rpl16_rps3	166	3	0.019	**trnI-CAU_ycf2**	210	20	0.238
rps7_rps12	54	1	0.019	**rbcL_accD**	2090	236	0.291

Average *p*-distance

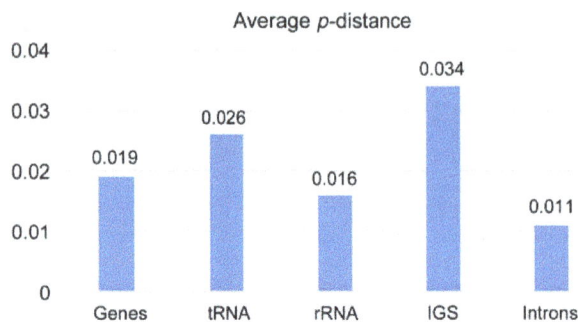

Figure 4. Average *p*-distance across five classes of genomic regions between two *Trillium*.

Table 3. Number of SSRs present in the three Parideae chloroplast genomes.

Taxon	*Paris verticillata*	*Trillium maculatum*	*Trillium tschonoskii*
Genome size	157 379	157 359	156 852
No. of SSRs			
A/T	133	117	127
C/G	5	4	6
AC/GT	3	3	3
AG/CT	17	19	18
AT/TA	32	35	32
AAG/CTT		2	2
AAT/ATT	6	10	3
ACT/AGT		2	1
ATC/GAT			1
AAAG/CTTT	–	–	1
AAAT/ATTT	3	4	4
AAGG/CCTT	1	1	1
AATC/GATT	1	1	1
AATG/CATT	1	1	3
AGAT/ATCT	1	1	1
ACTAT/ATAGT	1	–	1
AAAAT/ATTTT	–	1	1
AATAT/ATATT	–	2	1
AATATG/CATATT	–	1	–
AAAATC/GATTTT	–	1	–
ATATCC/GGATAT	–	–	1
AAAAAT/ATTTTT	–	–	2
AAGACT/AGTCTT	–	–	1
AACTAC/GTAGTT	–	–	1
AAAGAG/CTCTTT	–	–	1
Total	204	205	213

inversions occurred during diversification of the family, in which at least 20 occurred in *Cyphia*, and a minimum of 53 are now known in *Lobelia* (Knox 2014). Fabaceae are known to exhibit a number of unusual phenomena in their chloroplast genome: *Trifolium subterraneum* has undergone extensive genomic reconfiguration, including the loss of six genes and two introns and numerous gene order changes, attributable to 14–18 inversions (Cai *et al.* 2008).

Our results confirmed a single inversion in Melanthiaceae. It was remarkable that a single inversion of 3492 bp embedded four genes between *ndhC* and *rbcL* genes, which specifically occurred in the monophyletic subgenus *Phyllantherum* (Fig. 3). This event is thought to have occurred after the evolutionary divergence between subgenus *Phyllantherum* and subgenus *Trillium*. This new finding may be an effective molecular marker for classifying subgenera of the genus *Trillium*.

Diverse patterns of *trnI*-CAU duplication in Parideae

Gene duplication is an important process in organellar genome evolution. Most duplicated genes occur within the IR regions due to the mechanisms underlying IR expansion and contraction (Xiong *et al.* 2009). Gene duplication in plastid genome has been reported in tRNA genes (Hipkins *et al.* 1995; Vijverberg and Bachmann 1999; Schmickl *et al.* 2009) and in some protein-coding genes. Most of the duplications can be detected only in rearranged chloroplast genomes, as in grasses, legumes and conifers. Hipkins *et al.* (1995) compared the number of direct repeats between partially duplicated *trnY*-GUA and the complete *trnY*-GUA gene in *Pseudotsuga*. They found that the length-variable region in *Pseudotsuga* comprised imperfect tandem direct repeats based on the *trnY* gene sequence. Schmickl *et al.* (2009) used the 5′-*trnL*-UAA_*trnF*-GAA region for phylogeographic reconstructions, gene diversity calculations and phylogenetic analyses among the genera *Arabidopsis* and *Boechera*. The Cruciferous taxa are characterized by these pseudogenes in at least

four independent phylogenetic lineages. In addition, the tRNA gene as well as the coding gene could be confirmed by duplication events in *Jasminum* and *Menodora*, which have the duplicated *rbcL_psaI* region. Most chloroplast gene duplications outside of the IR involve tRNAs, as in the case of Oleaceae (Lee *et al.* 2007).

A total of 30–32 tRNA genes are present within the chloroplast genome of land plants (Tsudzuki *et al.* 1994; Vijverberg and Bachmann 1999), and they may be involved in chloroplast genome rearrangements through their secondary structure (Howe *et al.* 1988). These genes are dispersed throughout the genome, but five to eight

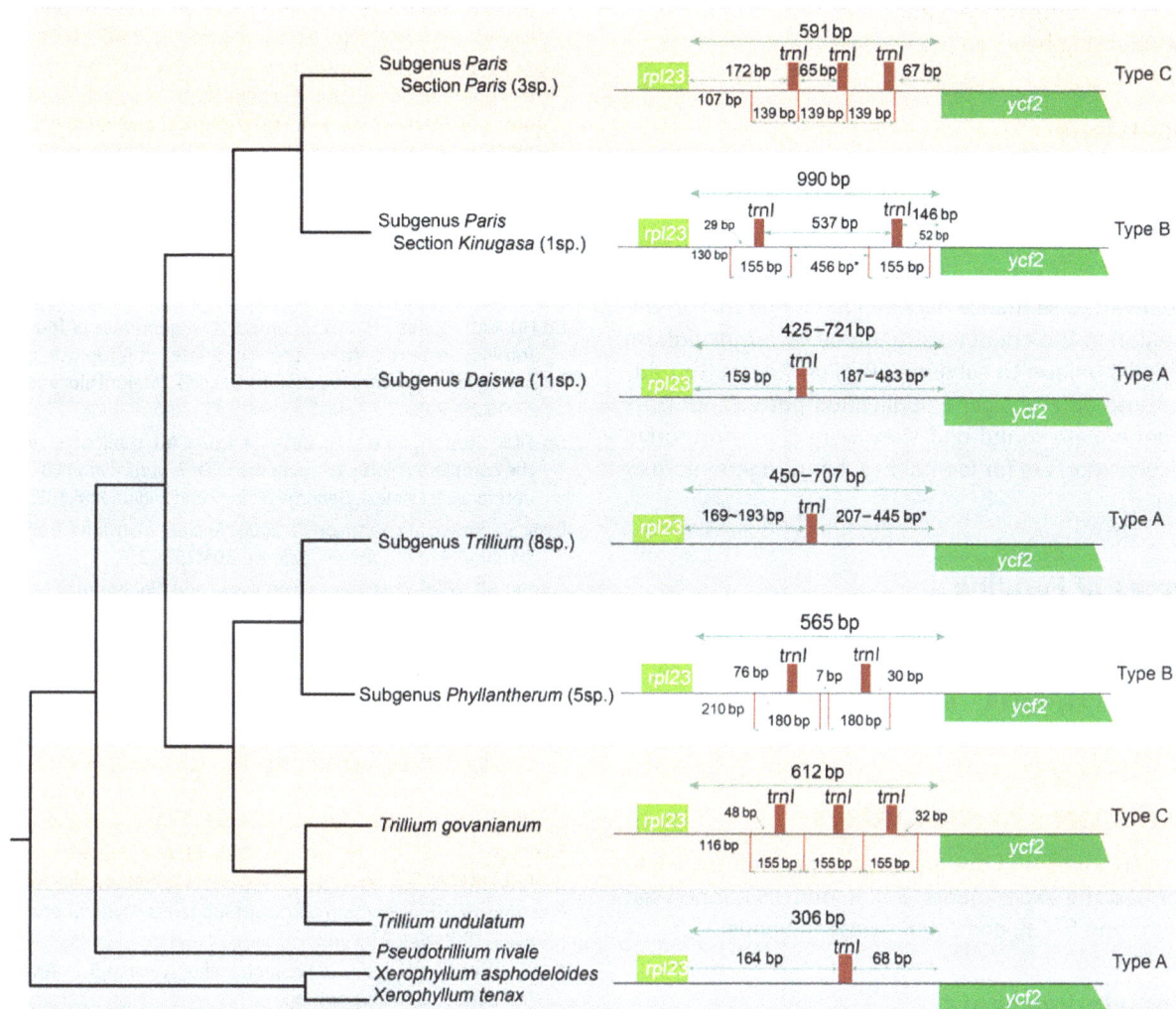

Figure 5. Summary of three types of *trnI*-CAU gene duplication in the tribe Parideae. *Including tandem repeats.

genes are located in the IR (Maréchal-Drouard *et al.* 1993). We found that three major types of *trnI*-CAU gene duplication are located between *rpl23* and *ycf2* at the IR of tribe Parideae (Fig. 5). Traditionally, Parideae included two genera, *Paris* and *Trillium*; however, *Trillium* was separated into two genera *Trillium* and *Pesudotrillium* in recent classifications (Farmer and Schilling 2002). Using the various duplication patterns of *trnI*-CAU in the IR region, the infrageneric circumscription of Parideae member was strongly supported. The type of *trnI*-CAU that had been discovered in *Xerophyllum*, *Pesudotrillium* and *T. undulatum* with one *trnI*-CAU between *rpl23* and *ycf2* was seen to be similar to the ancestor of Parideae (Type A, Fig. 5). This type was found also in most chloroplast genomes of Liliales (Liu *et al.* 2012; Bodin *et al.* 2013; Do *et al.* 2013; Kim and Kim 2013). It was modified in subgenus *Trillium* of *Trillium* and subgenus *Daiswa* of *Paris* to be extended by the tandem repeat between *trnI*-CAU and *ycf2*. Type B was found in subgenus *Phyllantherum* of

Trillium and section *Kinugasa* of subgenus *Paris* although section *Kinugasa* possessed the additional tandem repeat between *trnI*-CAU units. Type C, which was found in *T. govanianum* and section *Paris* of subgenus *Paris*, has three copies of the *trnI*-CAU gene. From the results, we suggested that duplicate events of *trnI*-CAU have occurred independently in the tribe Parideae of Melanthiaceae, and it provided useful information for determining the infrageneric circumscription. However, *T. govanianum*, which was classified into another genus *Trillidium* by Farmer and Schilling (2002) based on morphological characters and geographical distribution, was more similar to *Paris* than *Trillium*. Also, this result showed that the *trnI*-CAU gene duplication pattern of *T. govanianum* was more similar to *Paris* than *Trillium*. Interestingly, it was positioned at the same clade together with the North American species *T. undulatum* in the molecular phylogenetic tree (Farmer and Schilling 2002; S. C. Kim, J. S. Kim, W. C. Mark, F. F. Michael, J. H. Kim, unpublished data). Further studies are necessary to clarify the relationship between both species.

Conclusions

We analysed the complete chloroplast genomes of two species of *T. tschonoskii* (subgenus *Trillium*) and *T. maculatum* (subgenus *Phyllantherum*) to verify the specific feature in the genome level. As a result, we found a 3.4 kb inverted sequence between *ndhC* and *rbcL* in the LSC region in the chloroplast genome of *T. maculatum*, which was unique to subgenus *Phyllantherum*. In addition, three different gene duplication patterns of *trnI*-CAU gene were found and they were the informative molecular markers for identifying the infrageneric taxa of *Trillium*.

Sources of Funding

This work was supported by the National Research Foundation of Korea (NRF) Grant Foundation (MEST 2010-0029131) and Korea National Arboretum (KNA1-2-13,14-2).

Contributions by the Authors

J.-H.K conceived and designed the experiments, S.-C.K. performed the experiments, S.-C.K. and J.S.K. analysed the data and S.-C.K. and J.-H.K. wrote the paper.

Acknowledgements

The authors thank Dr Susan Farmer of Abraham Baldwin Agricultural College who kindly provided the *Trillium maculatum* for this study. Also we thank the editors and two reviewers for valuable comments and suggestions.

Supporting Information

The following additional information is available in the online version of this article –

   Table S1. Genes found in *Trillium tschonoskii* and *T. maculatum* chloroplast genomes.

   Table S2. The detailed list of insertion–deletion mutations between the chloroplast genomes of *T. tschonoskii* and *T. maculatum* in Parideae.

   Table S3. Sequences of *rpl23_ycf2* IGS among Parideae species.

Literature Cited

Bodin SS, Kim JS, Kim J-H. 2013. Complete chloroplast genome of *Chionographis japonica* (Willd.) Maxim. (Melanthiaceae): com-

parative genomics and evaluation of universal primers for Liliales. *Plant Molecular Biology Reporter* **31**:1407–1421.

Bryan GJ, McNicoll J, Ramsay G, Meyer RC, De Jong WS. 1999. Polymorphic simple sequence repeat markers in chloroplast genomes of Solanaceous plants. *Theoretical and Applied Genetics* **99**:859–867.

Cai Z, Guisinger M, Kim H-G, Ruck E, Blazier JC, Mcmurtry V, Kuehl JV, Boore J, Jansen RK. 2008. Extensive reorganization of the plastid genome of *Trifolium subterraneum* (Fabaceae) is associated with numerous repeated sequences and novel DNA insertions. *Journal of Molecular Evolution* **67**:696–704.

Do HD, Kim JS, Kim JH. 2013. Comparative genomics of four Liliales families inferred from the complete chloroplast genome sequence of *Veratrum patulum* O. Loes. (Melanthiaceae). *Gene* **530**:229–235.

Do HDK, Kim JS, Kim J-H. 2014. A *trnI_CAU* triplication event in the complete chloroplast genome of *Paris verticillata* M.Bieb. (Melanthiaceae, Liliales). *Genome Biology and Evolution* **6**:1699–1706.

Dyall SD, Brown MT, Johnson PJ. 2004. Ancient invasions: from endosymbionts to organelles. *Science* **304**:253–257.

Farmer SB. 2006. Phylogenetic analyses and biogeography of Trilliaceae. *Aliso* **22**:579–592.

Farmer SB, Schilling EE. 2002. Phylogenetic analyses of Trilliaceae based on morphological and molecular data. *Systematic Botany* **27**:674–692.

Flannery ML, Mitchell FJG, Coyne S, Kavanagh TA, Burke JI, Salamin N, Dowding P, Hodkinson TR. 2006. Plastid genome characterisation in *Brassica* and Brassicaceae using a new set of nine SSRs. *Theoretical and Applied Genetics* **113**:1221–1231.

Freeman JD. 1969. *A revisionary study of sessile-flowered* Trillium *L. (Liliaceae)*. PhD Dissertation, Vanderbilt University, Nashville, USA.

Freeman JD. 1975. Revision of *Trillium* subgenus *Phyllantherum* (Liliaceae). *Brittonia* **27**:1–62.

Gene Codes Corporation. 2012. *Sequencher (version 5.1)*. Ann Arbor, MI: Gene Codes Corporation.

Hipkins VD, Marshall KA, Neale DB, Rottmann WH, Strauss SH. 1995. A mutation hotspot in the chloroplast genome of a conifer (Douglas-fir: *Pseudotsuga*) is caused by variability in the number of direct repeats derived from a partially duplicated tRNA gene. *Current Genetics* **27**:572–579.

Howe CJ, Barker RF, Bowman CM, Dyer TA. 1988. Common features of three inversions in wheat chloroplast DNA. *Current Genetics* **13**: 343–349.

Jansen RK, Ruhlman TA. 2012. Plastid genomes of seed plants. In: Bock R, Knoop V, eds. *Genomics of chloroplasts and mitochondria*. Dordrecht, The Netherlands: Springer, 103–126.

Jansen RK, Raubeson LA, Boore JL, Depamphilis CW, Chumley TW, Haberle RC, Wyman SK, Alverson AJ, Peery R, Herman SJ, Fourcade HM, Kuehl JV, Mcneal JR, Leebens-Mack J, Cui L. 2005. Methods for obtaining and analyzing whole chloroplast genome sequences. *Methods in Enzymology* **395**:348–384.

Jansen RK, Cai Z, Raubeson LA, Daniell H, Depamphilis CW, Leebens-Mack J, Müller KF, Guisinger-Bellian M, Haberle RC, Hansen AK, Chumley TW, Lee SB, Peery R, Mcneal JR, Kuehl JV, Boore JL. 2007. Analysis of 81 genes from 64 plastid genomes resolves relationships in angiosperms and identifies genome-scale evolutionary patterns. *Proceedings of the National Academy of Sciences of the USA* **104**:19369–19374.

Kato H, Terauchi R, Utech FH, Kawano S. 1995. Molecular systematics of the Trilliaceae sensu lato as inferred from *rbcL* sequence data. *Molecular Phylogenetics and Evolution* **4**:184–193.

Katoh K, Misawa K, Kuma K, Miyata T. 2002. MAFFT: a novel method for rapid multiple sequence alignment based on fast Fourier transform. *Nucleic Acids Research* **30**:3059–3066.

Kearse M, Moir R, Wilson A, Stones-Havas S, Cheung M, Sturrock S, Buxton S, Cooper A, Markowitz S, Duran C, Thierer T, Ashton B, Meintjes P, Drummond A. 2012. Geneious Basic: an integrated and extendable desktop software platform for the organization and analysis of sequence data. *Bioinformatics* **28**:1647–1649.

Kim JS, Kim J-H. 2013. Comparative genome analysis and phylogenetic relationship of order Liliales insight from the complete plastid genome sequences of two Lilies (*Lilium longiflorum* and *Alstroemeria aurea*). *PLoS ONE* **8**:e68180.

Kim K-J, Choi K-S, Jansen RK. 2005. Two chloroplast DNA inversions originated simultaneously during the early evolution of the sunflower family (Asteraceae). *Molecular Biology and Evolution* **22**:1783–1792.

Knox EB. 2014. The dynamic history of plastid genomes in the Campanulaceae *sensu lato* is unique among angiosperms. *Proceedings of the National Academy of Sciences of the USA* **111**:11097–11102.

Kraemer L, Beszteri B, Gäebler-Schwarz S, Held C, Leese F, Mayer C, Pöehlmann K, Frickenhaus S. 2009. STAMP: extensions to the STADEN sequence analysis package for high throughput interactive microsatellite marker design. *BMC Bioinformatics* **10**:41.

Lee H-L, Jansen RK, Chumley TW, Kim K-J. 2007. Gene relocations within chloroplast genomes of *Jasminum* and *Menodora* (Oleaceae) are due to multiple, overlapping inversions. *Molecular Biology and Evolution* **24**:1161–1180.

Li R, Ma P-F, Wen J, Yi T-S. 2013. Complete sequencing of five Araliaceae chloroplast genomes and the phylogenetic implications. *PLoS ONE* **8**:e78568.

Liu J, Qi ZC, Zhao YP, Fu CX, Jenny Xiang QY. 2012. Complete cpDNA genome sequence of *Smilax china* and phylogenetic placement of Liliales—influences of gene partitions and taxon sampling. *Molecular Phylogenetics and Evolution* **64**:545–562.

Liu Y, Huo N, Dong L, Wang Y, Zhang S, Young HA, Gu YQ. 2013. Complete chloroplast genome sequences of Mongolia medicine *Artemisia frigida* and phylogenetic relationships with other plants. *PLoS ONE* **8**:e57533.

Lohse M, Drechsel O, Bock R. 2007. OrganellarGenomeDRAW (OGDRAW): a tool for the easy generation of high-quality custom graphical maps of plastid and mitochondrial genomes. *Current Genetics* **52**:267–274.

Maréchal-Drouard L, Weil JH, Dietrich A. 1993. Transfer RNAs and transfer RNA genes in plants. *Annual Review of Plant Biology* **44**:13–32.

Osaloo SK, Kawano S. 1999. Molecular systematics of Trilliaceae II. Phylogenetic analyses of *Trillium* and its allies using sequences of *rbcL* and *matK* genes of cpDNA and internal transcribed spacers of 18S–26S nrDNA. *Plant Species Biology* **14**:75–94.

Osaloo SK, Utech FH, Ohara M, Kawano S. 1999. Molecular systematics of Trilliaceae I. Phylogenetic analyses of *Trillium* using *matK* gene sequences. *Journal of Plant Research* **112**:35–49.

Provan J, Corbett G, McNicol JW, Powell W. 1997. Chloroplast DNA variability in wild and cultivated rice (*Oryza* spp.) revealed by polymorphic chloroplast simple sequence repeats. *Genome* **40**:104–110.

Provan J, Russell JR, Booth A, Powell W. 1999. Polymorphic chloroplast simple sequence repeat primers for systematic and population studies in the genus *Hordeum. Molecular Ecology* **8**:505–511.

Qian J, Song J, Gao H, Zhu Y, Xu J, Pang X, Yao H, Sun C, Li X, Li C, Liu J, Xu H, Chen S. 2013. The complete chloroplast genome sequence of the medicinal plant *Salvia miltiorrhiza. PLoS ONE* **8**:e57607.

Ravi V, Khurana JP, Tyagi AK, Khurana P. 2008. An update on chloroplast genome. *Plant Systematics and Evolution* **271**:101–122.

Rudall PJ, Stobart KL, Hong W-P, Conran JG, Furness CA, Kite GC, Chase MW. 2000. Consider the lilies: systematics of Liliales. In: Wilson KL, Morrison D, eds. *Monocots: systematics and evolution*. Collingwood: CSIRO Publishing, 347–359.

Schattner P, Brooks AN, Lowe TM. 2005. The tRNAscan-SE, snoscan and snoGPS web servers for the detection of tRNAs and snoRNAs. *Nucleic Acids Research* **33**:W686–W689.

Schmickl R, Kiefer C, Dobeš C, Koch MA. 2009. Evolution of *trnF*(GAA) pseudogenes in cruciferous plants. *Plant Systematics and Evolution* **282**:229–240.

Shendure J, Ji H. 2008. Next-generation DNA sequencing. *Nature Biotechnology* **26**:1135–1145.

Takhtajan A. 1997. *Diversity and classification of flowering plants*. New York: Columbia University Press.

Tamura K, Stecher G, Peterson D, Filipski A, Kumar S. 2013. MEGA6: molecular evolutionary genetics analysis version 6.0. *Molecular Biology and Evolution* **30**:2725–2729.

Tangphatsornruang S, Sangsrakru D, Chanprasert J, Uthaipaisanwong P, Yoocha T, Jomchai N, Tragoonrung S. 2010. The chloroplast genome sequence of mungbean (*Vigna radiata*) determined by high-throughput pyrosequencing: structural organization and phylogenetic relationships. *DNA Research* **17**:11–22.

The Angiosperm Phylogeny Group. 2009. An update of the Angiosperm Phylogeny Group classification for the orders and families of flowering plants: APG III. *Botanical Journal of the Linnean Society* **161**:105–121.

Tsudzuki J, Ito S, Tsudzuki T, Wakasugi T, Sugiura M. 1994. A new gene encoding tRNA^Pro (GGG) is present in the chloroplast genome of black pine: a compilation of 32 tRNA genes from black pine chloroplasts. *Current Genetics* **26**:153–158.

Vijverberg K, Bachmann K. 1999. Molecular evolution of a tandemly repeated *trnF*(GAA) gene in the chloroplast genomes of *Microseris* (Asteraceae) and the use of structural mutations in phylogenetic analyses. *Molecular Biology and Evolution* **16**:1329–1340.

Xiong AS, Peng RH, Zhuang J, Gao F, Zhu B, Fu XY, Yao QH. 2009. Gene duplication, transfer, and evolution in the chloroplast genome. *Biotechnology Advances* **27**:340–347.

Zhang Y-J, Ma P-F, Li D-Z. 2011. High-throughput sequencing of six bamboo chloroplast genomes: phylogenetic implications for temperate woody bamboos (Poaceae: Bambusoideae). *PLoS ONE* **6**:e20596.

Zomlefer WB, Williams NH, Whitten WM, Judd WS. 2001. Generic circumscription and relationships in the tribe Melanthieae (Liliales, Melanthiaceae), with emphasis on *Zigadenus*: evidence from ITS and *trnL-F* sequence data. *American Journal of Botany* **88**:1657–1669.

16

Genomic architecture of phenotypic divergence between two hybridizing plant species along an elevational gradient

Adrian C. Brennan[1,2,4]*, Simon J. Hiscock[3] and Richard J. Abbott[1]

[1] School of Biology, University of St Andrews, Harold Mitchell Building, St Andrews, Fife KY16 9TH, UK
[2] Estación Biológica de Doñana (EBD-CSIC), Avenida Américo Vespucio s/n, 41092 Sevilla, Spain
[3] School of Biological Sciences, University of Bristol, Woodland Road, Bristol BS8 1UG, UK
[4] Present address: School of Biological and Biomedical Sciences, University of Durham, South Road, Durham DH1 3LE, UK

Associate Editor: Diana Wolf

Abstract. Knowledge of the genetic basis of phenotypic divergence between species and how such divergence is caused and maintained is crucial to an understanding of speciation and the generation of biodiversity. The hybrid zone between *Senecio aethnensis* and *S. chrysanthemifolius* on Mount Etna, Sicily, provides a well-studied example of species divergence in response to conditions at different elevations, despite hybridization and gene flow. Here, we investigate the genetic architecture of divergence between these two species using a combination of quantitative trait locus (QTL) mapping and genetic differentiation measures based on genetic marker analysis. A QTL architecture characterized by physical QTL clustering, epistatic interactions between QTLs, and pleiotropy was identified, and is consistent with the presence of divergent QTL complexes resistant to gene flow. A role for divergent selection between species was indicated by significant negative associations between levels of interspecific genetic differentiation at mapped marker gene loci and map distance from QTLs and hybrid incompatibility loci. Within-species selection contributing to interspecific differentiation was evidenced by negative associations between interspecific genetic differentiation and genetic diversity within species. These results show that the two *Senecio* species, while subject to gene flow, maintain divergent genomic regions consistent with local selection within species and selection against hybrids between species which, in turn, contribute to the maintenance of their distinct phenotypic differences.

Keywords: Genetic differentiation; hybridization; phenotypic divergence; QTL architecture; QTL interactions; selection; speciation.

Introduction

Speciation commonly proceeds through genetic divergence between populations that ultimately become reproductively isolated from each other due to intrinsic and/or extrinsic breeding barriers (Orr and Turelli 2001; Coyne and Orr 2004; Smadja and Butlin 2011; Nosil and Feder 2012). Phenotypic trait divergence usually accompanies this process, often as a result of adaptation to different environments (Nosil 2012). Understanding how phenotypic trait divergence evolves between populations and is maintained between hybridizing species requires knowledge of the genetic basis of divergent traits and how selection acts on genes controlling these traits (Rieseberg *et al.* 2003; Lexer *et al.* 2005; Nosil *et al.* 2009; Nosil and Feder 2012).

* Corresponding author's e-mail address: a.c.brennan@durham.ac.uk

Quantitative trait locus (QTL) analysis is a powerful way of analysing the genetic basis of divergent traits between species (Rieseberg *et al.* 2003; Lexer *et al.* 2005; Bouck *et al.* 2007; Taylor *et al.* 2012; Rogers *et al.* 2013). It involves determination of the number and primary effects of QTLs, their genomic locations, the interactions between them (epistasis) and their effects across multiple traits (pleiotropy). The QTL architecture of divergent traits revealed by such analysis is likely to be shaped by divergent selection acting against relatively unfit recombinant hybrid phenotypes (Bierne *et al.* 2011; Servedio *et al.* 2011; Abbott *et al.* 2013; Yeaman 2013), especially where divergence between species occurs in the presence of interspecific gene flow (Via and West 2008; Nosil *et al.* 2009; Yeaman and Whitlock 2011; Yeaman 2013). This selective scenario could favour the evolution of QTL hotspots, epistasis and pleiotropy as effective means of preserving local adaptation despite gene flow (Whiteley *et al.* 2008; Gagnaire *et al.* 2013; Lindtke and Buerkle 2015). Alternatively, recombination and break-up of QTL complexes could be reduced by close physical proximity of QTLs (Yeaman and Whitlock 2011; Jones *et al.* 2012; Yeaman 2013) or recombination 'coldspots' such as near centromeres or chromosomal rearrangements (Turner *et al.* 2005; Kirkpatrick and Barton 2006; Lowry and Willis 2010; Twyford and Friedman 2015).

Complementary insights into the relationship between QTL architecture and divergent selection can be obtained by investigating genetic diversity and differentiation among mapped molecular marker loci (Rogers and Bernatchez 2007; Stinchcombe and Hoekstra 2008; Gompert *et al.* 2012; Renaut *et al.* 2012; Strasburg *et al.* 2012; Cruikshank and Hahn 2014). Heterogeneous differentiation across the genome is expected to result from divergent selection in the presence of gene flow (Wu 2001; Feder and Nosil 2010) and has been reported in several studies of ecologically divergent pairs of taxa (Turner *et al.* 2005; Rogers and Bernatchez 2007; Via and West 2008). However, such patterns of differentiation can be highly dependent on the biology and demographic histories of the focal taxa (Jones *et al.* 2012; Renaut *et al.* 2012), and their assessment must take account of genetic diversity both within and between focal taxa (Cruikshank and Hahn 2014).

Here, we present a quantitative genetic analysis of divergent traits between two diploid ($2n = 20$), short-lived perennial, self-incompatible, herbaceous species of *Senecio* (Asteraceae), *S. aethnensis* and *S. chrysanthemifolius*, which grow at elevations above 2000 m and below 1000 m, respectively, on Mount Etna, Sicily. Whereas *S. aethnensis* produces large flower heads (capitula) and fruits, and entire (spathulate) leaves, *S. chrysanthemifolius* has smaller flower heads and fruits, and highly dissected (pinnatisect) leaves.

The two species hybridize and form a hybrid zone at intermediate elevations on Mount Etna (James and Abbott 2005; Abbott and Brennan 2014). Although connected by hybrid populations, some barriers to interspecific gene flow are apparent in the field. For example, flowering times only partially overlap, with *S. chrysanthemifolius* flowering 6 weeks earlier (April–June) than *S. aethnensis* (July–September) (authors' personal observation). A previous analysis of the hybrid zone showed that leaf shape, flower head structure and fruit structure exhibited steeper clines and/or shifts in cline position relative to a molecular genetic cline (Brennan *et al.* 2009). This was attributed to both intrinsic and extrinsic environmental selection against hybrids.

An improved understanding of the level of genetic divergence between the two species and the importance of selection in driving genomic divergence recently came from a comparison of their transcriptomes (Chapman *et al.* 2013). This showed that genome-wide genetic differentiation between the species was low, with only 2.25 % of 8854 loci tested having been subject to divergent selection. Genetic maps for the two *Senecio* species based on segregation of molecular markers in F_2 mapping families (Brennan *et al.* 2014; Chapman *et al.* 2016) indicated that large genomic rearrangements were not a cause of reduced fitness in hybrids. However, many markers (~27 % of 127 maker loci tested, Brennan *et al.* 2014) exhibited significant transmission ratio distortion (TRD) in the F_2 family and clusters of TRD loci (TRDLs) were distributed across multiple linkage groups. This frequency of TRD was similar to that found in other crossing studies involving distinct 'species' (e.g. 49 and 33 % in *Mimulus* and *Iris*, respectively, Fishman *et al.* 2001; Taylor *et al.* 2012). Such extensive genomic incompatibility between the two species would be expected to affect the genetic structure of the hybrid zone on Mount Etna by limiting interspecific gene flow and promoting divergence across the genome. Chapman *et al.* (2016) further showed that loci exhibiting significant sequence or expression differentiation between the two species had a clustered distribution when placed on the map and several QTLs for species phenotypic differences coincided with these regions.

Here, we investigate further the genetic architecture of phenotypic trait differences and associated divergent selection acting on *S. aethnensis* and *S. chrysanthemifolius* by performing a QTL analysis of multiple quantitative traits that distinguish the two species. Our analysis examined additional traits and a larger mapping family relative to the recent study by Chapman *et al.* (2016), albeit with a reduced number of molecular marker loci. Our study aimed to determine the number and genomic locations of QTLs of relatively large effect controlling phenotypic differences and the extent of epistatic and pleiotropic effects of

QTLs that could limit introgression between the two species in the wild. We also conducted genetic differentiation outlier tests on mapped molecular markers in the two species to identify loci under divergent selection and test for associations between outlier loci and QTLs. In addition, we tested whether previously identified hybrid incompatibilities are associated with either QTLs for species differences or highly divergent loci as would be expected under divergent selection.

Methods

Samples

An F_2 mapping family (F_2AC) of a reciprocal cross between two cross-compatible F_1 progeny derived from a reciprocal cross between *S. aethnensis* (A) and *S. chrysanthemifolius* (C) was produced as described in Brennan *et al.* (2014) and used for QTL analysis. This family consisted of 100 individuals of known parental cytotype. For tests of selection based on genetic differentiation, seed was collected from two wild populations of *S. aethnensis* and three of *S. chrysanthemifolius* representing the elevational extremes of each species' range and also the source locations of the mapping family parents (NIC1 and PIC1) **[see Supporting Information—Table S1]**. Forty-two plants of each species, each representing a separately sampled maternal individual, were raised from this seed in a glasshouse at the same time and under the same conditions as F_2AC individuals.

Phenotype measurement

Twenty-five traits were measured on F_2AC parents and progeny, and also wild sampled individuals (see Brennan *et al.* 2009 for a description of traits measured). Extreme outlier values >3 standard deviations from the mean were removed from the datasets for progeny and wild samples of each species prior to analysis. Trait summary statistics were calculated and comparisons between wild sampled *S. aethnensis*, wild sampled *S. chrysanthemifolius* and the F_2AC mapping family were made using one-way analyses of variance and Mann–Whitney tests. Three traits—capitulum length, ray floret number and selfing rate—were dropped from further analysis after preliminary data exploration found that they showed extreme distributions that could not be satisfactorily resolved with data transformations. Remaining trait measurements were not transformed to become normally distributed before QTL analysis because (i) the expected density distributions of traits with additive effects contributed by multiple loci are not necessarily normally distributed, (ii) the significance of QTL logarithm of odds (LOD) scores can be adequately assessed with

data permutation and (iii) estimated sizes of QTL effects are more directly interpretable based on untransformed data (Churchill and Doerge 1994). Cross direction did not significantly influence any trait mean, so this was not required as a cofactor for QTL analysis. Independence between measured traits was examined using paired-trait Spearman correlations, and tests of their significance were performed separately for wild sampled *S. aethnensis*, wild sampled *S. chrysanthemifolius* and the F_2AC mapping family progeny leading to a subset of 13 highly independent traits being retained for QTL analysis. All tests were performed using R v2.13 software (R Development Core Team 2011).

DNA isolation and genotyping

DNA was extracted from each plant using the method described by Brennan *et al.* (2009). Plants were genotyped across 127 marker loci comprising 77 amplified fragment length polymorphisms (AFLPs), 8 simple sequence repeats (SSRs) and 42 expressed sequence tag (EST)-SSRs and indel molecular markers as described by Brennan *et al.* (2014). For ∼10 % of plants (randomly chosen), two independent DNA extracts were made to test for genotyping reliability.

Genetic mapping

A genetic map was constructed from the segregation of genetic markers in the F_2AC mapping family as described in Brennan *et al.* (2014) and Supplementary information. Genotype uncertainty due to scoring of dominant markers was accounted for by using the MapMaker genotype classes C (not a homozygote for the first parental allele) and D (not a homozygote for the second parental allele; Lincoln *et al.* 1993). The genetic map comprised 10 independent linkage groups with a total length of ∼400 cM. Transmission ratio distortion affected ∼27 % of mapped markers that were clustered into nine TRDLs. Sixty-five mapped loci were included in the QTL analysis after removing 39 loci that did not show F_2-like allelic segregation (i.e. each parent had an allele in common) and 23 loci that were located <0.5 cM from the nearest neighbouring marker and which therefore added little extra QTL mapping power.

Quantitative trait locus mapping and analysis

We analysed the data in the form of individual differences from the combined species mean, with the sign altered so that individuals that were more similar to *S. aethnensis* or *S. chrysanthemifolius* mean values were positive and negative, respectively. This data transformation preserved effect sizes in original units but had the added advantage of standardizing effect directions according to parental species across all traits. Comparisons with untransformed data showed that LOD scores (base 10 logarithm of odds) were largely unaffected by the transformation. Multiple interval mapping (MIM) was used to identify QTLs because this method has the advantage of simultaneously accounting for multiple QTLs and their interactions (Kao *et al.* 1999). Multiple interval mapping was performed with QTL cartographer v2.5.10 (Wang *et al.* 2011) using forward regression with a scanning interval of 3 cM and Bayesian information criterion (BIC-M0) model selection to determine the inclusion of extra QTL or QTL interaction parameters. Initial MIM models were then refined by testing indicated QTLs for significance according to BIC and adding additional QTLs until no further significant model improvement was achieved. Epistatic QTL interactions were also included if BIC was significantly improved. For comparison with MIM, composite interval mapping (CIM; Zeng 1994), a widely used QTL mapping method, was also performed and results obtained from this analysis, which did not differ greatly from those obtained with MIM, are presented in Supplementary information. The potential for TRDLs to influence the QTL results was tested using Spearman rank correlation tests of marker distance to nearest QTL peak against marker χ^2 test values for segregation distortion of genotypes, heterozygotes and parental alleles.

Multiple trait CIM (MtCIM) simultaneously analyses multiple trait data and can distinguish between linked QTLs and a single QTL affecting more than one trait through pleiotropy (MtCIM; Jiang and Zeng 1995). Multiple trait CIM analysis was performed using a scanning interval of 3 cM and automatic model selection using forward regression with five cofactor loci outside the test interval window of 10 cM. Significance of QTL LOD scores was tested with 1000 permutations of trait values (Churchill and Doerge 1994). A complementary test of the extent to which QTLs for different traits occupied the same genomic regions applied the 'sampling without replacement' method (Paterson 2002). Because the traits examined in this QTL dataset were selected to minimize covariance between them, spurious patterns of QTL coincidence generated by covariance were also assumed to be minimized, avoiding the need for additional statistical correction (Breitling *et al.* 2008). To perform the 'sampling without replacement' test, the genetic map was divided into smaller intervals of equal size corresponding to the mean QTL 2-LOD cM confidence interval of 16.5 cM with intervals chosen to be centred over each linkage group. This level of subdivision of the genetic map generates an optimal proportion of intervals occupied by a QTL for the purposes of this test (Paterson 2002), but the effect of using smaller interval sizes was also tested by repeating the test with 2, 4, 6, 8, 10, 12 and 14 cM interval sizes. A binary matrix describing the presence or absence of

QTLs for each trait within intervals was constructed and for each pair of traits, the probability of coincidence (*p*) was tested according to:

$$p = \binom{l}{m}\binom{n-l}{s-m}\bigg/\binom{n}{s}$$

where *n* is the number of intervals compared, *l* and *s* are the number of QTL intervals present in the samples with larger and smaller QTL counts, respectively, and *m* is the number of paired QTL interval matches present. To test whether QTL coincidence was greater than the null hypothesis of a random distribution of QTLs across the genetic map, the observed mean probability of QTL coincidence across paired-trait comparisons was compared against the distribution obtained from 1000 random permutations of QTL locations. The coincidence between TRDLs and QTLs was also investigated by including TRDL data in this analysis.

Genetic diversity analysis

Summary population genetic statistics were estimated for all mapped markers genotyped in wild samples of *S. aethnensis* and *S. chrysanthemifolius*. The population genetics software used included: Arlequin (Excoffier and Lischer 2010), GenAlEx v6.1 (Peakall and Smouse 2006) and HPrare (Kalinowsky 2005). The estimated statistics for AFLP and other dominantly scored markers were band presence frequency (*p*; assuming Hardy–Weinberg equilibrium), effective number of alleles (Ne), unbiased heterozygosity (UHe), allelic richness (Ar), private allelic richness (pAr), genetic differentiation among species (F_{ST}) and genotypic differentiation (Φ_{PT}). The same statistics, excluding *p* but including the minor allele frequency (MAF) and inbreeding coefficient (F_{IS}), were calculated for codominantly scored markers.

Patterns of differentiation across loci were investigated to detect both strongly and weakly differentiated outlier loci using BayeScan (Foll and Gaggiotti 2008), which employs Bayesian methods to estimate locus-specific differentiation and to evaluate its probability relative to population-level differentiation. Default starting parameter settings were used, except for a Monte Carlo Markov Chain size of 10 000, thinning interval of 50, ten pilot runs of 10 000 and an additional burn-in of 1 00 000. Outlier loci were identified based on $\log_{10}$ Bayes Factors values greater than one. Outlier analysis was performed with individuals classified according to both species and population. Initial runs suggested that loci with very low MAF were over-represented among outliers. To overcome this problem, only those loci with MAF >0.05 were included in final differentiation analyses, which were conducted separately on datasets comprising 64 codominant loci and 132 dominant loci.

The presence of 'genomic islands' of divergence was investigated by testing the genomic clustering of outlier markers with binomial tests that the observed number of neighbouring pairs of significantly selected loci was greater than the expected number of neighbouring paired selected loci given by the square of the observed frequency of selected loci. Genetic differentiation, measured as both F_{ST} and Φ_{PT}, was tested for an association with the genetic map distance to the nearest QTL peak and the nearest TRDL peak using Spearman rank correlation tests. Genetic differentiation was further tested for associations with local recombination rate, measured as the genetic map distance to the nearest mapped marker, and with genetic diversity within species, measured as each of UHe, Ar and MAF using Spearman rank correlation tests. Marker loci on linkage groups without QTLs were assigned large QTL distance values of 50 cM in order to include them as part of these association tests.

Results

Quantitative trait locus mapping and analysis

The two parent species, S. aethnensis and S. chrysanthemifolius, differed significantly for 22 of the 25 traits. The exceptions were flowering time, leaf number and selfed seed-set (Traits 1, 3 and 18) [see Supporting Information—Table S2 and Fig. S1]. We surmise that the lack of flowering time difference in the glasshouse compared with field observations reflects the importance of environmental conditions for the expression of this trait. For example, suitable growing conditions at the onset of spring start later in higher elevation S. aethnensis habitat than lower elevation S. chrysanthemifolius habitat. In summary, S. aethnensis differed from S. chrysanthemifolius in being shorter and less branched, possessing smaller, less dissected leaves (i.e. having entire or slightly lobed edges), and fewer but larger capitula that produced larger seed. Significant differences between the mean of the F_2AC family and those of one or both parent species were also evident for all traits apart from pollen viability and selfed seed-set (Traits 16 and 18) [see Supporting Information—Table S2]. The means of the F_2AC family for all traits were neither significantly higher nor lower than the means of both parents. Paired-trait correlations are summarized in Supporting Information—Table S3. Overall, 4.3, 2.7 and 11 % of pairs of traits were significantly correlated after correction for multiple testing among wild sampled S. aethnensis, wild sampled S. chrysanthemifolius and the F_2AC mapping family, respectively. Instances of non-independence between traits were reduced by dropping highly correlated traits

and traits used to calculate compound characters, leaving a subset of 13 independent traits for QTL analysis.

Significant QTLs for each trait were detected and characterized by LOD score, map position, two LOD confidence intervals, size of additive, dominance and epistatic effects, and percentage variance explained (PVE). A total of 29 significant QTLs were detected across the 13 traits examined with mean QTL effect size of 15 % (Table 1, Fig. 1). Quantitative trait loci were distributed across all major linkage groups except AC3 and AC6, with one to five QTLs detected for each trait (Fig. 1). The mean PVE of all identified QTLs per trait was 33.5 % (range = 10.0–69.8 %).

Four pairs of QTLs exhibited significant epistatic interaction effects with a mean PVE of 7.1 % (range = 5.1–10.1; Table 1). The MtCIM analysis of all traits identified three significant and three almost significant (within 1 LOD of the permutation threshold of 14.82 LOD) pleiotropic loci with multiple trait effects (Table 2) [see Supporting Information—Table S5]. These potential pleiotropic loci overlapped with the 2-LOD intervals of 14/29 of the individual trait QTLs, with up to four traits affected at each site (Table 1). Thus, 14 QTLs for eight traits exhibited pleiotropic effects. The 'sampling without replacement' method using the 16.5 cM interval size found four trait pairs, auricle width and pollen number, capitulum number and node length, capitulum number and flowering time, and node length and leaf dissection, that showed significantly coincident QTL locations [see Supporting Information—Table S6]. Sampling without replacement analyses using a range of shorter interval sizes found similar evidence for coincident QTL locations, but failed to find any previously identified TRDLs that were significantly coincident with trait QTLs [see Supporting Information—Table S6].

Genetic diversity analysis

Both species exhibited similar levels of genetic diversity, with the highest diversity recorded for anonymous SSRs, followed in turn by EST-SSRs, EST-indels and AFLPs, and other dominant markers [see Supporting Information—Tables S7 and S8]. Overall, inbreeding coefficients were not significantly different from zero in either species indicative of random mating ($F_{IS} = 0.02$ and 0.06 in S. aethnensis and S. chrysanthemifolius, respectively) [see Supporting Information—Table S7]. The two species were significantly genetically differentiated across all marker types with overall F_{ST} of 0.28 and 0.31 observed for dominant markers and codominant markers, respectively [see Supporting Information—Tables S7 and S8].

Bayesian analyses of species differentiation showed that 4.7 % of codominant markers, but 0 % of dominant markers, were divergent outliers and that the same

Table 1. Summary QTL results from a MIM analysis of a reciprocal F$_2$ S. aethnensis and S. chrysanthemifolius mapping family. Quantitative trait locus LG and position are the linkage group and maximum likelihood of odds score (LOD) cM position of significant QTLs identified from MIM. Quantitative trait locus cM interval is the 2-LOD cM interval around the maximum LOD value that is indicated in parentheses. The additive and dominance effects are in the same units as trait measures. Positive additive effects support the direction of the species difference and vice versa for negative effects, while positive dominance effects indicate that S. aethnensis alleles are dominant and vice versa for negative effects. The PVE is shown in parentheses. Epistatic interactions show the loci for each trait (numbered in the order they appear in the table) with significant interactions with the additional PVE shown in parentheses.

Trait ID number: trait (units)	QTL LG and position (cM)	QTL cM interval (max LOD)	Additive effect (PVE)	Dominant effect (PVE)	Total PVE	Epistatic interactions (PVE)
1: Time from first true leaf to flowering (days)	AC1; 41.3	36.6–44.5 (6.19)	−3.41 (5.7)	−2.31 (1.3)	7	2 × 3 (5.4)
	AC2; 9.3	6.3–11.3 (9.22)	−5.27 (14.3)	0.13 (0)	14.3	
	AC10A; 0	0–0.5 (15.35)	9.02 (46.2)	−3.56 (2.3)	48.5	
7: Primary stem midleaf auricle width (mm)	AC10B; 3	0–4.2 (2.21)	0.52 (7.3)	0.41 (2.7)	10	
8: Primary inflorescence capitulum number (count)	AC2; 11.3	0–14.2 (2.5)	1.29 (8.5)	0.96 (3.6)	12.1	
	AC5A; 0	0–6.9 (2.57)	−1.34 (9)	0.36 (0.4)	9.4	
	AC10A; 0	0–10 (4.89)	2.14 (20.7)	−0.43 (−0.1)	20.6	
9: Primary capitulum pedicel length (cm)	AC8A; 11.4	0–27.5 (2.5)	0.26 (11.7)	−0.07 (0.4)	12.1	
11: Primary capitulum disc diameter (mm)	AC4; 23.6	12.5–36.3 (3.52)	0.05 (6.4)	0.08 (9.8)	16.2	
	AC9; 0	0–15 (2.13)	−0.05 (6.2)	0.03 (1.2)	7.4	
15: Mean pollen number (per 3/40 florets)	AC10B; 3	0–4.2 (2.03)	7.79 (6.2)	−8.94 (3.8)	10	
16: Mean pollen viability (proportion)	AC1; 0.9	0–5.5 (5.4)	0.01 (−1.2)	−0.19 (25.9)	24.7	
19: Mean fruit length (mm)	AC1; 18	2.4–39.6 (2.89)	−0.21 (−5.8)	0.36 (26.1)	20.3	
20: Mean pappus length (mm)	AC1; 23	8.5–36.6 (4.6)	0.15 (5.9)	0.32 (12)	17.9	
	AC5B; 0	0–9.5 (2.09)	0.2 (5.6)	0.01 (0)	5.6	
	AC8A; 13.6	0–27.5 (3.43)	0.27 (10.6)	0.12 (1.3)	11.9	
21: Primary stem node length, height to leaf number ratio (cm)	AC4; 6	0–41.3 (2.8)	0.09 (8.6)	0.1 (2.9)	11.5	
	AC5A; 0	0–6.9 (6.37)	−0.12 (18)	−0.01 (0.1)	18.1	
	AC10A; 4.1	0–10 (11.57)	0.16 (30.1)	−0.11 (2)	32.1	
22: Branch number to leaf number (proportion)	AC4; 26.6	13.5–41.3 (4.3)	0.1 (9.9)	−0.17 (12.1)	22	
24: Primary capitulum ray display area (mm^2)	AC1; 15	6.5–25 (6.75)	4.69 (3.3)	3.37 (1.9)	5.2	1 × 3 (5.1)
	AC4; 30.3	4–41.3 (2.07)	6.9 (6.2)	0.25 (0.1)	6.3	1 × 4 (7.8)
	AC7A; 4.6	0–14.2 (6.45)	5.79 (2.1)	16.83 (7.1)	9.2	
	AC8A; 15.1	2.5–24.5 (5.5)	8.13 (8)	5.67 (−0.2)	7.8	
	AC10A; 2.8	0–8.6 (8.01)	15.43 (20.9)	0.54 (0.2)	21.1	
25: Primary stem midleaf dissection, perimeter to area ratio (per mm)	AC4; 3	0–41.3 (2.02)	0.04 (6.9)	0.04 (2.9)	9.8	2 × 3 (10.1)
	AC8A; 6	0–10 (7.29)	0.04 (7.7)	0.05 (6.3)	14	
	AC10A; 0	0–10 (9.01)	0.06 (21.1)	0.01 (0.5)	21.6	
	AC10B; 3	0–4.2 (2.6)	0.02 (3.9)	0.04 (4.4)	8.3	

percentages of each marker type were significantly convergent outliers (Table 3). When population information was included in these analyses, the tests were more sensitive and identified 7.8 and 5.3 % of significantly divergent codominant and dominant markers, respectively, and 4.7 and 0.8 % of significantly convergent codominant and dominant markers, respectively (Fig. 1, Table 3) **[see Supporting Information—Table S9]**. Significant

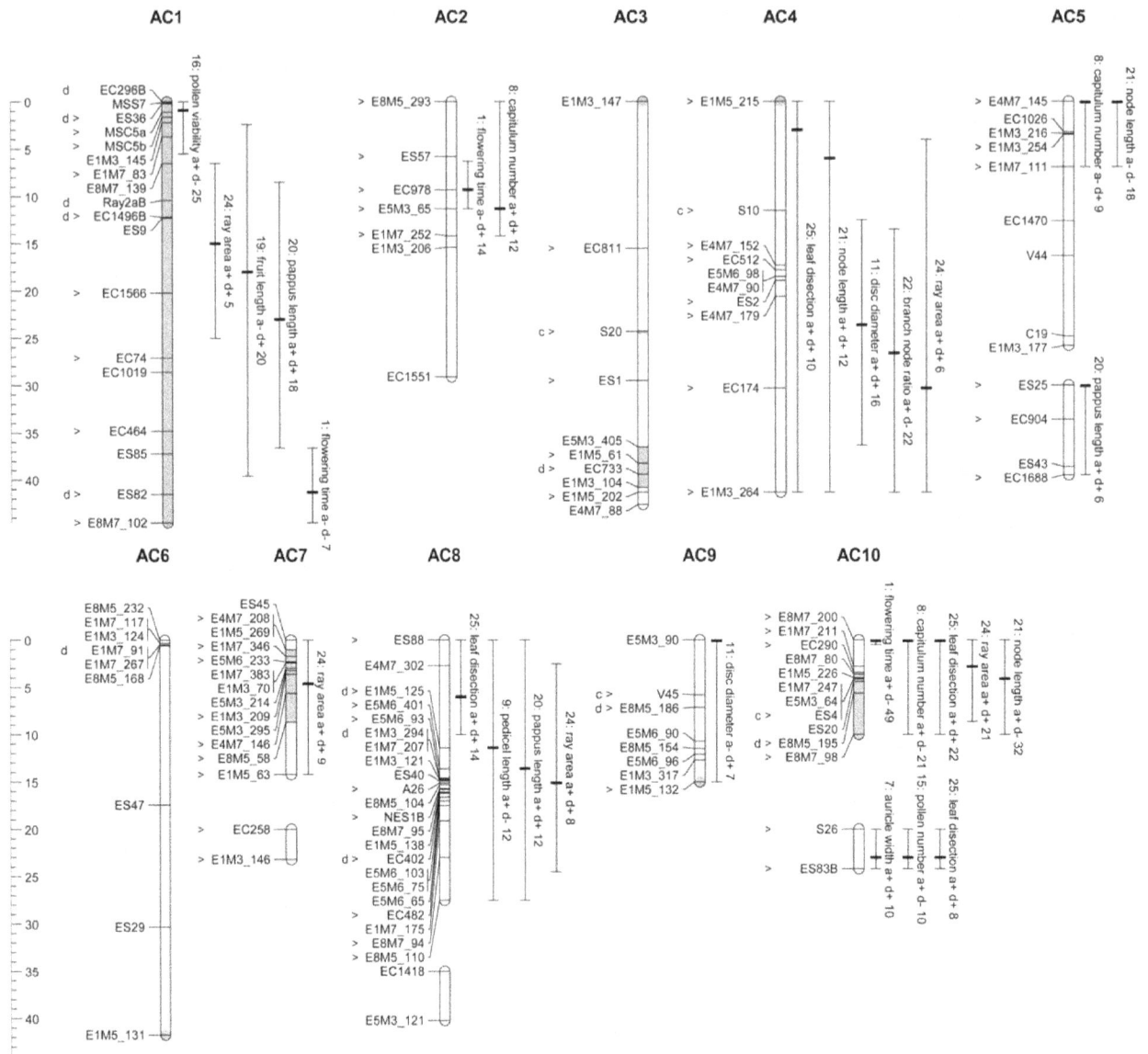

Figure 1. Genetic map of a reciprocal F₂ *S. aethnensis* and *S. chrysanthemifolius* mapping family showing quantitative trait loci identified by MIM and marker loci that were significantly divergent or convergent between species. Map distances in Kosambi centiMorgans are shown in the scale bar to the left of linkage groups. Linkage groups are represented by vertical bars with mapped locus positions indicated with horizontal lines. Weakly linked linkage groups (<4 LOD or >20 cM) that probably belong to the same chromosome are aligned vertically. Grey shading on linkage groups indicates regions exhibiting significant TRDLs. Locus names are listed to the left of linkage groups and mapped QTLs are listed to the right. 'c' or 'd' listed to the left of locus names indicates if locus was identified to be significantly convergent or divergent based on genetic differentiation analysis across sample populations, while the greater than symbol to the left of locus names indicates if the locus was included in QTL analysis. Quantitative trait loci were identified by MIM with significance testing by BIC model comparisons. Quantitative trait loci 2-LOD ranges are indicated with vertical lines with a bold horizontal line indicating the highest LOD score position. Quantitative trait locus summary information includes trait names, 'a' or 'd' each followed by '+' or '−' indicating additive or dominance effects and their direction of effect supporting or opposing the observed species difference, respectively, and the PVE.

outlier loci were distributed across most linkage groups of the genetic map (Fig. 1) and showed no evidence of clustering according to a one-way binomial test of an excess of neighbouring pairs of outlier markers ($P = 0.1754$). However, significant negative associations between measures of species differentiation for marker loci and the genetic map distance from the nearest QTL peak were

present (Fig. 2). Similarly, there was evidence for negative associations between marker gene differentiation and genetic map distance from the nearest TRDL (Fig. 2). A significant negative association between genetic differentiation and low recombination in the form of genetic map distance to closest neighbouring mapped locus was also found (Fig. 2). Also, significant negative

Table 2. Summary QTL results from a MtCIM analysis of a reciprocal F_2 S. aethnensis and S. chrysanthemifolius mapping family. Locus LG and peak cM are the linkage group and maximum likelihood odds score (LOD) cM position of a locus that affects the expression of multiple traits. Locus 2-LOD interval is the 2-LOD interval around the peak for the locus with the maximum LOD value indicated in parentheses. Overlapping quantitative trait loci from the MIM analysis are shown for comparison [see Table 1 and **Supporting Information—Table S4** for more details about QTLs].

Locus LG, peak cM position	Locus 2-LOD interval (peak LOD)	Overlapping QTL LOD intervals
AC1, 0.9	0–6.5 (14.52)	16: Mean poor pollen
AC1, 29.9	26.9–32.9 (14.35)	19: Mean fruit length
		20: Mean pappus length
AC2, 5.8	1.5–11.3 (14.1)	1: Time from first true leaf to flowering
		8: Primary inflorescence capitulum number
AC4, 3	1.8–4.2 (17.69)	21: Primary stem node length
		24: Primary capitulum ray display area
		25: Primary stem midleaf dissection
AC5A, 3	0–6.9 (17.24)	8: Primary inflorescence capitulum number
		21: Primary stem node length
AC10A, 8.6	3.4–10 (28.6)	8: Primary inflorescence capitulum number
		21: Primary stem node length
		24: Primary capitulum ray display area
		25: Primary stem midleaf dissection

Table 3. Summary of numbers of marker loci identified as significantly divergent or convergent outliers between S. aethnensis and S. chrysanthemifolius. Samples tested included all samples scored according to species (Species), populations (Populations) or data subsets of only S. aethnensis populations or only S. chrysanthemifolius populations. Only polymorphic loci with minor allele frequencies >0.05 were included in analyses. In the case of dominant loci, allele frequency was calculated assuming within-population Hardy–Weinberg equilibrium. Loci were considered significantly divergent or convergent with $\log_{10}$ Bayes Factor statistics >1.

Samples tested	No. codominant loci tested	No. dominant loci tested	No. codominant loci divergent (%)	No. codominant loci convergent (%)	No. dominant loci divergent (%)	No. dominant loci convergent (%)
Species	64	132	3 (4.7)	3 (4.7)	0 (0.0)	0 (0.0)
Populations	64	132	5 (7.8)	3 (4.7)	7 (5.3)	1 (0.8)
S. aethnensis	53	115	0 (0.0)	0 (0.0)	0 (0.0)	0 (0.0)
S. chrysanthemifolius	61	110	1 (1.6)	0 (0.0)	0 (0.0)	0 (0.0)

associations were present between genetic differentiation between species and the various intraspecific genetic diversity measures (Fig. 3). In general, all of these associations were stronger for codominant than for dominant markers.

Discussion

Quantitative trait locus architecture

Quantitative trait locus analysis identified 1–5 QTLs per trait and 29 QTLs in total for the 13 independent traits examined that distinguish the two Senecio species. In addition to resolving the primary effects of individual QTLs, MIM and MtCIM analyses provided evidence for epistatic interactions between four pairs of QTLs and

possible pleiotropic effects at six loci affecting eight traits (Tables 1 and 2). Sampling without replacement tests indicated that QTL map locations were significantly clustered across the genetic map, with significant physical associations evident for four trait pairs **[see Supporting Information—Table S5]**. Chapman et al. (2016) reported similar clustering of QTLs for species differences in an independent mapping study of S. aethnensis and S. chrysanthemifolius. However, their study did not investigate patterns of epistasis and pleiotropy. Regardless of whether the observed interactions between QTLs are due to epistasis or pleiotropy or physical linkage, they indicate that different traits are not genetically independent and that divergent selection acting on one trait would therefore also affect other traits.

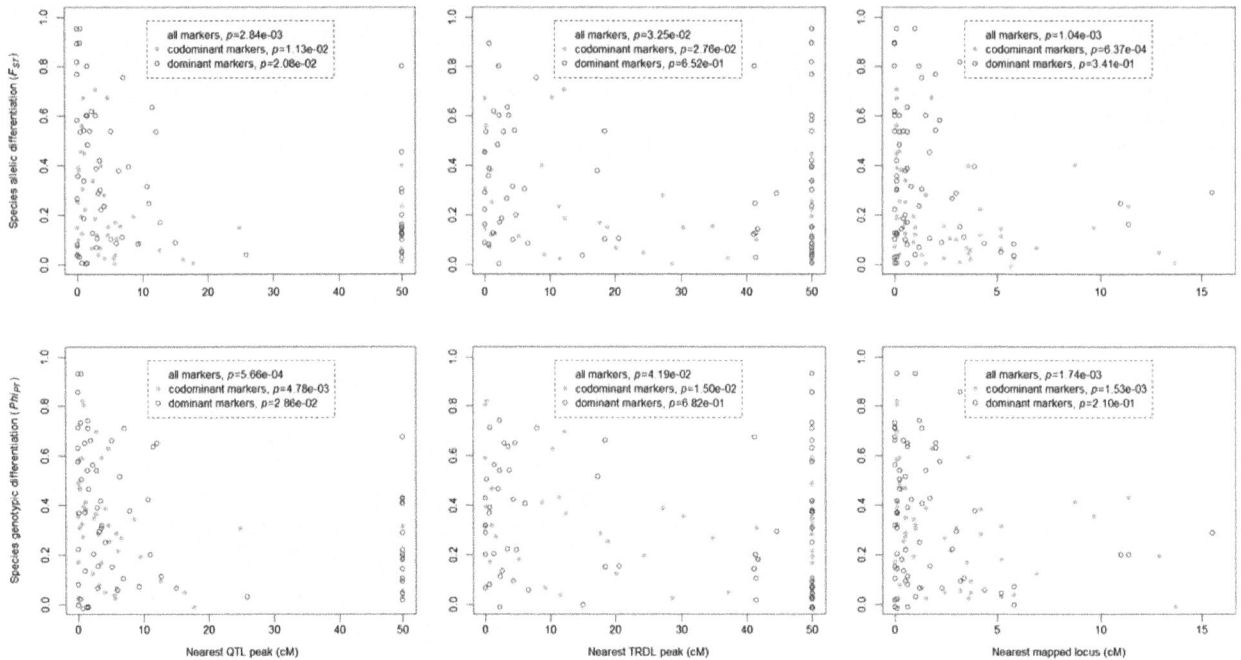

Figure 2. Relationships between genetic differentiation and genetic map distance to the nearest QTL peak, the nearest TRDL or the nearest mapped marker. Presented *P* values summarize Spearman rank correlation tests. All significant associations were negative. Sample sizes were 48 codominant loci and 63 dominant loci. Loci on linkage groups without a QTL or TRDL peak were assigned an unlinked genetic map distance of 50 cM.

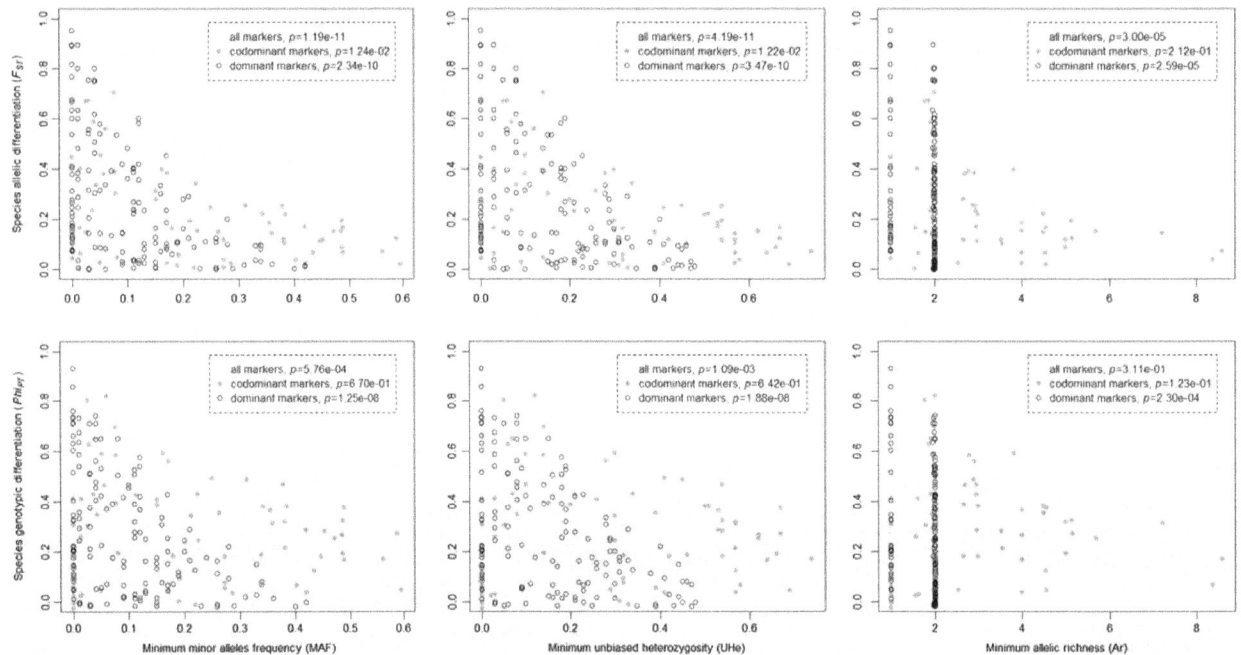

Figure 3. Relationships between genetic differentiation and genetic diversity of wild sampled *S. aethnensis* and *S. chrysanthemifolius*. Presented *P* values summarize Spearman rank correlation tests. All significant associations were negative. Sample sizes were 65 codominant loci and 145 dominant loci.

A QTL architecture involving extensive physical and epistatic interactions between QTLs, together with pleiotropic effects of individual QTLs, should limit introgression between the two *Senecio* species on Mount Etna since hybridization would tend to break up gene complexes that control the expression of adaptive phenotypes in each species (Fenster *et al.* 1997; Turelli *et al.* 2001). The complex genomic architecture of interspecific divergence revealed in *Senecio* might reflect the evolutionary outcome of selection for non-independence of different QTLs controlling traits under divergent selection (Kirkpatrick and Barton 2006; Nosil *et al.* 2009; Nosil and Feder 2012; Yeaman 2013). Limiting recombination seems to be the crucial factor permitting interacting QTLs to evolve into divergent co-adapted QTL complexes in the presence of gene flow. This can be achieved either through chromosomal rearrangement that causes recombination between rearranged regions to become deleterious (Feder *et al.* 2003; Lowry and Willis 2010; Twyford and Friedman 2015) or by evolution towards increased physical proximity (coincidence) through locally biased persistence, establishment or translocation of QTLs (Via and West 2008; Nosil *et al.* 2009; Yeaman 2013). Genetic mapping indicates that *S. aethnensis* and *S. chrysanthemifolius* are not distinguished by major genome rearrangements (Brennan *et al.* 2014), which instead emphasizes the importance of QTL coincidence for this system (our results and those of Chapman *et al.* 2016).

While TRDLs were not significantly coincident with QTLs for any trait according to the 'sampling without replacement' method, a QTL affecting pollen viability co-located with a TRDL of large effect in linkage group AC1 (Fig. 1). This finding is of interest as it adds to the result previously reported by Chapman *et al.* (2016) of co-localization of TRDLs with QTLs affecting F_2 hybrid necrosis. Hybrid incompatibilities, such as decreased F_2 pollen viability and hybrid necrosis and their associated TRDLs, are expected to limit introgression across large genomic regions allowing further divergence of these regions during speciation (Barton and Bengtsson 1986; Barton and de Cara 2009).

Non-random patterns of divergence across the genome

Levels of molecular genetic diversity were similar in wild samples of both *Senecio* species, while genetic differentiation between species was moderate. Genetic diversity decreased from estimates based on anonymous SSRs to EST-SSRs, to EST-indels to AFLPs, corresponding to the expected ability of each marker type to resolve allelic variation **[see Supporting Information—Tables S6 and S7]**. Low levels of genetic differentiation between the two species were also reported by Muir *et al.* (2013), Osborne *et al.* (2013) and Chapman *et al.* (2013), based on surveys of microsatellite and sequence variation. We identified a small percentage of loci that were either significantly divergent or convergent (up to 7.8 %) between species, dependent on the marker set analysed (Table 3) **[see Supporting Information—Table S9]**. This

value is slightly greater than the 2.25 % of outliers from a study of 8854 loci recorded by Chapman *et al.* (2013) based on a comparison of the two species' transcriptomes, but the two findings are probably within the bounds of error given the different numbers of loci examined. More discussion about the functions of significantly divergent or convergent loci is provided in the Supplementary information. Inevitably, the 196 marker loci for which patterns of differentiation were compared to detect significant divergence between species in the present study provide only a very coarse-grained perspective across the whole genome, and many of the true genetic targets of selection will not have been surveyed.

Reduced effective gene flow in the vicinity of selected loci is often used to explain significantly differentiated loci and 'islands of divergence' (Wu 2001; Feder and Nosil 2010). In support of this hypothesis, significant associations were found between interspecific genetic differentiation and genetic map distance to QTLs and TRDLs (Fig. 2). These associations were negative with more highly differentiated loci positioned closer to QTLs or TRDLs. These results support previous findings that selection against hybridization is important for maintaining species distinctiveness across the *Senecio* hybrid zone on Mount Etna (Brennan *et al.* 2009; Chapman *et al.* 2013, 2016). However, independently of gene flow, within-species directional selection can also generate the same pattern of divergence via species-specific reductions in diversity (Cruikshank and Hahn 2014). The latter is amplified when it occurs in regions of low recombination as it causes longer genomic regions to be affected by selection at linked markers. In accordance with these hypotheses and in agreement with the findings of Chapman *et al.* (2016), we also found evidence for intraspecific selection in the form of significant negative associations between interspecific differentiation and local recombination, and between interspecific differentiation and intraspecific genetic diversity (Figs 2 and 3). It is plausible that *S. aethnensis* and *S. chrysanthemifolius* experience distinct localized selection pressures related to the very different environments they occupy at different elevations on Mount Etna. Such within-species selection would be expected to reduce within-species genetic diversity in the genomic regions experiencing selection. These findings, therefore, suggest a role for environment-specific extrinsic selection in maintaining the cline with elevation on Mount Etna. While this pattern of diversity might also signal past periods of isolation facilitating divergence, other genetic studies suggest that gene flow between the two species has probably been continuous throughout their history (Chapman *et al.* 2013; Osborne *et al.* 2013; Filatov *et al.* 2016).

Conclusions

Our study shows that phenotypic divergence across the elevational gradient on Mount Etna involves divergence of multiple quantitative traits controlled by numerous interacting genes (QTLs). A breakdown in the complex genetic architecture of these traits following hybridization would be expected to reduce the fitness of most hybrid offspring and therefore contribute to introgression barriers between the two *Senecio* species. Our combined analyses of genetic differentiation, QTLs and TRDLs emphasize that divergence is non-randomly distributed across the genomes of these species and that both selection against hybrids between species and locally maladapted individuals within species will act to maintain phenotypic divergence between the two species in the face of gene flow.

Sources of Funding

The research was funded by a Natural Environment Research Council (NERC) Grant NE/D014166/1 to R.J.A. as principal investigator. A.C.B. was supported during part of the writing of this paper by funding from European Commission 7th Framework Programme Capacities Work Programme (FP7-REGPOT 2010-1), Grant No. 264125 Eco-Genes.

Contributions by the Authors

R.J.A., A.C.B. and S.J.H. designed the research. A.C.B. performed the experiments and analysis. A.C.B. wrote the first draft and A.C.B., R.J.A. and S.J.H. contributed to revisions.

Acknowledgements

We thank David Forbes for technical assistance and Ai-Lan Wang for help with measurement of quantitative traits. We thank the managing editor, Diana Wolf and anonymous reviewers for their suggestions to improve earlier versions of this paper.

Supporting Information

The following additional information is available in the online version of this article —

Table S1. Information on wild sampled populations of *S. aethnensis* and *S. chrysanthemifolius*.

Table S2. Summary quantitative trait results for *S. aethnensis*, *S. chrysanthemifolius* and a reciprocal F_2 *S. aethnensis* and *S. chrysanthemifolius* mapping family.

Table S3. Paired-trait correlations in (a) F_2AC progeny, (b) *Senecio aethnensis*, (c) *S. chrysanthemifolius* and (d) all three samples.

Table S4. Comparison of summary QTL results for a CIM and MIM analysis of a reciprocal F_2 *S. aethnensis* and *S. chrysanthemifolius* mapping family.

Table S5. Summary QTLs results from a MtCIM analysis compared with single trait QTL analyses of a reciprocal F_2 *S. aethnensis* and *S. chrysanthemifolius* mapping family.

Table S6. (a) 'Sampling without replacement' test results for paired-trait QTL coincidence, (b) permutation tests of overall paired-trait QTL coincidence using different QTL and TRDL datasets and genetic map interval sizes.

Table S7. Summary population genetic statistics for AFLPs and other dominantly scored molecular genetic markers from *S. aethnensis* and *S. chrysanthemifolius* samples.

Table S8. Summary population genetic statistics for codominantly scored molecular genetic markers from *S. aethnensis* and *S. chrysanthemifolius* samples.

Table S9. Expressed sequence tag loci showing evidence for divergent or convergent selection between *S. aethnensis* and *S. chrysanthemifolius*.

Figure S1. Boxplots summarizing quantitative trait results for *S. aethnensis*, *S. chrysanthemifolius* and a reciprocal F_2 mapping family that were included in the QTL analysis. Trait numbers in titles correspond to the trait numbering system of Table 1. Bold horizontal lines indicate median values. Boxes indicate 25–75 percentile range. Lines indicate the range of values within 1.5 times the upper and lower quartiles, respectively. Points indicate values more extreme than 1.5 times the upper and lower quartiles. Asterisks indicate the trait values of the mapping family parents. No mapping family parental values were available for flowering time as these individuals were vegetatively propagated for comparison with their progeny.

Figure S2. Genetic map of a reciprocal F_2 *S. aethnensis* and *S. chrysanthemifolius* mapping family showing quantitative trait loci identified by CIM and marker loci that were significantly divergent or convergent between species. Map distances in Kosambi centiMorgans are shown in the scale bar to the left of linkage groups. Linkage groups are represented by vertical bars with mapped locus positions indicated with horizontal lines. Weakly linked linkage groups (<4 LOD or >20 cM) that probably belong to the same chromosome are aligned vertically. Grey shading on linkage groups indicates regions exhibiting significant TRDLs. Locus names are listed to the left of linkage groups and mapped QTLs are listed to the right. 'c' or 'd' listed to the left of locus names indicates if that locus was identified as significantly convergent or divergent based on genetic differentiation analysis across sample populations, while greater than symbol to the left of locus names indicates if the locus was included in QTL analysis. Quantitative trait loci were identified by CIM with significance determined if the LOD score exceeded the 0.95 quantile of 1000 data permutations. Quantitative trait loci 2-LOD interval ranges are indicated with vertical lines with a bold horizontal line indicating the highest LOD score position. Quantitative trait locus summary information includes trait names, 'a' or 'd' each followed by '+' or '−' indicating additive or dominance effects and their direction of effect supporting or opposing the observed species difference, respectively, and the PVE.

Literature Cited

Abbott R, Albach D, Ansell S, Arntzen JW, Baird SJE, Bierrne N, Boughman J, Brelsford A, Buerkle CA, Buggs R, Butlin RK, Dieckmann U, Eroukhmanoff F, Grill A, Cahan SH, Hermansen JS, Hewitt G, Hudson AG, Jiggins C, Jones J, Keller B, Marczewski T, Mallet J, Martinez-Rodriguez P, Möst M, Mullen S, Nichols R, Nolte AW, Parisod C, Pfennig K, Rice AM, Ritchie MG, Seifert B, Smadja CM, Stelkens R, Szymura JM, Väinölä R, Wolf JBW, Zinner D. 2013. Hybridization and speciation. *Journal of Evolutionary Biology* **26**:229–246.

Abbott RJ, Brennan AC. 2014. Altitudinal gradients, plant hybrid zones and evolutionary novelty. *Philosophical Transactions of the Royal Society Series B* **369**:20130346.

Barton N, Bengtsson BO. 1986. The barrier to genetic exchange between hybridising populations. *Heredity* **57**:357–376.

Barton NH, De Cara MAR. 2009. The evolution of strong reproductive isolation. *Evolution* **63**:1171–1190.

Bierne N, Welch J, Loire E, Bonhomme F, David P. 2011. The coupling hypothesis: why genome scans may fail to map local adaptation genes. *Molecular Ecology* **20**:2044–2072.

Bouck A, Wessler SR, Arnold ML. 2007. QTL analysis of floral traits in Louisiana Iris hybrids. *Evolution* **61**:2308–2319.

Breitling R, Li Y, Tesson BM, Fu J, Wu C, Wiltshire T, Gerrits A, Bystrykh LV, De Haan G, Su AI, Jansen RC. 2008. Genetical genomics: spotlight on QTL hotspots. *PLoS Genetics* **4**:e1000232.

Brennan AC, Bridle JR, Wang A-L, Hiscock SJ, Abbott RJ. 2009. Adaptation and selection in the *Senecio* (Asteraceae) hybrid zone on Mount Etna, Sicily. *New Phytologist* **183**:702–717.

Brennan AC, Hiscock SJ, Abbott RJ. 2014. Interspecific crossing and genetic mapping reveal intrinsic genomic incompatibility between two *Senecio* species that form a hybrid zone on Mount Etna, Sicily. *Heredity* **113**:195–204.

Chapman MA, Hiscock SJ, Filatov DA. 2013. Genomic divergence during speciation driven by adaptation to altitude. *Molecular Biology and Evolution* **30**:2553–2567.

Chapman MA, Hiscock SJ, Filatov DA. 2016. The genomic bases of morphological divergence and reproductive isolation driven by ecological speciation in *Senecio* (Asteraceae). *Journal Evolutionary Biology* **29**:98–113.

Churchill GA, Doerge RW. 1994. Empirical threshold values for quantitative trait mapping. *Genetics* **138**:963–971.

Coyne JA, Orr HA. 2004. *Speciation*. Sunderland, MA: Sinauer Associates.

Cruikshank TE, Hahn MW. 2014. Reanalysis suggests that genomic islands of speciation are due to reduced diversity, not reduced gene flow. *Molecular Ecology* **23**:3133–3157.

Excoffier L, Lischer HEL. 2010. Arlequin suite ver 3.5: a new series of programs to perform population genetics analyses under Linux and Windows. *Molecular Ecology Resources* **10**:564–567.

Feder JL, Nosil P. 2010. The efficacy of divergence hitchhiking in generating genomic islands during ecological speciation. *Evolution* **64**:1729–1747.

Feder JL, Roethele JB, Filchak K, Niedbalski J, Romero-Severson J. 2003. Evidence for inversion polymorphism related to sympatric host race formation in the apple maggot fly, *Rhagoletis pomonella*. *Genetics* **163**:939–953.

Fenster CB, Galloway LF, Chao L. 1997. Epistasis and its consequences for the evolution of natural populations. *Trends in Ecology and Evolution* **12**:282–286.

Filatov DA, Osborne OG, Papadopulos AS. 2016. Demographic history of speciation in a *Senecio* altitudinal hybrid zone on Mt. Etna. *Molecular Ecology*, in press. doi:10.1111/mec.13618

Fishman L, Kelly AJ, Morgan E, Willis JH. 2001. A genetic map in the *Mimulus guttatus* species complex reveals transmission ratio distortion due to heterospecific interactions. *Genetics* **159**:1701–1716.

Foll M, Gaggiotti O. 2008. A genome-scan method to identify selected loci appropriate for both dominant and codominant markers: a Bayesian perspective. *Genetics* **180**:977–993.

Gagnaire P-A, Normandeau E, Pavey SA, Bernatchez L. 2013. Mapping phenotypic, expression and transmission ratio distortion QTL using RAD markers in the Lake Whitefish (*Coregonus clupeaformis*). *Molecular Ecology* **22**:3036–3048.

Gompert Z, Parchman TL, Buerkle CA. 2012. Genomics of isolation in hybrids. *Philosophical Transactions of the Royal Society Series B* **367**:439–450.

James JK, Abbott RJ. 2005. Recent, allopatric, homoploid hybrid speciation: the origin of *Senecio squalidus* (Asteraceae) in the British Isles from a hybrid zone on Mount Etna, Sicily. *Evolution* **59**:2533–2547.

Jiang C, Zeng ZB. 1995. Multiple trait analysis of genetic mapping for quantitative trait loci. *Genetics* **140**:1111–1127.

Jones FC, Grabherr MG, Chan YF, Russell P, Mauceli E, Johnson J, Swofford R, Pirun M, Zody MC, White S, Birney E, Searle S, Schmutz J, Grimwood J, Dickson MC, Myers RM, Miller CT, Summers BR, Knecht AK, Brady SD, Zhang H, Pollen AA, Howes T, Amemiya C, Broad Institute Genome Sequencing Platform & Whole Genome Assembly Team, Lander ES, Di Palma F, Lindblad-Toh K, Kingsley DM. 2012. The genomic basis of adaptive evolution in threespine sticklebacks. *Nature* **484**:55–61.

Kalinowsky ST. 2005. HP-RARE 1.0: a computer program for performing rarefaction on measures of allelic richness. *Molecular Ecology Notes* **5**:187–189.

Kao C-H, Zeng ZB, Teasdale RD. 1999. Multiple interval mapping for quantitative trait loci. *Genetics* **152**:1203–1216.

Kirkpatrick M, Barton N. 2006. Chromosome inversions, local adaptation and speciation. *Genetics* 173:419–434.

Lexer C, Rosenthal DM, Raymond O, Donovan LA, Rieseberg LH. 2005. Genetics of species differences in the wild annual sunflowers, *Helianthus annuus* and *H. petiolaris*. *Genetics* 169:2225–2239.

Lincoln SE, Daly MJ, Lander ES. 1993. *Constructing genetic linkage maps with MAPMAKER/EXP version 3.0: A tutorial and reference manual*, 3rd edn, Technical Report. Cambridge, MA: Whitehead Institute for Biomedical Research.

Lindtke D, Buerkle CA. 2015. The genetic architecture of hybrid incompatibilities and their effect on barriers to introgression in secondary contact. *Evolution* 69:1987–2004.

Lowry DB, Willis JH. 2010. A widespread chromosomal inversion polymorphism contributes to a major life-history transition, local adaptation, and reproductive isolation. *PLoS Biology* 8:e1000500.

Muir G, Osborne OG, Sarasa J, Hiscock SJ, Filatov DA. 2013. Recent ecological selection on regulatory divergence is shaping clinal variation in *Senecio* on Mount Etna. *Evolution* 67:3032–3042.

Nosil P. 2012. *Ecological speciation*. Oxford, UK: Oxford University Press.

Nosil P, Feder JL. 2012. Genomic divergence during speciation: causes and consequences. *Philosophical Transactions of the Royal Society Series B* 367:332–342.

Nosil P, Funk DJ, Ortíz-Barrientos D. 2009. Divergent selection and heterogeneous genomic divergence. *Molecular Ecology* 18: 375–402.

Orr HA, Turelli M. 2001. The evolution of postzygotic isolation: accumulating Dobzhansky-Muller incompatibilities. *Evolution* 55: 1085–1094.

Osborne OG, Batstone TE, Hiscock SJ, Filatov DA. 2013. Rapid speciation with gene flow following the formation of Mt. Etna. *Genome Biology and Evolution* 5:1704–1715.

Paterson AH. 2002. What has QTL mapping taught us about plant domestication? *New Phytologist* 154:591–608.

Peakall R, Smouse PE. 2006. GENALEX 6: genetic analysis in Excel. Population genetic software for teaching and research. *Molecular Ecology Notes* 6:288–295.

R Development Core Team. 2011. *R: a language and environment for statistical computing*. Vienna, Austria: R Foundation for Statistical Computing.

Renaut S, Maillet N, Normandeau E, Sauvage C, Derome N, Rogers SM, Bernatchez L. 2012. Genome-wide patterns of divergence during speciation: the lake whitefish case study. *Philosophical Transactions of the Royal Society Series B* 367:354–363.

Rieseberg LH, Raymond O, Rosenthal DM, Lai Z, Livingstone K, Nakazato T, Durphy JL, Schwarzbach AE, Donovan LA, Lexer C. 2003. Major ecological transitions in wild sunflowers facilitated by hybridization. *Science* 301:1211–1216.

Rogers SM, Bernatchez L. 2007. The genetic architecture of ecological speciation and the association with signatures of selection in natural lake whitefish (*Coregonus* sp. Salmonidae) species pairs. *Molecular Biology and Evolution* 24:1423–1438.

Rogers SM, Mee JA, Bowles E. 2013. The consequences of genomic architecture on ecological speciation in postglacial fishes. *Current Zoology* 59:53–71.

Servedio MR, Van Doorn GS, Kopp M, Frame AM, Nosil P. 2011. Magic traits in speciation: 'magic' but not rare? *Trends in Ecology and Evolution* 26:389–397.

Smadja CM, Butlin RK. 2011. A framework for comparing processes of speciation in the presence of gene flow. *Molecular Ecology* 20: 5123–5140.

Stinchcombe JR, Hoekstra HE. 2008. Combining population genomics and quantitative genetics: finding the genes underlying ecologically important traits. *Heredity* 100:158–170.

Strasburg JL, Sherman NA, Wright KM, Moyle LC, Willis JH, Rieseberg LH. 2012. What can patterns of differentiation across plant genomes tell us about adaptation and speciation? *Philosophical Transactions of the Royal Society Series B* 367: 364–373.

Taylor SJ, Rojas LD, Ho SW, Martin NH. 2012. Genomic collinearity and the genetic architecture of floral differences between the homoploid hybrid species *Iris nelsonii* and one of its progenitors, *Iris hexagona*. *Heredity* 110:63–70.

Turelli M, Barton NH, Coyne JA. 2001. Theory and speciation. *Trends in Ecology and Evolution* 16:330–343.

Turner TL, Hahn MW, Nuzhdin SV. 2005. Genomic islands of speciation in *Anopheles gambiae*. *PLoS Biology* 3:e285.

Twyford AD, Friedman J. 2015. Adaptive divergence in the monkey flower *Mimulus guttatus* is maintained by a chromosomal inversion. *Evolution* 69:1476–1486.

Via S, West J. 2008. The genetic mosaic suggests a new role for hitchhiking in ecological speciation. *Molecular Ecology* 17: 4334–4345.

Wang S, Basten CJ, Zeng Z-B. 2011. *Windows QTL Cartographer 2.5*. Raleigh, NC: Department of Statistics, North Carolina State University.

Whiteley AR, Derome N, Rogers SM, St-Cyr J, Laroche J, Labbe A, Nolte A, Renaut S, Jeukens J, Bernatchez L. 2008. The phenomics and expression quantitative trait locus mapping of brain transcriptomes regulating adaptive divergence in Lake Whitefish species pairs (*Coregonus* sp.). *Genetics* 180:147–164.

Wu C-I. 2001. The genic view of the process of speciation. *Journal of Evolutionary Biology* 14:851–865.

Yeaman S. 2013. Genomic rearrangements and the evolution of clusters of locally adaptive loci. *Proceedings of the National Academy of Sciences of the USA* 110:E1743–E1751.

Yeaman S, Whitlock MC. 2011. The genetic architecture of adaptation under migration–selection balance. *Evolution* 65: 1897–1911.

Zeng Z-B. 1994. Precision mapping of quantitative trait loci. *Genetics* 136:1457–1468.

17

Contrasting patterns of genetic variation in core and peripheral populations of highly outcrossing and wind pollinated forest tree species

Błażej Wójkiewicz*[1], Monika Litkowiec[1] and Witold Wachowiak[1,2]

[1] Institute of Dendrology, Polish Academy of Sciences, Parkowa 5, Kórnik 62-035, Poland
[2] Faculty of Biology, Adam Mickiewicz University, Institute of Environmental Biology, Umultowska 89, Poznań 61-614, Poland

Associate Editor: Kristina Hufford

Abstract. Gene flow tends to have a homogenising effect on a species' background genetic variation over large geographical areas. However, it is usually unknown to what extent the genetic structure of populations is influenced by gene exchange between core and peripheral populations that may represent stands of different evolutionary and demographic history. In this study, we looked at the patterns of population differentiation in Scots pine—a highly outcrossing and wind pollinated conifer species that forms large ecosystems of great ecological and economic importance in Europe and Asia. A set of 13 polymorphic nuclear microsatellite loci was analysed to infer the genetic relationships among 24 populations (676 individuals) from Europe and Asia Minor. The study included specimens from the primary continuous range and from isolated, marginal stands that are considered to be autochthonous populations representative of the species' putative refugial areas. Despite their presumably different histories, a similar level of genetic variation and no evidence of a population bottleneck was found across the populations. Differentiation among populations was relatively low (average $F_{ST} = 0.035$); however, the population structure was not homogenous, which was clearly evident from the allelic frequency spectra and Bayesian assignment analysis. Significant differentiation over short geographical distances was observed between isolated populations within the Iberian and Anatolian Peninsulas (Asia Minor), which contrasted with the absence of genetic differentiation observed between distant populations e.g., between central and northern Europe. The analysed populations were assigned to several groups that corresponded to the geographical regions of their occurrence. These results will be useful in genetics studies in Scots pine that aim to link nucleotide and phenotypic variation across the species distribution range and for development of sustainable breeding and management programs.

Keywords: Demographic history; genetic structure; glacial refugia; phylogeography; *Pinus sylvestris*; population history; recolonization.

* Corresponding author's e-mail address: bwojkiew@man.poznan.pl

Introduction

Demographic and evolutionary processes interplay to shape genetic variation that is crucial to maintaining species' adaptive responses to changing environments. Genetic variation among plant populations in the Northern Hemisphere has been shaped by range shifts and recolonization following the last glacial maximum (25–18 000 years ago) (Petit *et al.* 2003; Nosil and Feder 2013). The assessment of the genetic relationships among natural populations across the species distribution is important for tree management and for breeding and gene conservation programs, particularly in the face of ongoing environmental changes (Savolainen *et al.* 2011).

Forest trees are known to form large, wind pollinated populations and maintain a high level of genetic and phenotypic variation in comparison to other plant species (Petit and Hampe 2006). High diversity ensures that these long-lived organisms can survive and evolve under changing environmental conditions. Studies of historical processes such as population size fluctuations and geographical range shifts are needed to better understand the effect of demographic factors that influence background genetic variation and to effectively contrast neutral variation with that resulting from natural selection (Luikart *et al.* 2003; Li *et al.* 2012). However, currently available genetic data still lack sufficient information to fully reflect the usually complex demographic history of many forest tree species. For highly outcrossing and wind pollinated species, gene flow is supposed to have homogenising effects on the background, neutral genetic variation between populations over large geographical distances. However, its effect on the distribution of genetic variation between populations from core range vs. marginal populations of the species is not that clear, especially for long-lived temperate forest tree species.

In the present study, we focused on Scots pine (*Pinus sylvestris*), which is one of the most ecologically and economically important forest-forming tree species in Eurasia. Some former phylogeographic studies of this pine based on isozyme polymorphisms, organelle DNA and palynological records have shown that the most abundant populations of Scots pine survived the cold periods of the Pleistocene within southern Eurasia in the Iberian, Apennine and Balkan Peninsulas and in the Anatolian and Caucasus Mountains (Willis *et al.* 1998; Willis and van Andel 2004; Cheddadi *et al.* 2006; Naydenov *et al.* 2007; Pyhäjärvi *et al.* 2008; Soto *et al.* 2010; Buchovska *et al.* 2013). The Alps, the Carpathians and Moscow are considered refugial areas for other cold tolerant conifer species including Norway spruce (*Picea abies*) (Tollefsrud *et al.* 2008, 2009). Although there are strong indications regarding the location of some putative refugial stands, it is less clear how migration and gene flow have influenced the patterns of genetic variation at neutral gene markers across the present range of the species. Consequently, the relationships among the gene pools of populations from core and peripheral distribution are not well elucidated.

The aim of this study was to assess the genetic variation and structure of Scots pine populations from central and north Europe in relation to its marginal populations from the European and south-west Asiatic refugial areas based on the analysis of nuclear simple sequence repeat (*n*SSR) markers. Nuclear markers in pines are distributed by seeds and pollen and could potentially be dispersed at large geographical distances. Therefore, the identification of populations with divergent genetic backgrounds could suggest the existence of distinct populations that do not share a recent history. To our knowledge, this species has not yet been investigated on such a large geographical scale using this type of neutral marker. In this study, we used a set of 13 *n*SSR loci to examine genetic diversity and test for the existence of populations of different origin. We aimed to check if there is any difference between core and peripheral populations considering the presumably homogenising effect of long distance gene flow. Information regarding the genetic relationships among populations is particularly important to advance studies of the genetic architecture of observed variation in phenotypic traits that have been shaped by selection and local adaptation within the Scots pine distribution.

Methods

Plant material, DNA extraction and microsatellite genotyping

The study comprised 24 Scots pine populations (Fig. 1), 8 of which were from the core continuous species distribution from central and northern Europe. The rest of the populations analysed were collected from isolated, peripheral stands on the Iberian and Anatolian Peninsulas, the Massif Central of France, Scotland and the Balkans [**see Supporting Information—Table S1**]. The number of samples per population varied from 22 to 49 individuals with a total of 676 individuals (Table 1). Genomic DNA was extracted from needles using a CTAB protocol (Dumolin *et al.* 1995). The initial set of 22 nuclear microsatellite markers originally identified in pine species (Soranzo *et al.* 1998; Elsik *et al.* 2000; Liewlaksaneeyanawin *et al.* 2004; Sebastiani *et al.* 2012) were screened for their ability to provide repeatable, high quality results, sufficient polymorphism and unambiguous

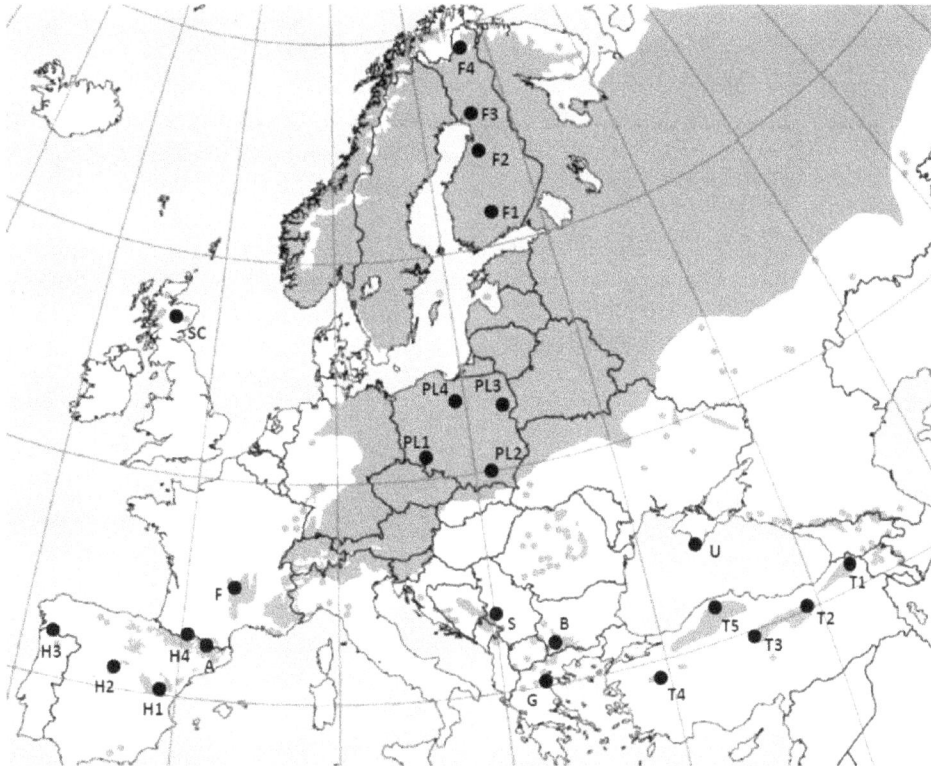

Figure 1. Geographic location of Scots pine populations included in this study. Acronyms and geographic coordinates of the populations are listed in **Supporting Information—Table S1**.

allele binding. The final set of loci used in this study included 13 nSSRs that provided high-quality amplification products. DNA amplification was carried out in three multiplex reactions including loci psyl2, psyl18, psyl25, psyl36, psyl42, psyl44 and psyl57 (multiplex 1); Spag7.14, PtTX2146, PtTX3107 and Spac11.4 (multiplex 2) and PtTX3025 and PtTX4011 (multiplex 3). The PCR for each sample was conducted in a total volume of 10 μL containing 5 μL of Qiagen Multiplex Master Mix (2×), 0.2 μL of primer mix (20 μM), 1 μL of Q-Solution (5×), 0.8 μL RNase-free water and 3 μL of DNA 84 template (approximately 10–20 ng). The following PCR amplification conditions were used: multiplex 1, initial denaturation at 95 °C for 15 min, 38 cycles of denaturation at 94 °C for 30 s, annealing at 57 °C for 90 s, extension at 72 °C for 90 s and final extension at 72 °C for 10 min; multiplex 2, initial denaturation at 95 °C for 15 min, 30 cycles of denaturation at 94 °C for 30 s, annealing at 55 °C for 90 s, extension at 72 °C for 90 s and final extension at 72 °C for 10 min; multiplex 3, initial denaturation at 95 °C for 15 min, 10 cycles of denaturation at 94 °C for 30 s, annealing at 60 °C for 40 s with temperature decreasing by 1 degree per cycle, extension at 72 °C for 90 s, 36 cycles of denaturation at 94 °C for 30 s, annealing at 50 °C for 40 s, extension at 72 °C for 90 s and final extension at 72 °C for 10 min. The fluorescently labelled PCR products, along with a size standard

(GeneScan 500 LIZ), were separated on a capillary sequencer ABI 3130 (Thermo Fisher Scientific, Waltham, USA). The allele's size was determined using GeneMapper software (ver. 4.0; Life Technologies).

Allelic diversity and within-population genetic variation

Genotypic disequilibrium between pairs of loci was tested at the single population level and across all populations with a Fisher's exact test using ARLEQUIN 3.11 (Excoffier and Lischer 2010). The allelic diversity of the studied loci and within-population genetic variation were estimated based on the following parameters: the number of alleles per locus (Al), the mean number of alleles per population (Np), the mean number of effective alleles per population (Ne), the mean number of private alleles per population (Pa), the observed heterozygosity (Ho) and the unbiased expected heterozygosity (He), all of which were computed using GenAlEx 6 (Peakall and Smouse 2006). In accordance with earlier studies that showed that microsatellites are known to be susceptible to genotyping errors (Guichoux et al. 2011), the null allele frequency for each loci was calculated using the EM algorithm with FreeNA software (Chapuis and Estoup 2007). We used FSTAT v 2.9.3 (Goudet 2001) to estimate gene diversity (Gd), rarefied allelic richness (A_{R22}) for a minimum sample size of 22 individuals and inbreeding coefficient values (F_{is}). The deviation of genotypic frequencies from Hardy–Weinberg equilibrium (HWE) were also identified utilizing the inbreeding coefficients (F_{is}; Weir and Cockerham 1984) with a correction for null alleles (F_{isNull}) for each population using the Bayesian method implemented in INEST 2.0 software (Chybicki 2015). The evaluation was performed using the IIM model with 100 000 MCMC iterations, storing every 100th value and with a burn-in period of 10 000. A Bayesian procedure based on the Deviance Information Criterion (DIC) was used to determine the statistical significance of the inbreeding component by comparing the full model with the random mating model (under the assumption $F_{is} = 0$).

Genetic differentiation between populations

To estimate the proportion of the total genetic variation due to differentiation among populations, an Analysis of Molecular Variance (AMOVA) based on two distance methods (F_{ST} and R_{ST}) was conducted using ARLEQUIN 3.11. Moreover, due to the presence of null alleles, the global, pairwise and within-geographic regions F_{ST} were calculated using FreeNA software. FreeNA applies the ENA (Excluding Null Alleles, F_{ST}ENA) correction method to effectively correct for the positive bias induced by the presence of null alleles in the F_{ST} estimation. Bootstrap 95 % confidence intervals (CI) were calculated for the global F_{ST}ENA values using 2000 replicates across the loci. The statistical significance of the F_{ST} values was verified with ARLEQUIN 3.11.

To evaluate the ability of the stepwise mutation model (SMM) to differentiate among populations and geographical regions, which in turn indicates whether phylogeographical structures exist, the computed F_{ST} and R_{ST} were compared. To test whether the difference between values of R_{ST} and permuted pR_{ST} (which corresponds to F_{ST}) was significant, the permutation test proposed by Hardy et al. (2003) was implemented in the program SpaGeDi 1.3d (Hardy and Vekemans 2002).

The genetic population structure (in the case of microsatellite markers) can arise due to isolation by distance (IBD), range expansions, diffusion of genes through space in migratory events and/or allelic surfing (Diniz-Filho et al. 2013). Because of that a Mantel test (1967) was applied to evaluate spatial processes driving populations structure by comparing the matrixes of pairwise geographic (logarithmic scale) and pairwise genetic (measured as $F_{ST}/(1-F_{ST})$) distances. The statistical significance of the correlation was calculated for all populations and sets of populations located along latitudinal and longitudinal transects using 9999 permutations with GenAlEx 6.

Population clustering and phylogenetic relationships

Principal Coordinates Analysis (PCoA) was applied to visualize the patterns of the genetic structure of the populations using a pairwise F_{ST}ENA matrix and GenAlEx 6 software. Phylogenetic relationships between the populations were investigated using POPTREEW (Takezaki et al. 2014). The phylogenetic tree was constructed from allele frequency data using the neighbour-joining (NJ) method. This method allows faithful depiction of genetic structure for some populations that have an isolation-by-distance population structure (Kalinowski 2009). Nei's standard genetic distance (D_{ST}) (Nei 1972) was chosen as a distance measure for the construction of the phylogeny. The statistical robustness of the branches was evaluated with 1000 bootstrap replicates.

The assignment of individuals and populations to genetically distinct groups was conducted using the Bayesian clustering method with the software STRUCTURE 2.3.4 (Pritchard et al. 2000; Falush et al. 2003; Hubisz et al. 2009). This program uses a Markov chain Monte Carlo (MCMC) algorithm to assign individuals to a given number of genetic clusters (K) without considering sampling origins and assuming that each cluster is in optimal Hardy–Weinberg (H–W) and linkage equilibrium (LE). The correlated allele frequencies and admixture model used allowed for mixed recent ancestry of individuals and assigned the proportion of the genome of each individual to the inferred clusters. Moreover, because all the microsatellite loci used in this study were affected by null alleles (see Results section), the recessive alleles option was chosen. Twenty independent runs were performed for each K, from $K = 1$ to 24, with burn-in lengths of 500 000 and 100 000 iterations. The probability distributions of the data (Ln$P[D]$) and the ΔK values (Evanno et al. 2005) were visualized in the STRUCTURE HARVESTER Web application (Earl and von Holdt 2011). Following Bayesian clustering, the hierarchical distribution of genetic variation was characterized using an analysis of molecular variance (AMOVA). A three-level AMOVA was conducted in ARLEQUIN 3.11 and significance was obtained via 10 000 random permutations.

Tests for genetic bottleneck model

The program Bottleneck v.1.2.02 (Cornuet and Luikart 1996) was used to evaluate whether the examined populations suffered a severe genetic bottleneck or had experienced recent reductions in their size. A Wilcoxon-signed rank test of heterozygosity excess was used to evaluate the significance of a potential bottleneck. The analysis was performed under three different models of microsatellite evolution: the SMM model, the infinite

Table 1. Genetic variation of Scots pine populations based on thirteen polymorphic nSSR loci. Nr and pop, population number and acronyms with reference to Fig. 1 and **Supporting Information — Table S1**, N, number of analyzed individuals; Np, the mean number of alleles per population; Ne, mean number of effective alleles per population; A_{R22}, allelic richness for a minimum sample size of 22 individuals; Pa, mean number of private alleles per population; Gd, gene diversity; Ho, observed heterozygosity; He, unbiased expected heterozygosity; F_{is}, inbreeding coefficient (*$P < 0.001$); F_{isNull}, inbreeding coefficient with null allele correction.

Nr.	Pop	N	Np	Ne	A_{R22}	Pa	Gd	Ho	He	F_{is}	F_{isNull}
1.	T1	25	5.46	3.39	5.19	0.00	0.53	0.477	0.515	0.09*	0.04 (0–0.12)
2.	T2	25	5.77	3.31	5.37	0.00	0.51	0.423	0.500	0.17*	0.07 (0–0.15)
3.	T3	25	5.54	2.91	5.33	0.08	0.52	0.456	0.507	0.12*	0.07 (0–0.17)
4.	T4	25	5.77	3.16	5.47	0.08	0.52	0.418	0.502	0.19*	0.08 (0–0.15)
5.	T5	25	6.23	3.41	5.83	0.23	0.53	0.483	0.521	0.09*	0.08 (0–0.15)
6.	U	25	4.69	3.10	5.59	0.15	0.50	0.377	0.516	0.28*	0.07 (0–0.13)
7.	G	31	6.62	3.45	5.85	0.00	0.54	0.494	0.536	0.09*	0.06 (0–0.06)
8.	B	25	6.31	3.62	5.87	0.00	0.54	0.490	0.530	0.09*	0.02 (0–0.08)
9.	S	26	6.15	3.49	5.68	0.00	0.53	0.481	0.515	0.08*	0.03 (0–0.13)
10.	H1	29	7.15	3.50	5.84	0.15	0.55	0.434	0.505	0.15*	0.06 (0–0.17)
11.	H2	31	7.00	3.25	5.27	0.15	0.49	0.370	0.439	0.18*	0.07 (0–0.15)
12.	H3	29	6.31	3.36	5.52	0.08	0.51	0.418	0.489	0.16*	0.07 (0–0.10)
13.	A	32	5.92	3.23	5.26	0.08	0.50	0.440	0.487	0.12*	0.05 (0–0.08)
14.	H4	32	5.92	3.25	5.31	0.00	0.49	0.456	0.481	0.07*	0.03 (0–0.06)
15.	FR	25	6.31	3.51	5.93	0.08	0.50	0.444	0.487	0.11*	0.02 (0–0.12)
16.	SC	39	6.31	3.01	5.34	0.23	0.50	0.437	0.491	0.12*	0.04 (0–0.10)
17.	PL1	33	6.54	3.84	6.21	0.00	0.52	0.467	0.496	0.07*	0.04 (0–0.10)
18.	PL2	45	5.85	3.70	5.66	0.00	0.51	0.456	0.536	0.12*	0.03 (0–0.08)
19.	PL3	22	5.46	3.22	5.28	0.08	0.45	0.425	0.483	0.14*	0.03 (0–0.08)
20.	PL4	28	6.23	3.59	5.65	0.00	0.48	0.409	0.475	0.13*	0.04 (0–0.07)
21.	F1	25	6.38	3.47	5.79	0.00	0.50	0.471	0.489	0.08*	0.02 (0–0.06)
22.	F2	25	6.38	3.69	5.85	0.08	0.50	0.491	0.490	0.02*	0.02 (0–0.05)
23.	F3	25	5.77	3.12	5.43	0.08	0.49	0.452	0.500	0.11*	0.03 (0–0.08)
24.	F4	24	5.77	3.16	5.42	0.00	0.48	0.458	0.470	0.06*	0.02 (0–0.07)
Mean		28.2	6.08	3.36	5.58	0.06	0.51	0.447	0.498	0.12*	0.04 (0–0.07)

allele model (IAM) and the two-phases model of mutation (TPM) with parameters of 30 % multiple-step mutations and 70 % single-step mutations. In addition, the distribution of allele frequencies over all loci was examined for a 'mode-shift', which might indicate a bottlenecked population rather than a stable population.

Results

Allelic diversity and within-population genetic variation

The 13 nuclear microsatellite loci investigated were polymorphic, providing a total of 160 size variants. There was no significant linkage disequilibrium between pairs of loci across all populations ($P > 0.01$). The number of alleles per locus ranged from 3 (psyl25) to 40 (Spag7.14), with an average of 12.3 [**see Supporting Information— Table S2**]. The estimated frequency of null alleles for most of the loci was moderate ($< 6\%$, but generally $< 2\%$) with the exceptions occurring at three loci: psyl18 (8.7 %), Spag7.14 (8.3 %) and PtTX3107 (16.2 %) [**see Supporting Information—Table S2**].

The basic statistics for genetic variation within the populations are summarized in Table 1. More than five alleles were observed in each population with an average Ap = 6.08. The allelic richness measures obtained based on a minimum of 22 samples (A_{R22}) were comparable in all investigated populations and ranged from 5.1 in the T1 population to 6.2 in the PL1 population. The mean effective number of alleles was Ae = 3.4. The lowest number of effective alleles was observed in the Tokat–Yıldızeli population from Turkey (T3, Ae = 2.9), whereas the highest number of effective alleles was found in the population from southwestern Poland (PL1, Ae = 3.8). Twenty private alleles were also detected among some of the studied populations, and their frequency ranged from Pa = 0.0 % to Pa = 23.1%. Similar levels of gene diversity were found in all populations, which ranged from 0.48 to 0.53. However, pine populations from the peripheral stands appear to show greater diversity (with an average of 0.52) in comparison with populations from central and northern Europe (with an average of 0.48). The level of the overall observed heterozygosity (Ho) per population (average = 0.45, range: 0.38–0.49) was similar for all populations and slightly lower than the level of expected heterozygosity (He) (average = 0.50, range: 0.44–0.54). The inbreeding coefficients ranged from 0.017 in the F2 population from Finland to 0.282 in the T5 population from Turkey, with an overall mean of 0.120. However, the inbreeding coefficients may be highly overvalued due to the presence of null alleles. Because of this and based on our previous results that showed that null alleles were present at all investigated loci [**see Supporting Information—Table S2**], we used the IIM approach (see M&M) to partition out their influence on F_{isNull} (inbreeding coefficients with null alleles correction) value. Recalculated values of inbreeding coefficients, taking into account the frequency of null alleles (F_{isNull}), were much lower than those obtained previously and ranged from 0.014 to 0.081 with an average of 0.045. The values of F_{isNull} and F_{is} were significantly different from zero in all populations. However, it seems that deficiency of heterozygotes is mostly due to the presence of null alleles. Moreover, it should be noted that the highest values of both F_{is} and F_{isNull} were observed in peripheral pine populations from Turkey and Spain.

Genetic differentiation between populations

The analysis of molecular variance (AMOVA) based on the number of different alleles (F_{ST}) and the sum of the squared size differences (R_{ST}) showed that differentiation between Scots pine populations was low but significant ($F_{ST} = 0.035$, $R_{ST} = 0.032$; $P < 0.0001$). The majority of the variance was found within populations (Table 2). The global F_{ST} values, estimated both with and without the ENA correction, was $F_{ST} = 0.035$ and $F_{ST}ENA = 0.037$, respectively. The similarity of these values implies that the presence of null alleles is not a significant factor

affecting the level of genetic differentiation. The results of genetic differentiation among populations within the studied geographical regions are presented in Table 2. The greatest differentiation was found between populations from Turkey and Spain, whereas the within-region F_{ST} values obtained for the Balkans, Poland and Finland were not statistically significant. Most of the pairwise F_{ST} population values were significant ($P < 0.001$) [**see Supporting Information—Table S3**]. The greatest difference (0.11) was between the T4 population from Çatacık in Turkey and the H2 population from the Sierra de Neila in Spain, and the lowest ($F_{ST} < 0.01$) was between the PL2 population located in southern Poland and the PL4 population located in northern Poland.

The permutation test, by which global R_{ST} was compared against the distribution of 10 000 pR_{ST} values, did not detect a significant difference between these parameters ($R_{ST} = 0.032$; p$R_{ST} = 0.025$; CIpR_{ST} 95 % = 0.01–0.04; p_{H1}: $R_{ST} > pR_{ST} = 0.164$). This suggests an absence of phylogeographic structure and that gene flow is high compared with the mutation rate. However, the results of the Mantel test correlation between genetic distance and the logarithm of geographic distance among all *P. sylvestris* populations indicated that the genetic diversity is structured in geographic space ($R = 0.39$, $P < 0.05$). The strongest correlation was found along a longitudinal transect among populations from Anatolia and the Iberian Peninsula ($R = 0.73$, $P < 0.05$). The spatial genetic structure among all populations was rather weak, whereas only 15 % ($R_2 = 0.152$) of the genetic divergence was explained by geographical distance. The strong structure (53 % of the genetic divergence could be explained by geographical distance) found among south peripheral populations is most likely due to the presence of geographical barriers along Mediterranean basin transects which intensify the effects of the process of isolation by distance (IBD).

Population clustering and phylogenetic relationships

The genetic structure of the Scots pine populations based on the pairwise $F_{ST}ENA$ matrix is illustrated in the PCoA plot (Fig. 2). The east–west subdivision of the southernmost peripheral populations was clearly shown by the first coordinate, which explained more than 25 % of the variation. The second variable (second coordinate), which was responsible for more than 20 % of the total variation, separated the pine populations from central and northern Europe from those in southern Eurasia. Moreover, according to the F_{ST} values obtained for each geographical region (Table 2), populations from both Turkey and Spain formed a much more heterogeneous

Table 2. Analysis of molecular variance (AMOVA) at 13 nSSR loci. (a) Assuming no population structure; (b) among populations within geographical regions; (c) assuming population structure as defined by allelic frequency spectra and Bayesian assignment tests (Table 3).

Source of variation		d.f.	Sum of squares	Variance components	Percentage of variation	P
(a)						
Among populations	F_{ST}	23	240.861	0.12172	3.55	< 0.0001
	R_{ST}		48 995.78	25.16	3.50	< 0.0001
Within populations	F_{ST}	652	4318.50	3.31	96.45	< 0.0001
	R_{ST}		903 431.90	693.43	96.50	< 0.0001
(b)						
Turkey	F_{ST}	5	46.48	0.11	3.25	< 0.0001
		100	954.49	3.37	96.75	< 0.0001
Spain	F_{ST}	4	34.31	0.08	2.38	< 0.0001
		124	977.59	3.29	97.62	< 0.0001
Balkans	F_{ST}	2	8.47	0.01	0.22	0.9984
		51	556.43	3.49	99.78	< 0.0001
Poland	F_{ST}	3	18.59	0.03	1.13	0.2424
		95	812.06	3.24	98.87	< 0.0001
Finland	F_{ST}	3	12.60	0.01	0.49	0.9918
		74	622.43	3.21	99.51	< 0.0001
(c)						
Among groups	F_{ST}	5	120.93	0.08	2.51	< 0.0001
Among populations within groups		18	109.37	0.04	1.36	< 0.0001
Among individuals within populations		652	2327.52	0.35	10.64	< 0.0001
Within individuals		676	1932.00	2.85	85.48	< 0.0001

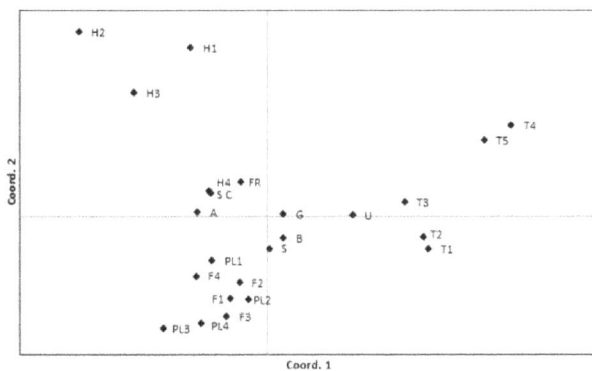

Figure 2. Principal Coordinate Analysis (PCoA) based on pairwise population F_{ST}ENA values. Acronyms of Scots pine populations are listed in **Supporting Information—Table S1**.

group in comparison with populations from the Balkans and central and northern Europe.

The relationships between populations are illustrated in **Supporting Information—Figure S1**, which shows the

phylogenetic tree based on the NJ method [**see Supporting Information—Figure S1**].

The Bayesian assignment of samples obtained with STRUCTURE indicated that three clusters ($K = 3$) provide the most probably representation of the overall genetic structure of the analysed Scots pine populations. The admixture proportions of each of the three gene pools (clusters) were estimated and differed among populations (Fig. 3). The highest frequency of cluster 1 (indicated in red) was found in populations from Turkey (T1, T2, T3, T4 and T5), with an average of 77.6 %, whereas cluster 2 (indicated in green) was most frequent in populations from Poland and Finland (PL1, PL2, PL3, PL4, F1, F2, F3 and F4), with a mean of 61.5%. Three populations from Spain (H1, H2 and H3) showed the highest frequencies of cluster 3 (indicated in blue; mean value of 69.1%). Pines from Ukraine, Balkans, north of Spain, Massif Central in France and Scotland were classified as mixed populations. However, the gene pool of the population from Ukraine exhibited a predominance of cluster 1,

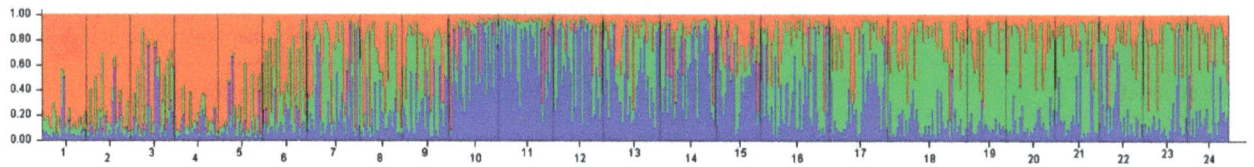

Figure 3. The STRUCTURE assignment of each individual to inferred genetic clusters (marked in different colours). The study populations are separated by thick black lines. Population numbers (1–24) are listed in Table 1 and **Supporting Information—Table S1**.

Table 3. Genetic differentiation (F_{ST}) between groups of Scots pine populations according to population grouping analysis and geographical location of populations (*$P < 0.001$).

	1	2	3	4	5
1. Turkey and Ukraine	0.000				
2. Greece, Serbia and Bulgaria	0.021*	0.000			
3. Southern Spain (H1, H2 and H3)	0.063*	0.039*	0.000		
4. Northern Spain (H4), Andorra and France	0.035*	0.017*	0.025*	0.000	
5. Scotland	0.041*	0.021*	0.029*	0.010*	0.000
6. Poland and Finland	0.035*	0.009*	0.042*	0.014*	0.016*

whereas populations from the Balkans showed a predominance of cluster 2. Gene pools from north of the Iberian Peninsula (A, H4) and the population from France exhibited a higher frequency of cluster 3. Clusters 2 and 3 contained the dominant frequencies in the gene pool of Scottish pine populations, with a higher frequency of cluster 2.

Based on these findings and the pairwise F_{ST} matrix [**see Supporting Information—Table S3**], an AMOVA analysis was conducted between six groups of populations. The largest group consisted of the populations from Turkey and Ukraine (Group 1). The next largest groups included populations from the Balkans (Group 2), southern Spain (Group 3), north of the Iberian Peninsula (populations H4 and A) and the population from the Massif Central in France (Group 4). Two other groups were formed by the population from Scotland (Group 5) and the populations from central and northern Europe (Group 6). The AMOVA results and the pairwise F_{ST} values between these groups of populations are presented in Tables 2 and 3, respectively. Significant genetic differentiation over short geographical distances was observed between the populations within Spain and Turkey (Table 2) but not between populations from the Balkans, Poland or Finland. The highest degree of genetic differentiation was found between southern Spain (Group 3) and Turkey and Ukraine (Group 1) and between those geographical areas and the rest of the populations (Table 3; **see Supporting Information—Table S3**).

Tests for genetic bottleneck model

Despite the fact that some of the investigated populations were characterized by lower genetic diversity, no evidence for a recent genetic bottleneck or reductions in effective population size under the two-phased mutation model (TPM), infinite allele mutation model (IAM) or stepwise mutation model (SMM) was detected. Moreover, the mode-shift test, based on the frequency distribution of alleles, demonstrated a typical L-shaped mode for all populations, which is consistent with a non-bottleneck model.

Discussion

Genetic variation and differentiation

In our study, we evaluated the neutral genetic variation and genetic structure of Scots pine populations from the core and peripheral distribution in Europe and Asia Minor based on variation in nSSR markers. All the 13 microsatellite loci studied were polymorphic with a mean of 12.3 alleles per locus. Considerably lower numbers of alleles were found at the psyl loci (an average of 6.1) compared with the Spag and PtTX loci (an average of 30.5 and 14.0, respectively). The most variable loci were Spag7.14, Spac11.4 and PtTX2146. Therefore, these loci appear to be the most informative for population genetics studies of *P. sylvestris* that require a high resolution of markers for fine-scale analysis. Consistent with earlier studies on the genetic variation of conifer species, we found that the analysed loci contained some level of null alleles (Liewlaksaneeyanawin *et al.* 2004; Belletti *et al.* 2012; Sebastiani *et al.* 2012). However, as evidenced by our dataset, they did not significantly affect the observed patterns of variation across the studied populations.

Our data show that most of the genetic variation is located within the Scots pine populations. The level of genetic variation, as measured by basic statistics, was high and very similar across populations (Table 1) which is typical to woody species (Hamrick *et al.* 1992). The levels of observed heterozygosity were only slightly less than the values of unbiased expected heterozygosity and were primarily due to the presence of null alleles. These findings indicate that there is no significant allele frequency disequilibrium in any of the studied Scots pine populations. There were, however, some noticeable differences in the parameters describing the within-population genetic variation of peripheral populations vs. populations from the core distribution. In general, the peripheral populations had higher frequencies of private alleles, with the exception of the Balkan populations. An excess of private alleles was also observed in the peripheral Scots pine populations from Spain and Italy based on an analysis of allozyme and *n*SSR loci (Belletti *et al.* 2012; Prus-Głowacki *et al.* 2012). Those populations had also higher inbreeding coefficients in comparison with populations from the Balkans, Poland and Finland. The results of this study indicate isolation and probably a limited effective population size in the peripheral populations from Turkey (T2, T3, T4 and T5) and the south of Spain (H1, H2 and H3). The stronger isolation of these pine populations from the south regions apparently restricts gene flow and seems to cause a more distinct pattern of geographical variation in these studied regions. This conclusion is supported by the results of the Mantel test correlation analysis, which is based on genetic and geographical distance. We found that there was strong spatial structure and overall among-population differentiation between southern peripheral populations. This high differentiation most likely results from action of the processes of isolation-by-distance and genetic drift, which are most effective in isolated stands. It must be pointed, however, that the isolation did not have a strong negative impact on the level of within population genetic variation found in the peripheral pine populations. Similarly relationship was already demonstrated in the case of other plant species living in isolation, as for example in the Cheddar pink (Putz *et al.* 2015). Our results also show contrasting patterns of genetic variation between populations located within distinct geographical areas. Populations within the Iberian and Anatolian Peninsulas are separated by relatively short geographical distances but showed much higher divergence compared with populations from northern Europe and the Balkans, which are separated by several thousand kilometres. In some previous studies, relatively high genetic diversity ($F_{ST} = 0.058$) was found within the Scots pine natural range in Italy (Belletti *et al.* 2012) as compared to differentiation between populations from the Scandinavian region ($F_{ST} = 0.02$, Karhu *et al.*1996). Similarly, a high level of differentiation between *P. sylvestris* populations from mountain regions in Spain in relation to other European populations has been found (Prus-Głowacki and Stephan 1993; Prus-Głowacki *et al.* 2003). Moreover, a high level of differentiation between populations of Norway spruce from Bulgarian mountains in comparison to populations from other regions of natural distribution of this species in Europe was noted by Tollefsrud *et al.* (2008). Mountain regions of southern Europe are assumed to have provided particularly suitable habitats for species survival during last glacial period, because they allowed species to respond to climate oscillations using the altitudinal gradient (Hewitt 1996). Our results support the hypothesis of a possible long-lasting isolation of pinewoods in separate Iberian refugia (Rubiales *et al.* 2010) and some pine populations in separate refugial areas within the Anatolian Peninsula. Moreover these findings suggest that the homogenizing effect of gene flow via pollen dispersal on gene pool variation across geographical ranges may be limited.

The genetic relationships between populations

Although only a small percentage of the total variation was due to differentiation among populations (approximately 4%), we found clear signals of population sub-structuring at *n*SSR loci within the Scots pine distribution. Based on the analysis of allelic frequency spectra and population assignment methods, we could distinguish several groups of populations from distinct geographical locations. Five groups of populations are clearly represented at the PCoA plot where Scottish population is grouped with the populations from north of Iberia Peninsula and Massif Central in France. However taking into account the admixture proportions of each of the three gene pools in the analysed populations and the result of phylogenic relationship analysis, we decided to separate the population from Scotland as a distinct group because it could not be clearly attributed to any of the designated previously groups of populations. As a result, the AMOVA analysis was conducted between six groups of populations. The east–west subdivision was clearly shown between southernmost groups of populations from Turkey (Group 1) and Spain (Group 3). The differentiation between the Iberian and Anatolian populations was also the greatest based on the morphological and anatomical traits of needles from Scots pines in the Mediterranean Basin (Jasińska *et al.* 2014). Phenotypic and genetic differentiation among populations from these regions support the isolation of *P. sylvestris* in the southern European and Anatolian mountain regions and suggest that these populations are characterized by a different population history. In our study, the populations from Turkey and the south of Spain were significantly different from populations in the Balkans (Group 2), north of the Iberian Peninsula and the Massif Central (Group 4), Scotland (Group 5) and Poland and Finland (Group 6). In some previous mitochondrial DNA studies, unique haplotypes not observed in other parts of Europe were detected in Spanish and Turkish populations (Cheddadi *et al.* 2006; Naydenov *et al.* 2007; Pyhäjärvi *et al.* 2008). Our data show a divergence between populations within the Iberian Peninsula and indicate that southern populations have distinct gene pools compared with northern Spain and Andora populations. The latter population has probably contributed to the recolonization of Europe (or was admixed) after the last glacial maximum, as evidenced by genetic variation similar to populations from Massif Central and the Balkans and minor differentiation from populations within the continuous distribution. The subdivision of populations from the Iberian Peninsula has also been suggested by the analysis of biochemical and molecular markers (Prus-Głowacki *et al.* 2003, 2012; Robledo-Arnuncio *et al.* 2005).

The position of the Scottish populations is not straightforward. Scottish populations showed some similarity to groups from northern Spain and the Massif Central and populations from central and northern Europe. Previous pollen, allozyme and monoterpene studies suggested a west/east population subdivision within Scotland, reflecting different origins of the populations contributing to the postglacial colonization (Birks 1989; Kinloch *et al.* 1986; Sinclair *et al.* 1998). Based on this finding, two hypotheses have been put forth to explain this phenomenon. Kinloch *et al.* (1986) suggested that the western Scotland pine populations are relicts and might have survived the last glacial maximum in some region of the British Isles because they show relatively little genetic affinity to pines from the continuous range. The eastern populations were likely admixed with populations of continental origin or derived from that distribution. Scottish populations also show high levels of genetic polymorphism at nuclear gene loci and patterns of allelic frequency incompatible with a simple model of population expansion from mainland locations (Wachowiak *et al.* 2011). Given predictions regarding the extent of the British ice sheet during the last glaciation (Clark *et al.* 2012) and

the geographic distribution of some unique mitochondrial haplotypes, it appears more probable that colonization of the Scottish Highlands occurred from at least two different sources (Sinclair *et al.* 1998; Soranzo *et al.* 2000). Our data show that the composition of the gene pool of the Scottish population has characteristics in common with populations from both central Europe and north of the Iberian Peninsula. We also found high frequencies of private alleles in this population. However, our data are too limited to provide any definitive answer regarding the source of the probable Scots pine admixture in the Scottish Highlands.

The Scots pine populations from the continuous range in Poland and Finland were characterized by the most homogeneous gene pools, and differentiation between populations from those regions was not statistically significant. The low level of genetic differentiation between Scots pine populations from central and northern Europe was also observed for isozymes (Prus-Głowacki *et al.* 2012; Sannikov and Petrova 2012) and nucleotide diversity at nuclear gene loci (Wachowiak *et al.* 2014). These results imply free gene exchange among those populations and that they probably share a common postglacial history. Moreover, our data show that populations from the Balkans share gene pools with populations from central and northern Europe. More challenging is the identification of putative source populations for the recolonization of pines following the last glacial maximum. None of the analysed populations in our study showed exceptionally high genetic variation or evidence for recent bottlenecks following the patterns of nucleotide sequence variation at nuclear gene loci (Pyhäjärvi *et al.* 2007; Wachowiak *et al.* 2009; Kujala and Savolainen 2012). Taking into account findings about genetic consequences of glacial isolation and postglacial colonization on the genetic structure of other cold tolerant tree species e.g. silver birch (*Betula pendula*) and Norway spruce (*Picea abies*) which at present co-occur with Scots pine across large parts of its natural habitat, we can conclude that Europe most likely was reoccupied by at least three main waves of recolonization. The source populations could have their origins in the regions of Alps, Balkans and east refugia with origin at intermediate latitude in Moscow region (Palmé *et al.* 2003; Willis and van Andel 2004; Maliouchenko *et al.* 2007; Tollefsrud *et al.* 2008; Parducci *et al.* 2012). These several distinct lineages of colonization might have mixed in central Europe as suggested based on colonization routes of Norway spruce (Dering and Lewandowski 2009). However, according to results obtained by Parducci *et al.* (2012) which suggest that conifer trees might have also survived the last glaciation in the ice-free refugia of Scandinavia, it appears that detailed studies of populations from Scandinavia and the eastern distribution in Asia, including regions of Moscow (Buchovska *et al.* 2013), are needed to test the possible recolonization trajectories of the species to the central Europe.

Conclusions

Our data show that isolated populations from southern Eurasia show a high genetic divergence over a short geographical distance, which contrasts with the pattern of variation in populations from the northern part of Europe. A clear subdivision was found between populations from distinct parts of the Eurasian distribution. With the exception of the southernmost stands, populations from the central and northern parts of the studied distribution range showed some genetic similarity that may reflect their shared postglacial history or effective admixture between populations of different origin. High-resolution *mt*DNA markers dispersed by seeds across small geographic areas in pines would be needed to verify migration trajectories and the location of source populations for the species' postglacial recolonization.

Sources of Funding

This work was financially supported by the Polish National Science Centre (DEC-2012/05/E/NZ9/03476).

Contributions by the Authors

W.W. and B.W. designed and conceptualized the study. B.W. and M.L. collected and analysed the data. B.W. wrote the manuscript; W.W. assisted in drafting the manuscript; B.W. and W.W. critically reviewed and revised the manuscript for content; all authors read and approved the final manuscript.

Acknowledgements

We thank Krystyna Boratyńska and Jacek Oleksyn for providing some of the plant materials used in the study. The authors would like to express their thanks to Weronika Żukowska, the editor and two anonymous reviewers for constructive comments on the previous version of the article.

Supporting Information

The following additional information is available in the online version of this article —

Table S1. Populations of *Pinus sylvestris* used in this study.

Table S2. Descriptive statistics for the thirteen nuclear microsatellite loci used in this study. Al, number of alleles; $A_{NullFreq}$, mean frequency of null alleles.

Table S3. Pairwise F_{ST}ENA matrix for 24 Scots pine populations (*p < 0.001).

Figure S1. Phylogenetic tree of 24 Scots pine populations based on Nei's standard genetic distance (D_{ST}) at 13 *n*SSR loci (1000 bootstraps) using the neighbour-joining method (NJ).

Literature Cited

Belletti P, Ferrazzini D, Piotti A, Monteleone I, Ducci F. 2012. Genetic variation and divergence in Scots pine (*Pinus sylvestris* L.) within its natural range in Italy. *European Journal of Forest Research* **131**:1127–1138.

Birks HJB. 1989. Holocene isochrones maps and patterns of tree-spreading in the British Isles. *Journal of Biogeography* **16**:503–540.

Buchovska J, Danusevicius D, Baniulis D, Stanys V, Siksnianiene JB, Kavaliauskas D. 2013. The location of the northern glacial refugium of Scots Pine based on mitochondrial DNA markers. *Baltic Forestry* **19**:2–12.

Chapuis MP, Estoup A. 2007. Microsatellite null alleles and estimation of population differentiation. *Molecular Biology and Evolution* **24**:621–631.

Cheddadi R, Vendramin GG, Litt T, François L, Kageyama M, Lorentz S, Laurent J-M, de Beaulieu J-L, Sadori L, Jost A, Lunt D. 2006. Imprints of glacial refugia in the modern genetic diversity of *Pinus sylvestris*. *Global Ecology and Biogeography* **15**:271–282.

Chybicki I. 2015. INEST 2.0 [Computer Software]. Retrieved from http://www.ukw.edu.pl/pracownicy/strona/igor_chybicki/soft ware_ukw/ last accessed: 18 August, 2016.

Clark CD, Hughes ALC, Greenwood SL, Jordan C, Sejrup HP. 2012. Pattern and timing of retreat of the last British-Irish Ice Sheet. *Quaternary Science Reviews* **44**:112–146.

Cornuet J-M, Luikart G. 1996. Description and power analysis of two tests for detecting recent population bottlenecks from allele frequency data. *Genetics* **144**:2001–2014.

Dering M, Lewandowski A. 2009. Finding the meeting zone: Where have the northern and southern ranges of Norway spruce overlapped? *Forest Ecology and Management* **259**:229–235.

Diniz-Filho JAF, Soares TN, Lima JS, Dobrovolski R, Landeiro VL, de Campos Telles MP, Rangel TF, Bini LM. 2013. Mantel test in population genetics. *Genetics and Molecular Biology* **36**:475–485.

Dumolin S, Demesure B, Petit RJ. 1995. Inheritance of chloroplast and mitochondrial genomes in pedunculate oak investigated with an efficient PCR method. *Theoretical and Applied Genetics* **91**:1253–1256.

Earl DA, von Holdt BM. 2011. STRUCTURE HARVESTER: a website and program for visualizing STRUCTURE output and implementing the Evanno method. *Conservation Genetics Resources* **4**:359–361.

Elsik CG, Minihan VT, Scarpa AM, Hall SE, Williams CG. 2000. Low-copy microsatellite markers for *Pinus taeda* L. *Genome* **43**:550–555.

Evanno G, Regnaut S, Goudet J. 2005. Detecting the number of clusters of individuals using the software STRUCTURE: a simulation study. *Molecular Ecology* **14**:2611–2620.

Excoffier L, Lischer HEL. 2010. Arlequin ver. 3.0: a new series of programs to perform population genetics analyses under Linux and Windows. *Molecular Ecology Resources* **10**:564–567.

Falush D, Stephens M, Pritchard JK. 2003. Inference of population structure using multilocus genotype data: linked loci and correlated allele frequencies. *Genetics* **164**:1567–1587.

Goudet J. 2001. FSTAT, a program to estimate and test gene diversities and fixation indices (version 2.9.3). Available from http://www2.unil.ch/popgen/softwares/fstat.htm; last accessed: 18 August, 2016. Updated from Goudet (1995).

Guichoux E, Lagache E, Wagner S, Chaumeil P, Léger P, Lepais O, Lepoittevin C, Malausa T, Revardel E, Salin F, Petit RJ. 2011. Current trends in microsatellite genotyping. *Molecular Ecology Resources* **11**:591–611.

Hamrick JL, Godt MJW, Sherman-Broyles SL. 1992. Factors influencing levels of genetic diversity in woody plant species. *New Forests* **6**:95–124.

Hardy OJ, Vekemans X. 2002. SPAGeDi: a versatile computer program to analyse spatial genetic structure at the individual or population levels. *Molecular Ecology Notes* **2**:618–620.

Hardy OJ, Charbonnel N, Fréville H, Heuertz M. 2003. Microsatellite allele sizes: a simple test to assess their significance on genetic differentiation. *Genetics* **163**:1467–1482.

Hewitt GM. 1996. Some genetic consequences of ice ages, and their role in divergence and speciation. *Biological Journal of the Linnean Society* **58**:247–276.

Hubisz MJ, Falush D, Stephens M, Pritchard JK. 2009. Inferring weak population structure with the assistance of sample group information. *Molecular Ecology Resources* **9**:1322–1332.

Jasińska AK, Boratyńska K, Dering M, Sobierajska KI, Ok T, Romo A, Boratyński A. 2014. Distance between south-European and south-west Asiatic refugia areas involved morphological differentiation: *Pinus sylvestris* case study. *Plant Systematics and Evolution* **300**:1487–1502.

Kalinowski ST. 2009. How well do evolutionary trees describe genetic relationships among populations? *Heredity* **102**: 506–513.

Karhu A, Hurme P, Karjalainen M, Karvonen P, Karkkainen K, Neale D, Savolainen O. 1996. Do molecular markers reflect patterns of differentiation in adaptive traits of conifers? *Theoretical and Applied Genetics* **93**:215–221.

Kinloch BB, Westfall RD, Forrest GI. 1986. Caledonian Scots pine: origins and genetic structure. *New Phytologist* **104**:703–729.

Kujala ST, Savolainen O. 2012. Sequence variation patterns along a latitudinal cline in Scots pine (*Pinus sylvestris*): signs of clinal adaptation? *Tree Genetics & Genomes* **8**:1451–1467.

Liewlaksaneeyanawin C, Ritland CE, El-Kassaby YA, Ritland K. 2004. Single-copy, species-transferable microsatellite markers developed from loblolly pine ESTs. *Theoretical and Applied Genetics* **109**:361–369.

Li J, Li H, Jakobsson M, Li S, Sjodin P, Lascoux M. 2012. Joint analysis of demography and selection in population genetics: where do we stand and where could we go? *Molecular Ecology* **21**:28–44.

Luikart G, England PR, Tallmon D, Jordan S, Taberlet P. 2003. The power and promise of population genomics: from genotyping to genome typing. *Genetics* **4**:981–994.

Maliouchenko O, Palmé AE, Buonamici A, Vendramin GG, Lascoux M. 2007. Comparative phylogeography and population structure of European *Betula* species, with particular focus on *B. pendula* and *B. pubescens*. *Journal of Biogeography* **34**:1601–1610.

Mantel N. 1967. The detection of disease clustering and a generalized regression approach. *Cancer Research* **27**:209–220.

Naydenov K, Senneville S, Beaulieu J, Tremblay F, Bousquet J. 2007. Glacial variance in Eurasia: mitochondrial DNA evidence from Scots pine for a complex heritage involving genetically distinct refugia at mid-northern latitudes and Asia Minor. *BMC Evolutionary Biology* **7**:233.

Nei M. 1972. Genetic distance between populations. *The American Naturalist* **106**:283–392.

Nosil P, Feder JL. 2013. Genome evolution and speciation: toward quantitative descriptions of pattern and process. *Evolution* **67**: 2461–2467.

Palmé AE, Su Q, Rautenberg A, Manni F, Lascoux M. 2003. Postglacial recolonization and cpDNA variation of silver birch, *Betula pendula*. *Molecular Ecology* **12**:201–212.

Parducci L, Jørgensen T, Tollefsrud MM, Elverland E, Alm T, Fontana SL, Bennett KD, Haile J, Matetovici I, Suyama Y, Edwards ME, Andersen K, Rasmussen M, Boessenkool S, Coissac E, Brochmann C, Taberlet P, Houmark-Nielsen M, Larsen NK, Orlando L, Gilbert MTP, Kjær KH, Alsos IG, Willerslev E. 2012. Glacial Survival of Boreal Trees in Northern Scandinavia. *Science* **335**:1083–1086.

Peakall R, Smouse PE. 2006. GENALEX 6: genetic analysis in Excel. Population genetic software for teaching and research. *Molecular Ecology Notes* **6**:288–295.

Petit RJ, Aguinagalde I, de Beaulieu J-L, Bittkau C, Brewer S, Cheddadi R, Ennos R, Fineschi S, Grivet D, Lascoux M, Mohanty A, Muller-Starck G, Demesure-Musch B, Palme A, Martin JP, Rendell S, Vendramin GG. 2003. Glacial refugia: hotspots but not melting pots of genetic diversity. *Science* **300**:1563–1565.

Petit RJ, Hampe A. 2006. Some evolutionary consequences of being a tree. *Annual Review of Ecology, Evolution, and Systematics* **37**: 187–214.

Pritchard JK, Stephens M, Donnelly P. 2000. Inference of population structure using multilocus genotype data. *Genetics* **155**:945–959.

Prus-Głowacki W, Stephan BR. 1993. Genetic variation of *Pinus sylvestris* from Spain in relations to other European populations. *Silvae Genetica* **43**:7–14.

Prus-Głowacki W, Stephan BR, Bujas E, Alia R, Marciniak A. 2003. Genetic differentiation of autochthonous populations of *Pinus sylvestris* (*Pinaceae*) from the Iberian Peninsula. *Plant Systematics and Evolution* **239**:55–66.

Prus-Głowacki W, Urbaniak L, Bujas E, Curtu AL. 2012. Genetic variation of isolated and peripheral populations of *Pinus sylvestris* (L.) from glacial refugia. *Flora* **207**:150–158.

Putz CM, Schmid C, Reisch C. 2015. Living in isolation – population structure, reproduction, and genetic variation of the endangered plant species *Dianthus gratianopolitanus* (Cheddar pink). *Ecology and Evolution* **5**:3610–3621.

Pyhäjärvi T, García-Gil MR, Knürr T, Mikkonen M, Wachowiak W, Savolainen O. 2007. Demographic history has influenced nucleotide diversity in European *Pinus sylvestris* populations. *Genetics* **177**:1713–1724.

Pyhäjärvi T, Salmela MJ, Savolainen O. 2008. Colonization routes of *Pinus sylvestris* inferred from distribution of mitochondrial DNA variation. *Tree Genetics & Genomes* **4**:247–254.

Robledo-Arnuncio JJ, Collada C, Alía R, Gil L. 2005. Genetic structure of montane isolates of *Pinus sylvestris* L. in Mediterranean refugial area. *Journal of Biogeography* **32**:595–605.

Rubiales JM, García-Amorena I, Hernández L, Génova M, Martínez F, Gómez Manzaneque F, Morla C. 2010. Late Quaternary dynamics of pinewoods in the Iberian Mountains. *Review of Palaeobotany and Palynology* **162**:476–491.

Sannikov SN, Petrova IV. 2012. Phylogenogeography and genotaxonomy of *Pinus sylvestris* L. populations. *Russian Journal of Ecology* **4**:273–280.

Savolainen O, Kujala SJ, Sokol C, Pyhäjärvi T, Avia K, Knürr T, Kärkkäinen K, Hicks S. 2011. Adaptive potential of northernmost tree populations to climate change, with emphasis on Scots pine (*Pinus sylvestris* L.). *Journal of Heredity* **102**:526–536.

Sebastiani F, Pinzauti F, Kujala ST, González-Martínez SC, Vendramin GG. 2012. Novel polymorphic nuclear microsatellite markers for *Pinus sylvestris* L. *Conservation Genetics Resources* **4**:231–234.

Sinclair WT, Morman JD, Ennos RA. 1998. Multiple origins for Scots pine (*Pinus sylvestris* L.) in Scotland: evidence from mitochondrial DNA variation. *Heredity* **80**:233–240.

Soranzo N, Alia R, Provan J, Powell W. 2000. Patterns of variation at mitochondrial sequence-tagged-sites locus provides new insights into the postglacial history of European *Pinus sylvestris* populations. *Molecular Ecology* **9**:1205–1211.

Soranzo N, Provan J, Powell W. 1998. Characterization of microsatellite loci in *Pinus sylvestris* L. *Molecular Ecology* **7**: 1247–1263.

Soto A, Robledo-Arnuncio JJ, González-Martínez SC, Smouse PE, Alia R. 2010. Climatic niche and neutral genetic diversity of the six Iberian pine species: a retrospective and prospective view. *Molecular Ecology* **19**:1346–1409.

Takezaki N, Nei M, Tamura K. 2014. POPTREEW: web version of POPTREE for constructing population trees from allele frequency data and computing other population statistics. *Molecular Biology and Evolution* **31**:1622–1624.

Tollefsrud MM, Kissling R, Gugerli F, Johnsen Ø, Skrøppa T, Cheddadi R, Van der Knaap WO, Latałowa M, Terhürne-Berson R, Litt T, Geburek T, Brochmann C, Sperisen C. 2008. Genetic consequences of glacial survival and postglacial colonization in Norway spruce: combined analysis of mitochondrial DNA and fossil pollen. *Molecular Ecology* **17**:4134–4150.

Tollefsrud MM, Sønstebø JH, Brochmann C, Johnsen Ø, Skrøppa T, Vendramin GG. 2009. Combined analysis of nuclear and mitochondrial markers provide new insight into the genetic structure of North European *Picea abies*. *Heredity* **102**:549–562.

Wachowiak W, Balk PA, Savolainen O. 2009. Search for nucleotide diversity patterns of local adaptation in dehydrins and other cold-related candidate genes in Scots pine (*Pinus sylvestris* L.). *Tree Genetics & Genomes* **5**:117–132.

Wachowiak W, Salmela MJ, Ennos RA, Iason G, Cavers S. 2011. High genetic diversity at the extreme range edge: nucleotide variation at nuclear loci in Scots pine (*Pinus sylvestris* L.) in Scotland. *Heredity* **106**:775–787.

Wachowiak W, Wójkiewicz B, Cavers S, Lewandowski A. 2014. High genetic similarity between Polish and North European Scots pine (*Pinus sylvestris* L.) populations at nuclear gene loci. *Tree Genetics & Genomes* **10**:1015–1025.

Weir BS, Cockerham CC. 1984. Estimating F-Statistics for the analysis of population structure. *Evolution* **38**:1358–1370.

Willis KJ, van Andel TH. 2004. Trees or no trees? The environments of central and eastern Europe during the Last Glaciation. *Quaternary Science Reviews* **23**:2369–2387.

Willis KJ, Bennet KD, Birks JB. 1998. The late quaternary dynamics of pine in Europe. In: Richardson DM, ed. *Ecology and biogeography of Pinus*. Cambridge: Cambridge University Press, 107–119.

18

Genetic structure of colline and montane populations of an endangered plant species

Tiphaine Maurice[1,2,3] Diethart Matthies[4], Serge Muller[1,5] and Guy Colling*[2]

[1] Université de Lorraine, CNRS UMR 7360, Laboratoire Interdisciplinaire des Environnements Continentaux (LIEC), rue du Général Delestraint, F-57070 Metz, France
[2] Musée National d'Histoire Naturelle, Population Biology and Evolution, 25 rue Münster, L-2160 Luxembourg, Luxembourg
[3] Fondation Faune Flore, 24 rue Münster, L-2160 Luxembourg, Luxembourg
[4] Philipps-Universität, Fachbereich Biologie, Pflanzenökologie, D-35032 Marburg, Germany
[5] Muséum national d'Histoire naturelle, UMR 7205 ISYEB, CNRS, Université Pierre-et-Marie-Curie, EPHE, Sorbonne Universités, CP 39, 16 rue Buffon, F-75005 Paris, France

Associate Editor: Philippine Vergeer

Abstract. Due to land-use intensification, lowland and colline populations of many plants of nutrient-poor grasslands have been strongly fragmented in the last decades, with potentially negative consequences for their genetic diversity and persistence. Populations in mountains might represent a genetic reservoir for grassland plants, because they have been less affected by land-use changes. We studied the genetic structure and diversity of colline and montane Vosges populations of the threatened perennial plant *Arnica montana* in western central Europe using AFLP markers. Our results indicate that in contrast to our expectation even strongly fragmented colline populations of *A. montana* have conserved a considerable amount of genetic diversity. However, mean seed mass increased with the proportion of polymorphic loci, suggesting inbreeding effects in low diversity populations. At a similar small geographical scale, there was a clear IBD pattern for the montane Vosges but not for the colline populations. However, there was a strong IBD-pattern for the colline populations at a large geographical scale suggesting that this pattern is a legacy of historical gene flow, as most of the colline populations are today strongly isolated from each other. Genetic differentiation between colline and montane Vosges populations was strong. Moreover, results of a genome scan study indicated differences in loci under selection, suggesting that plants from montane Vosges populations might be maladapted to conditions at colline sites. Our results suggest caution in using material from montane populations of rare plants for the reinforcement of small genetically depauperate lowland populations.

Keywords: AFLP; altitude; clonality; conservation genetics; fragmentation; genome scan.

Introduction

Nutrient-poor grasslands in lowland areas have been strongly fragmented during the last decades due to changes in land-use, nutrient enrichment through fertilizers or the cessation of traditional agricultural practices (Ratcliffe 1984; Bignal and McCracken 1996). As a consequence, many formerly common grassland species have been reduced to small and isolated populations (see

*Corresponding author's e-mail address: guy.colling@mnhn.lu

Fischer and Matthies 1998; Kéry et al. 2000; Colling and Matthies 2006). These populations face an increased risk of extinction because of their higher sensitivity to environmental, demographic and genetic stochasticity (Young et al. 1996; Matthies et al. 2004). Small and isolated populations are threatened through a loss of genetic diversity due to genetic drift and reduced gene flow (Young et al. 1996; Jacquemyn et al. 2009), because the loss of genetic variation and increased inbreeding are expected to lead to lower fitness of individual plants and a reduced ability of the populations to respond to environmental changes (Kéry and Matthies 2004; Ouborg et al. 2006; Walisch et al. 2012).

To enhance the chances of survival of small and isolated populations, it has been suggested to artificially augment threatened populations by introducing seeds from extant large populations to increase their size and genetic diversity (Ingvarsson 2001; Hufford and Mazer 2003; Tallmon et al 2004). European mountains could represent a genetic reservoir for plants of nutrient-poor grasslands as the intensification of land-use affecting lowland areas since several decades has only recently begun in mountain areas (Fischer and Wipf 2002; Peter et al. 2009). However, environmental conditions at higher altitudes such as low temperatures, a short growing season, strong winds, high irradiance, low air pressure and variation in the persistence of snow cover (Körner 2007) impose strong environmental constraints that can lead to marked genetic differences among plant populations along altitudinal gradients (Parker et al. 2003; Montesinos-Navarro et al. 2011). If populations at higher altitudes were locally adapted, this could present a problem for management measures such as the reinforcement of lowland populations with seeds from mountain populations, because the transplants could be maladapted (Vergeer et al. 2004; McKay et al. 2005). Moreover, crossings between strongly differentiated genotypes could result in outbreeding depression in the offspring (Hufford and Mazer 2003; Galloway and Etterson 2005; Walisch et al. 2012).

We studied the population genetic structure of the endangered long-lived grassland species Arnica montana, a characteristic species of acid nutrient-poor grasslands in Central Europe, in the colline Ardennes-Eifel and Hunsrück regions and the nearest montane region, the Vosges mountains. A. montana has strongly declined in lowland and colline regions and is now considered to be endangered in many parts of Europe (Korneck et al. 1996; Colling 2005; Kestemont 2010). Its decline has been attributed to the deterioration of habitat quality due to changes in land use, increased use of fertilizer and aerial deposition of nitrogen (Fennema 1992; Vergeer et al. 2005; Maurice et al. 2012). The remaining colline and lowland populations, but also populations in some mountain ranges are fragmented (Kahmen and Poschlod 2000;

Luijten et al. 2000), while in other mountain ranges like the Vosges A. montana is still rather common and even harvested (Ellenberger 1998; Schnitzler and Muller 1998).

We asked the following questions: (1) Do the genetic diversity and genetic structure of the colline Ardennes-Eifel and Hunsrück populations and montane Vosges populations of A. montana differ, and in particular, (2) Are the colline populations genetically depauperate? (3) Are there differences between colline and Vosges populations in AFLP loci?

Methods

Study species

Arnica montana (Asteraceae) is a long-lived perennial plant that produces large rosettes from a rhizome. The species is restricted to Europe (Hultén and Fries 1996). A. montana can form dense mats that may consist of several different genotypes, and without genetic analyses it is not possible to distinguish individual genets (Luijten et al. 1996). A. montana has a sporophytic self-incompatibility system (Luijten et al. 2002). The large orange-yellow flowerheads of A. montana produce achenes (hereafter called seeds), which are wind-dispersed. Although the seeds of A. montana are small (mass c. 1.3 mg) and possess a pappus, their dispersal is very limited (Luijten et al. 1996; Strykstra et al. 1998). A. montana is an important source of pharmaceutical compounds (Lyss et al. 1997; Klaas et al. 2002) and the species is still harvested in some mountain regions due to difficulties in cultivating it (Delabays and Mange 1991; Mardari et al. 2015).

Study area and sampling procedure

To study the genetic structure of A. montana in Western-Central Europe, samples were taken in 30 populations of different sizes in three neighbouring geographical regions: (1) The colline region of the Ardennes (Belgium), the Oesling (Luxembourg) and the Eifel (Germany), (2) the colline region of the Hunsrück (Germany) with the neighbouring Pays de Bitche region (France) and (3) the montane belt of the French Vosges mountains, which is the mountain range nearest to the colline study area (Fig. 1 and Table 1). Apart from altitude, the colline and montane populations differ in the composition of the vegetation and, based on Ellenberg indicator values calculated from the vegetation data, in soil moisture, but not in soil reaction and soil nutrients (Maurice et al. 2012).

The studied 20 colline populations (281–633 m a.s.l.) represent a large part of the extant populations in the area. Although some large populations still occur in the colline region of the Ardennes-Eifel, many of the extant populations are small due to small habitat size and

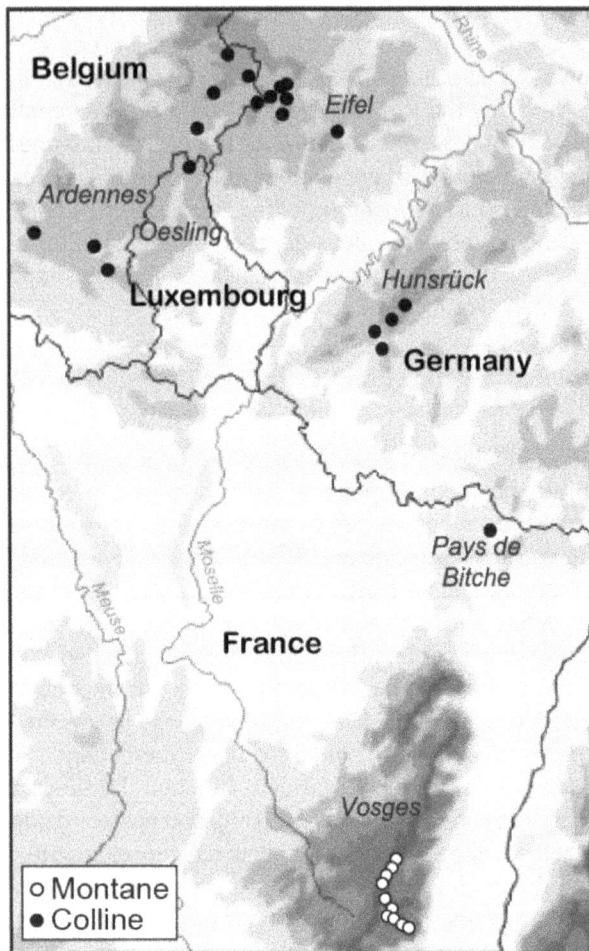

Figure 1. Map showing the location of the studied montane populations of *A. montana* in the Vosges mountains (open circles) and of the colline populations (filled circles) in the Ardennes–Eifel, Hunsrück and Pays de Bitche regions.

low habitat quality, in particular high nutrient levels which are known to be detrimental to *A. montana* (Vergeer *et al.* 2005, Maurice *et al.* 2012). The geographical distance between most extant populations in the colline area was large due to the intense habitat fragmentation, and ranged from 0.7 km to 190.7 km (median = 77.9 km).

Ten montane populations were sampled in the French Vosges mountains (1175–1268 m a.s.l., Fig. 1). The populations were sampled in the part of the Vosges with the highest density of populations of *A. montana*. The geographical distance between montane populations was much smaller than that between colline populations (0.7–17.3 km; median = 8.0 km).

Population sizes were estimated as the total number of rosettes (ramets) per population as it is not possible to distinguish individual genets in the field due to the clonal growth of *A. montana*. Based on the results of a former study of the structure of *A. montana* populations

(Maurice *et al.* 2012), ramet population size was calculated by dividing the number of flowering stems by the proportion of flowering rosettes per population.

In June 2007, we collected one fresh leaf from each of 20 rosettes in each population along transects of 20 m length. Within each transect, we recorded the distances among the sampled plants. Because *A. montana* is a clonal species, the minimum distance between the sampled rosettes was 1 m to avoid sampling of the same genetic individual twice. In very small populations, the number of sampled rosettes was less than 20 (Table 1). The leaf samples were immediately stored in silica gel and kept at room temperature until DNA extraction. In several populations, we had problems with the PCR-reaction, probably due to the high content of secondary metabolites in the leaf tissue of *A. montana* (Ekenäs *et al.* 2009) and a lower number of samples were used for the genetic analyses (Table 1).

DNA extraction, purification and AFLP analysis

A 96 wells DNeasy kit extraction (Qiagen®) was performed on 10 mg of dried leaf tissue for 494 samples after grinding (Retsch MM200, Retsch, Haan, Germany). Extracted DNA was purified from secondary metabolites with a ChargeSwitch® gDNA Plant Kit (Invitrogen®). A further purification step was done by electrophoresis on a 2 % agarose gel (90 V, 200 mA, 45 min) in 10X TBE buffer UltraPure (Invitrogen®, Tris 1 mM, Boric Acid 0.9 mM, EDTA 0.01 mM). Extraction of the samples from the gel was done using a QIAquick® 96 PCR Purification Kit (Qiagen®).

DNA (0.1 µg) was digested at 37 °C for 2 h using *Eco*RI and *Mse*I (1.3 U, Invitrogen®), 0.65 µL of 10X REact® 1 Buffer (Invitrogen®, 50 mM Tris-HCl, 10 mM MgCl$_2$) and 0.65 µL of 10X REact® 3 Buffer (Invitrogen®, 50 mM Tris-HCl, 10 mM MgCl2, 100 mM NaCl) in a final volume of 10 µL. Endonucleases were inactivated by 15 min at 70 °C. Adaptor ligation was achieved by adding 12.48 µL of Adapter/Ligation Solution (Invitrogen®, *Eco*RI/*Mse*I adapters, 0.4 mM ATP, 10 mM Tris-HCl, 10 mM Mg-acetate, 50 mM K-acetate) and 0.52 µL of T4 DNA ligase (1 U/µL) before incubation for 2 h at 20 °C. Ligation solution was diluted to 1:6 and 4 µL were used to perform the pre-amplification by adding 16 µL of pre-amp primer mix (Invitrogen®), 2 µL of 10X *pfu* buffer with MgSO$_4$ (Fermentas®) and 1 U of *pfu* DNA polymerase (Fermentas®) in a final volume of 24 µL.

Polymerase chain reaction was performed using a thermocycler (iCycler, Bio-Rad Laboratories). A first cycle was performed at 94 °C for 3 min, then 20 cycles were performed at 94 °C for 30 s, 56 °C for 60 s and 72 °C for 60 s, with a final cycle of 5 min at 72 °C. Three primer combinations with distinct polymorphic loci were

Table 1. Characteristics of the 30 studied populations of *A. montana*. Region, geographical region (see text for details); pop. Name, population name; pop. size, population size calculated as total number of rosettes (see methods for details); *n*, number of rosettes analysed genetically. In some populations (in italics) a lower number of rosettes was analysed due to PCR problems (see text for details). P, proportion of polymorphic loci; H_e, Nei's expected heterozygosity based on allele frequencies calculated by the square root method, assuming Hardy–Weinberg equilibrium; CI, STRUCTURE cluster ID: AE, Ardennes–Eifel; H, Hunsrück; V, Vosges mountains. The proportion of individuals assigned to the clusters is indicated as subscript.

Region	Pop. name	Alt. (m)	Pop. size	Latitude (° North)	Longitude (° East)	*n*	P (%)	H_e	CI
Ardennes – Eifel	A-Bas	449	11	49.833	5.606	11	60.9	0.236	$AE_{1.00}$
	A-Jus	448	340	49.888	5.549	20	72.4	0.236	$AE_{1.00}$
	A-Roc	418	670	49.920	5.238	20	72.4	0.239	$AE_{1.00}$
	A-Tho	496	270	50.239	6.001	9	76.2	0.229	$AE_{1.00}$
	A-Em	609	150	50.326	6.405	6	65.9	0.222	$AE_{1.00}$
	A-Els	591	18000	50.454	6.259	20	84.0	0.240	$AE_{1.00}$
	A-Sch	457	1100	50.225	6.049	20	86.7	0.248	$AE_{1.00}$
	A-Kap	485	110	50.311	6.142	11	73.4	0.252	$AE_{1.00}$
	O-Lux	516	260	50.146	6.036	15	74.4	0.247	$AE_{1.00}$
	E-Auf	537	710	50.364	6.534	20	83.5	0.236	$AE_{0.95}$
	E-Asb	538	10000	50.364	6.524	20	78.2	0.205	$AE_{1.00}$
	E-Leu	598	4000	50.349	6.484	20	74.4	0.211	$AE_{0.95}$
	E-Ste	584	7600	50.329	6.558	20	74.2	0.206	$AE_{1.00}$
	E-Man	538	2100	50.291	6.540	20	75.9	0.221	$AE_{1.00}$
	E-Dau	481	2300	50.240	6.826	20	81.7	0.232	$AE_{1.00}$
Hunsrück	H-Abe	496	570	49.657	7.094	20	61.4	0.174	$H_{1.00}$
	H-Bra	558	1200	49.574	7.003	13	46.6	0.164	$H_{1.00}$
	H-Eis	448	4200	49.612	7.049	19	55.6	0.168	$H_{1.00}$
	H-Otz	410	1300	49.601	6.979	20	62.2	0.173	$H_{1.00}$
	P-Bit	281	300	49.066	7.528	20	72.2	0.210	$AE_{0.90}$
Vosges	V-Fer	1225	600	48.052	7.020	18	68.9	0.198	$V_{1.00}$
	V-Cha	1251	5000	48.049	7.012	7	72.9	0.206	$V_{1.00}$
	V-Hon	1243	260	48.040	7.008	18	68.2	0.203	$V_{1.00}$
	V-Sch	1225	1400	48.034	6.996	17	68.7	0.209	$V_{1.00}$
	V-Her	1223	1500	47.986	6.980	5	54.1	0.201	$V_{1.00}$
	V-Hah	1223	2200	47.942	7.022	14	60.9	0.194	$V_{1.00}$
	V-Ste	1214	170	47.931	7.018	20	60.2	0.182	$V_{1.00}$
	V-Mar	1175	2800	47.924	7.034	18	57.9	0.183	$V_{1.00}$
	V-Moo	1218	4300	47.907	7.075	16	58.1	0.189	$V_{1.00}$
	V-Haa	1268	450	47.904	7.093	17	59.1	0.187	$V_{1.00}$

selected for the selective amplification: E-CTA/M-ACA, E-CTA/M-AAC and E-CAA/M-ACA (Invitrogen). To assess reproducibility of the primer combinations, three different and independent individuals were repeated three times from the same DNA extraction for each combination. The mean reproducibility values for the three combinations were reasonably high (86.1–93.0 %). Amplifications were performed using 5 µL of 1:6 diluted pre-amplification reaction, adding 0.4 µL of dNTPs (10 mM), 2 µL of *pfu* buffer with $MgSO_4$, 0.4 U of *pfu* DNA polymerase, 1 µL of *Eco*RI primers (100 mM) and 1 µL *Mse*I primer (50 mM) to a total volume of 20 µL.

Amplifications were programmed for 1 cycle at 94 °C for 2 min, 10 cycles consisting of 20 s at 94 °C and 30 s at 66 °C and 2 min at 72 °C. The 66 °C annealing temperature of the 10 cycles was subsequently reduced by 1 °C every cycle, and continued at 56 °C for the remaining 20 cycles, with a final hold at 60 °C for 30 min.

Capillary electrophoresis of all samples was performed with the selective amplification products of AFLP on an automated 48-capillary DNA sequencer (MegaBACE™ 500, GE Healthcare). Samples were prepared for analysis by diluting the final amplified product to 1:10. All samples included 1 μL of MegaBACE ET550-R DNA size standard (GE Healthcare) diluted at 1:6. Samples were run for 75 min using GT Dye Set 2 [ET-Rox, FAM, NED, HEX].

Data analysis

The fragments amplified by AFLP primers were visualized using MegaBACE Fragment profiler v1.2 (GE Healthcare) and manually scored as either present (1) or absent (0). Fragments with lengths between 60 and 500 base pairs were included in the analysis. Estimates of allelic frequencies were computed using the square root method of the null homozygote frequency assuming Hardy–Weinberg equilibrium, as implemented in the program AFLP-SURV V1.0 (Vekemans et al. 2002). Genetic diversity within populations was estimated as the proportion of polymorphic loci at the 5 % level, and as Nei's expected gene diversity (H_e) that averages expected heterozygosity of the marker loci (Lynch and Milligan 1994). To test for effects of genetic drift on reproduction, we correlated mean mass of seeds (available for 17 of the populations; see Maurice et al. 2012) with the proportion of polymorphic loci and H_e in the populations.

The genetic structure within and among populations was analysed on the basis of AFLP allele frequencies using the square-root method implemented in AFLP-SURV V1.0 assuming that the populations were in Hardy–Weinberg equilibrium, to calculate an overall F_{ST} value following the treatment by Lynch and Milligan (1994) with 1000 permutations. We also performed a separate analysis with AFLP-SURV for each altitude class.

The genetic structure of A. montana was studied at the landscape level using a Bayesian clustering method to infer population structure and assign individuals to geographical regions, as implemented in STRUCTURE V2.3 (Pritchard et al. 2000) which allows the analysis of dominant data (Falush et al. 2007). We used a model of no population admixture for the ancestry of the individuals without prior information about the regional membership of the populations and assumed that the allele frequencies are correlated within populations. We conducted a series of 11 independent runs for each value of K (the number of clusters) between 1 and 30 in order to quantify the amount of variation of the likelihood of each K. We found that a length of the burn-in and Markov chain Monte Carlo (MCMC) of 10 000 each was sufficient. Longer burn-in or MCMC did not significantly change the results. The model choice criterion implemented in STRUCTURE to detect the K most appropriate to describe the data is an estimate of the posterior probability of the data for a given K, Pr(X|K) (Pritchard et al. 2000). This value is called 'Ln P(D)' in STRUCTURE, which we refer to as $L(K)$ afterwards. An ad hoc quantity based on the second-order rate of change of the likelihood function with respect to K (ΔK) did show a clear peak at the true value of K (Evanno et al. 2005). We calculated $\Delta K = m(|L(K+1) - 2L(K) + L(K-1)|)/SD[L(K)]$ where m is the mean and SD the standard deviation. The best estimate of K was defined by the model giving the highest probability of the data, with a peak in the ΔK graph, and which also gave consistent results over multiple runs. Finally, the runs of the STRUCTURE simulation were aligned using the FullSearch option with the cluster matching and permutation program CLUMPP V1.1.2 (Jakobsson and Rosenberg 2007).

A hierarchical analysis of molecular variance (AMOVA) was used to partition the genetic variability among colline and montane Vosges populations, populations within population groups and individuals as implemented in GenAlEX 6.41 (Peakall and Smouse 2006). The variance components from the analysis were used to estimate Φ-statistics which are similar to F-statistics (Excoffier et al. 1992).

We identified non-neutral markers with the program BAYESCAN (Foll et al. 2008), removed them from the dataset, and ran a second AMOVA. The false discovery rate (FDR) in BAYESCAN was set to 0.001 (see Foll et al. 2008). The method used by BAYESCAN 2.01 was found to be robust against deviations from the island model and yielded very few false positives in all simulations in a recent study comparing several methods for detecting markers under selection (De Mita et al. 2013).

We obtained the following bioclimatic variables for each study site (representative of 1950–2000) in a gridsize of about one square kilometre (30 arc seconds) from the Worldclim database version 1.4. (Hijmans et al. 2005; www.worldclim.org): annual mean temperature, temperature seasonality, temperature annual range, temperature of driest quarter, and annual precipitation. Because these variables were intercorrelated, we identified two principal components (PCs) by PCA with varimax rotation (SPSS 19.0). PC ALTI explained 45.6 % of the variation and was highly correlated with annual precipitation (r = 0.97) and mean annual temperature (r = −0.96), indicating that PC ALTI corresponded to a

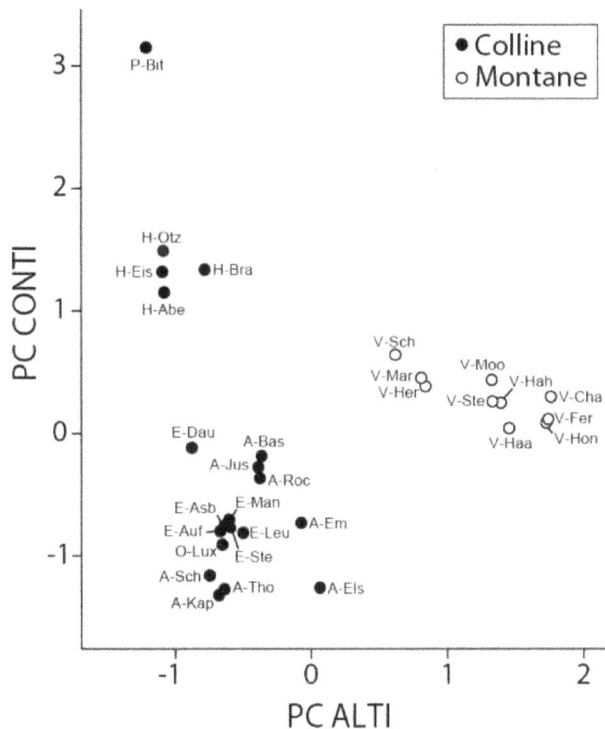

Figure 2. Principal component analysis of five bioclimatic variables extracted for 30 populations of *A. montana* from the Worldclim database version 1.4. (Hijmans *et al.* 2005; www.worldclim.org): annual mean temperature, temperature seasonality, temperature annual range, temperature of driest quarter, and annual precipitation. The first factor PC ALTI was highly correlated with annual precipitation ($r = 0.97$) and mean annual temperature ($r = -0.96$), indicating that PC ALTI corresponded to a climatic gradient related to altitude. PC CONTI was highly correlated with temperature seasonality ($r = 0.976$) and temperature annual range ($r = 0.96$); indicating that PC CONTI corresponded to a gradient in continentality. For abbreviations of population names see Table 1.

climatic gradient related to altitude. PC CONTI explained a further 33.5 % of the variation and was highly correlated with temperature seasonality ($r = 0.98$) and temperature annual range ($r = 0.96$); indicating that PC CONTI corresponded to a gradient in continentality.

We then studied the relationship between the frequency of the identified non-neutral markers and PC ALTI and PC CONTI with a generalized linear model with a logit link and a quasibinomial error distribution (see Crawley 2009), using the glm package of R version 3.0.1. To correct for spatial autocorrelation, we included latitude and longitude in the model. Moreover, the first two components of a PCA ordination of the neutral AFLP loci were also added to the model to correct for the genetic structure present in the neutral model. McFadden's Pseudo R^2 was estimated as the ratio among the log-likelihood of the model of interest and the log-likelihood of the null model.

Pairwise Φst genetic distances among (1) all pairs of populations, (2) separately for the colline and the montane Vosges populations and (3) for a subset of the colline populations whose geographical distances were similar to those of the montane Vosges populations were related to geographical distances and the significance of the relationships tested with a Mantel–Test implemented in GenAlEX (1000 permutations).

Results

Climatic characteristics of the study sites

The clusters identified by the PCA-analysis of the bioclimatic variables corresponded well to the three geographical regions (Table 1 and Fig. 2). A first cluster consisted of the four populations of the Hunsrück region and the one Pays de Bitche population characterized by a warm climate with relatively low precipitation, high annual temperature range and high temperature seasonality. The second cluster corresponded to the populations of the Ardennes–Eifel region characterized by lower temperature and precipitation. Our results thus indicate that the Hunsrück populations grow in a climate different from that of the Ardennes–Eifel populations, although they are at the same altitudinal level (Table 1). The third cluster corresponded to the ten populations of the Vosges mountains characterized by a cold and wet climate with relatively high temperature seasonality.

Genetic diversity within populations

Using three primer combinations, 399 AFLP bands were scored with no private bands specific to a population. All 494 individuals had a unique multilocus genotype. The proportion of polymorphic loci (PPL) in the 30 populations ranged from 46.6 to 86.7 % (Table 1). The proportion of polymorphic loci varied strongly among regions ($F_{2,27} = 14.55, P < 0.001$) and was higher in the Ardennes–Eifel region (75.1 %) than in the Hunsrück (59.2 %) and Vosges region (62.4 %). The proportion of polymorphic loci in the montane Vosges populations was lower than in the colline populations (62.4 % vs. 71.2 %; $F_{1,28} = 6.19, P < 0.05$).

The mean value for Nei's genetic diversity within populations (H_e) assuming Hardy–Weinberg equilibrium was 0.210. Genetic diversity was 10.2 % lower in montane Vosges ($H_e = 0.195$) than in colline populations ($H_e = 0.217, F_{1,28} = 5.86, P < 0.05$), and differed significantly among the three geographical regions Ardennes–Eifel, Hunsrück and the Vosges mountains ($F_{2,27} = 35.16, P < 0.001$). Among the colline regions, mean genetic diversity of the populations in the Hunsrück region was significantly lower than that of the populations of the Ardennes–Eifel region ($H_e = 0.178$ vs. $H_e = 0.231, F_{1,18}$

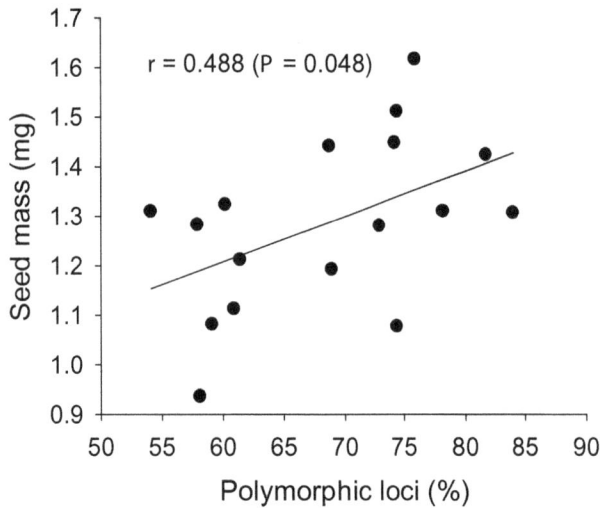

Figure 3. The relationship between mean seed mass in 17 populations of A. montana and the proportion of polymorphic loci.

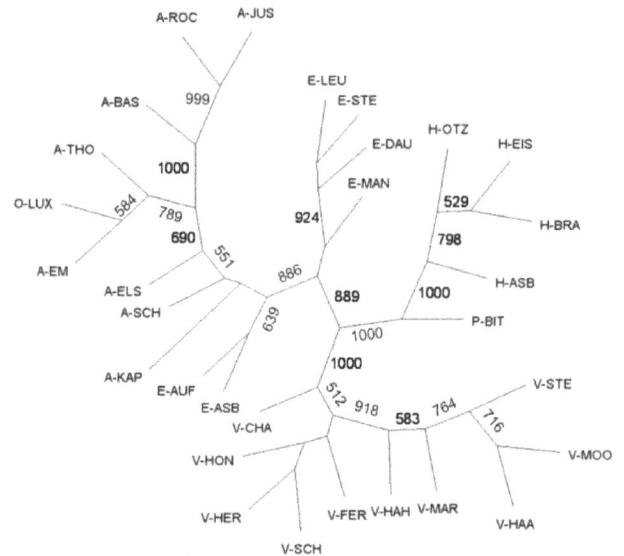

Figure 4. Neighbour-joining tree of 30 populations of A. montana based on Nei's genetic distances derived from AFLP markers. Numbers near the branches indicate bootstrap values above 500 of 1000 bootstraps. For abbreviations of population names see Table 1.

$= 42.3$, $P < 0.001$). In multiple regressions relating the measures of genetic diversity to altitude and (log)population size, separately for the three regions, Nei's gene diversity was not significantly related to the two explanatory variables in any of the regions. However, adjusted for the effects of altitude, the number of polymorphic loci significantly increased with population size in the Ardennes-Eifel region ($\beta = 0.83$, $t = 3.65$, $P < 0.01$). Seed mass increased with the proportion of polymorphic loci (Fig. 3), but not with H_e ($r = 0.287$, $P = 0.265$).

Population genetic structure

Using the modal value of ΔK rather than the maximum value of L(K) allowed us to identify with STRUCTURE several groups corresponding to the uppermost hierarchical level of partitioning between populations. The highest modal value of ΔK was at $K = 3$, corresponding to the number of geographical regions. There was a nearly complete correspondence between the clusters identified by STRUCTURE and the three geographical regions (Table 1). A first cluster consisted of the populations of the Ardennes–Eifel region and the Pays de Bitche population, which showed some admixture between the Ardennes–Eifel and the Hunsrück regions. The second cluster corresponded to the four populations of the Hunsrück region. The third cluster corresponded to the ten populations of the Vosges mountains. The proportion of membership of the individuals of the populations in each of the three identified clusters ranged from 0.901 to 1.000 (Table 1). The neighbour-joining tree based on Nei's genetic distance revealed a clustering pattern similar to the clusters identified by STRUCTURE (Table 1).

However, the population with lowest elevation in the Pays de Bitche region (P-Bit; 281 m a.s.l.), was more related to the Hunsrück region in the neighbour-joining tree (Fig. 4). Furthermore, the neighbour-joining tree indicated that populations from the south-west of Belgium (A-Jus, A-Bas and A-Roc), and from the north of Luxemburg and neighbouring E-Belgium (A-Tho, A-Lux and A-Em), formed two separate sub-groups within the Ardennes and Oesling region in concordance with the geographical position of the sampled populations (Figs 1 and 4).

The estimate of overall F_{ST} obtained by AFLP-Surv assuming Hardy–Weinberg equilibrium was lower (0.122 ± 0.11) than the value obtained by the AMOVA ($\Phi_{ST} = 0.159$). Results of the AMOVA showed that there was a significant genetic differentiation between the Vosges and the colline region (8.2 %) and among the populations within the groups (7.8 % of total variation), although the largest part of the total genetic variation was due to differences between plants within populations (84 %, Table 2).

A separate analysis with AFLP-SURV for the colline and the montane Vosges populations indicated that the genetic differentiation among colline populations ($F_{ST} = 0.12$) was much higher than that among Vosges populations ($F_{ST} = 0.004$). Overall genetic differentiation between all 30 populations (pairwise Φ_{ST}) was related to their geographic distance (Mantel test, $r = 0.44$, $P < 0.01$). This isolation by distance pattern was much stronger for the colline populations (Mantel test, $r = 0.61$, $P < 0.001$;

Table 2. Summary of analysis of molecular variance based on AFLP-analysis of 494 *A. montana* individuals from 30 populations. The genetic variation was partitioned between colline and populations in the Vosges mountains.

Source	df	Variance component	Variance (%)	P
Between Vosges and colline populations	1	5.32	8.2	< 0.001
Among populations within population groups	28	5.03	7.8	< 0.001
Within populations	464	54.45	84.0	< 0.001

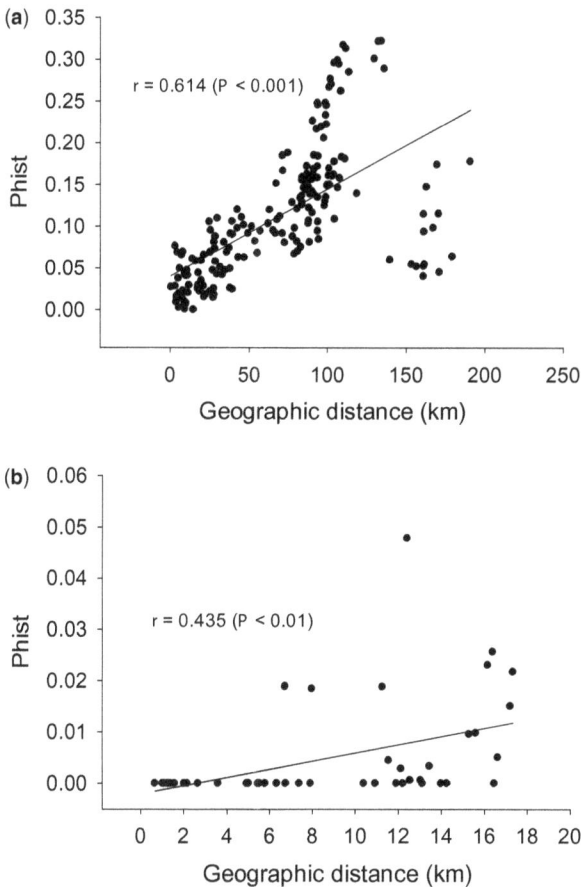

(a)

r = 0.614 (P < 0.001)

Phist

Geographic distance (km)

(b)

r = 0.435 (P < 0.01)

Phist

Geographic distance (km)

Figure 5. The relationship between the genetic and geographical distance between (a) 20 pairs of colline populations and (b) 10 pairs of montane populations of *A. montana*.

Fig. 5a) than for the Vosges populations (Mantel test, r = 0.44, P < 0.01; Fig. 5b). However, in a subset of pairs of colline populations whose geographical distances were similar to those separating the Vosges populations (< 27 km), the IBD pattern among colline populations was much weaker (r = 0.21 P = 0.18).

Putative selective loci

Using the program BAYESCAN 2.01, 63 loci (15.8 %) were identified as outliers with FDR values below 0.001. Divergence of 49 loci (77.8 %) was higher and that of 14 (22.2 %) significantly lower than under a neutral expectation indicating that directional selection occurred at a higher frequency than stabilizing selection.

Multiple logistic regressions revealed that of the 63 putative selective loci, 44 showed a significant ($P < 0.05$) relationship with one or two of the principal components derived from a PCA of bioclimatic variables. After correcting for spatial autocorrelation and for the genetic structure of neutral loci two loci were significantly related to climatic PCs in analyses of deviance, suggesting that these loci may be adaptive and their frequency related to climatic conditions (Table 3). However, other environmental factors like soil conditions that were not studied, but vary among populations could also be responsible for the differences in the frequency of putatively adaptive loci.

A second AMOVA, using a reduced data set with the 63 non-neutral molecular markers removed, resulted in lower Φ_{ST} values than the analysis using the complete dataset ($\Phi_{ST} = 0.12$ vs. $\Phi_{ST} = 0.16$). This was also the case when the variation was partitioned among the colline and the montane Vosges populations ($\Phi_{ST} = 0.06$ vs. $\Phi_{ST} = 0.08$), suggesting that not only genetic drift but also divergent selection has influenced the genetic differentiation among the colline and montane Vosges populations.

Discussion

Genetic diversity within populations

We found that large *A. montana* populations still exist at both altitudinal levels. Although colline populations are much more isolated than the montane Vosges populations, even most small colline populations have conserved a considerable amount of genetic diversity. Due to their isolation, current gene flow among most colline populations is probably very low, but the effects of genetic drift in small populations are not yet very pronounced, as there was no clear relationship between genetic diversity and population size. However, we only could estimate the number of rosettes in the populations, and the relationship between the number of genets and rosettes in this clonal plant is not known and might vary strongly among populations. A prevailing assumption has long been that sexual recruitment is rare in clonal plants implying low genetic diversity, but an increasing number of studies indicate that populations of clonal plants may maintain considerable amounts of genetic diversity (Holderegger *et al.* 1998; Bengtsson 2003; Pluess and Stöcklin 2004).

Table 3. Intercepts and regression coefficients from multiple logistic regression analyses of the relationship between the frequency of two putative loci under selection in populations of A. montana and two principal components describing bioclimatic variables. PC ALTI corresponded to a climatic gradient related to altitude (annual precipitation and mean annual temperature) and PC CONTI to a gradient in continentality (temperature seasonality and temperature annual range), to correct for spatial autocorrelation latitude and longitude were included in the model. To correct for the genetic structure present in the neutral model the first two components of a PCA ordination (DIM1 and DIM2) of the neutral AFLP loci were also added to the model. *, $P < 0.05$; **, $P < 0.01$. Bp, fragment size expressed as number of base-pairs. McFadden's pseudo R^2 for the models is also indicated (see text for details).

Dependent variable (locus)	bp	R^2	Intercept	Explanatory variable	Estimate	t-value	
E-CTA/M-ACA							
B1-37	96	0.891	52.167	PC ALTI	−0.237	−0.534	
				PC CONTI	1.132	2.958	**
				lat	−1.005	−1.472	
				long	−0.206	−0.327	
				DIM1	0.001	0.012	
				DIM2	−0.115	−1.673	
E-CTA/M-AAC							
B2-30	89	0.933	105.417	PC ALTI	1.231	2.466	*
				PC CONTI	1.941	3.371	**
				lat	−1.983	−1.533	
				long	−1.247	−1.516	
				DIM1	−0.001	−0.020	
				DIM2	0.024	0.266	

Our finding of considerable genetic variation in the populations is in agreement with the situation of A. montana in the Rhön mountains (Kahmen and Poschlod 2000). Fragmented populations of long-lived plant species like A. montana may conserve high levels of genetic diversity for a long time, especially if the surviving plants are remnants of formerly large, well-connected populations (Honnay and Bossuyt 2005; Beatty et al. 2008). In contrast, in the Dutch populations of A. montana studied by Luijten et al. (2000) genetic variation was very low. In the Netherlands, fragmentation of A. montana populations is far more pronounced and many of the populations studied by Luijten et al. (2000) were very small. However, comparing the results from our AFLP study to those of Luijten et al. (2000) and Kahmen and Poschlod (2000) is difficult because the types of markers differ (Garcia et al. 2004; Nybom 2004).

We found a positive relationship between mean seed mass and the proportion of polymorphic loci in 17 populations of A. montana, but no positive relationship with population size estimated by the number of rosettes. This suggests inbreeding effects in low diversity populations. Similarly, in the Netherlands, several components of fitness were significantly related to population size in A. montana (Luijten et al. 2000).

Genetic differentiation among populations

The analysis of the genetic population structure revealed significant genetic differentiation between the Vosges and the colline populations and among populations within regions. Overall, the genetic differentiation among the studied A. montana populations was moderate (AMOVA, $F_{ST} = 0.16$) in comparison to that of other species studied using dominant markers (Nybom 2004). Genetic differentiation among the colline populations was higher than among the montane Vosges populations, but this was due to the greater distances among the studied colline populations, as genetic distance increased with geographical distance. This is in accordance with an isolation by distance model (IBD) where geographically closer populations are connected more efficiently by gene flow (Lowe et al. 2004). While at a similar small scale, there was a clear IBD pattern for the montane Vosges but not for the colline populations, at a large scale, there was a strong IBD-pattern for the colline populations. The significant IBD pattern at the large scale is likely to be a legacy of historical gene flow, whereas the lack of IBD at shorter ranges is most likely caused by random genetic drift after fragmentation reduced gene flow more recently. At the beginning of the 20th century, large areas of the region were still covered by heathland and nutrient-poor grassland communities (Hoyois 1949; Dumont 1979), which were suitable habitats for *A. montana* (Maurice *et al.* 2012). Moreover, populations of long-lived plants like *A. montana* may be buffered against the effects of fragmentation due to their long generation times (Honnay and Bossuyt 2005; Beatty *et al.* 2008, Walisch *et al.* 2015).

The STRUCTURE analysis indicated three groups of populations that were separated genetically from each other: those in the Vosges mountains, in the Hunsrück and in the Ardennes-Eifel region. The three groups of populations have probably been isolated for a long time, because the old Rhenish massif of the Hunsrück is separated from the Ardennes–Eifel region by the deep Moselle river valley, and the two colline groups from the Vosges mountains by the Saar river valley. As *A. montana* is a characteristic plant of open heathlands and nutrient-poor acidic grasslands (Maurice *et al.* 2012), the nutrient-rich riparian habitats of the Mosel and Saar river valleys could have isolated the groups of populations. Overall, the results suggest that the current population structure of *A. montana* can be described as regional ensembles of populations with more recent historical gene flow within regions. However, it is likely that at least some part of the genetic differences among regions is due to differences in allele frequencies of non-neutral markers.

Non-neutral markers

The genome scan study of the 30 populations of *Arnica montana* showed that after controlling for spatial autocorrelation and patterns of neutral variation, two AFLP-loci strongly correlated with the two bioclimatic principal components representing a climatic gradient with altitude (annual precipitation and mean annual temperature) and a gradient in continentality (temperature seasonality, temperature annual range), which suggests that these molecular markers may be under directional selection. These results could indicate that populations of *A. montana* are adapted to the local climatic conditions. Although this is only correlative evidence and the observed pattern could also be due other environmental factors, local adaptation in response to local climatic conditions has frequently been found in plants (Becker *et al.*, 2006; Leinonen *et al.*, 2009).

Conclusions

Our results indicate that in contrast to our expectation even strongly fragmented colline populations of *A. montana* have conserved a considerable amount of genetic diversity. In the short term, habitat destruction and deterioration, and not genetic erosion, are the strongest threats to both colline and montane populations. For colline populations, eutrophication through aerial deposition of nitrogen and influx from neighbouring fertilised fields negatively affects habitat suitability for *A. montana* (Maurice *et al.* 2012).

However, without suitable management measures, populations will continue to decrease in size (Maurice *et al.* 2012) and lose genetic diversity due to random genetic drift as already seen in the Netherlands (Luijten *et al.* 2000) and affect population persistence in the long term. In order to preserve actual genetic diversity, suitable management measures aimed at reducing eutrophication and increasing the size of small populations are necessary. Management measures such as turf cutting could enhance seedling recruitment in small colline *A. montana* populations (Knapp 1953, Vergeer *et al.* 2005) and thus preserve genetic variability.

The strong genetic differentiation found between colline and montane Vosges populations probably precludes the use of plants from montane populations for the reintroduction of *A. montana* or the reinforcement of populations in the lowlands. Moreover, results of a genome scan study indicated differences in loci under selection, suggesting that plants from montane Vosges populations could be maladapted to conditions at colline sites. There could also be a considerable risk of outbreeding depression. Our results suggest caution in using material from montane populations of rare plants for reinforcement of small genetically depauperate lowland populations.

Sources of Funding

The project was supported by the National Research Fund of Luxembourg (Ref. BFR07 TR-PHD-054) and by the Musée National d'Histoire Naturelle, Luxembourg.

Contributions by the Authors

T.M., G.C., D.M. and S.M. designed the study. T.M. carried out the practical work. T.M., G.C. and D.M. analysed the data, T.M., G.C., D.M. and S.M. wrote the manuscript.

Acknowledgements

An anonymous reviewer made useful comments on an earlier version of the manuscript. We are grateful to Sylvie Hermant for help in the laboratory. We would like to thank Gaëtan Botin from Natagora in Belgium, and Gerd Ostermann and Steffen Caspari in Germany for their help to locate populations. We thank Struktur- und Genehmigungsdirektion Nord (Rheinland-Pflanz) in Germany, the Département de la Nature et des Forêts in

Belgium and the Ministère du Développement durable et des Infrastructures, Luxembourg for permission to collect samples of the legally protected A. montana.

Literature Cited

Beatty GE, Peter M, McEvoy PM, Sweeney O, Provan J. 2008. Range-edge effects promote clonal growth in peripheral populations of the one-sided wintergreen Orthilia secunda. Diversity and Distributions 14:546–555.

Becker U, Colling G, Dostal P, Jakobsson A, Matthies D. 2006. Local adaptation in the monocarpic perennial Carlina vulgaris at different spatial scales across Europe. Oecologia 150:506–518.

Bengtsson BO. 2003. Genetic variation in organisms with sexual and asexual reproduction. Journal of Evolutionary Biology 16:189–199.

Bignal EM, McCracken DI. 1996. Low-intensity farming systems in the conservation of the countryside. Journal of Applied Ecology 33:413–424.

Colling G. 2005. Red list of the vascular plants of Luxembourg. Luxembourg, Ferrantia.

Colling G, Matthies D. 2006. Effects of habitat deterioration on population dynamics and extinction risk of an endangered, long-lived perennial herb (Scorzonera humilis). Journal of Ecology 94:959–972.

Crawley MJ. 2009. The R book. Chichester: John Wiley & Sons.

De Mita S, Thuillet AC, Gay L, Ahmadi N, Manel S, Ronfort J, Vigouroux Y. 2013. Detecting selection along environmental gradients: analysis of eight methods and their effectiveness for outbreeding and selfing populations. Molecular Ecology 22:1383–1399.

Delabays N, Mange N. 1991. La culture d'Arnica montana L.: aspects agronomiques et phytosanitaires. Revue Suisse De Viticulture, Arboriculture Et Horticulture 23:313–319.

Dumont JM. 1979. Les anciennes prairies à Colchicum autumnale du plateau des tailles (Belgique). Bulletin Du Jardin Botanique National De Belgique 49:121–138.

Ekenäs C, Rosén J, Wagner S, Merfort I, Backlund A, Andreasen K. 2009. Secondary chemistry and ribosomal DNA data congruencies in Arnica (Asteraceae). Cladistics 25:78–92.

Ellenberger A. 1998. Assuming responsibility for a protected plant: WELEDA's endeavour to secure the firm's supply of Arnica montana. In: TRAFFIC-Europe eds: Medicinal plant trade in Europe: conservation and supply. Proceedings of the first international symposium on the conservation of medicinal plants in trade in Europe. TRAFFIC-Europe, Brussels, pp 127–130.

Evanno G, Regnaut S, Goudet J. 2005. Detecting the number of clusters of individuals using the software STRUCTURE: a simulation study. Molecular Ecology 14:2611–2620.

Excoffier L, Smouse PE, Quattro JM. 1992. Analysis of molecular variance inferred from metric distances among DNA haplotypes: application to human mitochondrial DNA restriction data. Genetics 131:479–491.

Falush D, Stephens M, Pritchard JK. 2007. Inference of population structure using multilocus genotype data: dominant markers and null alleles. Molecular Ecology Notes 7:574–578.

Fennema F. 1992. SO_2 and NH_3 deposition as possible causes for the extinction of Arnica montana L. Water, Air, & Soil Pollution 62:325–336.

Fischer M, Matthies D. 1998. Effects of population size on performance in the rare plant Gentianella germanica. Journal of Ecology 86:195–204.

Fischer M, Wipf S. 2002. Effect of low-intensity grazing on the species-rich vegetation of traditionally mown subalpine meadows. Biological Conservation 104:1–11.

Foll M, Beaumont MA, Gaggiotti O. 2008. An approximate Bayesian computation approach to overcome biases that arise when using Amplified Fragment Length Polymorphism markers to study population structure. Genetics 179:927–939.

Galloway LF, Etterson JR. 2005. Population differentiation and hybrid success in Campanula americana: geography and genome size. Journal of Evolutionary Biology 18:81–89.

Garcia AAF, Benchimol LL, Barbosa AMM, Geraldi IO, Souza CL Jr, de Souza AP. 2004. Comparison of RAPD, RFLP, AFLP and SSR markers for diversity studies in tropical maize inbred lines. Genetics and Molecular Biology 27:579–588.

Hijmans RJ, Cameron SE, Parra JL, Jones PG, Jarvis A. 2005. Very high resolution interpolated climate surfaces for global land areas. International Journal of Climatology 25:1965–1978.

Holderegger R, Stehlik I, Schneller JJ. 1998. Estimation of the relative importance of sexual and vegetative reproduction in the clonal woodland herb Anemone nemorosa. Oecologia 117:105–107.

Honnay O, Bossuyt B. 2005. Prolonged clonal growth: escape route or route to extinction? Oikos 108:427–432.

Hoyois G. 1949. L'Ardenne et l'Ardennais. Tome 1. Gembloux: J. Duculot.

Hufford KM, Mazer SJ. 2003. Plant ecotypes: genetic differentiation in the age of ecological restoration. Trends in Ecology & Evolution 18:147–155.

Hultén E, Fries M. 1996. Atlas of the North-European vascular plants. North of the Tropic of Cancer. Königsstein: Koeltz Scientific Books.

Ingvarsson PK. 2001. Restoration of genetic variation lost – the genetic rescue hypothesis. Trends in Ecology & Evolution 16:62–63.

Jacquemyn H, Brys R, Adriaens D, Honnay O, Roldan-Ruiz I. 2009. Effects of population size and forest management on genetic diversity and structure of the tuberous orchid Orchis mascula. Conservation Genetics 10:161–168.

Jakobsson M, Rosenberg NA. 2007. CLUMPP: a cluster matching and permutation program for dealing with label switching and multimodality in analysis of population structure. Bioinformatics 23:1801–1806.

Kahmen S, Poschlod P. 2000. Population size, plant performance, and genetic variation in the rare plant Arnica montana L. in the Rhön, Germany. Basic and Applied Ecology 1:43–51.

Kéry M, Matthies D, Spillmann HH. 2000. Reduced fecundity and offspring performance in small populations of the declining grassland plants Primula veris and Gentiana lutea. Journal of Ecology 88:17–30.

Kéry M, Matthies D. 2004. Reduced fecundity in small populations of the rare plant Gentianopsis ciliata (Gentianaceae). Plant Biology 6:683–688.

Kestemont B. 2010. A red list of Belgian threatened species. Brussels: Statistics Belgium.

Klaas CA, Wagner G, Laufer S, Sosa S, Della Loggia R, Bomme U, Pahl HL, Merfort I. 2002. Studies on the anti-inflammatory activity of phytopharmaceuticals prepared from *Arnica* flowers. *Planta Medica* **68**:385–391.

Knapp R. 1953. Über die natürliche Verbreitung von *Arnica montana* L. und ihre Entwicklungsmöglichkeit auf verschiedenen Böden. *Berichte Der Deutschen Botanischen Gesellschaft* **66**: 168–179.

Korneck D, Schnittler M, Vollmer L. 1996. Rote Liste der Farn- und Blütenpflanzen (Pteridophyta et Spermatophyta) Deutschlands. *Schriftenreihe Für Vegetationskunde* **28**:21–187.

Körner C. 2007. The use of 'altitude' in ecological research. *Trends in Ecology & Evolution* **22**:569–574.

Leinonen PH, Sandring S, Quilot B, Clauss MJ, Mitchell-Olds T, Ågren J, Savolainen O. 2009. Local adaptation in European populations of *Arabidopsis lyrata* (Brassicaceae). *American Journal of Botany* **96**:1129–1137.

Lowe AJ, Harris SA, Ashton P. 2004. *Ecological genetics. Design, analysis and application.* Oxford: Blackwell.

Luijten SH, Oostermeijer JGB, Leeuwen NC, Den Nijs HCM. 1996. Reproductive success and clonal genetic structure of the rare *Arnica montana* (Compositae) in The Netherlands. *Plant Systematics and Evolution* **201**:15–30.

Luijten SH, Dierick A, Oostermeijer JGB, Raijmann LEL, Den Nijs HCM. 2000. Population size, genetic variation, and reproductive success in a rapidly declining, self-incompatible perennial *Arnica montana* in the Netherlands. *Conservation Biology* **14**:1776–1787.

Luijten SH, Kéry M, Oostermeijer JGB, Den Nijs HCM. 2002. Demographic consequences of inbreeding and outbreeding in *Arnica montana*: a field experiment. *Journal of Ecology* **90**: 593–603.

Lynch M, Milligan BG. 1994. Analysis of population genetic structure with RAPD markers. *Molecular Ecology* **3**:91–99.

Lyss G, Schmidt TJ, Merfort I, Pahl HL. 1997. Helenalin, an anti-inflammatory sesquiterpene lactone from *Arnica*, selectively inhibits transcription factor NF-κB. *Biological Chemistry* **378**: 951–961.

Mardari C, Dănilă D, Bîrsan C, Balaeş T, Ştefanache C, Tănase C. 2015. Plant communities with *Arnica montana* in natural habitats from the central region of Romanian Eastern Carpathians. *Journal of Plant Development* **22**:95–105.

Matthies D, Bräuer I, Maibom W, Tscharntke T. 2004. Population size and the risk of local extinction: empirical evidence from rare plants. *Oikos* **105**:481–488.

Maurice T, Colling G, Muller S, Matthies D. 2012. Habitat characteristics, stage structure and reproduction of colline and montane populations of the threatened species *Arnica montana*. *Plant Ecology* **213**:831–842.

McKay JK, Christian CE, Harrison S, Rice KJ. 2005. "How local is local?" – A review of practical and conceptual issues in the genetics of restoration. *Restoration Ecology* **13**:432–440.

Montesinos-Navarro A, Wig J, Picó FX, Tonsor SJ. 2011. *Arabidopsis thaliana* populations show clinal variation in a climatic gradient associated with altitude. *New Phytologist* **189**:282–294.

Nybom H. 2004. Comparison of different nuclear DNA markers for estimating intraspecific genetic diversity in plants. *Molecular Ecology* **13**:1143–1155.

Ouborg NJ, Vergeer P, Mix C. 2006. The rough edges of the conservation genetics paradigm for plants. *Journal of Ecology* **94**:1233–1248.

Parker IM, Rodriguez J, Loik ME. 2003. An evolutionary approach to understanding the biology of invasions: local adaptation and general-purpose genotypes in the weed *Verbascum thapsus*. *Conservation Biology* **17**:59–72.

Peakall R, Smouse PE. 2006. GENALEX 6: genetic analysis in Excel. Population genetic software for teaching and research. *Molecular Ecology Notes* **6**:288–295.

Peter M, Gigon A, Edwards P, Lüscher A. 2009. Changes over three decades in the floristic composition of nutrient-poor grasslands in the Swiss Alps. *Biodiversity and Conservation* **18**:547–567.

Pluess AR, Stöcklin J. 2004. Population genetic diversity of the clonal plant *Geum reptans* (Rosaceae) in the Swiss Alps. *American Journal of Botany* **91**:2013–2021.

Pritchard JK, Stephens M, Donnelly P. 2000. Inference of population structure using multilocus genotype data. *Genetics* **155**:945–959.

Ratcliffe DA. 1984. Post-medieval and recent changes in British vegetation: the culmination of human influence. *New Phytologist* **98**:73–100.

Schnitzler A, Muller S. 1998. Towards an ecological basis for the conservation of subalpine heath-grassland on the upper ridges of the Vosges. *Journal of Vegetation Science* **9**:317–326.

Strykstra RJ, Pegtel DM, Bergsma A. 1998. Dispersal distance and achene quality of the rare anemochorous species *Arnica montana* L.: implications for conservation. *Acta Botanica Neerlandica* **47**:45–56.

Tallmon DA, Luikart G, Waples RS. 2004. The alluring simplicity and complex reality of genetic rescue. *Trends in Ecology & Evolution* **19**:489–496.

Vekemans X, Beauwens T, Lemaire M, Roldan-Ruiz I. 2002. Data from amplified fragment length polymorphism (AFLP) markers show indication of size homoplasy and of a relationship between degree of homoplasy and fragment size. *Molecular Ecology* **11**:139–151.

Vergeer P, Sonderen E, Ouborg NJ. 2004. Introduction strategies put to the test: local adaptation versus heterosis. *Conservation Biology* **18**:812–821.

Vergeer P, Berg LJL, Roelofs JGM, Ouborg NJ. 2005. Single-family versus multi-family introductions. *Plant Biology* **7**:509–515.

Walisch TJ, Colling G, Poncelet M, Matthies D. 2012. Effects of inbreeding and interpopulation crosses on performance and plasticity of two generations of offspring of a declining grassland plant. *American Journal of Botany* **99**:1300–1313.

Walisch TJ, Matthies D, Hermant S, Colling G. 2015. Genetic structure of *Saxifraga rosacea* subsp. *sponhemica*, a rare endemic rock plant of Central Europe. *Plant Systematics and Evolution* **301**:251–263.

Young A, Boyle T, Brown T. 1996. The population genetic consequences of habitat fragmentation for plants. *Trends in Ecology & Evolution* **11**:413–418.

PERMISSIONS

All chapters in this book were first published in AOB PLANTS, by Oxford University Press; hereby published with permission under the Creative Commons Attribution License or equivalent. Every chapter published in this book has been scrutinized by our experts. Their significance has been extensively debated. The topics covered herein carry significant findings which will fuel the growth of the discipline. They may even be implemented as practical applications or may be referred to as a beginning point for another development.

The contributors of this book come from diverse backgrounds, making this book a truly international effort. This book will bring forth new frontiers with its revolutionizing research information and detailed analysis of the nascent developments around the world.

We would like to thank all the contributing authors for lending their expertise to make the book truly unique. They have played a crucial role in the development of this book. Without their invaluable contributions this book wouldn't have been possible. They have made vital efforts to compile up to date information on the varied aspects of this subject to make this book a valuable addition to the collection of many professionals and students.

This book was conceptualized with the vision of imparting up-to-date information and advanced data in this field. To ensure the same, a matchless editorial board was set up. Every individual on the board went through rigorous rounds of assessment to prove their worth. After which they invested a large part of their time researching and compiling the most relevant data for our readers.

The editorial board has been involved in producing this book since its inception. They have spent rigorous hours researching and exploring the diverse topics which have resulted in the successful publishing of this book. They have passed on their knowledge of decades through this book. To expedite this challenging task, the publisher supported the team at every step. A small team of assistant editors was also appointed to further simplify the editing procedure and attain best results for the readers.

Apart from the editorial board, the designing team has also invested a significant amount of their time in understanding the subject and creating the most relevant covers. They scrutinized every image to scout for the most suitable representation of the subject and create an appropriate cover for the book.

The publishing team has been an ardent support to the editorial, designing and production team. Their endless efforts to recruit the best for this project, has resulted in the accomplishment of this book. They are a veteran in the field of academics and their pool of knowledge is as vast as their experience in printing. Their expertise and guidance has proved useful at every step. Their uncompromising quality standards have made this book an exceptional effort. Their encouragement from time to time has been an inspiration for everyone.

The publisher and the editorial board hope that this book will prove to be a valuable piece of knowledge for researchers, students, practitioners and scholars across the globe.

LIST OF CONTRIBUTORS

Susan J. Mazer and Scott A. Hodges
Ecology, Evolution and Marine Biology, University of California, Santa Barbara, CA, USA

Kristina M. Hufford
Ecology, Evolution and Marine Biology, University of California, Santa Barbara, CA, USA
Present address: Ecosystem Science and Management, University of Wyoming, Laramie, WY, USA

Renuka Agrawal and Rajesh Tandon
Laboratory of Cellular and Molecular Cytogenetics, Department of Botany, University of Delhi, Delhi 110007, India

Nitin Agrawal
Cluster Innovation Centre, University of Delhi, Delhi 110007, India

Soom Nath Raina
Present address: Amity Institute of Biotechnology, Amity University, Sector 125, Noida 201303, Uttar Pradesh, India

Sylvain Pineau and Pierre Huynh
LUNAM Université, Groupe Ecole Supérieure d'Agriculture, UPSP Légumineuses, Ecophysiologie Végétale, Agroécologie, 55 rue Rabelais, BP 30748, F-49007 Angers Cedex 01, France

Romain Barillot
LUNAM Université, Groupe Ecole Supérieure d'Agriculture, UPSP Légumineuses, Ecophysiologie Végétale, Agroécologie, 55 rue Rabelais, BP 30748, F-49007 Angers Cedex 01, France
Present address: INRA, Centre de Versailles-Grignon, U.M.R. INRA/AgroParisTech Environnement et Grandes Cultures, 78850 Thiverval-Grignon, France

Abraham J. Escobar-Gutiérrez and Didier Combes
INRA, UR4 P3F, Equipe Ecophysiologie des plantes fourragères, Le Chêne – RD 150, BP 6, F-86600 Lusignan, France

Diana E. Wolf and Naoki Takebayashi
Department of Biology and Wildlife, Institute of Arctic Biology, University of Alaska Fairbanks, 311 Irving I, Fairbanks, AK 99775-7000, USA

Janette A. Steets
Department of Biology and Wildlife, Institute of Arctic Biology, University of Alaska Fairbanks, 311 Irving I, Fairbanks, AK 99775-7000, USA

Department of Botany, Oklahoma State University, 301 Physical Sciences, Stillwater, OK 74078-3013, USA

Gary J. Houliston
Department of Biology and Wildlife, Institute of Arctic Biology, University of Alaska Fairbanks, 311 Irving I, Fairbanks, AK 99775-7000, USA
Landcare Research, Gerald St, Lincoln 7608, New Zealand

Lidia Poggio, María Florencia Realini, María Florencia Fourastié, Ana María García and Graciela Esther González
Instituto de Ecología, Genética y Evolución (IEGEBA)-Consejo Nacional de Investigaciones Científicas y Técnicas (CONICET) and Laboratorio de Citogenética y Evolución (LaCyE), Departamento de Ecología, Genética y Evolución, Facultad de Ciencias Exactas y Naturales, Universidad de Buenos Aires, Ciudad Autónoma de Buenos Aires, Argentina

Danielle R. Begley-Miller
Department of Ecosystem Science and Management, The Pennsylvania State University, University Park, PA 16802, USA

AndrewL. Hipp, Bethany H. Brown and Marlene Hahn
The Morton Arboretum, Lisle, IL 60532, USA

Thomas P. Rooney
Department of Biological Sciences, Wright State University, Dayton, OH 45435, USA

Paramvir Singh Ahuja and Ram Kumar Sharma
Biotechnology Division, CSIR–Institute of Himalayan Bioresource Technology, Palampur, 176061 Himachal Pradesh, India

Akshay Nag
Biotechnology Division, CSIR–Institute of Himalayan Bioresource Technology, Palampur, 176061 Himachal Pradesh, India Academy for Scientific and Innovative Research (AcSIR), CSIR–Institute of Himalayan Bioresource Technology, Palampur, 176061 Himachal Pradesh, India

Fu Qin Wu, Shi Kang Shen, Xin Jun Zhang and Yue Hua Wang
Present address: School of Life Sciences, Yunnan University, Kunming No. 2, Green Lake North Road, Kunming, Yunnan 650091, The People's Republic of China

Wei Bang Sun
Kunming Botanical Garden, Kunming Institute of Botany, Chinese Academy of Sciences, Kunming 650201, The People's Republic of China

G. Danny Esselink and M. J. M. Smulders
Wageningen UR Plant Breeding, Wageningen University and Research Center, NL-6700 AJWageningen, The Netherlands

F. Bongers
Center for Ecosystem Studies, Forest Ecology and Forest Management Group,Wageningen University and Research Center, NL-6700 AAWageningen, The Netherlands

A. B. Addisalem
Wageningen UR Plant Breeding, Wageningen University and Research Center, NL-6700 AJWageningen, The Netherlands
Center for Ecosystem Studies, Forest Ecology and Forest Management Group,Wageningen University and Research Center, NL-6700 AAWageningen, The Netherlands
Wondo Genet College of Forestry and Natural Resources, Shashemene, Ethiopia

Jane C. Stout, Paul A. Egan, Maeve Harbourne and Trevor R. Hodkinson
School of Natural Sciences and Trinity Centre for Biodiversity Research, Trinity College Dublin, Dublin 2, Ireland

Karl J. Duffy
School of Natural Sciences and Trinity Centre for Biodiversity Research, Trinity College Dublin, Dublin 2, Ireland
School of Life Sciences, University of KwaZulu-Natal, Private Bag X01, Scottsville, Pietermaritzburg 3209, South Africa

Rachel S. Welt
Department of Biological Sciences, Fordham University, Bronx, New York, NY 10458, USA
Department of Herpetology, American Museum of Natural History, New York, NY 10024, USA

Amy Litt
The New York Botanical Garden, Bronx, New York, NY 10458, USA
Botany and Plant Sciences, UC Riverside, Riverside, CA 92521, USA

Steven J. Franks
Department of Biological Sciences, Fordham University, Bronx, New York, NY 10458, USA

The New York Botanical Garden, Bronx, New York, NY 10458, USA

Sayantan Panda and Avinash Kamble
Department of Botany, Savitribai Phule Pune University, Ganeshkhind, Pune 411007, India

Dhiraj Naik
Department of Environmental Sciences, Indian Institute of Advanced Research, Koba Institutional Area, Gandhinagar 382007, India

Laura Zoratti and Hely Häggman
Department of Genetics and Physiology, University of Oulu, FI-90014 Oulu, Finland

Luisa Palmieri
Department of Food Quality and Nutrition, Research and Innovation Center, Fondazione Edmund Mach, Via E. Mach, 1-38010 San Michele all'Adige (TN), Italy

Laura Jaakola
Department of Arctic and Marine Biology, UiT The Arctic University of Norway, Climate Laboratory, 9037 Tromsø, Norway
Norwegian Institute of Bioeconomy Research, NIBIO Holt, A°s, Norway

Abdul Qayyum
Genomics Lab, Department of Plant Breeding and Genetics, Faculty of Agricultural Sciences and Technology, Bahauddin Zakariya University, Multan 60000, Pakistan

Muhammad Ali Abid and Waqas Malik
Genomics Lab, Department of Plant Breeding and Genetics, Faculty of Agricultural Sciences and Technology, Bahauddin Zakariya University, Multan 60000, Pakistan
Biotechnology Research Institute, Chinese Academy of Agricultural Sciences, 100081 Beijing, China

Rui Zhang, Chengzhen Liang and Sandui Guo
Biotechnology Research Institute, Chinese Academy of Agricultural Sciences, 100081 Beijing, China

Azra Yasmeen
Department of Agronomy, Faculty of Agricultural Sciences and Technology, Bahauddin Zakariya University, Multan 60000, Pakistan

Sang-Chul Kim, Jung Sung Kim and Joo-Hwan Kim
Department of Life Science, Gachon University, Seongnamdaero 1342, Seongnam-si, Gyeonggi-do 461-701, Korea

Richard J. Abbott
School of Biology, University of St Andrews, Harold Mitchell Building, St Andrews, Fife KY16 9TH, UK

Simon J. Hiscock
School of Biological Sciences, University of Bristol, Woodland Road, Bristol BS8 1UG, UK

Adrian C. Brennan
School of Biology, University of St Andrews, Harold Mitchell Building, St Andrews, Fife KY16 9TH, UK
Estación Biológica de Doñana (EBD-CSIC), Avenida Américo Vespucio s/n, 41092 Sevilla, Spain
School of Biological and Biomedical Sciences, University of Durham, South Road, Durham DH1 3LE, UK

Błażej Wójkiewicz and Monika Litkowiec
Institute of Dendrology, Polish Academy of Sciences, Parkowa 5, Kórnik 62-035, Poland

Witold Wachowiak
Institute of Dendrology, Polish Academy of Sciences, Parkowa 5, Kórnik 62-035, Poland
Faculty of Biology, Adam Mickiewicz University, Institute of Environmental Biology, Umultowska 89, Poznań 61-614, Poland

Tiphaine Maurice
Universitéde Lorraine, CNRS UMR 7360, Laboratoire Interdisciplinaire des Environnements Continentaux (LIEC), rue du Général Delestraint, F-57070 Metz, France

Musée National d'Histoire Naturelle, Population Biology and Evolution, 25 rue Münster, L-2160 Luxembourg, Luxembourg
Fondation Faune Flore, 24 rue Münster, L-2160 Luxembourg, Luxembourg

Guy Colling
Musée National d'Histoire Naturelle, Population Biology and Evolution, 25 rue Münster, L-2160 Luxembourg, Luxembourg

Diethart Matthies
Philipps-Universität, Fachbereich Biologie, Pflanzenökologie, D-35032 Marburg, Germany

Serge Muller
Universitéde Lorraine, CNRS UMR 7360, Laboratoire Interdisciplinaire des Environnements Continentaux (LIEC), rue du Général Delestraint, F-57070 Metz, France
Muséum national d'Histoire naturelle, UMR 7205 ISYEB, CNRS, Université Pierre-et-Marie-Curie, EPHE, Sorbonne Universités, CP 39, 16 rue Buffon, F-75005 Paris, France

Index

www.ingramcontent.com/pod-product-compliance
Lightning Source LLC
Chambersburg PA
CBHW080253230326
41458CB00097B/4440

9 781682 867556